Absolute CM-Periods

Mathematical
Surveys
and
Monographs

Volume 106

Absolute CM-Periods

Hiroyuki Yoshida

American Mathematical Society

2000 *Mathematics Subject Classification.* Primary 11G15, 11Fxx, 11M41, 11R42, 33B99; Secondary 11J89, 11Y99, 30E15.

For additional information and updates on this book, visit
www.ams.org/bookpages/surv-106

Library of Congress Cataloging-in-Publication Data

Yoshida, Hiroyuki, 1947–
Absolute CM-periods / Hiroyuki Yoshida.
p. cm. — (Mathematical surveys and monographs, ISSN 0076-5376 ; v. 106)
Includes bibliographical references and index.
ISBN 0-8218-3453-3 (alk. paper)
1. Abelian varieties. 2. Automorphic forms. 3. Functions, Zeta. I. Title. II. Mathematical surveys and monographs ; no. 106.

QA564.Y67 2003
516.3′53—dc22 2003062762

∞ The paper used in this book is acid-free and falls within the guidelines
established to ensure permanence and durability.
Visit the AMS home page at `http://www.ams.org/`

10 9 8 7 6 5 4 3 2 1 08 07 06 05 04 03

TABLE OF CONTENTS

Preface vii

Notation and Terminology ix

Introduction 1

Chapter I. MULTIPLE GAMMA FUNCTION AND ITS GENERALIZATIONS 13
§1. Basic integral representations 13
§2. Shintani's formulas 17
§3. The second derivative of $\zeta(s, A, x)$ at $s = 0$ 24
§4. An asymptotic expansion of the multiple gamma function 34
Exercises 41

Chapter II. THE STARK–SHINTANI CONJECTURE 43
§1. Stark's general conjecture 43
§2. Transition to a more precise conjecture 48
§3. Shintani's formulas for the partial zeta function 52
§4. An example 54
Exercises 58

Chapter III. ABSOLUTE CM–PERIODS 61
§1. Shimura's period symbol p_K 61
§2. The case of abelian fields 65
§3. Conjectures on absolute CM–periods 74
§4. Numerical examples 94
§5. Further investigations on the invariant $X(c)$ 108
§6. Numerical examples (continued) 126
Exercises 133

Chapter IV. EXPLICIT CONE DECOMPOSITIONS AND APPLICATIONS 135
§1. A special decomposition of a higher dimensional cube 135
§2. Topological preparations 139
§3. A sufficient condition for a cone decomposition 144
§4. Examples 147
§5. Explicit cone decompositions for index finite subgroups 149
§6. Applications 161
Exercises 166

Chapter V. APPLICATIONS OF A LIMIT FORMULA OF KRONECKER'S TYPE 169
§1. A limit formula of Kronecker's type 169
§2. A generalization of the exact Chowla–Selberg formula 179
§3. L-functions of orders of an algebraic number field 189
§4. Toward the reciprocity law for the h-function 197
§5. A connection of automorphic forms with group cohomology 207
Exercises 213

Appendix I. EISENSTEIN SERIES ON $GL(2)$ 215
§1. Eisenstein series on $GL(2)$ 215
§2. Calculations of local integrals 222
§3. The functional equation of Eisenstein series 234
§4. Eisenstein series of class 1 243

Appendix II. ON HIGHER DERIVATIVES OF L-FUNCTIONS 249
§1. A search for new invariants 249
§2. The case of the second derivatives of L–functions 253

Appendix III. TRANSCENDENTAL PROPERTY OF CM–PERIODS 269
Exercises 274

References 275

Index 281

PREFACE

The central theme of this book is an invariant attached to an ideal class of a totally real algebraic number field. Using this invariant, we can construct the Stark-Shintani units and the periods of abelian varieties with complex multiplication simultaneously. These periods were studied by Shimura and he crystallized them as the period symbol. The highlight of this book is the construction of the absolute CM-period symbol and the description of the remarkable properties of this symbol.

I planned to write this book in 1997 after getting a substantial part of the results in chapters I, II, III and V. My motivation was twofold: One was to present the results and conjectures in a systematic and precise form convenient for further research. Second was to organize the material to make it accessible to graduate students providing many illustrative examples. I benefitted from lecturing on the subject at several universities during these years, and the experience is incorporated into the presentation of this book. I would like to thank Professors Jacques Tilouine and Michael Harris who arranged my stay in Paris for two months in 1999, during which time I could draft chapters II and III. Finally I would like to thank Professor Goro Shimura for his extremely useful advice and encouragement.

Kyoto, June 2003. Hiroyuki Yoshida

NOTATION AND TERMINOLOGY

For a set S, $|S|$ denotes the cardinality of S. If S is a disjoint union of a family of sets T_λ, $\lambda \in \Lambda$, we express this fact by $S = \sqcup_{\lambda\in\Lambda} T_\lambda$. For $x \in \mathbf{R}$, $[x]$ denotes the largest integer which does not exceed x. By $\mathbf{R}_+$, we denote the set of all positive real numbers. For a, $b \in \mathbf{C}$, we write $a \sim b$ if $b \neq 0$ and a/b is an algebraic number. By $\mathfrak{H}$, we denote the complex upper half plane. For a subset X of $\mathbf{R}^n$, we put $\partial X = X \setminus X^\circ$, where X° denotes the set of interior points of X.

We denote by $\zeta(s)$ the Riemann zeta function. Euler's constant is denoted by γ. For a function $f(x)$ and a nonnegative integer p, $f^{(p)}(x)$ denotes the p-th derivative of $f(x)$ with the convention $f^{(0)}(x) = f(x)$. This notation will be used only in chapter I and appendix II in an obvious context. For a function $f(s)$ which is meromorphic in a neighborhood of $a \in \mathbf{C}$, $\mathrm{Res}_{s=a} f(s)$ denotes the residue of $f(s)$ at a.

For a finite group G, $\widehat{G}$ denotes the set of all equivalence classes of irreducible representations of G on vector spaces over $\mathbf{C}$. For a subgroup H of G and a representation ψ of H, the induced representation from ψ is denoted by $\mathrm{Ind}_H^G \psi$. For an associative ring R with unit, $R^\times$ denotes the group of all invertible elements of R. By $M(m, n, R)$, we denote the set of all $m \times n$-matrices with entries in R. We abbreviate $M(n, n, R)$ to $M(n, R)$ and set $GL(n, R) = M(n, R)^\times$. For two fields F_1 and F_2 contained in a field K, $F_1 \vee F_2$ denotes the composite field of F_1 and F_2.

We fix an algebraic closure $\overline{\mathbf{Q}}$ of $\mathbf{Q}$ in $\mathbf{C}$. By an algebraic number field, we understand an algebraic extension of $\mathbf{Q}$ of finite degree contained in $\overline{\mathbf{Q}}$. We denote by ρ the complex conjugation.

Let F be a global field, i.e., an algebraic number field or a function field of dimension 1 with a finite field as the field of constants. For a place v of F, F_v denotes the completion of F at v. For $a \in F_v$, $|a|_v$ denotes the absolute value of a, i.e., we have $d(ax) = |a|_v dx$ for a Haar measure dx on F_v. By F_A and $F_A^\times$, we denote the adele ring and the idele group of F respectively. By F_∞ (resp. $F_\infty^\times$), we denote the infinite part of F_A (resp. $F_A^\times$). By $(F_A)_f$ (resp. $(F_A^\times)_f$), we denote the finite part of F_A (resp. $F_A^\times$). For an algebraic group G defined over F, G_A denotes the adelization of G and G_∞ denotes the infinite part of G_A. For a place v of F, G_v denotes the group of F_v-rational points of G. By a Hecke character of $F_A^\times$, we mean a continuous homomorphism of $F_A^\times$ into $\mathbf{C}^\times$ which is trivial on $F^\times$. For $x \in F_A$ and a place v of F, x_v denotes the v-component of x. The finite part of x is denoted by x_f. For $x \in F_A^\times$, $|x|_A$ denotes the idele norm of x

Let F be an algebraic number field. The ring of integers and the group of units of F are denoted by $\mathfrak{O}_F$ and E_F respectively. We denote by E_F^+ the group of all totally positive units of F. The group of roots of unity contained in F is denoted by W_F. We put $w_F = |W_F|$. The regulator, the different, the absolute discriminant and the class number of F are denoted by R_F, $\mathfrak{d}_F$, D_F and h_F respectively. We denote by $I(F)$ the ideal group of F. For an integral ideal $\mathfrak{f}$ of F, $I_{\mathfrak{f}}(F)$ denotes the ideal group of F modulo $\mathfrak{f}$, i.e., the group of all fractional ideals relatively prime to $\mathfrak{f}$. For $x \in F_A^\times$, $\mathrm{div}(x)$ denotes the fractional ideal $\prod_v \mathfrak{p}_v^{a_v}$, where v extends over all finite places of F, $\mathfrak{p}_v$ is the prime ideal corresponding to v and $a_v = \mathrm{ord}_v x_v$ with the normalized additive valuation ord_v at v. By $r_1(F)$ and $r_2(F)$, we denote the number of real places and of imaginary places of F respectively. We denote by J_F the set of all isomorphisms of F into $\mathbf{C}$ and by I_F the free abelian group generated by J_F. For $a \in F$, $a \gg 0$ means that a is totally positive. We denote the maximal abelian extension of F in $\overline{\mathbf{Q}}$ by F_{ab}. For $a \in F_A^\times$, $[a, F] \in \mathrm{Gal}(F_{ab}/F)$ denotes the image of a under the Artin map. For an abelian extension L of F of finite degree and a fractional ideal $\mathfrak{a}$ of F, which is relatively prime to the conductor of L, $(\frac{L/F}{\mathfrak{a}})$ denotes the Artin symbol.

Let K be an algebraic number field which is an extension of F of finite degree. By $\mathrm{Res}_{K/F}$, we denote the restriction homomorphism from I_K to I_F. By $\mathrm{Inf}_{K/F}$, we denote the homomorphism from I_F to I_K such that, for $\sigma \in J_F$, $\mathrm{Inf}_{K/F}(\sigma)$ is the sum of all elements of J_K whose restrictions to F coincide with σ. The norm map and the trace map from K to F are denoted by $N_{K/F}$ and by $\mathrm{Tr}_{K/F}$ respectively. The relative discriminant (resp. the relative different) of K over F is denoted by $D(K/F)$ (resp. $\mathfrak{d}_{K/F}$).

By a CM-field, we understand a totally imaginary quadratic extension of a totally real algebraic number field. For a CM-filed K, $\Phi \in I_K$ is called a CM-type of K if $\Phi = \sum_{i=1}^n \sigma_i \in I_K$ with $\sigma_i \in J_K$ and if $\Phi + \Phi\rho$ is the sum of all elements in J_K. We often identify Φ with the set of isomorphisms $\{\sigma_1, \sigma_2, \cdots, \sigma_n\}$ or with the representation of K into $M(n, \mathbf{C})$ sending $a \in K$ to the diagonal matrix with a^{σ_1}, a^{σ_2}, $\ldots$, a^{σ_n} on the diagonal entries. A point $z = (z_1, z_2, \ldots, z_n) \in \mathfrak{H}^n$ is called a CM-point if for a CM-field K, a CM-type $\Phi = \{\sigma_1, \sigma_2, \ldots, \sigma_n\}$ and for $a \in K$, $z_i = a^{\sigma_i}$, $1 \leqq i \leqq n$ holds.

INTRODUCTION

In number theory, various kinds of "class invariants" were introduced and studied. One of the most remarkable is the invariant studied by Stark-Shintani, which is the exponential of the derivative at $s = 0$ of the partial zeta function of a class c. This invariant is conjectured to give a unit of an abelian extension of a number field and should give an answer to Hilbert's 12th problem. The aim of this book is to describe a more enlarged and coherent picture by introducing Shimura's CM-periods into our consideration of class invariants.

To explain the contents of this book in a proper perspective, we start with a brief historical survey on the subjects. Let F be an algebraic number field of finite degree. One of the most famous formula in number theory, due to Dirichlet-Dedekind, is the analytic class number formula[1]:

$$\lim_{s\to 1}(s-1)\zeta_F(s) = \frac{2^{r_1}(2\pi)^{r_2}h_F R_F}{w_F|D_F|^{1/2}}. \tag{1}$$

Here $\zeta_F(s)$ is the Dedekind zeta function of F, r_1 (resp. r_2) denotes the number of real (resp. imaginary) infinite places of F; h_F, R_F, D_F denote the class number, the regulator, the discriminant of F respectively; w_F denotes the number of roots of unity contained in F. When K is an abelian extension of F, the calculation of the relative class number h_K/h_F times R_K/R_F can be reduced to that of $L(1,\chi)$ for the non-trivial characters of $\mathrm{Gal}(K/F)$. In fact, Kummer performed the calculation for the cyclotomic fields and derived many interesting results on class numbers and units.

On the other hand, Kronecker applied his limit formula to the study of algebraic number fields. Let F be an imaginary quadratic field, c be an ideal class of F and $\zeta_F(s,c)$ be the partial zeta function of the class c. His formula gives[2]

$$\zeta_F(s,c) = \frac{2\pi}{w_F|D_F|^{1/2}}\left(\frac{1}{s-1} + \gamma(c) + O(s-1)\right), \qquad s \to 1, \tag{2}$$

$$\gamma(c) = \log\frac{1}{2|D_F|^{1/2}} + 2\gamma - \log\Im(w) - \log|\eta(w)|^4. \tag{3}$$

Here we take an ideal $\mathfrak{A}$ in the class c and write $\mathfrak{A} = \mathbf{Z}\omega_1 + \mathbf{Z}\omega_2$, $\Im(\omega_1/\omega_2) > 0$, $w = \omega_1/\omega_2$; η is the Dedekind eta function and γ is Euler's constant.

Next let us briefly recall Hecke's work. In his dissertation, Hecke generalized classical theory of complex multiplication to CM-fields of degree 4 using Hilbert

[1] cf. [Ded], III, p. 176.

[2] cf. [Kr], IV, p. 494–495 for his limit formula. In the form (2), (3), see [Ded], II, p. 227.

modular functions of two variables. This work of Hecke was generalized[3] to the theory of complex multiplication of abelian varieties by Shimura-Taniyama-Weil around 1955. By the work of Shimura during 1960-1980, we have now a good perspective on the zeta functions and periods of abelian varieties with complex multiplication together with their relations to modular forms (cf. [Sh12]). As a related but a different approach to Hilbert's 12th problem, Hecke tried to generalize Kronecker's limit formula to algebraic number fields of higher degree. He was interested in the special function which appears in the constant term of the Laurent expansion of the partial zeta function of an algebraic number field. He expected that such a function would give an analogue of the η-function, hence should play a key role in constructing abelian extensions of a number field. He investigated this problem from various points of view. We refer the reader to papers No. 3, 10, 15, 17, 20, 21 in his Werke. A part of Hecke's work is presented together with Kronecker's work by the master hand of Siegel [Si2]. In 1970, Asai [As] published a generalization of Kronecker's limit formula to an arbitrary algebraic number field. Also in this year, Stark [St1] published a conjecture on the general nature of $L(1,\chi)$. This conjecture was made very precise by the work of Stark-Shintani in 1970's.

Let us explain their research. Using the functional equation, we see that (1) is equivalent to

$$(1') \qquad \zeta_F(s) \sim -\frac{h_F R_F}{w_F} s^{r_1+r_2-1}, \qquad s \to 0.$$

Let K be a Galois extension of F of finite degree. Put $G = \mathrm{Gal}(K/F)$ and let $\widehat{G}$ denote the set of all irreducible characters of G. We have

$$(4) \qquad \zeta_K(s) = \prod_{\chi \in \hat{G}} L(s,\chi,K/F)^{\dim \chi}$$

where $L(s,\chi,K/F)$ denotes the Artin L-function associated to χ. We set

$$L(s,\chi,K/F) = c(\chi)s^{r(\chi)} + O(s^{r(\chi)+1}), \qquad s \to 0$$

with $c(\chi) \in \mathbf{C}^\times$ and $0 \leqq r(\chi) \in \mathbf{Z}$. Stark defined a generalized regulator $R(\chi)$ and conjectured that

$$(5) \qquad \left(\frac{c(\chi)}{R(\chi)}\right)^\sigma = \frac{c(\chi^\sigma)}{R(\chi^\sigma)}$$

for every $\sigma \in \mathrm{Aut}(\mathbf{C})$. Here $\mathrm{Aut}(\mathbf{C})$ denotes the automorphism group of $\mathbf{C}$. The generalized regulator gives a decomposition of R_K; hence (5) can be regarded as a factorization of the right-hand side of (1′) according to (4). The generalized regulator $R(\chi)$ is the determinant of a matrix of size $r(\chi)$ whose entries are linear combinations of the logarithms of the absolute values of units of K with algebraic coefficients. Thus it was expected that the investigation of $L'(0,\chi)$ when $r(\chi) = 1$ would yield very interesting results on the units of K.

To explain the conjecture more explicitly, let F be a totally real algebraic number field of degree n. Let $\mathfrak{O}_F$ be the ring of all integers of F and $E_F = \mathfrak{O}_F^\times$ be the group of units of F. Let $\infty_1, \infty_2, \ldots, \infty_n$ denote all the infinite places of F. Let K be an abelian extension of finite degree of F such that ∞_1 is unramified and

[3]This word is not completely precise. See the introduction of [Sh12] for comments on this work of Hecke.

∞_i, $2 \leqq i \leqq n$ are ramified in K. Then the conductor of K as a class field over F is of the form $\mathfrak{f}\infty_2 \cdots \infty_n$ with an integral ideal $\mathfrak{f}$ of F. For $\sigma \in \mathrm{Gal}(K/F)$, let

$$\zeta(s,\sigma) = \sum_{\mathfrak{A},(\frac{K/F}{\mathfrak{A}})=\sigma} N(\mathfrak{A})^{-s}$$

be the partial zeta function attached to σ. Here $(\frac{K/F}{\mathfrak{A}})$ denotes the Artin symbol. We regard F as a subfield of $\mathbf{R}$ and assume that ∞_1 corresponds to the identical embedding of F into $\mathbf{R}$. Then the Stark conjecture states

CONJECTURE S. *Assume $n \geqq 2$. Then there exists a unit ϵ of K such that*

$$\exp(-2\zeta'(0,\sigma)) = \epsilon^{\sigma} \qquad \textit{for every} \quad \sigma \in \mathrm{Gal}(K/F).$$

In chapter II, we will explain a method to derive Conjecture S from (5) in detail.[4]. This conjecture was discovered independently by Shintani when F is a real quadratic field. Furthermore he obtained an explicit formula in terms of the multiple gamma function for the derivative of the partial zeta function at $s = 0$ for an arbitrary totally real field. Since his result is of central importance for our investigation, let us describe it in some detail.

Let $\omega = (\omega_1, \omega_2, \dots, \omega_r)$, $\omega_i > 0$ for $1 \leqq i \leqq r$ and $x > 0$. We define the r-ple Riemann zeta function by

$$\zeta_r(s,\omega,x) = \sum_{\Omega = m_1\omega_1 + m_2\omega_2 + \cdots + m_r\omega_r} (x+\Omega)^{-s}.$$

Here $(m_1, m_2, \dots, m_r)$ extends over all r-tuples of non-negative integers. By a method which goes back to Riemann, Barnes showed that $\zeta_r(s,\omega,x)$ can be continued meromorphically to the whole s-plane and holomorphic at $s = 0$; furthermore we have an elementary formula for $\zeta_r(0,\omega,x)$:

$$\begin{aligned} &\zeta_r(0,(\omega_1,\omega_2,\dots,\omega_r),\sum_{k=1}^{r}\omega_k x_k) \\ &=(-1)^r \sum_{l_1,\dots,l_r \geqq 0,\, l_1+\cdots+l_r=r} \omega_1^{l_1-1}\omega_2^{l_2-1}\cdots\omega_r^{l_r-1} \frac{B_{l_1}(x_1)}{l_1!}\frac{B_{l_2}(x_2)}{l_2!}\cdots\frac{B_{l_r}(x_r)}{l_r!}, \end{aligned} \tag{6}$$

where $B_n(x)$ denotes the n-th Bernoulli polynomial. We put

$$-\log \rho_r(\omega) = \lim_{x \to +0}\left\{ \frac{\partial}{\partial s}\zeta_r(s,\omega,x)\Big|_{s=0} + \log x \right\},$$

$$\frac{\partial}{\partial s}\zeta_r(s,\omega,x)\Big|_{s=0} = \log\frac{\Gamma_r(x,\omega)}{\rho_r(\omega)}.$$

The function $\Gamma_r(x,\omega)$ is the r-ple gamma function introduced by Barnes ([Bar2]). If $r = 1$, we have

$$\frac{\Gamma_1(x,\omega)}{\rho_1(\omega)} = \frac{1}{\sqrt{2\pi}}\Gamma(\frac{x}{\omega})\exp((\frac{x}{\omega}-\frac{1}{2})\log\omega).$$

---

[4]Strictly speaking, a conjecture with $0 < e \in \mathbf{Z}$ in place of 2 will be derived.

Shintani introduced a further generalization of the r-ple Riemann zeta function. Let $A = (a_{ij})$ be an $n \times r$-matrix with $a_{ij} > 0$ for all i and j. For ${}^t x = (x_1, x_2, \ldots, x_r)$, $x_i \geqq 0$, $1 \leqq i \leqq r$, $x \neq 0$ and $s \in \mathbf{C}$, we define

$$\zeta(s, A, x) = \sum_{m_1, \ldots, m_r = 0}^{\infty} \prod_{i=1}^{n} \left\{ \sum_{j=1}^{r} a_{ij}(m_j + x_j) \right\}^{-s}.$$

For $1 \leqq i \leqq n$, let $A^{(i)}$ denote the i-th row of A. By definition, we have

$$\zeta(s, A^{(i)}, x) = \zeta_r(s, (a_{i1}, a_{i2} \ldots, a_{ir}), \sum_{j=1}^{r} a_{ij} x_j).$$

We can show that $\zeta(s, A, x)$ can be continued meromorphically to the whole s-plane and holomorphic at $s = 0$. Shintani obtained that ([Shi3])

$$\zeta(0, A, x) = \frac{1}{n} \sum_{i=1}^{n} \zeta(0, A^{(i)}, x), \tag{7}$$

$$\begin{aligned} \frac{\partial}{\partial s} \zeta(s, A, x) \bigg|_{s=0} &= \sum_{i=1}^{n} \log \frac{\Gamma_r(\sum_{j=1}^{r} a_{ij} x_j, (a_{i1}, \ldots, a_{ir}))}{\rho_r((a_{i1}, \ldots, a_{ir}))} \\ &+ \frac{(-1)^r}{n} \sum_{l} C_l(A) \prod_{j=1}^{r} \frac{B_{l_j}(x_j)}{l_j!}. \end{aligned} \tag{8}$$

Here in (8), the last term is "the correction term" and we refer the reader to chapter I, §2.3 for the precise definition; we simply note that it is of elementary nature. A remarkable feature of (8) is that

$$\begin{aligned} \zeta'(0, A, x) &= \sum_{i=1}^{n} \zeta'(0, A^{(i)}, x) \\ &+ \frac{1}{n} \sum_{1 \leqq i, k \leqq n, i \neq k} \left[\zeta'(0, \begin{pmatrix} A^{(i)} \\ A^{(k)} \end{pmatrix}, x) - \zeta'(0, A^{(i)}, x) - \zeta'(0, A^{(k)}, x) \right]. \end{aligned} \tag{8'}$$

Here a prime denotes the derivative with respect to s. The second term of the right-hand side represents the correction term. In chapter I, we will give proofs of these results together with a generalization to the second derivative of $\zeta(s, A, x)$ at $s = 0$. In analogy to (8′), we have[5]

[5]The formula (8′) (resp. (9)) is tautological when $n = 2$ (resp. $n = 2, 3$) without the explicit expressions of the correction terms. But to write the formulas in these forms is suggestive.

$$
\begin{aligned}
\zeta''(0,A,x) &= n\sum_{i=1}^{n}\zeta''(0,A^{(i)},x)\\
&+\frac{3-n}{n}\sum_{1\leqq i,k\leqq n, i\neq k}\left[\zeta''(0,\begin{pmatrix}A^{(i)}\\A^{(k)}\end{pmatrix},x)-2\zeta''(0,A^{(i)},x)-2\zeta''(0,A^{(k)},x)\right]\\
&+\frac{1}{2n}\sum_{1\leqq i,k,p\leqq n, i\neq k, k\neq p, p\neq i}\left[\zeta''(0,\begin{pmatrix}A^{(i)}\\A^{(k)}\\A^{(p)}\end{pmatrix},x)-3\zeta''(0,A^{(i)},x)\right.\\
&\qquad\left.-3\zeta''(0,A^{(k)},x)-3\zeta''(0,A^{(p)},x)\right].
\end{aligned}
\tag{9}
$$

Here the second and the third terms have the meaning of "correction terms" and can be written explicitly. In chapter I, we will also prove an asymptotic expansion of the multiple gamma function.

For r linearly independent vectors $v_1, v_2, \ldots, v_r \in \mathbf{R}^n$, put

$$
C(v_1,v_2,\ldots,v_r)=\left\{\sum_{i=1}^{r}t_iv_i \;\middle|\; t_1,t_2,\ldots,t_r>0\right\}
$$

and call $C(v_1,v_2,\ldots,v_r)$ an r-dimensional open simplicial cone with basis v_1, v_2 , $\ldots$, v_r. Let σ_1, σ_2, $\ldots$, σ_n be all the isomorphisms of F into $\mathbf{R}$. For $x\in F$, we put $x^{(i)}=x^{\sigma_i}$. We embed F into $\mathbf{R}^n$ by

$$
F\ni x\longrightarrow (x^{(1)},x^{(2)}\ldots,x^{(n)})\in\mathbf{R}^n.
$$

Let $\mathbf{R}_+$ be the group of all positive numbers and E_F^+ be the group of all totally positive units of F. A theorem of Shintani states that there exist finitely many open simplicial cones C_j, $j\in J$ with basis in $\mathfrak{O}_F$ such that

$$
\mathbf{R}_+^n=\sqcup_{\epsilon\in E_F^+}\epsilon\,(\sqcup_{j\in J}C_j). \tag{10}
$$

Let $r(j)$ be the dimension of C_j and put

$$
C_j=C(v_{j1},v_{j2},\ldots,v_{jr(j)}),\qquad v_{j1},v_{j2},\ldots,v_{jr(j)}\in\mathfrak{O}_F.
$$

For $z\in C_j$, we define the coordinate $x_i(z)\in\mathbf{R}_+$ by

$$
z=x_1(z)v_{j1}+x_2(z)v_{j2}+\ldots+x_{r(j)}(z)v_{jr(j)}.
$$

Put ${}^t x(z)=(x_1(z),x_2(z),\ldots,x_{r(j)}(z))$. For a fractional ideal $\mathfrak{a}$ of F, we put

$$
R(C_j,\mathfrak{a})=\{z\in\mathfrak{a}\cap C_j \mid 0<x_1(z),x_2(z),\ldots,x_{r(j)}(z)\leqq 1,\ {}^t x(z)\in\mathbf{Q}^{r(j)}\}.
$$

Now let $\mathfrak{f}$ be an integral ideal of F and let $C_{\mathfrak{f}}$ denote the ideal class group of F defined modulo $\mathfrak{f}\infty_1\infty_2\cdots\infty_n$. Let h_0 be the class number of F in the narrow sense and let $\mathfrak{a}_1$, $\mathfrak{a}_2$, $\ldots$, $\mathfrak{a}_{h_0}$ be integral ideals which represent narrow ideal classes. For $c\in C_{\mathfrak{f}}$, take $\mathfrak{a}_\mu$ so that c and $\mathfrak{a}_\mu\mathfrak{f}$ belong to the same narrow ideal class and put

$$
R(C_j,c)=\{z\in R(C_j,(\mathfrak{a}_\mu\mathfrak{f})^{-1}) \mid (z)\mathfrak{a}_\mu\mathfrak{f}=c \text{ in } C_{\mathfrak{f}}\}.
$$

This is a finite set. To C_j, we associate an $n \times r(j)$-matrix A_j by $A_j = (v_{jl}^{\sigma_i})$ (the (i,l)-component of A_j is $v_{jl}^{\sigma_i}$). Then we have

$$\zeta_F(s,c) = N(\mathfrak{a}_\mu \mathfrak{f})^{-s} \sum_{j \in J} \sum_{z \in R(C_j,c)} \zeta(s, A_j, x(z)). \tag{11}$$

For $1 \leqq i \leqq n$, let $A_j^{(i)}$ denote the i-th row of A_j. Now we have

$$\zeta_F(0,c) = \frac{1}{n} \sum_{j \in J} \sum_{z \in R(C_j,c)} \sum_{i=1}^{n} \zeta(0, A_j^{(i)}, x(z)) \in \mathbf{Q}, \tag{12}$$

$$\begin{aligned} \zeta_F'(0,c) = \sum_{j \in J} \sum_{z \in R(C_j,c)} & \Bigg[\sum_{i=1}^{n} \zeta'(0, A_j^{(i)}, x(z)) \\ & + \frac{1}{n} \sum_{1 \leqq i,k \leqq n} \left\{ \zeta'(0, \begin{pmatrix} A_j^{(i)} \\ A_j^{(k)} \end{pmatrix}, x(z)) - \zeta'(0, A_j^{(i)}, x(z)) - \zeta'(0, A_j^{(k)}, x(z)) \right\} \Bigg] \\ & - \log N(\mathfrak{a}_\mu \mathfrak{f}) \zeta_F(0,c). \end{aligned} \tag{13}$$

The first formula gives a closed formula for $\zeta_F(0,c)$ by (6) and the second formula gives an expression of $\zeta_F'(0,c)$ by the multiple gamma function by (8).

Next let K be a CM-field of degree $2n$. Let ρ denote the complex conjugation. We denote by J_K the set of all isomorphisms of K into $\mathbf{C}$ and by I_K the free abelian group generated by J_K. Shimura introduced the period symbol $p_K : I_K \times I_K \longrightarrow \mathbf{C}^\times/\overline{\mathbf{Q}}^\times$ which is linear with respect to both of the arguments; when Φ is a CM-type of K, $p_K(\sigma, \Phi)$, $\sigma \in \Phi$ coincides with a geometric period of an abelian variety with complex multiplication. The period symbol p_K plays a fundamental role in the analysis of the behavior of arithmetic automorphic forms at CM-points. For the convenience of notation, we regard $p_K(\xi, \eta)$ as an element of $\mathbf{C}^\times$ well defined modulo $\overline{\mathbf{Q}}^\times$. For a, $b \in \mathbf{C}$, we write $a \sim b$ if $b \neq 0$ and $a/b \in \overline{\mathbf{Q}}$.

Assume that K is normal over $\mathbf{Q}$. Put $G = \mathrm{Gal}(K/\mathbf{Q})$. A representation η of G is called *odd* (resp. *even*) if $\eta(\rho) = -\mathrm{id}$ (resp. $\eta(\rho) = \mathrm{id}$). Let $\widehat{G}$ (resp. $\widehat{G}_-$) denote the set of all equivalence classes of irreducible (resp. irreducible odd) representations of G. When K is abelian over $\mathbf{Q}$, we can show that

$$p_K(\mathrm{id}, \sigma) \sim \pi^{-\mu(\sigma)/2} \prod_{\eta \in \hat{G}_-} \exp\Big(\frac{\eta(\sigma)}{[K:\mathbf{Q}]} \frac{L'(0,\eta)}{L(0,\eta)}\Big), \qquad \sigma \in G. \tag{14}$$

Here $\mu(\sigma) = 1$ (resp. -1, resp. 0) if $\sigma = 1$ (resp. $\sigma = \rho$, resp. $\sigma \neq 1, \rho$). This result is essentially due to Anderson [An]. We give a proof in chapter III, §2. The idea is to consider the factors of the jacobian variety of the Fermat curve $x^n + y^n = 1$. The geometric periods of this factor can be expressed by the gamma function and we can obtain sufficient information on $p_K(\mathrm{id}, \sigma)$ in terms of the gamma function[6]. When

[6]This is the main reason of the success in the abelian case. For $p_K(\sigma, \tau)$, K being a non-abelian CM-field, we do not know any geometric objects which can supply a connection of CM-periods to the multiple gamma function.

K is an imaginary quadratic field of discriminant $-d$, (14) gives the Chowla-Selberg formula:

$$\pi p_K(\mathrm{id},\mathrm{id})^2 \sim \prod_{a=1}^{d-1}\Gamma(\frac{a}{d})^{w_K\chi(a)/2h_K} = d\exp(\frac{L'(0,\chi)}{L(0,\chi)}), \tag{15}$$

where χ is the Dirichlet character corresponding to K. As a natural generalization of (14), we conjecture

CONJECTURE A. *Let K be a CM-field normal over $\mathbf{Q}$. Let c be a conjugacy class of G. Then*

$$\prod_{\sigma\in c} p_K(\mathrm{id},\sigma) \sim \pi^{-\mu(c)/2}\prod_{\eta\in\hat{G}_-}\exp(\frac{|c|\chi_\eta(\sigma)}{[K:\mathbf{Q}]}\frac{L'(0,\eta)}{L(0,\eta)}). \tag{16}$$

Here $\mu(c)=1$ (resp. -1, resp. 0) if $c=\{1\}$ (resp. $c=\{\rho\}$, resp. $c\neq\{1\},\{\rho\}$); χ_η is the character of η and we denote $L(s,\chi_\eta,K/\mathbf{Q})$ as $L(s,\eta)$.

This conjecture is essentially equivalent to a conjecture of Colmez [Co]; the present formulation is due to the author [Y3]. Conjecture A provides an expression of $\prod_{\sigma\in c} p_K(\mathrm{id},\sigma)$ in terms of the multiple gamma function but not of the individual $p_K(\mathrm{id},\sigma)$. To get a reasonable formula on $p_K(\mathrm{id},\sigma)$, we must factorize the right-hand side of (16) suitably.

Now let F be a totally real algebraic number field of degeree n and K be an abelian extension of F of finite degree. We assume that K is a CM-field. Set $G=\mathrm{Gal}(K/F)$. Let $J_F=\{\sigma_1,\sigma_2,\dots,\sigma_n\}$ as before and we use the same letter σ_i for any of its extensions to J_K. From Conjecture A, we can derive a relation

$$\prod_{i=1}^{n} p_K(\sigma_i,\tau\sigma_i) \sim \pi^{-n\mu(\tau)/2}\prod_{\omega\in\hat{G}_-}\exp(\frac{\omega(\tau)}{|G|}\frac{L'(0,\omega)}{L(0,\omega)}) \tag{17}$$

for $\tau\in G$. By the properties of the period symbol p_K, the left hand side mod $\overline{\mathbf{Q}}^\times$ does not depend on the choice of σ_i. We also have $p_K(\sigma_i,\tau\sigma_i)\sim p_{K^{\sigma_i}}(\mathrm{id},\sigma_i^{-1}\tau\sigma_i)$. Let $\tilde{\mathfrak{f}}\infty_1\infty_2\cdots\infty_n$ be the conductor of K as a class field over F, $\tilde{\mathfrak{f}}$ being an integral ideal of F. For an integral ideal $\mathfrak{f}$ of F, let $(\widehat{G}_-)_{\mathfrak{f}}$ denote the set of all characters of $\widehat{G}_-$ whose conductors are equal to $\mathfrak{f}\infty_1\infty_2\cdots\infty_n$. Then (17) can be written as

$$\begin{aligned}&\prod_{i=1}^{n} p_{K^{\sigma_i}}(\mathrm{id},\sigma_i^{-1}\tau\sigma_i)\\ &\quad\sim \pi^{-n\mu(\tau)/2}\exp(\frac{1}{|G|}\sum_{\mathfrak{f}|\tilde{\mathfrak{f}}}\sum_{\omega\in(\hat{G}_-)_{\mathfrak{f}}}\frac{\omega(\tau)}{L(0,\omega)}\sum_{c\in C_{\mathfrak{f}}}\omega(c)\zeta_F'(0,c)).\end{aligned} \tag{18}$$

The formula (13) for $\zeta_F'(0,c)$ has the form as the sum of "conjugate terms"; in view of this, we decompose $\zeta_F'(0,c)$ naturally. Regard F as a subfield of $\mathbf{R}$ and choose $\sigma_1=\mathrm{id}$. For $c\in C_{\mathfrak{f}}$, we put

$$G(c)=\sum_{j\in J}\sum_{z\in R(C_j,c)}\zeta'(0,A_j^{(1)},x(z)),$$

$$V(c)=\sum_{j\in J}\sum_{z\in R(C_j,c)}\left(\frac{1}{n}\sum_{k=2}^{n}v_{1,k}^{(j)}-\frac{1}{n^2}\sum_{1\leqq i,k\leqq n, i\neq k}v_{i,k}^{(j)}\right),$$

$$v_{i,k}^{(j)}=\zeta'(0,\begin{pmatrix}A_j^{(i)}\\A_j^{(k)}\end{pmatrix},x(z))-\zeta'(0,A_j^{(i)},x(z))-\zeta'(0,A_j^{(k)},x(z)),$$

$$W(c)=-\frac{1}{n}\log N(\mathfrak{a}_\mu\mathfrak{f})\sum_{j\in J}\sum_{z\in R(C_j,c)}\zeta(0,A_j^{(1)},x(z)),$$

$$X(c)=G(c)+V(c)+W(c),$$

$$g_K(\mathrm{id},\tau)=\pi^{-\mu(\tau)/2}\exp(\frac{1}{|G|}\sum_{\mathfrak{f}|\tilde{\mathfrak{f}}}\sum_{\omega\in(\hat{G}_-)_\mathfrak{f}}\frac{\omega(\tau)}{L(0,\omega)}\sum_{c\in C_\mathfrak{f}}\omega(c)X(c)).$$

The class invariant $X(c)$ depends on the choice of $\{C_j\}_{j\in J}$ and $\mathfrak{a}_\mu$. We write it as $X(c;\{C_j\}_{j\in J},\mathfrak{a}_\mu)$ and $g_K(\mathrm{id},\tau)$ as $g_K(\mathrm{id},\tau;\{C_j\}_{j\in J},\{\mathfrak{a}_\mu\})$ when we have to be precise. We have the decomposition

$$\sum_{i=1}^{n}X(c^{\sigma_i};\{C_j^{\sigma_i}\}_{j\in J},\mathfrak{a}_\mu^{\sigma_i})=\zeta_F'(0,c). \tag{19}$$

CONJECTURE B. $p_K(\mathrm{id},\tau)\sim g_K(\mathrm{id},\tau)$ *for* $\tau\in\mathrm{Gal}(K/F)$.

We should like to stress the importance of the factorization process described above; it suggests the possibility of further generalizations and also Conjecture B is considerably deeper than Conjecture A.

We can test Conjecture B numerically using a theorem of Shimura. Let L be a CM-field, Φ be a CM-type of L and $\mathfrak{q}$ be an integral ideal of L. Let $I_\mathfrak{q}(L)$ denote the group of all fractional ideals of L relatively prime to $\mathfrak{q}$. Let λ be a character of $I_\mathfrak{q}(L)$ such that

$$\lambda((\alpha))=\prod_{\sigma\in\Phi}(\alpha^{\sigma\rho}/|\alpha^\sigma|)^{t_\sigma}\qquad\text{if}\quad\alpha\equiv1\mod^\times\mathfrak{q},$$

where t_σ, $\sigma\in\Phi$ are non-negative integers; λ is called a Grössencharacter of L. Let m be an integer such that $m\equiv t_\sigma \mod 2$ and $-t_\sigma<m\leqq t_\sigma$ for every $\sigma\in\Phi$. Then a theorem of Shimura states that

$$L(m/2,\lambda)\sim\pi^{e/2}p_L(\sum_{\sigma\in\Phi}t_\sigma\cdot\sigma,\ \Phi), \tag{20}$$

where $e=m[L:\mathbf{Q}]/2+\sum_{\sigma\in\Phi}t_\sigma$.

By virtue of this theorem, we can choose Grössencharacters $\lambda_1,\lambda_2,\ldots,\lambda_q$ of K and integers $\epsilon_1,\epsilon_2,\ldots,\epsilon_q,m$ so that

$$\prod_{i=1}^{q}L(m/2,\lambda_i)^{\epsilon_i}\sim\pi^A\prod_{i=1}^{r}p_{K^{\sigma_i}}(\mathrm{id},\sigma_i^{-1}\tau_i\sigma_i)^{a_i}$$

holds with $A\in2^{-1}\mathbf{Z}$, $a_1,a_2,\ldots,a_r\in\mathbf{Z}$, $\tau_1,\tau_2,\ldots,\tau_r\in\mathrm{Gal}(K/F)$. Then we expect

CONJECTURE C. *For $\sigma \in \mathrm{Aut}(\mathbf{C})$, we have*

$$(\frac{\prod_{i=1}^{q} L(m/2, \lambda_i)^{\epsilon_i}}{\pi^A \prod_{i=1}^{r} g_{K^{\sigma_i}}(\mathrm{id}, \sigma_i^{-1}\tau_i\sigma_i; \{C_j^{\sigma_i}\}_{j\in J}, \{\mathfrak{a}_\mu^{\sigma_i}\})^{a_i}})^\sigma$$
$$= \zeta(\sigma) \cdot \frac{\prod_{i=1}^{q} L(m/2, \lambda_i^\sigma)^{\epsilon_i}}{\pi^A \prod_{i=1}^{r} g_{K^{\sigma_i\sigma}}(\mathrm{id}, \sigma^{-1}\sigma_i^{-1}\tau_i\sigma_i\sigma; \{C_j^{\sigma_i\sigma}\}_{j\in J}, \{\mathfrak{a}_\mu^{\sigma_i\sigma}\})^{a_i}}$$

with a root of unity $\zeta(\sigma)$.

We can show that the validity of Conjectures B and C does not depend on the choice of $\{C_j\}_{j\in J}$ and $\{\mathfrak{a}_\mu\}$. We have defined $X(c)$ carefully to get this consistency. Conjecture C gives more precise information than Conjecture B and we can use it for the numerical test. Since $g_K(\mathrm{id}, \tau; \{C_j\}_{j\in J}, \{\mathfrak{a}_\mu\})$ has at most n "conjugates" under the action of $\mathrm{Aut}(\mathbf{C})$, we call it *an absolute CM-period.* In chapter III, §4, we will discuss several convincing examples. We will see that there is a close relation between $\zeta(\sigma)$ and $V(c)$.

The building block to construct $g_K(\mathrm{id}, \tau)$ is the class invariant $X(c)$. The formula (19) shows that the invariant $X(c)$ contains all information about the Stark-Shintani units besides CM-periods. By the detailed investigation in chapter III, §5, we get the following insight about the nature of $X(c)$. Let F be a real quadratic field and $C_\mathfrak{f}$ be the ideal class group modulo $\mathfrak{f}\infty_1\infty_2$ with an integral ideal $\mathfrak{f}$ of F. Take $\nu_1, \nu_2 \in \mathfrak{O}_F$ so that $\nu_1 \equiv 1 \mod \mathfrak{f}$, $\nu_2 \equiv 1 \mod \mathfrak{f}$, $\nu_1^{(1)} < 0$, $\nu_1^{(2)} > 0$, $\nu_2^{(1)} > 0$, $\nu_2^{(2)} < 0$. Let $s_i \in C_\mathfrak{f}$ be the class of (ν_i), $i = 1, 2$. Then we can prove that $X(c) + X(cs_2) = a(c) \log \epsilon$ with $a(c) \in \mathbf{Q}$ where ϵ is the fundamental unit of F. On the other hand, $\exp(-2(X(c) + X(cs_1))$ is closely related to a Stark-Shintani unit. Assuming the Stark-Shintani conjecture, we see that the transcendantal degree of the field generated over $\mathbf{Q}$ by four quantities

$$\exp(X(c)), \quad \exp(X(cs_1)), \quad \exp(X(cs_2)), \quad \exp(X(cs_1s_2))$$

is at most 1.[7] We feel that the approach in chapter III, §5 gives a hint for a possible proof of the Stark-Shintani conjecture.

In chapter IV, we will first study geometric construction of a cone decomposition (10). For example, we will show that for a suitable index finite subgroup E of E_F^+ in place of E_F^+, we can take $C_j = C(v_{j1}, v_{j2}, \cdots, v_{jr(j)})$ with $v_{ji} \in E$, $1 \leqq i \leqq r(j)$ for all j. As an application of this result, we will prove that

$$W(c) = -\frac{1}{n} \log N(\mathfrak{a}_\mu \mathfrak{f}) \zeta_F(0, c), \tag{21}$$

$$V(c) = \sum_{i=1}^{m} a_i \log \epsilon_i \qquad \text{with} \quad a_i \in \tilde{F},\ \epsilon_i \in E_{\tilde{F}}^+ \tag{22}$$

where $\tilde{F}$ is the normal closure of F over $\mathbf{Q}$. We expect that $\tilde{F}$ can be replaced by F in (22); (21) is equivalent to $\sum_{j\in J} \sum_{z\in R(C_j,c)} \zeta(0, A_j^{(1)}, x(z)) \in \mathbf{Q}$ (cf. (12)).

[7]As similar phenomenon where transcendental and algebraic invariants appear simultaneously, we remind the reader that the values of the η-function at CM-points can be described by CM-periods while a suitable quotient of them gives a unit; cf. also chapter III, the ending remark of §5 for a simple example.

In chapter V, we will study the relation of Conjecture B with the limit formula of Kronecker's type. For simplicity, we assume, in this introduction, that the class number of F is 1. Let $\mathfrak{H}$ denote the complex upper half plane. Let $\mathbf{a}$ denote the set of all archimedean places of F. For $x = (x_1, x_2, \dots, x_n) \in \mathbf{R}_+^n$ and $s \in \mathbf{C}$, we put $x^{s\mathbf{a}} = \prod_{i=1}^n x_i^s$. Put $V = (\mathfrak{O}_F \oplus \mathfrak{O}_F) \backslash \{(0,0)\}$. For $z = (z_1, z_2, \dots z_n) \in \mathfrak{H}^n$, $s \in \mathbf{C}$, we define a standard Eisenstein series $E(z,s)$ by

$$E(z,s) = \sum_{(c,d) \in V/E_F} y^{s\mathbf{a}} |cz+d|^{-2s\mathbf{a}}.$$

Here $cz + d = (c^{(1)}z_1 + d^{(1)}, c^{(2)}z_2 + d^{(2)}, \dots, c^{(n)}z_n + d^{(n)})$, $y = (y_1, y_2, \dots, y_n)$, $y_k = \Im(z_k)$. As a function of s, $E(z,s)$ can be continued meromorphically to the whole s-plane, holomorphic except for the simple pole at $s = 1$. In the usual manner, $SL(2,F)$ acts on $\mathfrak{H}^n$. We have $E(\gamma z, s) = E(z,s)$ for $\gamma \in SL(2, \mathfrak{O}_F)$. The limit formula of Kronecker's type is

$$\begin{aligned} E(z,s) = \frac{2^{n-2}\pi^n R_F}{D_F} \Bigg[\frac{1}{s-1} - 2n\log 2 + \frac{D_F^{1/2}\gamma_F}{2^{n-2}R_F} \\ + h(z) - \log y^{\mathbf{a}} \Bigg] + O(s-1), \qquad s \to 1. \end{aligned} \tag{23}$$

Here γ_F is generalized Euler's constant defined by

$$\zeta_F(s) = \frac{2^{n-1}R_F}{D_F^{1/2}} \cdot \frac{1}{s-1} + \gamma_F + O(s-1), \qquad s \to 1$$

and $h(z)$ is a certain real analytic function on $\mathfrak{H}^n$ with the automorphic property such that $h(z) - \log y^{\mathbf{a}}$ is invariant under $SL(2, \mathfrak{O}_F)$; $y = y(z)$ is the imaginary part of z as above. The function $h(z)$ is closely related to a holomorphic function studied by Hecke which has a very complicated automorphic property (Werke No. 20, p. 398). If $n = 1$, we have

$$h(z) = -\log |\eta(z)|^4$$

where $\eta(z)$ is the Dedekind η-function. Let K be a CM-field such that $[K : F] = 2$. Let χ be the Hecke character of F which corresponds to the quadratic extension K/F. Let $L(s,\chi)$ denote the L-function attached to χ. As a special case of Conjecture A, we have

$$\exp\left(\frac{L'(0,\chi)}{L(0,\chi)}\right) \sim \pi^n \prod_{\sigma \in J_K} p_K(\sigma,\sigma). \tag{24}$$

Let $\mathfrak{A}_1, \mathfrak{A}_2, \dots, \mathfrak{A}_{h_K}$ be integral ideals of K representing ideal classes of K. We write $\mathfrak{A}_i = \mathfrak{O}_F\omega_1^{(i)} \oplus \mathfrak{O}_F\omega_2^{(i)}$ and put $w_i = \omega_1^{(i)}/\omega_2^{(i)}$. Choose a CM-type Φ_i of K so that $\Im(w_i^\sigma) > 0$ for every $\sigma \in \Phi_i$. Then we have

$$\begin{aligned} \frac{L'(0,\chi)}{L(0,\chi)} = n\gamma + n\log 4\pi - \frac{1}{2}\log|D_K| - \frac{D_F^{1/2}\gamma_F}{2^{n-1}R_F} \\ - \frac{1}{h_K}\sum_{i=1}^{h_K} \Big(h(w_i) - \log \prod_{\sigma \in \Phi_i} \Im(w_i^\sigma)\Big). \end{aligned} \tag{25}$$

If $F = \mathbf{Q}$, (25) gives

$$\exp(\frac{L'(0,\chi)}{L(0,\chi)}) \sim \pi(\prod_{i=1}^{h_K} |\eta(w_i)|^4)^{1/h_K} \sim \pi p_K(\mathrm{id},\mathrm{id})^2$$

since $|\eta(w_i)|^4 \sim p_K(\mathrm{id},\mathrm{id})^2$, $1 \leqq i \leqq h_K$; thus the Chowla-Selberg formula (15) follows.

Comparing (25) with (24), we should be able to obtain arithmetical information of $h(z)$ at CM-points. In chapter V, we will investigate this problem in great detail taking account of the Stark conjecture; we will see the source of difficulty in the case $n \geqq 2$.

In appendix I, we will give a general theory of Eisenstein series on $GL(2)$ over an arbitrary global field. In particular, we will prove analytic continuation and the functional equation of Eisenstein series.

In view of the simple form of Conjecture A and the factorization process of $\zeta'_F(0,c)$ employed to define $X(c)$, it is tempting to make similar constructions starting from the second derivative $\zeta''_F(0,c)$. In fact, by (9) and (11), we can express $\zeta''_F(0,c)$ using a natural generalization of the multiple gamma function. In appendix II, we will investigate the problem to define the class invariant $X^{(2)}(c)$ for $c \in C_{\mathfrak{f}}$ so that $\sum_{i=1}^{n} X^{(2)}(c^{\sigma_i}) = \zeta''_F(0,c)$ holds in analogy to (19). Then we will define a symbol $g_K^{(2)}(\mathrm{id},\tau)$ by a similar procedure to that defines $g_K(\mathrm{id},\tau)$. We will end with a speculation on a property of the symbol $g_K^{(2)}$.

In appendix III, we assemble a few remarks on the transcendence property of the periods, for the convenience of the reader.

CHAPTER I

MULTIPLE GAMMA FUNCTION AND ITS GENERALIZATIONS

In §1, we will derive contour integral representations for the multiple Riemann zeta function and for the multiple gamma function. The results are due to E. W. Barnes [Bar1], [Bar2]. We also define a generalization of the multiple gamma function by considering a higher derivative of the multiple Riemann zeta function at $s = 0$. In §2, we consider a generalization of the multiple Riemann zeta function. We will derive the formulas of the value and the derivative at $s = 0$ for this zeta function. The results are due to T. Shintani [Shi1], [Shi2], [Shi3]. In §3, we will derive the formula of the second derivative at $s = 0$ for the zeta function studied in §2. In §4, we will study the asymptotic expansion of the multiple gamma function due to Barnes. We will give an estimate for the remainder term of the asymptotic expansion under a certain condition.

1. Basic integral representations

1.1. Let r be a natural number and let $\omega = (\omega_1, \omega_2, \dots, \omega_r) \in \mathbf{R}_+^r$, $x > 0$. We define the r-ple Riemann zeta function by

$$\zeta_r(s, \omega, x) = \sum_{\Omega = m_1\omega_1 + m_2\omega_2 + \cdots + m_r\omega_r} (x + \Omega)^{-s}. \tag{1.1}$$

Here $(m_1, m_2, \dots, m_r)$ extends over all r-tuples of non-negative integers. Let us show that the series in (1.1) converges locally uniformly when $\Re(s) > r$. Put $\sigma = \Re(s)$. Since

$$\Big| \sum_{\Omega = m_1\omega_1 + m_2\omega_2 + \cdots + m_r\omega_r} (x + \Omega)^{-s} \Big| \leqq \sum_{\Omega = m_1\omega_1 + m_2\omega_2 + \cdots + m_r\omega_r} (x + \Omega)^{-\sigma},$$

it suffices to prove that the series on the right-hand side converges locally uniformly when $s = \sigma > r$. First assume $r = 1$. Then, for $\sigma > 1$, we have

$$\sum_{m_1=0}^{\infty} (x + m_1\omega_1)^{-\sigma} = x^{-\sigma} + \sum_{m_1=1}^{\infty} (x + m_1\omega_1)^{-\sigma} \leqq x^{-\sigma} + \int_0^{\infty} (x + \omega_1 t)^{-\sigma} dt.$$

We obtain

$$\sum_{m_1=0}^{\infty} (x + m_1\omega_1)^{-\sigma} \leqq \left\{ \frac{1}{x} + \frac{1}{\omega_1(\sigma - 1)} \right\} x^{-\sigma+1}, \qquad \sigma > 1. \tag{1.2}$$

Put

$$C_r(x) = \prod_{i=1}^{r} \left\{ \frac{1}{x} + \frac{1}{\omega_i(\sigma - i)} \right\} \qquad \text{for} \quad \sigma > r,$$

and we are going to show

$$\sum_{\Omega=m_1\omega_1+m_2\omega_2+\cdots+m_r\omega_r}(x+\Omega)^{-\sigma} \leqq C_r(x)x^{-\sigma+r}, \qquad \sigma > r \tag{1.3}$$

by induction on r. Suppose that (1.3) holds. Then, for $\sigma > r+1$, we have

$$\begin{aligned}&\sum_{\Omega=m_1\omega_1+m_2\omega_2+\cdots+m_r\omega_r+m_{r+1}\omega_{r+1}}(x+\Omega)^{-\sigma}\\ &\leqq C_r(x+m_{r+1}\omega_{r+1})\sum_{m_{r+1}=0}^{\infty}(x+m_{r+1}\omega_{r+1})^{-\sigma+r}\end{aligned}$$

using (1.3). Since $C_r(x+m_{r+1}\omega_{r+1}) \leqq C_r(x)$, we can majorize the series on the right-hand side by

$$C_r(x)\left\{\frac{1}{x}+\frac{1}{\omega_{r+1}(\sigma-r-1)}\right\}x^{-\sigma+r+1} = C_{r+1}(x)x^{-\sigma+r+1}$$

using (1.2). Thus we have proved (1.3) and the assertion on the convergence of (1.1) follows.

We see that $\zeta_1(s,1,x) = \sum_{m=0}^{\infty}(x+m)^{-s}$ coincides with the Hurwitz zeta function and $\zeta_1(s,1,1)=\zeta(s)$.

1.2. Let L denote the contour which starts from $+\infty$, runs on the real axis, encircling the origin once counter-clockwise on the circle of a small radius ϵ with the center at 0, runs the real axis and returns to $+\infty$. We choose the branch of $\log(-z)$ on L so that it takes real values when $z<0$. We put $(-z)^{s-1} = \exp((s-1)\log(-z))$. Then, if $\Re(s) > r$, we have

$$\begin{aligned}&\int_L \frac{e^{-xz}}{\prod_{k=1}^{r}(1-e^{-\omega_k z})}(-z)^{s-1}dz\\ &=2\sqrt{-1}\sin((s-1)\pi)\int_0^{\infty}\frac{e^{-xt}}{\prod_{k=1}^{r}(1-e^{-\omega_k t})}t^{s-1}dt\\ &=-2\sqrt{-1}\sin\pi s\,\Gamma(s)\sum_{\Omega}(x+\Omega)^{-s}\end{aligned}$$

where Ω extends as in (1.1). Here notice the difference of $(-z)^{s-1}$ on $L\cap\mathbf{R}_+$ when z is going or returning way viewed from $+\infty$; also notice that it is permissible to schrink the small circle if $\epsilon < 2\pi/\omega_k$ for all k. Hence we obtain

$$\zeta_r(s,\omega,x) = \frac{\sqrt{-1}\Gamma(1-s)}{2\pi}\int_L \frac{e^{-xz}}{\prod_{k=1}^{r}(1-e^{-\omega_k z})}(-z)^{s-1}dz. \tag{1.4}$$

The contour integral of this type has the origin in Riemann's celebrated paper [Ri]. As a simpler case, we have

$$x^{-s} = \frac{\sqrt{-1}\Gamma(1-s)}{2\pi}\int_L(-z)^{s-1}e^{-xz}dz. \tag{1.5}$$

The integral $\int_L\cdots$ in (1.4) converges absolutely for all s and the convergence is locally uniform with respect to s. This is the main virtue to obtain a contour integral representation for an analytic function. In particular, we see that $\zeta_r(s,\omega,x)$ can be continued meromorphically to the whole s-plane, holomorphic except for possible simple poles at $s = 1, 2, \cdots, r$. It is also legitimate to differentiate arbitrary times with respect to s under the integral sign. Let i be a positive integer. We obtain

$$\begin{aligned}(1.6)\qquad &(\frac{\partial}{\partial s})^i \zeta_r(s,\omega,x)\Big|_{s=0} \\ &= \sum_{j=0}^{i} \binom{i}{j} (-1)^j \Gamma^{(j)}(1) \frac{1}{2\pi\sqrt{-1}} \int_L \frac{e^{-xz}\{\log(-z)\}^{i-j}}{z\prod_{k=1}^{r}(1-e^{-\omega_k z})} dz.\end{aligned}$$

In particular, we have

$$\zeta_r(0,\omega,x) = \frac{1}{2\pi\sqrt{-1}} \int_L \frac{e^{-xz}}{\prod_{k=1}^{r}(1-e^{-\omega_k z})} \frac{dz}{z}, \tag{1.7}$$

$$\frac{\partial}{\partial s}\zeta_r(s,\omega,x)\Big|_{s=0} = \frac{1}{2\pi\sqrt{-1}} \int_L \frac{e^{-xz}}{\prod_{k=1}^{r}(1-e^{-\omega_k z})} \frac{\log(-z)}{z} dz + \gamma\,\zeta_r(0,\omega,x), \tag{1.8}$$

since $\Gamma'(1) = -\gamma$. Here γ denotes Euler's constant. Differentiating the formula (1.5) i times with respect to s at $s = 0$, we have

$$-\frac{\sqrt{-1}}{2\pi} \sum_{j=0}^{i} \binom{i}{j} (-1)^j \Gamma^{(j)}(1) \int_L e^{-xz}\{\log(-z)\}^{i-j} \frac{dz}{z} = (-\log x)^i. \tag{1.9}$$

We define

$$-\log \rho_r^{[i]}(\omega) = \lim_{x\to+0} \left[(\frac{\partial}{\partial s})^i \zeta_r(s,\omega,x)\Big|_{s=0} + (-1)^{i+1}(\log x)^i \right]. \tag{1.10}$$

By (1.6) and (1.9), we have

$$\begin{aligned}&(\frac{\partial}{\partial s})^i \zeta_r(s,\omega,x)\Big|_{s=0} + (-1)^{i+1}(\log x)^i \\ &= \frac{\sqrt{-1}}{2\pi} \sum_{j=0}^{i} \binom{i}{j} (-1)^j \Gamma^{(j)}(1) \int_L e^{-xz}\{\log(-z)\}^{i-j} \left\{ \frac{1-\prod_{k=1}^{r}(1-e^{-\omega_k z})}{\prod_{k=1}^{r}(1-e^{-\omega_k z})} \right\} \frac{dz}{z}.\end{aligned}$$

Since $\dfrac{1-\prod_{k=1}^{r}(1-e^{-\omega_k z})}{z\prod_{k=1}^{r}(1-e^{-\omega_k z})}$ is rapidly decreasing on L for $|z| \longrightarrow \infty$, the integral converges for $x \longrightarrow +0$. Hence the limit in (1.10) exists and we obtain

$$\begin{aligned}(1.11)\qquad \log \rho_r^{[i]}(\omega) =& \frac{\sqrt{-1}}{2\pi} \sum_{j=0}^{i} \binom{i}{j} (-1)^j \Gamma^{(j)}(1) \\ &\times \int_L \{\log(-z)\}^{i-j} \left\{ \frac{1-\prod_{k=1}^{r}(1-e^{-\omega_k z})}{\prod_{k=1}^{r}(1-e^{-\omega_k z})} \right\} \frac{dz}{z}.\end{aligned}$$

1.3. Now we define a generalization of the r-ple gamma function by

$$\log \frac{\Gamma_r^{[i]}(x,\omega)}{\rho_r^{[i]}(\omega)} = (\frac{\partial}{\partial s})^i \zeta_r(s,\omega,x)\Big|_{s=0}. \tag{1.12}$$

The r-ple gamma function introduced by Barnes is the case $i = 1$. In this case, we put

$$\Gamma_r(x,\omega) = \Gamma_r^{[1]}(x,\omega), \qquad \rho_r(\omega) = \rho_r^{[1]}(\omega).$$

For $1 \leqq l \leqq r$, put

$$\check{\omega}(l) = (\omega_1, \dots, \omega_{l-1}, \omega_{l+1}, \dots, \omega_r).$$

By definition, we obviously have

$$\zeta_r(s, \omega, x) - \zeta_r(s, \omega, x + \omega_l) = \zeta_{r-1}(s, \check{\omega}(l), x)$$

if $r \geqq 2$. Hence we obtain the difference equation

$$\log \frac{\Gamma_r^{[i]}(x + \omega_l, \omega)}{\rho_r^{[i]}(\omega)} - \log \frac{\Gamma_r^{[i]}(x, \omega)}{\rho_r^{[i]}(\omega)} = -\log \frac{\Gamma_{r-1}^{[i]}(x, \check{\omega}(l))}{\rho_{r-1}^{[i]}(\check{\omega}(l))}. \tag{1.13}$$

If $r = 1$, (1.13) holds with $-(-\log x)^i$ on the right-hand side. By (1.6), we have an expression by contour integrals:

$$\log \frac{\Gamma_r^{[i]}(x, \omega)}{\rho_r^{[i]}(\omega)} = \frac{1}{2\pi\sqrt{-1}} \sum_{j=0}^{i} \binom{i}{j} (-1)^j \Gamma^{(j)}(1) \int_L \frac{e^{-xz}\{\log(-z)\}^{i-j}}{z \prod_{k=1}^r (1 - e^{-\omega_k z})} dz. \tag{1.14}$$

In the case $i = 1$, (1.14) can be written as

$$\log \frac{\Gamma_r(x, \omega)}{\rho_r(\omega)} = \frac{1}{2\pi\sqrt{-1}} \int_L \frac{e^{-xz}\{\log(-z) + \gamma\}}{z \prod_{k=1}^r (1 - e^{-\omega_k z})} dz. \tag{1.15}$$

This is the contour integral representation obtained by Barnes. Though Barnes studied $\Gamma_r(x, \omega)$ and $\rho_r(\omega)$ separately in great detail, we will have little occasions to treat them separately.

We define the Bernoulli polynomials by

$$\frac{e^{xt}}{e^t - 1} = \sum_{m=0}^{\infty} \frac{B_m(x)}{m!} t^{m-1}.$$

We have

$$B_0(x) = 1, \qquad B_1(x) = x - \frac{1}{2}, \qquad B_2(x) = x^2 - x + \frac{1}{6},$$
$$B_3(x) = x^3 - \frac{3}{2}x^2 + \frac{1}{2}x, \qquad B_4(x) = x^4 - 2x^3 + x^2 - \frac{1}{30}, \qquad \dots .$$

Suppose that $x = \sum_{k=1}^r \omega_k x_k > 0$ with $x_k \in \mathbf{R}$. By (1.7), we see that $\zeta_r(0, \omega, x)$ is equal to the constant term when we expand $\dfrac{\exp(-\sum_{k=1}^r \omega_k x_k z)}{\prod_{k=1}^r (1 - \exp(-\omega_k z))}$ into the Laurent series of z. This is equal to

$$\sum_{l_1, \cdots, l_r \geqq 0,\, l_1 + \cdots + l_r = r} \frac{B_{l_1}(1 - x_1)}{l_1!} \frac{B_{l_2}(1 - x_2)}{l_2!} \cdots \frac{B_{l_r}(1 - x_r)}{l_r!} \omega_1^{l_1 - 1} \omega_2^{l_2 - 1} \cdots \omega_r^{l_r - 1}.$$

Since $B_m(1 - x) = (-1)^m B_m(x)$, we obtain

THEOREM 1.1.

$$\begin{aligned} &\zeta_r(0, (\omega_1, \omega_2, \dots, \omega_r), \sum_{k=1}^r \omega_k x_k) \\ =&(-1)^r \sum_{l_1, \dots, l_r \geqq 0,\, l_1 + \cdots + l_r = r} \omega_1^{l_1 - 1} \omega_2^{l_2 - 1} \cdots \omega_r^{l_r - 1} \frac{B_{l_1}(x_1)}{l_1!} \frac{B_{l_2}(x_2)}{l_2!} \cdots \frac{B_{l_r}(x_r)}{l_r!}. \end{aligned}$$

1.4. Let us digress to the case $i = 1$, $r = 1$. For $x > 0$, define the Hurwitz zeta function by

$$\zeta(s, x) = \sum_{m=0}^{\infty} (x+m)^{-s}.$$

By Theorem 1.1, we have

$$\zeta(0, x) = \frac{1}{2} - x. \tag{1.16}$$

It is well known that ([WW], p. 271)

$$\frac{\partial}{\partial s}\zeta(s, x)\bigg|_{s=0} = \log\Gamma(x) - \frac{1}{2}\log(2\pi). \tag{1.17}$$

We will have occasions to use (1.17) in Chapter III. Let us give a simple proof of (1.17). Put $f(x) = \frac{\partial}{\partial s}\zeta(s, x)\bigg|_{s=0}$. As noted below (1.13), we have

$$f(x+1) - f(x) = \log x. \tag{1.18}$$

Differentiating (1.15) under the integral sign, we have

$$\frac{d^2}{dx^2}f(x) = \frac{1}{2\pi\sqrt{-1}}\int_L \frac{e^{-xz}z\{\log(-z)+\gamma\}}{1-e^{-z}}dz = \int_0^{\infty}\frac{e^{-xz}z}{1-e^{-z}}dz > 0.$$

By the well known characterization of the gamma function by (1.18) and the convex property, we have $f(x) = \log\Gamma(x) + f(1)$, $x > 0$. To finish the proof, it suffices to show that $\zeta'(0) = -\frac{1}{2}\log(2\pi)$. This fact follows easily from the functional equation of $\zeta(s)$ (cf. V, §1.1). Since

$$\zeta_1(s, \omega, x) = \omega^{-s}\zeta(s, \frac{x}{\omega}),$$

we find

$$\frac{\Gamma_1(x, \omega)}{\rho_1(\omega)} = \frac{1}{\sqrt{2\pi}}\Gamma(\frac{x}{\omega})\exp((\frac{x}{\omega} - \frac{1}{2})\log\omega). \tag{1.19}$$

2. Shintani's formulas

2.1. Let $A = (a_{ij})$, $1 \leqq i \leqq n$, $1 \leqq j \leqq r$ be an $n \times r$-matrix. We assume $a_{ij} > 0$ for all i and j. Let x be a column vector such that ${}^t x = (x_1, x_2, \dots, x_r)$, $x_i \geqq 0$ for all i and that $x \neq 0$. For $s \in \mathbf{C}$, we define

$$\zeta(s, A, x) = \sum_{m_1, \dots, m_r = 0}^{\infty} \prod_{i=1}^{n}\{\sum_{j=1}^{r} a_{ij}(m_j + x_j)\}^{-s}. \tag{2.1}$$

This series converges if $\Re(s) > r/n$. Note that the r-ple Riemann zeta function is included as the special case $n = 1$:

$$\zeta(s, (a_{11}, a_{12}, \dots, a_{1r}), x) = \zeta_r(s, (a_{11}, a_{12}, \dots, a_{1r}), \sum_{j=1}^{r} a_{1j}x_j).$$

First let us describe a contour integral expression of $\zeta(s, A, x)$ due to Shintani. When $\Re(s) > 0$, we have

$$\int_0^\infty t_i^{s-1} \exp\{-(\sum_{j=1}^r a_{ij}(m_j + x_j))t_i\} dt_i = \Gamma(s)(\sum_{j=1}^r a_{ij}(m_j + x_j))^{-s}.$$

Assume $\Re(s) > r/n$. Then we have

$$\Gamma(s)^n \prod_{i=1}^n (\sum_{j=1}^r a_{ij}(m_j + x_j))^{-s}$$
$$= \int_0^\infty \int_0^\infty \cdots \int_0^\infty (t_1 t_2 \cdots t_n)^{s-1} \exp\{-\sum_{i=1}^n (\sum_{j=1}^r a_{ij}(m_j + x_j))t_i\} dt_1 dt_2 \cdots dt_n.$$

For $t = (t_1, t_2, \ldots, t_n)$, put

$$\text{(2.2)} \qquad \begin{aligned} L_j(t) &= \sum_{i=1}^n a_{ij} t_i \qquad (1 \leqq j \leqq r), \\ g(t) &= \prod_{j=1}^r \frac{\exp((1 - x_j)L_j(t))}{\exp(L_j(t)) - 1}. \end{aligned}$$

Then

$$\Gamma(s)^n \zeta(s, A, x) = \int_0^\infty \int_0^\infty \cdots \int_0^\infty g(t)(t_1 t_2 \cdots t_n)^{s-1} dt_1 dt_2 \cdots dt_n$$

For $1 \leqq i \leqq n$, put

$$D_i = \{(t_1, t_2, \ldots, t_n) \in \mathbf{R}_+^n \mid t_1, \ldots, t_{i-1}, t_{i+1}, \ldots, t_n \leqq t_i\}.$$

Then we get

$$\text{(2.3)} \qquad \Gamma(s)^n \zeta(s, A, x) = \sum_{i=1}^n \int_{D_i} g(t)(t_1 t_2 \cdots t_n)^{s-1} dt_1 dt_2 \cdots dt_n.$$

On D_i, we change variables $y = t_i$, $u_k = t_k/t_i = t_k/y$ for $1 \leqq k \leqq n$, $k \neq i$. Then we obtain

$$\int_{D_i} g(t)(t_1 t_2 \cdots t_n)^{s-1} dt_1 dt_2 \cdots dt_n = \int_0^\infty \int_0^1 \cdots \int_0^1 g(yu) \prod_{l \neq i} u_l^{s-1} y^{ns-1} \prod_{l \neq i} du_l dy$$

where $u = (u_1, u_2, \ldots, u_n)$, $u_i = 1$.

For a small positive number $\epsilon > 0$, let $I(\epsilon, +\infty)$ (resp. $I(\epsilon, 1)$) denote the contour which starts from $+\infty$ (resp. 1), runs the real axis, encircling the origin once conter-clockwise on the circle of radius ϵ with the center at 0, runs the real axis and returns to $+\infty$ (resp. 1). The contour $I(\epsilon, +\infty)$ was denoted by L in §1. As in §1, we choose the branch of $\log(-z)$ on $I(\epsilon, +\infty)$ and on $I(\epsilon, 1)$ so that it takes real values when $z < 0$ and put $(-z)^{s-1} = \exp((s-1)\log(-z))$. Since

$$\exp((ns-1)(\log y + \pi i)) - \exp((ns-1)(\log y - \pi i)) = -2\sqrt{-1}\sin(ns\pi) y^{ns-1}$$

for $y > 0$, we can change the above integral to the contour integral

$$\begin{aligned}&\int_{D_i} g(t)(t_1t_2\cdots t_n)^{s-1}dt_1dt_2\cdots dt_n\\ =&\frac{(-1)^n}{(2\sqrt{-1})^n(\sin\pi s)^{n-1}\sin ns\pi}\\ &\times\int_{I(\epsilon,+\infty)}\int_{I(\epsilon,1)^{n-1}} g(yu)\prod_{l\neq i}(-u_l)^{s-1}(-y)^{ns-1}\prod_{l\neq i}du_l dy\end{aligned}$$

when ϵ is sufficiently small. To see this, first notice that we can choose $\delta > 0$ so that if $0 < \epsilon < \delta$ and if $|y| = \epsilon$, $|u_l| = \epsilon$ for all $l \neq i$, then $0 < |L_j(yu)| < 2\pi$ for $1 \leqq j \leqq r$. This fact follows from $u_i = 1$. Therefore the value of the contour integral does not depend on ϵ when $\epsilon < \delta$. Next let C_ϵ denote the circle of radius ϵ with the center at 0. Since $|g(yu)| \leqq c\epsilon^{-r}$ with a constant c independent of $\epsilon < \delta$, we see that

$$\lim_{\epsilon\to+0}\int_{C_\epsilon^n} g(yu)\prod_{l\neq i}(-u_l)^{s-1}(-y)^{ns-1}\prod_{l\neq i}du_l dy = 0.$$

This justifies the transformation to the contour integral. Hereafter ϵ denotes such a small positive number.

We have

$$\begin{aligned}&\Gamma(s)^{-n}(-2\sqrt{-1})^{-n}(\sin\pi s)^{1-n}(\sin n\pi s)^{-1}\\ =&(-1)^n(2\pi\sqrt{-1})^{-n}\Gamma(1-s)^n\sin\pi s(\sin n\pi s)^{-1}.\end{aligned}$$

On D_i, we put $g(tu) = g_i(t,u)$. Then

$$g_i(t,u) = \prod_{j=1}^{r}\frac{\exp(t(1-x_j)L_j(u))}{\exp(tL_j(u))-1}. \tag{2.4}$$

Therefore, writing the variable y as t, we obtain

$$\begin{aligned}&\zeta(s,A,x) = (-1)^n\Gamma(1-s)^n\sin\pi s(\sin n\pi s)^{-1}\\ &\times\sum_{i=1}^{n}\frac{1}{(2\pi\sqrt{-1})^n}\int_{I(\epsilon,+\infty)}\int_{I(\epsilon,1)^{n-1}} g_i(t,u)\prod_{l\neq i}(-u_l)^{s-1}(-t)^{ns-1}\prod_{l\neq i}du_l dt.\end{aligned} \tag{2.5}$$

The integrals converge absolutely for all s and the convergence is locally uniform with respect to s. In particular, $\zeta(s,A,x)$ can be continued meromorphically to the whole s-plane, holomorphic except for possible poles at $s = 1, 2, \ldots, [r/n]$ and $s = t/n$, $t \in \mathbf{Z}$, $t \leqq r$, $n \nmid t$.

2.2. By (2.5), we have

$$\zeta(0,A,x) = \frac{1}{n}\cdot\frac{1}{(2\pi\sqrt{-1})^n}\sum_{i=1}^{n}\int_{I(\epsilon,+\infty)}\int_{I(\epsilon,1)^{n-1}} g_i(t,u)\prod_{l\neq i}u_l^{-1}t^{-1}\prod_{l\neq i}du_l dt. \tag{2.6}$$

By residue calculus, we see that

$$\frac{1}{(2\pi\sqrt{-1})^{n-1}}\int_{I(\epsilon,1)^{n-1}} g_i(t,u)\prod_{l\neq i}u_l^{-1}du_l = g_i(t,0).$$

Therefore we obtain

$$\zeta(0,A,x)=\frac{1}{2n\pi\sqrt{-1}}\sum_{i=1}^{n}\int_{I(\epsilon,+\infty)}\frac{\exp(-t\sum_{j=1}^{r}a_{ij}x_j)}{\prod_{j=1}^{r}(1-\exp(-ta_{ij}))}\frac{dt}{t}.$$

By (1.7), we have

$$\zeta(0,A,x)=\frac{1}{n}\sum_{i=1}^{n}\zeta_r(0,(a_{i1},a_{i2},\dots,a_{ir}),\sum_{j=1}^{r}a_{ij}x_j). \tag{2.7}$$

By Theorem 1.1, we obtain the following theorem.

THEOREM 2.1.

$$\zeta(0,A,x)=\frac{1}{n}\sum_{i=1}^{n}\zeta_r(0,(a_{i1},a_{i2},\dots,a_{ir}),\sum_{j=1}^{r}a_{ij}x_j)$$

$$=\frac{(-1)^r}{n}\sum_{i=1}^{n}\sum_{l_1,\dots,l_r\geqq 0,l_1+\dots+l_r=r}a_{i1}^{l_1-1}a_{i2}^{l_2-1}\cdots a_{ir}^{l_r-1}\frac{B_{l_1}(x_1)}{l_1!}\frac{B_{l_2}(x_2)}{l_2!}\cdots\frac{B_{l_r}(x_r)}{l_r!}.$$

2.3. We see that $\Gamma(1-s)^n\sin\pi s(\sin n\pi s)^{-1}$ is holomorphic at $s=0$. Hence differentiating (2.5) at $s=0$ under the integral sign, we obtain

$$\begin{aligned}&\frac{\partial}{\partial s}\zeta(s,A,x)\Big|_{s=0}\\&=\gamma\cdot\frac{1}{(2\pi\sqrt{-1})^n}\sum_{i=1}^{n}\int_{I(\epsilon,+\infty)}\int_{I(\epsilon,1)^{n-1}}g_i(t,u)\prod_{l\neq i}u_l^{-1}t^{-1}\prod_{l\neq i}du_l dt\\&+\frac{1}{n}\frac{1}{(2\pi\sqrt{-1})^n}\sum_{i=1}^{n}\int_{I(\epsilon,+\infty)}\int_{I(\epsilon,1)^{n-1}}g_i(t,u)\\&\qquad\times\{\sum_{l\neq i}\log(-u_l)+n\log(-t)\}\prod_{l\neq i}u_l^{-1}du_l t^{-1}dt.\end{aligned}$$

For $1\leqq i\leqq n$, we put

$$\begin{aligned}J_i=\frac{1}{(2\pi\sqrt{-1})^n}&\int_{I(\epsilon,+\infty)}\int_{I(\epsilon,1)^{n-1}}g_i(t,u)\\&\times\{\sum_{l\neq i}\log(-u_l)+n\log(-t)\}\prod_{l\neq i}u_l^{-1}du_l t^{-1}dt.\end{aligned}$$

By (2.6), we obtain

$$\frac{\partial}{\partial s}\zeta(s,A,x)\Big|_{s=0}=n\gamma\zeta(0,A,x)+\frac{1}{n}\sum_{i=1}^{n}J_i. \tag{2.8}$$

We have

$$\begin{aligned}J_i&=\frac{n}{(2\pi\sqrt{-1})^n}\int_{I(\epsilon,+\infty)}\int_{I(\epsilon,1)^{n-1}}g_i(t,u)\prod_{l\neq i}u_l^{-1}du_l\frac{\log(-t)}{t}dt\\&+\frac{1}{(2\pi\sqrt{-1})^n}\sum_{k\neq i}\int_{I(\epsilon,+\infty)}\int_{I(\epsilon,1)^{n-1}}g_i(t,u)\log(-u_k)\prod_{l\neq i}u_l^{-1}du_l t^{-1}dt.\end{aligned}$$

Integrating with respect to $u_1, u_2, \dots, u_{i-1}, u_{i+1}, \dots, u_n$, we obtain

$$\frac{1}{(2\pi\sqrt{-1})^n}\int_{I(\epsilon,+\infty)}\int_{I(\epsilon,1)^{n-1}} g_i(t,u)\prod_{l\neq i} u_l^{-1}du_l \frac{\log(-t)}{t}dt$$
$$=\frac{1}{2\pi\sqrt{-1}}\int_{I(\epsilon,+\infty)} \frac{\exp(-t\sum_{j=1}^r a_{ij}x_j)}{\prod_{j=1}^r(1-\exp(-ta_{ij}))}\frac{\log(-t)}{t}dt.$$

By (1.7) and (1.15), this is equal to

$$\log\frac{\Gamma_r(\sum_{j=1}^r a_{ij}x_j,(a_{i1},\dots,a_{ir}))}{\rho_r((a_{i1},\dots,a_{ir}))} - \gamma\zeta_r(0,(a_{i1},\dots,a_{ir}),\sum_{j=1}^r a_{ij}x_j).$$

Integrating with respect to u_j, $j \neq i, k$, we obtain

$$\frac{1}{(2\pi\sqrt{-1})^n}\int_{I(\epsilon,+\infty)}\int_{I(\epsilon,1)^{n-1}} g_i(t,u)\log(-u_k)\prod_{l\neq i} u_l^{-1}du_l t^{-1}dt$$
$$(2.9)\qquad =\frac{1}{(2\pi\sqrt{-1})^2}\int_{I(\epsilon,+\infty)}\int_{I(\epsilon,1)} g_i(t,(0,\dots,u_k,\dots 1,0,\dots,0)) \times\frac{\log(-u_k)}{u_k}\frac{1}{t}du_k dt.$$

We have

$$g_i(t,(0,\dots,u_k,\dots 1,0,\dots,0)) = \prod_{j=1}^r \frac{\exp((1-x_j)t(a_{ij}+a_{kj}u_k))}{\exp(t(a_{ij}+a_{kj}u_k))-1}.$$

Put

$$(2.10)\qquad g_{ik}(t,u) = \prod_{j=1}^r \frac{\exp((1-x_j)t(a_{ij}+a_{kj}u))}{\exp(t(a_{ij}+a_{kj}u))-1}.$$

Then the integral (2.9) can be written as

$$\frac{1}{(2\pi\sqrt{-1})^2}\int_{I(\epsilon,+\infty)}\int_{I(\epsilon,1)} g_{ik}(t,u)\frac{\log(-u)}{u}\frac{1}{t}dudt.$$

We see easily that

$$(2.11)\qquad \int_{I(\epsilon,1)}\frac{\log(-u)}{u}du = 0.$$

Hence this integral is equal to

$$\frac{1}{(2\pi\sqrt{-1})^2}\int_{I(\epsilon,+\infty)}\int_{I(\epsilon,1)} \{g_{ik}(t,u)-g_{ik}(t,0)\}\frac{\log(-u)}{u}\frac{1}{t}dudt.$$

Therefore we obtain

$$(2.12)\qquad \begin{aligned} J_i &= n \log \frac{\Gamma_r(\sum_{j=1}^r a_{ij}x_j, (a_{i1}, \dots, a_{ir}))}{\rho_r((a_{i1}, \dots, a_{ir}))} \\ &- n\gamma\zeta_r(0, (a_{i1}, a_{i2}, \dots, a_{ir}), \sum_{j=1}^r a_{ij}x_j) \\ &+ \frac{1}{(2\pi\sqrt{-1})^2} \sum_{k\neq i} \int_{I(\epsilon,+\infty)} \int_{I(\epsilon,1)} \{g_{ik}(t,u) - g_{ik}(t,0)\} \frac{\log(-u)}{u} \frac{1}{t} du dt. \end{aligned}$$

We see

$$\begin{aligned} g_{ik}(t,u) - g_{ik}(t,0) &= \prod_{j=1}^r \sum_{l_j=0}^\infty \frac{B_{l_j}(1-x_j)}{l_j!}((a_{ij} + a_{kj}u)t)^{l_j-1} \\ &- \prod_{j=1}^r \sum_{l_j=0}^\infty \frac{B_{l_j}(1-x_j)}{l_j!}(a_{ij}t)^{l_j-1}. \end{aligned}$$

Considering the residue, we obtain

$$\begin{aligned} &\frac{1}{2\pi\sqrt{-1}} \int_{I(\epsilon,+\infty)} \{g_{ik}(t,u) - g_{ik}(t,0)\} \frac{1}{t} dt \\ &= \sum_{l_1,\dots,l_r \geqq 0, l_1+\dots+l_r=r} \prod_{j=1}^r \frac{B_{l_j}(1-x_j)}{l_j!} (\prod_{j=1}^r (a_{ij} + a_{kj}u)^{l_j-1} - \prod_{j=1}^r (a_{ij})^{l_j-1}). \end{aligned}$$

Since every a_{ij} is non-zero, the rational function

$$u^{-1}(\prod_{j=1}^r (a_{kj}u + a_{ij})^{l_j-1} - \prod_{j=1}^r (a_{ij})^{l_j-1})$$

is holomorphic at $u = 0$. We see easily that

$$\begin{aligned} &\frac{1}{2\pi\sqrt{-1}} \int_{I(\epsilon,1)} (\prod_{j=1}^r (a_{ij} + a_{kj}u)^{l_j-1} - \prod_{j=1}^r (a_{ij})^{l_j-1}) \frac{\log(-u)}{u} du \\ &= \int_0^1 (\prod_{j=1}^r (a_{ij} + a_{kj}u)^{l_j-1} - \prod_{j=1}^r (a_{ij})^{l_j-1}) \frac{du}{u}. \end{aligned}$$

For $l = (l_1, l_2, \dots, l_r)$, $0 \leqq l_i \in \mathbf{Z}$, $1 \leqq j,k \leqq n$, $j \neq k$, we put

$$(2.13)\qquad \begin{aligned} C_{l,j,k}(A) &= \int_0^1 \{ \prod_{m=1}^r (a_{jm} + a_{km}u)^{l_m-1} - \prod_{m=1}^r a_{jm}^{l_m-1} \} \frac{du}{u}, \\ C_l(A) &= \sum_{(j,k)} C_{l,j,k}(A). \end{aligned}$$

In the second formula, (j,k) extends over all ordered pairs of integers such that $1 \leqq j,k \leqq n$, $j \neq k$. Then we obtain

$$
\begin{aligned}
J_i =& n \log \frac{\Gamma_r(\sum_{j=1}^r a_{ij}x_j, (a_{i1}, \dots, a_{ir}))}{\rho_r((a_{i1}, \dots, a_{ir}))} \\
&-n\gamma\zeta_r(0, (a_{i1}, a_{i2}, \dots, a_{ir}), \sum_{j=1}^r a_{ij}x_j) \\
&+(-1)^r \sum_l \sum_{k\neq i} C_{l,i,k}(A) \prod_{j=1}^r \frac{B_{l_j}(x_j)}{l_j!}.
\end{aligned} \tag{2.14}
$$

Here $l = (l_1, l_2, \dots, l_r)$ extends over all r-tuples of non-negative integers such that $l_1 + l_2 + \dots + l_r = r$. By (2.7), (2.8) and (2.14), we obtain the following theorem.

THEOREM 2.2.

$$
\begin{aligned}
\frac{\partial}{\partial s}\zeta(s, A, x)\Big|_{s=0} =& \sum_{i=1}^n \log \frac{\Gamma_r(\sum_{j=1}^r a_{ij}x_j, (a_{i1}, \dots, a_{ir}))}{\rho_r((a_{i1}, \dots, a_{ir}))} \\
&+ \frac{(-1)^r}{n} \sum_l C_l(A) \prod_{j=1}^r \frac{B_{l_j}(x_j)}{l_j!}.
\end{aligned}
$$

Here $l = (l_1, l_2, \dots, l_r)$ extends over all r-tuples of non-negative integers such that $l_1 + l_2 + \dots + l_r = r$ and $C_l(A)$ is defined by (2.13).

Theorems 2.1 and 2.2 were obtained by Shintani.

2.4. Clearly $C_l(A)$ can be expressed as an elementary function of entries of A. We prove the following result ([Shi3]) for the use in succeeding chapters.

LEMMA 2.3. *Put*

$$
T_1(l) = \{m \mid 1 \leqq m \leqq r,\ l_m > 0\}, \qquad T_2(l) = \{m \mid 1 \leqq m \leqq r,\ l_m = 0\}.
$$

We abbreviate $T_1(l)$ and $T_2(l)$ to T_1 and T_2 respectively. If

$$
a_{kp}a_{jq} - a_{kq}a_{jp} \neq 0 \quad \text{for every} \quad p, q \in T_2,\ p \neq q, \tag{2.15}
$$

then we have

$$
C_{l,j,k}(A) = -\sum_{p\in T_2} \frac{1}{a_{jp}a_{kp}} \frac{\prod_{i\in T_1}(a_{ji}a_{kp} - a_{jp}a_{ki})^{l_i - 1}}{\prod_{q\in T_2, q\neq p}(a_{jq}a_{kp} - a_{jp}a_{kq})} \log(1 + \frac{a_{kp}}{a_{jp}}), \tag{2.16}
$$

$$
C_{l,j,k}(A) + C_{l,k,j}(A) = \sum_{p\in T_2} \frac{1}{a_{jp}a_{kp}} \frac{\prod_{i\in T_1}(a_{ji}a_{kp} - a_{jp}a_{ki})^{l_i - 1}}{\prod_{q\in T_2, q\neq p}(a_{jq}a_{kp} - a_{jp}a_{kq})} \log \frac{a_{jp}}{a_{kp}}. \tag{2.17}
$$

PROOF. We have

$$\sum_{m\in T_1}(l_m-1)=|T_2|. \tag{2.18}$$

Since $a_{km}u+a_{jm}$ are mutually prime for $m\in T_2$ by the assumption (2.15), we can write

$$\begin{aligned}&\frac{\prod_{m\in T_1}(a_{km}u+a_{jm})^{l_m-1}-\prod_{m=1}^{r}a_{jm}^{l_m-1}\prod_{m\in T_2}(a_{km}u+a_{jm})}{u\prod_{m\in T_2}(a_{km}u+a_{jm})}\\ &=\sum_{m\in T_2}\frac{r_m}{a_{km}u+a_{jm}}\end{aligned}$$

with $r_m\in\mathbf{R}$, in view of (2.18). Considering the asymptotic behaviors of both sides for $u\longrightarrow -a_{jm}/a_{km}$, we easily obtain

$$r_m=-\frac{1}{a_{jm}}\frac{\prod_{i\in T_1}(a_{km}a_{ji}-a_{ki}a_{jm})^{l_i-1}}{\prod_{q\in T_2,q\neq m}(a_{km}a_{jq}-a_{kq}a_{jm})}.$$

Now (2.16) follows from

$$\int_0^1\frac{du}{a_{km}u+a_{jm}}=\frac{1}{a_{km}}\log(1+\frac{a_{km}}{a_{jm}}),$$

and (2.17) follows from (2.16) and (2.18).

3. The second derivative of $\zeta(s,A,x)$ at $s=0$

3.1. Now we are going to consider higher derivatives of $\zeta(s,A,x)$ at $s=0$. Put

$$G(s)=\Gamma(1-s)^n\sin\pi s(\sin n\pi s)^{-1}.$$

By (2.5), we obtain

$$\begin{aligned}(\frac{\partial}{\partial s})^p\zeta(s,A,x)\Big|_{s=0}&=\sum_{q=0}^{p}\binom{p}{q}G^{(p-q)}(0)\\ &\times\frac{1}{(2\pi\sqrt{-1})^n}\sum_{i=1}^{n}\int_{I(\epsilon,+\infty)}\int_{I(\epsilon,1)^{n-1}}g_i(t,u)\{\sum_{l\neq i}\log(-u_l)+n\log(-t)\}^q\\ &\qquad\prod_{l\neq i}u_l^{-1}du_lt^{-1}dt.\end{aligned} \tag{3.1}$$

We can decompose the term for $q=2$ into the sum of three integrals:
(3.2)

$$\begin{aligned}&\frac{1}{(2\pi\sqrt{-1})^n}\int_{I(\epsilon,+\infty)}\int_{I(\epsilon,1)^{n-1}} g_i(t,u)\{\sum_{l\neq i}\log(-u_l)+n\log(-t)\}^2\prod_{l\neq i}u_l^{-1}du_l t^{-1}dt\\ =&\frac{n^2}{(2\pi\sqrt{-1})^n}\int_{I(\epsilon,+\infty)}\int_{I(\epsilon,1)^{n-1}} g_i(t,u)\prod_{l\neq i}u_l^{-1}du_l\frac{(\log(-t))^2}{t}dt\\ +&\frac{2n}{(2\pi\sqrt{-1})^n}\sum_{k\neq i}\int_{I(\epsilon,+\infty)}\int_{I(\epsilon,1)^{n-1}} g_i(t,u)\log(-u_k)\prod_{l\neq i}u_l^{-1}du_l\frac{\log(-t)}{t}dt\\ +&\frac{1}{(2\pi\sqrt{-1})^n}\sum_{k_1,k_2\neq i}\int_{I(\epsilon,+\infty)}\int_{I(\epsilon,1)^{n-1}} g_i(t,u)\log(-u_{k_1})\log(-u_{k_2})\prod_{l\neq i}u_l^{-1}du_l\frac{dt}{t}.\end{aligned}$$

3.2. The first integral can be integrated with respect to $u_1, u_2, \ldots, u_{i-1}, u_{i+1}, \ldots, u_n$. We see that it equals

$$\begin{aligned}&\frac{n^2}{2\pi\sqrt{-1}}\int_{I(\epsilon,+\infty)} g_i(t,0)\frac{(\log(-t))^2}{t}dt\\ =&\frac{n^2}{2\pi\sqrt{-1}}\int_{I(\epsilon,+\infty)}\frac{\exp(-t\sum_{j=1}^r a_{ij}x_j)}{\prod_{j=1}^r(1-\exp(-ta_{ij}))}\frac{(\log(-t))^2}{t}dt.\end{aligned}$$

Put

$$G_p=\log\frac{\Gamma_r^{[p]}(x,\omega)}{\rho_r^{[p]}(\omega)},\qquad \gamma_q=\Gamma^{(q)}(1),$$

$$I_{p-q}=\frac{1}{2\pi\sqrt{-1}}\int_{I(\epsilon,+\infty)}\frac{e^{-xt}(\log(-t))^{p-q}}{\prod_{j=1}^r(1-\exp(\omega_j t))}\frac{dt}{t}.$$

Then (1.6) and (1.12) can be written as

$$G_p=\sum_{q=0}^{p}\binom{p}{q}(-1)^q\gamma_q I_{p-q} \tag{3.3}$$

with $G_0=\zeta_r(0,x,\omega)$. We easily have

$$I_2=G_2-2\gamma G_1+(2\gamma^2-\gamma_2)G_0.$$

Hence we obtain

$$\begin{aligned}&\frac{1}{2\pi\sqrt{-1}}\int_{I(\epsilon,+\infty)}\frac{e^{-xt}(\log(-t))^2}{\prod_{j=1}^r(1-\exp(-\omega_j t))}\frac{dt}{t}\\ =&\log\frac{\Gamma_r^{[2]}(x,\omega)}{\rho_r^{[2]}(\omega)}-2\gamma\log\frac{\Gamma_r(x,\omega)}{\rho_r(\omega)}+(2\gamma^2-\gamma_2)\zeta_r(0,x,\omega).\end{aligned} \tag{3.4}$$

Here $\Gamma_r(x,\omega) = \Gamma_r^{[1]}(x,\omega)$ is the r-ple gamma function. Therefore the first term of the right-hand side of (3.2) can be expressed as

$$
\begin{aligned}
&\frac{n^2}{(2\pi\sqrt{-1})^n}\int_{I(\epsilon,+\infty)}\int_{I(\epsilon,1)^{n-1}} g_i(t,u)\prod_{l\neq i} u_l^{-1}du_l \frac{(\log(-t))^2}{t}dt \\
&= n^2\Bigg[\log\frac{\Gamma_r^{[2]}(\sum_{j=1}^r a_{ij}x_j,(a_{i1},a_{i2},\dots,a_{ir}))}{\rho_r^{[2]}((a_{i1},a_{i2},\dots,a_{ir}))} \\
&\quad - 2\gamma\log\frac{\Gamma_r(\sum_{j=1}^r a_{ij}x_j,(a_{i1},a_{i2},\dots,a_{ir}))}{\rho_r((a_{i1},a_{i2},\dots,a_{ir}))} \\
&\quad + (2\gamma^2-\gamma_2)\zeta_r(0,(a_{i1},a_{i2},\dots,a_{ir}),\sum_{j=1}^r a_{ij}x_j)\Bigg].
\end{aligned}
\tag{3.5}
$$

This term gives the main contribution. Below we are going to calculate "correction terms" and will attain the final result in §3.5. For a further explanation of the meaning of the calculation, see §3.6.

3.3. Next we consider the second term of the right-hand side of (3.2). Integrating with respect to $u_1, u_2, \dots, \check{u}_k, \dots, \check{u}_i, \dots, u_n$, we obtain

$$
\begin{aligned}
&\frac{1}{(2\pi\sqrt{-1})^n}\int_{I(\epsilon,+\infty)}\int_{I(\epsilon,1)^{n-1}} g_i(t,u)\log(-u_k)\prod_{l\neq i} u_l^{-1}du_l\frac{\log(-t)}{t}dt \\
&=\frac{1}{(2\pi\sqrt{-1})^2}\int_{I(\epsilon,+\infty)}\int_{I(\epsilon,1)} g_i(t,(0,\dots,0,u_k,0,\dots,0,1,0,\dots,0)) \\
&\qquad\qquad \times\frac{\log(-u_k)}{u_k}\frac{\log(-t)}{t}du_k dt \\
&=\frac{1}{(2\pi\sqrt{-1})^2}\int_{I(\epsilon,+\infty)}\int_{I(\epsilon,1)} g_{ik}(t,u)\frac{\log(-u)}{u}\frac{\log(-t)}{t}dudt \\
&=\frac{1}{(2\pi\sqrt{-1})^2}\int_{I(\epsilon,+\infty)}\int_{I(\epsilon,1)} (g_{ik}(t,u)-g_{ik}(t,0))\frac{\log(-u)}{u}\frac{\log(-t)}{t}dudt,
\end{aligned}
$$

using (2.11). By (2.10), (1.7) and (1.15), we have

$$
\begin{aligned}
&\frac{1}{(2\pi\sqrt{-1})}\int_{I(\epsilon,+\infty)} g_{ik}(t,u)\frac{\log(-t)}{t}dt \\
&=\log\frac{\Gamma_r(\sum_{j=1}^r(a_{kj}u+a_{ij})x_j,(a_{k1}u+a_{i1},a_{k2}u+a_{i2},\dots,a_{kr}u+a_{ir}))}{\rho_r((a_{k1}u+a_{i1},a_{k2}u+a_{i2},\dots,a_{kr}u+a_{ir}))} \\
&\quad -\gamma\zeta_r(0,(a_{k1}u+a_{i1},a_{k2}u+a_{i2},\dots,a_{kr}u+a_{ir}),\sum_{j=1}^r(a_{kj}u+a_{ij})x_j).
\end{aligned}
$$

Therefore we obtain

$$\frac{1}{(2\pi\sqrt{-1})^2}\int_{I(\epsilon,+\infty)}\int_{I(\epsilon,1)}(g_{ik}(t,u)-g_{ik}(t,0))\frac{\log(-u)}{u}\frac{\log(-t)}{t}dudt$$
$$=\int_0^1\left[\log\frac{\Gamma_r(\sum_{j=1}^r(a_{kj}u+a_{ij})x_j,(a_{k1}u+a_{i1},a_{k2}u+a_{i2},\dots,a_{kr}u+a_{ir}))}{\rho_r((a_{k1}u+a_{i1},a_{k2}u+a_{i2},\dots,a_{kr}u+a_{ir}))}\right.$$
$$\left.-\log\frac{\Gamma_r(\sum_{j=1}^r a_{ij}x_j,(a_{i1},a_{i2},\dots,a_{ir}))}{\rho_r((a_{i1},a_{i2},\dots,a_{ir}))}\right]\frac{du}{u}$$
$$-\gamma\int_0^1\left[\zeta_r(0,(a_{k1}u+a_{i1},a_{k2}u+a_{i2},\dots,a_{kr}u+a_{ir}),\sum_{j=1}^r(a_{kj}u+a_{ij})x_j)\right.$$
$$\left.-\zeta_r(0,(a_{i1},a_{i2},\dots,a_{ir}),\sum_{j=1}^r a_{ij}x_j)\right]\frac{du}{u}.$$

For $\mathfrak{a}=(a_1,a_2,\dots,a_r)$, $\mathfrak{b}=(b_1,b_2,\dots,b_r)\in\mathbf{R}^r$ such that a_i, $b_j>0$ for all i and j and ${}^t\mathfrak{X}=(x_1,x_2,\dots,x_r)\in\mathbf{R}^r$, we put

$$\Lambda_r(\mathfrak{a},\mathfrak{b},\mathfrak{X})=\int_0^1\left[\log\frac{\Gamma_r((u\mathfrak{a}+\mathfrak{b})\mathfrak{X},u\mathfrak{a}+\mathfrak{b})}{\rho_r(u\mathfrak{a}+\mathfrak{b})}-\log\frac{\Gamma_r(\mathfrak{b}\mathfrak{X},\mathfrak{b})}{\rho_r(\mathfrak{b})}\right]\frac{du}{u}. \tag{3.6}$$

By Theorem 1.1, we have

$$\int_0^1\left[\zeta_r(0,(a_{k1}u+a_{i1},a_{k2}u+a_{i2},\dots,a_{kr}u+a_{ir}),\sum_{j=1}^r(a_{kj}u+a_{ij})x_j)\right.$$
$$\left.-\zeta_r(0,(a_{i1},a_{i2},\dots,a_{ir}),\sum_{j=1}^r a_{ij}x_j)\right]\frac{du}{u}$$
$$=(-1)^r\sum_l\int_0^1\{\prod_{j=1}^r(a_{kj}u+a_{ij})^{l_j-1}-\prod_{j=1}^r a_{ij}^{l_j-1}\}\frac{du}{u}\times\prod_{j=1}^r\frac{B_{l_j}(x_j)}{l_j!}$$
$$=(-1)^r\sum_l C_{l,i,k}(A)\prod_{j=1}^r\frac{B_{l_j}(x_j)}{l_j!}.$$

Here $l=(l_1,l_2,\dots,l_r)$ extends over all r-tuples of non-negative integers such that $l_1+l_2+\dots+l_r=r$. Hereafter in this section, $\sum_l$ will always have this meaning. Therefore concerning the second term of the right-hand side of (3.2), we obtain

$$\frac{2n}{(2\pi\sqrt{-1})^n}\sum_{k\neq i}\int_{I(\epsilon,+\infty)}\int_{I(\epsilon,1)^{n-1}}g_i(t,u)\log(-u_k)\prod_{l\neq i}u_l^{-1}du_l\frac{\log(-t)}{t}dt$$
$$=2n\sum_{k\neq i}\left[\Lambda_r((a_{k1},a_{k2},\dots,a_{kr}),(a_{i1},a_{i2},\dots,a_{ir}),{}^t(x_1,x_2,\dots,x_r))\right. \tag{3.7}$$
$$\left.-(-1)^r\gamma\sum_l C_{l,i,k}(A)\prod_{j=1}^r\frac{B_{l_j}(x_j)}{l_j!}\right].$$

3.4. Now we consider the third term of the right-hand side of (3.2). We divide the summation $\sum_{k_1,k_2\neq i}$ as $\sum_{k_1,k_2\neq i,k_1\neq k_2}+\sum_{k_1=k_2=k,k\neq i}$. First assume $k_1\neq k_2$. Then integrating with respect to u_1, u_2, ..., $\check{u}_{k_1}$, ..., $\check{u}_{k_2}$, ..., $\check{u}_i$, ..., u_n, we obtain

(3.8)
$$\begin{aligned}&\frac{1}{(2\pi\sqrt{-1})^n}\sum_{k_1,k_2\neq i}\int_{I(\epsilon,+\infty)}\int_{I(\epsilon,1)^{n-1}}g_i(t,u)\log(-u_{k_1})\log(-u_{k_2})\prod_{l\neq i}u_l^{-1}du_l\frac{dt}{t}\\&=\frac{1}{(2\pi\sqrt{-1})^3}\sum_{k_1,k_2\neq i}\int_{I(\epsilon,+\infty)}\int_{I(\epsilon,1)^2}g_{i,k_1,k_2}(t,u_{k_1},u_{k_2})\frac{\log(-u_{k_1})}{u_{k_1}}\frac{\log(-u_{k_2})}{u_{k_2}}\frac{dt}{t},\end{aligned}$$

where

$$\begin{aligned}&g_{i,k_1,k_2}(t,u_{k_1},u_{k_2})=g_i(t,(0,\dots,0,u_{k_1},0,\dots,0,u_{k_2},0,\dots,1,\dots,0))\\&=\prod_{j=1}^{r}\frac{\exp(t(1-x_j)(a_{ij}+a_{k_1j}u_{k_1}+a_{k_2j}u_{k_2}))}{\exp(t(a_{ij}+a_{k_1j}u_{k_1}+a_{k_2j}u_{k_2}))-1}.\end{aligned}\tag{3.9}$$

According as the decomposition

$$\begin{aligned}&g_{i,k_1,k_2}(t,u_{k_1},u_{k_2})=g_{i,k_1,k_2}(t,0,0)\\&+(g_{i,k_1,k_2}(t,u_{k_1},0)-g_{i,k_1,k_2}(t,0,0))+(g_{i,k_1,k_2}(t,0,u_{k_2})-g_{i,k_1,k_2}(t,0,0))\\&+(g_{i,k_1,k_2}(t,u_{k_1},u_{k_2})-g_{i,k_1,k_2}(t,u_{k_1},0)-g_{i,k_1,k_2}(t,0,u_{k_2})+g_{i,k_1,k_2}(t,0,0)),\end{aligned}$$

we decompose the integral on the second line of (3.8) as $I_1+I_2+I_3+I_4$. By (2.11), we see immediately that $I_1=I_2=I_3=0$. Since

$$g_{i,k_1,k_2}(t,u_{k_1},u_{k_2})=\prod_{j=1}^{r}\sum_{l_j=0}^{\infty}\frac{B_{l_j}(1-x_j)}{l_j!}((a_{ij}+a_{k_1j}u_{k_1}+a_{k_2j}u_{k_2})t)^{l_j-1},$$

we obtain

$$\begin{aligned}&\frac{1}{(2\pi\sqrt{-1})}\int_{I(\epsilon,+\infty)}g_{i,k_1,k_2}(t,u_{k_1},u_{k_2})\frac{dt}{t}\\&=(-1)^r\sum_{l}\prod_{j=1}^{r}(a_{ij}+a_{k_1j}u_{k_1}+a_{k_2j}u_{k_2})^{l_j-1}\frac{B_{l_j}(x_j)}{l_j!}.\end{aligned}$$

Therefore we have

$$\begin{aligned}I_4=&\frac{1}{(2\pi\sqrt{-1})^3}\int_{I(\epsilon,+\infty)}\int_{I(\epsilon,1)^2}(g_{i,k_1,k_2}(t,u_{k_1},u_{k_2})-g_{i,k_1,k_2}(t,u_{k_1},0)\\&-g_{i,k_1,k_2}(t,0,u_{k_2})+g_{i,k_1,k_2}(t,0,0))\frac{\log(-u_{k_1})}{u_{k_1}}\frac{\log(-u_{k_2})}{u_{k_2}}du_{k_1}du_{k_2}\frac{dt}{t}\\=&\frac{(-1)^r}{(2\pi\sqrt{-1})^2}\sum_{l}\int_{I(\epsilon,1)^2}\{\prod_{j=1}^{r}(a_{ij}+a_{k_1j}u_{k_1}+a_{k_2j}u_{k_2})^{l_j-1}-\prod_{j=1}^{r}(a_{ij}+a_{k_1j}u_{k_1})^{l_j-1}\\&-\prod_{j=1}^{r}(a_{ij}+a_{k_2j}u_{k_2})^{l_j-1}+\prod_{j=1}^{r}a_{ij}^{l_j-1}\}\frac{\log(-u_{k_1})}{u_{k_1}}\frac{\log(-u_{k_2})}{u_{k_2}}du_{k_1}du_{k_2}\times\prod_{j=1}^{r}\frac{B_{l_j}(x_j)}{l_j!}\\=&(-1)^r\sum_{l}C_{l,i,k_1,k_2}(A)\prod_{j=1}^{r}\frac{B_{l_j}(x_j)}{l_j!},\end{aligned}$$

where we put

$$\begin{aligned}(3.10)\qquad & C_{l,i,k_1,k_2}(A) \\ =& \int_0^1\int_0^1 \Big\{ \prod_{j=1}^r (a_{ij} + a_{k_1 j}u_1 + a_{k_2 j}u_2)^{l_j-1} - \prod_{j=1}^r (a_{ij} + a_{k_1 j}u_1)^{l_j-1} \\ & - \prod_{j=1}^r (a_{ij} + a_{k_2 j}u_2)^{l_j-1} + \prod_{j=1}^r a_{ij}^{l_j-1}\Big\} \frac{du_1}{u_1}\frac{du_2}{u_2}.\end{aligned}$$

Next assume $k_1 = k_2 = k$. Integrating with respect to $u_1, u_2, \dots, \check{u}_k, \dots, \check{u}_i, \dots, u_n$, we obtain

$$\begin{aligned}&\frac{1}{(2\pi\sqrt{-1})^n}\int_{I(\epsilon,+\infty)}\int_{I(\epsilon,1)^{n-1}} g_i(t,u)(\log(-u_k))^2 \prod_{l\neq i} u_l^{-1}du_l \frac{dt}{t} \\ =&\frac{1}{(2\pi\sqrt{-1})^2}\int_{I(\epsilon,+\infty)}\int_{I(\epsilon,1)} g_{ik}(t,u)\frac{(\log(-u))^2}{u}du\frac{dt}{t}.\end{aligned}$$

Integrating with respect to t first, we obtain

$$\begin{aligned}&\frac{1}{(2\pi\sqrt{-1})^2}\int_{I(\epsilon,+\infty)}\int_{I(\epsilon,1)} (g_{ik}(t,u) - g_{ik}(t,0))\frac{(\log(-u))^2}{u}du\frac{dt}{t} \\ =&\frac{1}{2\pi\sqrt{-1}}(-1)^r\sum_l \int_{I(\epsilon,1)} \{\prod_{j=1}^r (a_{ij} + a_{kj}u)^{l_j-1} - \prod_{j=1}^r a_{ij}^{l_j-1}\}\frac{(\log(-u))^2}{u}du \\ &\times \prod_{j=1}^r \frac{B_{l_j}(x_j)}{l_j!} \\ =&(-1)^r\sum_l \int_0^1 \{\prod_{j=1}^r (a_{ij} + a_{kj}u)^{l_j-1} - \prod_{j=1}^r a_{ij}^{l_j-1}\}\frac{2\log u}{u}du \times \prod_{j=1}^r \frac{B_{l_j}(x_j)}{l_j!} \\ =&(-1)^r\sum_l 2C'_{l,i,k}(A))\prod_{j=1}^r \frac{B_{l_j}(x_j)}{l_j!},\end{aligned}$$

where we put

$$(3.11)\qquad C'_{l,i,k}(A) = \int_0^1 \{\prod_{j=1}^r (a_{ij} + a_{kj}u)^{l_j-1} - \prod_{j=1}^r a_{ij}^{l_j-1}\}\frac{\log u}{u}du.$$

Since

$$\int_{I(\epsilon,1)} \frac{(\log(-u))^2}{u}du = \frac{2}{3}(\pi\sqrt{-1})^3,$$

we get

$$\begin{aligned}&\frac{1}{(2\pi\sqrt{-1})^2}\int_{I(\epsilon,+\infty)}\int_{I(\epsilon,1)} g_{ik}(t,0)\frac{(\log(-u))^2}{u}du\frac{dt}{t} \\ =&\frac{1}{3}(\pi\sqrt{-1})^2\frac{1}{(2\pi\sqrt{-1})}\int_{I(\epsilon,1)} g_{ik}(t,0)\frac{dt}{t} \\ =&-\frac{1}{3}\pi^2\zeta_r(0,(a_{i1},a_{i2},\dots,a_{ir}),\sum_{j=1}^r a_{ij}x_j).\end{aligned}$$

Summing up the calculations above, we obtain

$$
\begin{aligned}
&\frac{1}{(2\pi\sqrt{-1})^n}\int_{I(\epsilon,+\infty)}\int_{I(\epsilon,1)^{n-1}} g_i(t,u)\{\sum_{l\neq i}\log(-u_l)+n\log(-t)\}^2 \\
&\qquad\qquad\qquad\qquad \prod_{l\neq i} u_l^{-1}du_l t^{-1}dt \\
&= n^2\left[\log\frac{\Gamma_r^{[2]}(\sum_{j=1}^r a_{ij}x_j,(a_{i1},a_{i2},\dots,a_{ir}))}{\rho_r^{[2]}((a_{i1},a_{i2},\dots,a_{ir}))}\right. \\
&\quad \left. -2\gamma\log\frac{\Gamma_r(\sum_{j=1}^r a_{ij}x_j,(a_{i1},a_{i2},\dots,a_{ir}))}{\rho_r((a_{i1},a_{i2},\dots,a_{ir}))}\right] \\
&\quad +2n\sum_{k\neq i}\left[\Lambda_r((a_{k1},a_{k2},\dots,a_{kr}),(a_{i1},a_{i2},\dots,a_{ir}),{}^t(x_1,x_2,\dots,x_r))\right. \\
&\quad \left. -(-1)^r\gamma\sum_l C_{l,i,k}(A)\prod_{j=1}^r\frac{B_{l_j}(x_j)}{l_j!}\right] \\
&\quad +\sum_{k_1,k_2\neq i,k_1\neq k_2}\left[(-1)^r\sum_l C_{l,i,k_1,k_2}(A)\prod_{j=1}^r\frac{B_{l_j}(x_j)}{l_j!}\right] \\
&\quad +\sum_{k\neq i}\left[(-1)^r\sum_l 2C'_{l,i,k}(A)\prod_{j=1}^r\frac{B_{l_j}(x_j)}{l_j!}\right] \\
&\quad +\left[n^2(2\gamma^2-\gamma_2)-\frac{1}{3}(n-1)\pi^2\right]\zeta_r(0,(a_{i1},a_{i2},\dots,a_{ir}),\sum_{j=1}^r a_{ij}x_j).
\end{aligned}
\tag{3.12}
$$

3.5. We see easily that

$$
G(0)=\frac{1}{n},\qquad G'(0)=\gamma,\qquad G''(0)=\gamma_2+(n-1)\gamma^2+\frac{n-1}{3n}\pi^2.
\tag{3.13}
$$

We have

$$
\gamma_2=\gamma^2+\frac{\pi^2}{6}.
\tag{3.14}
$$

To see this, we use the formula ([WW], p. 250)

$$
\frac{d^2}{ds^2}\log\Gamma(s)=\sum_{n=0}^{\infty}\frac{1}{(s+n)^2}
$$

which holds except for $0\geqq s\in\mathbf{Z}$. Putting $s=1$, we get $\Gamma''(1)=\Gamma'(1)^2+\zeta(2)$.

Using (3.13), (3.14) and (2.14), we get

$$
\begin{aligned}
&\sum_{q=0}^{2}\binom{2}{q}G^{(2-q)}(0)\\
&\times\frac{1}{(2\pi\sqrt{-1})^n}\int_{I(\epsilon,+\infty)}\int_{I(\epsilon,1)^{n-1}}g_i(t,u)\{\sum_{l\neq i}\log(-u_l)+n\log(-t)\}^q\\
&\qquad\prod_{l\neq i}u_l^{-1}du_l t^{-1}dt\\
=&n\log\frac{\Gamma_r^{[2]}(\sum_{j=1}^{r}a_{ij}x_j,(a_{i1},a_{i2},\dots,a_{ir}))}{\rho_r^{[2]}((a_{i1},a_{i2},\dots,a_{ir}))}\\
&+2\sum_{k\neq i}\Lambda_r((a_{k1},a_{k2},\dots,a_{kr}),(a_{i1},a_{i2},\dots,a_{ir}),{}^t(x_1,x_2,\dots,x_r))\\
&+\frac{(-1)^r}{n}\Bigg[\sum_{k_1,k_2\neq i,k_1\neq k_2}\sum_{l}C_{l,i,k_1,k_2}(A)\prod_{j=1}^{r}\frac{B_{l_j}(x_j)}{l_j!}\\
&\qquad+\sum_{k\neq i}\sum_{l}2C'_{l,i,k}(A))\prod_{j=1}^{r}\frac{B_{l_j}(x_j)}{l_j!}\Bigg]\\
&+\frac{(n-1)\pi^2}{6}\zeta_r(0,(a_{i1},a_{i2},\dots,a_{ir}),\sum_{j=1}^{r}a_{ij}x_j).
\end{aligned}\tag{3.15}
$$

Thus we obtain the following theorem:

THEOREM 3.1.

$$
\begin{aligned}
&(\frac{\partial}{\partial s})^2\zeta(s,A,x)\Big|_{s=0}\\
=&n\sum_{i=1}^{n}\log\frac{\Gamma_r^{[2]}(\sum_{j=1}^{r}a_{ij}x_j,(a_{i1},a_{i2},\dots,a_{ir}))}{\rho_r^{[2]}((a_{i1},a_{i2},\dots,a_{ir}))}\\
&+2\sum_{1\leqq i,k\leqq n,k\neq i}\Lambda_r((a_{k1},a_{k2},\dots,a_{kr}),(a_{i1},a_{i2},\dots,a_{ir}),{}^t(x_1,x_2,\dots,x_r))\\
&+\frac{(-1)^r}{n}\sum_{1\leqq i,k_1,k_2\leqq n,k_1\neq k_2,k_2\neq i,i\neq k_1}\sum_{l}C_{l,i,k_1,k_2}(A)\prod_{j=1}^{r}\frac{B_{l_j}(x_j)}{l_j!}\\
&+\frac{2(-1)^r}{n}\sum_{1\leqq i,k\leqq n,k\neq i}\sum_{l}C'_{l,i,k}(A)\prod_{j=1}^{r}\frac{B_{l_j}(x_j)}{l_j!}\\
&+\sum_{i=1}^{n}\frac{(n-1)\pi^2}{6}\zeta_r(0,(a_{i1},a_{i2},\dots,a_{ir}),\sum_{j=1}^{r}a_{ij}x_j).
\end{aligned}
$$

Here $l=(l_1,l_2,\dots,l_r)$ extends over all r-tuples of non-negative integers such that $l_1+l_2+\cdots+l_r=r$; Λ_r, $C_{l,i,k_1,k_2}(A)$ and $C'_{l,i,k}(A)$ are defined by (3.6), (3.10), (3.11) *respectively.*

3.6. There are interesting features in our calculations in §2 and §3. First considerable cancellations took place before reaching the final results. In this respect, we note that the integral representation (2.5) is slightly different from that of Shintani due to the choice of the branch of $\log(-z)$; (2.5) has the advantage that $(\frac{\partial}{\partial s})^p \zeta(s, A, x)\big|_{s=0}$ is manifestly real. Next take $0 < f \in \mathbf{Z}$. From $A \in M(n, r, \mathbf{R})$ as above, we define $B \in M(fn, r, \mathbf{R})$ by ${}^tB = ({}^tA\,{}^tA \cdots {}^tA)$. Then, by definition, we have $\zeta(s, B, x) = \zeta(fs, A, x)$. Therefore

$$(3.16)\qquad (\frac{\partial}{\partial s})^p \zeta(s, B, x)\Big|_{s=0} = f^p (\frac{\partial}{\partial s})^p \zeta(s, A, x)\Big|_{s=0}$$

for $0 \leqq p \in \mathbf{Z}$. We can check easily that Theorems 2.1 and 2.2 are consistent with (3.16).

Let us write the formula of Theorem 3.1 in the form by which (3.16) is obvious. For $1 \leqq i \leqq n$, let $A^{(i)}$ denote the i-th row of A. For $1 \leqq i, k \leqq n$, put

$$y_{i,k} = (\frac{\partial}{\partial s})^2 \zeta(s, \begin{pmatrix} A^{(i)} \\ A^{(k)} \end{pmatrix}, x)\Big|_{s=0} - 2(\frac{\partial}{\partial s})^2 \zeta(s, A^{(i)}, x)\Big|_{s=0} - 2(\frac{\partial}{\partial s})^2 \zeta(s, A^{(k)}, x)\Big|_{s=0}.$$

For $1 \leqq i, k, p \leqq n$, put

$$\begin{aligned} z_{i,k,p} = (\frac{\partial}{\partial s})^2 \zeta(s, \begin{pmatrix} A^{(i)} \\ A^{(k)} \\ A^{(p)} \end{pmatrix}, x)\Big|_{s=0} &- 3(\frac{\partial}{\partial s})^2 \zeta(s, A^{(i)}, x)\Big|_{s=0} \\ &- 3(\frac{\partial}{\partial s})^2 \zeta(s, A^{(k)}, x)\Big|_{s=0} - 3(\frac{\partial}{\partial s})^2 \zeta(s, A^{(p)}, x)\Big|_{s=0}. \end{aligned}$$

Applying Theorem 3.1 to A, $\begin{pmatrix} A^{(i)} \\ A^{(k)} \end{pmatrix}$, $\begin{pmatrix} A^{(i)} \\ A^{(k)} \\ A^{(p)} \end{pmatrix}$, we find

$$(3.17)\qquad \begin{aligned} (\frac{\partial}{\partial s})^2 \zeta(s, A, x)\Big|_{s=0} &= n \sum_{i=1}^{n} (\frac{\partial}{\partial s})^2 \zeta(s, A^{(i)}, x)\Big|_{s=0} + \frac{3-n}{n} \sum_{1 \leqq i,k \leqq n, i \neq k} y_{i,k} \\ &+ \frac{1}{2n} \sum_{1 \leqq i,k,p \leqq n, i \neq k, k \neq p, p \neq i} z_{i,k,p}. \end{aligned}$$

Applying (3.17) to $\begin{pmatrix} A^{(i)} \\ A^{(i)} \end{pmatrix}$ and $\begin{pmatrix} A^{(i)} \\ A^{(i)} \\ A^{(i)} \end{pmatrix}$, we find $y_{i,i} = z_{i,i,i} = 0$ for $1 \leqq i \leqq n$ in view of (3.16). Next applying (3.17) to $\begin{pmatrix} A^{(i)} \\ A^{(k)} \end{pmatrix}$ and $\begin{pmatrix} A^{(i)} \\ A^{(k)} \\ A^{(i)} \\ A^{(k)} \end{pmatrix}$ and using (3.16) with $f = 2$, we obtain

$$(3.18)\qquad z_{i,i,k} + z_{k,k,i} = 2(y_{i,k} + y_{k,i}), \qquad 1 \leqq i, k \leqq n.$$

By (3.18), we have

$$\sum_{1 \leqq i,k,p \leqq n} z_{i,k,p} = \sum_{1 \leqq i,k,p \leqq n, i \neq k, k \neq p, p \neq i} z_{i,k,p} + 6 \sum_{1 \leqq i,k \leqq n, i \neq k} y_{i,k}.$$

Therefore (3.17) can be written as

$$(\frac{\partial}{\partial s})^2\zeta(s,A,x)\Big|_{s=0} = n\sum_{i=1}^{n}(\frac{\partial}{\partial s})^2\zeta(s,A^{(i)},x)\Big|_{s=0} - \sum_{1\leqq i,k\leqq n} y_{i,k} + \frac{1}{2n}\sum_{1\leqq i,k,p\leqq n} z_{i,k,p}. \tag{3.19}$$

In the form (3.19), the relation (3.16) is manifest.

It is remarkable that only the r-ple gamma function is needed in Theorem 2.2, except for elementary factors, to express $(\frac{\partial}{\partial s})\zeta(s,A,x)|_{s=0}$. In Theorem 3.1, the occurence of $\log\dfrac{\Gamma_r^{[2]}(x,\omega)}{\rho_r^{[2]}(\omega)}$ is inevitable, but one more new function Λ_r is needed, which may be viewed as an unsatisfactory point. To clarify the nature of Λ_r, suppose $r=1$. Then, by definition and (1.19), we get

$$\begin{aligned}\Lambda_1(a,b,x) &= \int_0^1\left[\log\frac{\Gamma_1((ua+b)x,ua+b)}{\rho_1((ua+b))} - \log\frac{\Gamma_1(bx,b)}{\rho_1(b)}\right]\frac{du}{u}\\ &= (x-1/2)\int_0^1(\log(ua+b)-\log b)\frac{du}{u} = (x-1/2)f(a/b),\end{aligned}$$

where we put

$$f(k) = \int_0^1 \log(1+ku)\frac{du}{u}, \qquad k>0.$$

We have $f'(k) = \log(1+k)/k$. Hence $\frac{d}{dk}(f(k)+f(1/k)) = \log k/k$. Using $f(1) = \pi^2/12$, we get

$$f(k) + f(\frac{1}{k}) = \frac{1}{2}(\log k)^2 + \frac{\pi^2}{6}.$$

(Using the dilogarithm function $\mathrm{Li}_2(x)$, we can write $f(k) = -\mathrm{Li}_2(-k)$ and this is a well known functional equation for the dilogarithm function.) Therefore we obtain

$$\Lambda_1(a,b,x) + \Lambda_1(b,a,x) = (x-\frac{1}{2})\left\{\frac{1}{2}\log(\frac{b}{a})^2 + \frac{\pi^2}{6}\right\}. \tag{3.20}$$

For $r>1$, it is an open question whether $\Lambda_r(\mathfrak{a},\mathfrak{b},\mathfrak{X}) + \Lambda_r(\mathfrak{b},\mathfrak{a},\mathfrak{X})$ can be expressed elementarily or not.

3.7. Suggested by (3.19), we can make the following prediction on the nature of $(\frac{\partial}{\partial s})^p\zeta(s,A,x)\Big|_{s=0}$.

CONJECTURE 3.2. *There exist constants $c_s^{(p)}$, $1\leqq s\leqq p$ independent of A and n such that*

$$\begin{aligned}(\frac{\partial}{\partial s})^p\zeta(s,A,x)\Big|_{s=0} &= n^{p-1}\sum_{i=1}^{n}(\frac{\partial}{\partial s})^p\zeta(s,A^{(i)},x)\Big|_{s=0}\\ &+ \sum_{1\leqq s\leqq p} c_s^{(p)} n^{p-s-1}\sum_{1\leqq i_1,i_2,\dots,i_{s+1}\leqq n} z_{i_1,i_2,\dots,i_{s+1}},\end{aligned} \tag{3.21}$$

where

$$z_{i_1,i_2,\dots,i_{s+1}} = (\frac{\partial}{\partial s})^p \zeta(s, \begin{pmatrix} A^{(i_1)} \\ A^{(i_2)} \\ \vdots \\ A^{(i_{s+1})} \end{pmatrix}, x)\Big|_{s=0} - (s+1)^{p-1} \sum_{j=1}^{s+1} (\frac{\partial}{\partial s})^p \zeta(s, A^{(i_j)}, x)\Big|_{s=0}.$$

In the form of (3.21), it is evident that (3.16) is satisfied. Theorem 2.2 and (3.19) give

$$c_1^{(1)} = 1, \qquad c_1^{(2)} = -1, \quad c_2^{(2)} = \frac{1}{2}.$$

If the conjecture is true, the constant $c_s^{(p)}$ must satisfy several consistency conditions. For example, if $A^{(1)}$ is multiplied by t, $t > 0$, then $(\frac{\partial}{\partial s})^p \zeta(s, \begin{pmatrix} A^{(i_1)} \\ A^{(i_2)} \\ \vdots \\ A^{(i_{s+1})} \end{pmatrix}, x)\Big|_{s=0}$ changes to $t^{-ms}(\frac{\partial}{\partial s})^p \zeta(s, \begin{pmatrix} A^{(i_1)} \\ A^{(i_2)} \\ \vdots \\ A^{(i_{s+1})} \end{pmatrix}, x)\Big|_{s=0}$, where m is the number of indices j such that $i_j = 1$, $1 \leqq j \leqq s+1$. Using this fact, we can determine $c_s^{(p)}$ without difficulty. The result is

$$c_s^{(p)} = \frac{(-1)^{p+s}}{s!(p-s)!}, \qquad 1 \leqq s \leqq p. \tag{3.22}$$

4. An asymptotic expansion of the multiple gamma function

In this section, we will describe an asymptotic expansion of the multiple gamma function due to Barnes. We will further give an explicit estimation of the remainder term as promised in [Y4].

4.1. Our basis of investigation is the formula (1.15). Let

$$\frac{(-1)^r z}{\prod_{k=1}^r (1 - e^{-\omega_k z})} = \frac{(-1)^r A_r}{z^{r-1}} + \frac{(-1)^{r-1} A_{r-1}}{z^{r-2}} + \cdots - A_1 + \sum_{n=1}^{\infty} \frac{{}_rB_n}{n!} z^n \tag{4.1}$$

be the Laurent expansion at $z = 0$. The coefficients ${}_rB_n$, A_1, $\dots$, A_r are symmetric rational functions of ω_1, ω_2, $\dots$, ω_r. For a positive integer N, we put

$$g_N(z) = \frac{(-1)^r z}{\prod_{k=1}^r (1 - e^{-\omega_k z})} - (\sum_{l=1}^r \frac{(-1)^l A_l}{z^{l-1}} + \sum_{n=1}^N \frac{{}_rB_n}{n!} z^n).$$

Then we have

$$\begin{aligned} \log \frac{\Gamma_r(x,\omega)}{\rho_r(\omega)} &= \sum_{l=1}^r (-1)^{l+r} A_l \frac{1}{2\pi\sqrt{-1}} \int_L \frac{e^{-xz}\{\log(-z) + \gamma\}}{z^{l+1}} dz \\ &+ \sum_{n=1}^N \frac{(-1)^r {}_rB_n}{n!} \frac{1}{2\pi\sqrt{-1}} \int_L z^{n-2} e^{-xz} \{\log(-z) + \gamma\} dz \\ &+ \frac{(-1)^r}{2\pi\sqrt{-1}} \int_L \frac{g_N(z) e^{-xz}\{\log(-z)+\gamma\}}{z^2} dz. \end{aligned} \tag{4.2}$$

For $m \geqq 0$, we have

$$\frac{1}{2\pi\sqrt{-1}}\int_L z^m e^{-xz}\{\log(-z)+\gamma\}dz = \int_0^\infty z^m e^{-xz}dz.$$

Hence we obtain

$$\frac{1}{2\pi\sqrt{-1}}\int_L z^m e^{-xz}\{\log(-z)+\gamma\}dz = m!\,x^{-m-1}, \qquad m \geqq 0. \tag{4.3}$$

By (1.5), we have

$$\frac{1}{2\pi\sqrt{-1}}\int_L (-z)^{s-1}e^{-xz}dz = -x^{-s}\frac{1}{\Gamma(1-s)}.$$

Differentiating with respect to s, we obtain

$$\frac{1}{2\pi\sqrt{-1}}\int_L (-z)^{s-1}\log(-z)e^{-xz}dz = x^{-s}\log x\cdot\frac{1}{\Gamma(1-s)} - x^{-s}\cdot\frac{\Gamma'(1-s)}{\Gamma(1-s)^2}. \tag{4.4}$$

Let $0 \leqq m \in \mathbf{Z}$. Putting $s=-m$ in (4.4), we get

$$\begin{aligned}&\frac{1}{2\pi\sqrt{-1}}\int_L z^{-m-1}e^{-xz}\log(-z)dz\\ &=\frac{(-1)^{m+1}}{\Gamma(m+1)}x^m\log x + \frac{(-1)^m}{\Gamma(m+1)^2}\Gamma'(m+1)x^m.\end{aligned} \tag{4.5}$$

We see easily that

$$\frac{\Gamma'(m+1)}{\Gamma(m+1)} = 1+\frac{1}{2}+\cdots+\frac{1}{m}-\gamma \qquad (m \geqq 1)$$

and

$$\frac{1}{2\pi\sqrt{-1}}\int_L z^{-m-1}e^{-xz}dz = \frac{(-x)^m}{\Gamma(m+1)}$$

by the residue calculus. By (4.5), we get

$$\begin{aligned}&\frac{1}{2\pi\sqrt{-1}}\int_L z^{-m-1}e^{-xz}\{\log(-z)+\gamma\}dz\\ &=\frac{(-1)^{m+1}}{m!}x^m\log x + \frac{(-1)^m}{m!}(1+\frac{1}{2}+\cdots+\frac{1}{m})x^m\end{aligned} \tag{4.6}$$

if $m \geqq 1$. If $m=0$, the left hand side is equal to $-\log x$. Inserting (4.3) and (4.6) into (4.2), we obtain

$$\begin{aligned}\log\frac{\Gamma_r(x,\omega)}{\rho_r(\omega)} &= (-1)^{r-1}\sum_{m=1}^{r}\frac{A_m}{m!}x^m(\log x - 1 - \frac{1}{2} - \cdots - \frac{1}{m})\\ &\quad + (-1)^{r-1}{}_rB_1\log x + (-1)^r\sum_{n=1}^{N-1}\frac{{}_rB_{n+1}}{n(n+1)}\frac{1}{x^n} + R_N,\end{aligned} \tag{4.7}$$

where

$$R_N = \frac{(-1)^r}{2\pi\sqrt{-1}}\int_L \frac{g_N(z)e^{-xz}\{\log(-z)+\gamma\}}{z^2}dz. \tag{4.8}$$

Since $g_N(z)/z^2$ is holomorphic at $z=0$, we have

$$R_N = (-1)^r \int_0^\infty \frac{g_N(z)}{z^2} e^{-xz} dz. \tag{4.9}$$

We clearly have

$$g_N(z) = O(z^N), \quad z \longrightarrow +\infty, \qquad g_N(z) = O(z^{N+1}), \quad z \longrightarrow 0.$$

Hence there exists a positive constant c_N such that

$$|g_N(z)| \leqq c_N z^{N+1}. \tag{4.10}$$

Then we get

$$|R_N| \leqq (N-1)!\, c_N \cdot \frac{1}{x^N}. \tag{4.11}$$

The asymptotic expansion (4.7), (4.11) was obtained by Barnes [Bar2].

REMARK. In a similar way, we can obtain an asymptotic expansion of $\log \dfrac{\Gamma_r^{[2]}(x,\omega)}{\rho_r^{[2]}(\omega)}$. (See Exercises 3.)

4.2. Now we are going to estimate the remainder term R_N more explicitly. Put

$$f(z) = \frac{(-1)^r}{z\prod_{k=1}^r (1-e^{-\omega_k z})} - \Big(\sum_{l=1}^r \frac{(-1)^l A_l}{z^{l+1}} + {}_rB_1 z^{-1}\Big), \tag{4.12}$$

$$f_N(z) = \Big(\frac{d}{dz}\Big)^{N-1} f(z). \tag{4.13}$$

Clearly $(\frac{d}{dz})^{N-1}(\frac{g_N(z)}{z^2}) = f_N(z)$ and it is not difficult to check that $f_N(z) = O(z^{-N})$ for $z \to +\infty$. Hence, by (4.9) and using integration by parts repeatedly, we get

$$R_N = (-1)^r \cdot \frac{1}{x^{N-1}} \int_0^\infty f_N(z) e^{-xz} dz. \tag{4.14}$$

Therefore if

$$|f_N(a)| \leqq c_N, \qquad a \geqq 0 \tag{4.15}$$

for some positive constant c_N, we have

$$|R_N| \leqq c_N x^{-N}. \tag{4.16}$$

Hereafter we assume that $\omega_j/\omega_k \notin \mathbf{Q}$ if $j \neq k$. Let $0 \leqq a < R$ and let C_R be the circle of radius R with the center at the origin. We assume that $f(z)$ has no poles on C_R. Then by residue calculus, we have

$$\begin{aligned}
&\frac{1}{2\pi\sqrt{-1}}\int_{C_R}\frac{f(z)}{(z-a)^N}dz=\frac{f_N(a)}{(N-1)!}\\
(4.17)\qquad &+(-1)^r\sum_{k=1}^{r}\{\sum_{n\in\mathbf{Z},0<|n|\leqq\omega_k R/2\pi}\frac{1}{2n\pi\sqrt{-1}}\\
&\cdot\prod_{j=1,j\neq k}^{r}\frac{1}{(1-\exp(-2n\pi\sqrt{-1}\omega_j/\omega_k))}(\frac{2n\pi\sqrt{-1}}{\omega_k}-a)^{-N}\}.
\end{aligned}$$

LEMMA 4.1. *Let $\mu_1, \mu_2, \ldots, \mu_r$ be positive numbers and let $0<\delta<1/2r$. For any $R_0>0$, there exists a real number $R>R_0$ which satisfies that*

$$\delta\leqq R\mu_i-[R\mu_i]\leqq 1-\delta,\qquad 1\leqq i\leqq r.$$

PROOF. Let $t>0$ and put $n=[t\mu_i]$. If $\delta>t\mu_i-[t\mu_i]$, then we have $n\leqq t\mu_i<n+\delta$, $\dfrac{n}{\mu_i}\leqq t<\dfrac{n}{\mu_i}+\dfrac{\delta}{\mu_i}$. If $t\mu_i-[t\mu_i]>1-\delta$, then we have $n+1-\delta<t\mu_i<n+1$, $\dfrac{n+1}{\mu_i}-\dfrac{\delta}{\mu_i}<t<\dfrac{n+1}{\mu_i}$. Put

$$I_i(n)=\left[\frac{n}{\mu_i},\frac{n}{\mu_i}+\frac{\delta}{\mu_i}\right),\qquad J_i(n)=\left(\frac{n+1}{\mu_i}-\frac{\delta}{\mu_i},\frac{n+1}{\mu_i}\right)$$

for $1\leqq i\leqq r$. Assume that the assertion does not hold. Then we have

$$(R_0,x)\subset\cup_{i=1}^{r}\cup_{n=0}^{\infty}(I_i(n)\cup J_i(n))$$

for any $x>R_0$. If $n\geqq\mu_i x$, then $I_i(n)\cap(R_0,x)=\emptyset$, $J_i(n)\cap(R_0,x)=\emptyset$. Hence we have

$$(R_0,x)\subset\cup_{i=1}^{r}\cup_{n=0}^{[\mu_i x]}(I_i(n)\cup J_i(n))$$

for any $x>R_0$. The measure of the set of the right-hand side does not exceed $\displaystyle\sum_{i=1}^{r}\frac{2\delta}{\mu_i}([\mu_i x]+1)\leqq 2r\delta x+\sum_{i=1}^{r}\frac{2\delta}{\mu_i}$. Since $\delta<1/2r$, this leads to a contradiction when x is sufficiently large. This completes the proof.

By Lemma 4.1, we can choose a sequence of positive real numbers $\{R_m\}_{m=1}^{\infty}$ so that $\lim_{m\to\infty}R_m=\infty$,

$$\frac{1}{4r}\leqq\frac{\omega_k R_m}{2\pi}-\left[\frac{\omega_k R_m}{2\pi}\right]\leqq 1-\frac{1}{4r},\qquad 1\leqq k\leqq r.$$

Keeping the assumption $N\geqq 1$, we are going to show that

$$\lim_{m\to\infty}\frac{1}{2\pi\sqrt{-1}}\int_{C_{R_m}}\frac{f(z)}{(z-a)^N}dz=0. \tag{4.18}$$

It suffices to show that there exists a positive constant c, which does not depend on k and m, such that

$$|1-e^{-\omega_k z}|^{-1}\leqq c\qquad\text{for}\quad z\in C_{R_m}. \tag{4.19}$$

We may assume that $\omega_k R_m > 1$, $1 \leqq k \leqq r$. We choose $1 > \epsilon > 0$ so that $\epsilon^2 < \pi/4r$. Let $z \in C_{R_m}$ and put $z = R_m e^{i\theta}$, $\theta \in \mathbf{R}$. Since $|1 - e^w| \geqq |e^{\Re(w)} - 1|$ for $w \in \mathbf{C}$, we have

$$(4.20) \qquad |1 - e^{-\omega_k z}| \geqq |e^{-\omega_k R_m \cos\theta} - 1|.$$

If $\cos\theta \geqq \dfrac{\epsilon}{\omega_k R_m}$, then $|1-e^{-\omega_k z}| \geqq 1 - e^{-\epsilon}$. If $\cos\theta \leqq -\dfrac{\epsilon}{\omega_k R_m}$, then $|1-e^{-\omega_k z}| \geqq e^{\epsilon} - 1 \geqq 1 - e^{-\epsilon}$. Assume $|\cos\theta| \leqq \dfrac{\epsilon}{\omega_k R_m}$. Since ω_k is real, we may assume that $\sin\theta \geqq 0$. We have

$$\sqrt{1 - (\frac{\epsilon}{\omega_k R_m})^2} \leqq \sin\theta \leqq 1.$$

Hence

$$\omega_k R_m - \frac{\epsilon^2}{\omega_k R_m} \leqq \omega_k R_m \sin\theta \leqq \omega_k R_m.$$

Put $l = \left[\dfrac{\omega_k R_m}{2\pi}\right]$. Since

$$2\pi l + \frac{\pi}{2r} \leqq \omega_k R_m \leqq 2\pi l + 2\pi - \frac{\pi}{2r},$$

we get

$$2\pi l + \frac{\pi}{4r} \leqq \omega_k R_m \sin\theta \leqq 2\pi l + 2\pi - \frac{\pi}{2r}.$$

We have

$$1 - e^{-\omega_k z} = (1 - e^{-\omega_k R_m \cos\theta} \cos(\omega_k R_m \sin\theta)) + e^{-\omega_k R_m \cos\theta} \sin(\omega_k R_m \sin\theta)\sqrt{-1}.$$

If $2\pi l + \dfrac{\pi}{2} \leqq \omega_k R_m \sin\theta \leqq 2\pi l + \dfrac{3\pi}{2}$, then $\Re(1 - e^{-\omega_k z}) \geqq 1$. If $2\pi l + \dfrac{\pi}{4r} \leqq \omega_k R_m \sin\theta < 2\pi l + \dfrac{\pi}{2}$ or $2\pi l + \dfrac{3\pi}{2} < \omega_k R_m \sin\theta \leqq 2\pi l + 2\pi - \dfrac{\pi}{2r}$, then $|\sin(\omega_k R_m \sin\theta)| \geqq \sin\dfrac{\pi}{4r}$. Hence we get $|1 - e^{-\omega_k z}| \geqq e^{-\epsilon}\sin\dfrac{\pi}{4r}$. Thus we have established (4.19). By (4.17) and (4.18), we obtain

$$(4.21) \qquad \begin{aligned} \frac{f_N(a)}{(N-1)!} &= \lim_{m\to\infty} (-1)^{r-1} \sum_{k=1}^{r} \Bigg\{ \sum_{n\in\mathbf{Z}, 0<|n|\leqq \omega_k R_m/2\pi} \frac{1}{2n\pi\sqrt{-1}} \\ &\cdot \prod_{j=1, j\neq k}^{r} \frac{1}{(1 - \exp(-2n\pi\sqrt{-1}\omega_j/\omega_k))} (\frac{2n\pi\sqrt{-1}}{\omega_k} - a)^{-N} \Bigg\}. \end{aligned}$$

4.3. The following lemma is well known as Liouville's theorem. (cf. Hardy-Wright, [HW], p. 161.) We will give a proof for the sake of completeness.

LEMMA 4.2. *Let ψ be an algebraic number of degree $l \geqq 2$ and let*

$$f(X) = a_0 X^l + a_1 X^{l-1} + \cdots + a_l, \qquad a_i \in \mathbf{Z}, \quad a_0 > 0$$

be the irreducible polynomial over $\mathbf{Z}$ such that $f(\psi) = 0$. Let $\psi = \psi^{(1)}$, $\psi^{(2)}$, ..., $\psi^{(l)}$ be the conjugates of ψ over $\mathbf{Q}$. Put

$$c = \frac{1}{a_0} \prod_{i=2}^{l} (|\psi^{(i)} - \psi| + 1)^{-1}.$$

Then we have

$$|\psi - \frac{m}{n}| > c \cdot \frac{1}{|n|^l} \qquad \textit{for} \quad n, m \in \mathbf{Z},\ n \neq 0.$$

PROOF. Take n, $m \in \mathbf{Z}$, $n \neq 0$. Clearly $0 \neq n^l f(m/n) \in \mathbf{Z}$. Hence we have $|f(m/n)| \geqq 1/|n|^l$. Since

$$f(X) = a_0(X - \psi)(X - \psi^{(2)}) \cdots (X - \psi^{(l)}),$$

we have

$$a_0 |\frac{m}{n} - \psi||\frac{m}{n} - \psi^{(2)}| \cdots |\frac{m}{n} - \psi^{(l)}| \geqq \frac{1}{|n|^l}.$$

Hence we get

$$|\psi - \frac{m}{n}| \geqq \frac{1}{a_0} \prod_{i=2}^{l} |\psi^{(i)} - \frac{m}{n}|^{-1} \cdot \frac{1}{|n|^l}. \tag{4.22}$$

Notice that the assertion holds if $|\psi - \frac{m}{n}| \geqq \frac{1}{|n|}$ since $c < 1$. We may assume that $|\psi - \frac{m}{n}| < \frac{1}{|n|}$. Then we have

$$|\psi^{(i)} - \frac{m}{n}| \leqq |\psi^{(i)} - \psi| + |\psi - \frac{m}{n}| < |\psi^{(i)} - \psi| + \frac{1}{|n|} \leqq |\psi^{(i)} - \psi| + 1$$

and the assertion follows from (4.22).

Hereafter we assume that $\omega_j/\omega_k \in \overline{\mathbf{Q}}$ for every j, k. By Lemma 4.2, there exist $c_{j,k} > 0$ and $0 < l \in \mathbf{Z}$ such that

$$|\frac{\omega_j}{\omega_k} - \frac{m}{n}| \geqq c_{j,k} \cdot \frac{1}{|n|^l} \qquad \text{for} \quad n, m \in \mathbf{Z},\ n \neq 0 \tag{4.23}$$

when $j \neq k$. We may assume that $c_{j,k} \leqq 1/4$. Let us show that

$$\begin{aligned} &\left| \prod_{j=1, j\neq k}^{r} \frac{1}{(1 - \exp(-2n\pi\sqrt{-1}\omega_j/\omega_k))} \right| \\ &\leqq 4^{-(r-1)} \prod_{j=1, j\neq k}^{r} (c_{j,k})^{-1} \cdot |n|^{(r-1)(l-1)}, \qquad 0 \neq n \in \mathbf{Z}. \end{aligned} \tag{4.24}$$

Put $\theta = n\omega_j/\omega_k$. By (4.23), we have

$$|\theta - m| \geqq c_{j,k}/|n|^{l-1} \qquad \text{for every} \quad m \in \mathbf{Z}.$$

Hence we get

$$c_{j,k}/|n|^{l-1} \leqq \theta - [\theta] \leqq 1 - c_{j,k}/|n|^{l-1}.$$

If $1/4 \leqq \theta - [\theta] \leqq 3/4$, then $|\exp(2\pi i\theta) - 1| \geqq 1$. Assume that $c_{j,k}/|n|^{l-1} \leqq \theta - [\theta] \leqq 1/4$ or $3/4 \leqq \theta - [\theta] \leqq 1 - c_{j,k}/|n|^{l-1}$. Then we have

$$|\exp(2\pi i\theta) - 1| \geqq \sin(2\pi \cdot \frac{c_{j,k}}{|n|^{l-1}}) \geqq 4c_{j,k}/n^{l-1}.$$

Hence we obtain (4.24).

By (4.21) and (4.24), we have

$$\left|\frac{f_N(a)}{(N-1)!}\right| \leqq \sum_{k=1}^{r} 4^{-(r-1)} \prod_{j=1, j\neq k}^{r} (c_{j,k})^{-1} \times 2\sum_{n=1}^{\infty} \frac{1}{2n\pi}\left(\frac{\omega_k}{2n\pi}\right)^N n^{(r-1)(l-1)}$$

if $N \geqq (r-1)(l-1)+1$. Therefore

$$\left|\frac{f_N(a)}{(N-1)!}\right| \leqq \frac{2}{4^{r-1}(2\pi)^{N+1}}\left[\sum_{k=1}^{r} \omega_k^N \prod_{j=1, j\neq k}^{r} (c_{j,k})^{-1}\right]\zeta(N+1-(r-1)(l-1)). \tag{4.25}$$

Summing up, we have proved

THEOREM 4.3. *Assume that* $\omega_j/\omega_k \in \overline{\mathbf{Q}} - \mathbf{Q}$ *if* $j \neq k$. *Then* (4.23) *holds with* $0 < c_{j,k} \leqq 1/4$ *and* $0 < l \in \mathbf{Z}$. *Let* $N \geqq (r-1)(l-1)+1$ *and put*

$$c_N = \frac{2(N-1)!}{4^{r-1}(2\pi)^{N+1}}\left[\sum_{k=1}^{r} \omega_k^N \prod_{j=1, j\neq k}^{r} (c_{j,k})^{-1}\right]\zeta(N+1-(r-1)(l-1)).$$

Then (4.15) *holds and the remainder term* R_N *of Barnes' asymptotic expansion* (4.7) *is subject to the estimate*

$$|R_N| \leqq c_N x^{-N}.$$

EXAMPLE 4.4. Let us compare the estimate given in Theorem 4.3 with that given in [WW], p. 252. Whittaker-Watson define (cf. [WW], p. 127) the Bernoulli numbers by

$$\frac{t}{e^t-1} = 1 - \frac{t}{2} + \frac{B_1}{2!}t^2 - \frac{B_2}{4!}t^4 + \cdots. \tag{4.26}$$

This definition is different from ours but in this example we use (4.26). We consider the case $r=1$, $\omega_1 = 1$. We have

$$A_1 = 1, \quad {}_1B_1 = -\frac{1}{2}, \quad {}_1B_n = (-1)^{n/2}B_{n/2} \quad \text{if } n \text{ is even, } n \geqq 2,$$
$${}_1B_n = 0 \quad \text{if } n \text{ is odd, } n \geqq 3.$$

By (1.19), we have

$$\frac{\Gamma_1(x,(1))}{\rho_1((1))} = \frac{1}{\sqrt{2\pi}}\Gamma(x). \tag{4.27}$$

The asymptotic expansion (4.7) can be written as

$$\log\Gamma(x) = (x-\frac{1}{2})\log x - x + \frac{1}{2}\log 2\pi + \sum_{m=1}^{n}\frac{(-1)^{m-1}B_m}{(2m-1)2m}\frac{1}{x^{2m-1}} + R_{2n+1}, \tag{4.28}$$

taking $N = 2n+1$, $n \geqq 0$. By Theorem 4.3, we have

$$c_{2n+1} = 2(2n)!(2\pi)^{-(2n+2)}\zeta(2n+2).$$

It is well known that

$$\zeta(2n+2) = 2^{2n+1}\pi^{2n+2}\cdot\frac{B_{n+1}}{(2n+2)!}.$$

Hence we obtain

$$c_{2n+1} = \frac{B_{n+1}}{(2n+1)(2n+2)}, \qquad |R_{2n+1}| \leqq \frac{B_{n+1}}{(2n+1)(2n+2)} \frac{1}{x^{2n+1}}.$$

This estimate is as sharp as that given in [WW], p. 252.

EXAMPLE 4.5. We illustrate the estimate given by Theorem 4.3 by a numerical example. Let $f(X) = X^3 - 4X - 1$; $f(X)$ is irreducible over $\mathbf{Q}$ and has three real roots $\rho = 2.11490\ldots$, $\rho' = -0.25410\ldots$, $\rho'' = -1.86080\ldots$. Let $F = \mathbf{Q}(\rho)$; F is a totally real cubic field. Let $\epsilon = \rho^2$, $\eta = \rho + 2$. Then ϵ and η generate E_F^+. Let us consider the asymptotic expansion (4.7) for $\log \dfrac{\Gamma_3(x, (1, \epsilon, \epsilon\eta))}{\rho_3((1, \epsilon, \epsilon\eta))}$. We will need numerical values of this function in chapter III, §4.2 to study an example.

Put $\omega_1 = 1$, $\omega_2 = \epsilon$, $\omega_3 = \epsilon\eta$. The constants $c_{j,k}$ can be calculated easily by Lemma 4.2 taking $l = 3$. We can take

$$\begin{aligned} c_{2,1}^{-1} &= 10.87\ldots, & c_{3,1}^{-1} &= 365.07\ldots, & c_{1,2}^{-1} &= 17.32\ldots, \\ c_{3,2}^{-1} &= 16.76\ldots, & c_{1,3}^{-1} &= 29.65\ldots, & c_{2,3}^{-1} &= 10.55\ldots. \end{aligned}$$

Hence, for $N \geqq 5$, we have

$$c_N \leqq \frac{(N-1)!}{8(2\pi)^{N+1}} \{3970 + 291\epsilon^N + 314(\epsilon\eta)^N\} \zeta(N-3).$$

For $x \geqq 1000$ and $N = 25$, we get $|R_N| \leqq 2 \times 10^{-39}$, by using $\zeta(N-3) \leqq 1 + 2^{-21}$.

Exercises

1. Using (1.4), show the following.
   (a) At $s = r$, $\zeta_r(s, \omega, x)$ has a simple pole with residue $\dfrac{1}{(r-1)!\,\omega_1 \cdots \omega_r}$.
   (b) When $r \geqq 2$, $\zeta_r(s, \omega, x)$ is holomorphic at $s = r - 1$ if and only if $x = \frac{1}{2}\sum_{k=1}^{r} \omega_k$.
2. Show that, for some x_1, x_2, b_1, $b_2 > 0$, we have

$$\zeta_r(ns, (b_1, \ldots, b_1), x_1) < \zeta(s, A, x) < \zeta_r(ns, (b_2, \ldots, b_2), x_2), \qquad s > r/n.$$

   Deduce that $\zeta(s, A, x)$ has a simple pole at $s = r/n$.
3. Show the asymptotic expansion

$$\begin{aligned} \log \frac{\Gamma_r^{[2]}(x, \omega)}{\rho_r^{[2]}(\omega)} &= (-1)^r \sum_{m=1}^{r} \frac{A_m}{m!} x^m \Big\{ (\log x)^2 - 2(1 + \frac{1}{2} + \cdots + \frac{1}{m}) \log x \\ &\quad + (1 + \frac{1}{2} + \cdots + \frac{1}{m})^2 + (1 + \frac{1}{4} + \cdots + \frac{1}{m^2}) \Big\} + (-1)^r {}_rB_1 (\log x)^2 \\ &\quad + 2(-1)^r \sum_{n=1}^{N-1} \frac{{}_rB_{n+1}}{n(n+1)} \frac{1}{x^n} \Big\{ (1 + \frac{1}{2} + \cdots + \frac{1}{n-1}) - \log x \Big\} + R_N, \end{aligned}$$

$$\begin{aligned} R_N &= 2(-1)^r \int_0^\infty \frac{g_N(t)}{t^2} e^{-xt} (\log t + \gamma) dt, \\ R_N &= O(x^{-N} \log x), \qquad x \to +\infty. \end{aligned}$$

Here $N \geqq 1$ and the term $(1+\frac{1}{2}+\cdots+\frac{1}{n-1})$ should be dropped when $n=1$.

CHAPTER II

THE STARK-SHINTANI CONJECTURE

In this chapter, we will study a beautiful and important conjecture of Stark–Shintani. We will borrow ideas from Tate [T] at several places.

In §1, we will state Stark's general conjecture concerning the leading term of the Taylor expansion of Artin's L-function at $s = 0$. This conjecture is formulated using the generalized regulator. We will compute it explicitly for the one dimensional character. In §2, we will bring the conjecture to a more concrete form, by considering an abelian extension of a totally real number field. In §3, we will prove Shintani's formulas which express the value and the derivative at $s = 0$ of a partial zeta function of a totally real number field. In §4, we will discuss an example.

As interesting recent literatures, which were brought to the author's attention after finishing this chapter, we refer the reader to Rubin [Ru] and Darmon [Da1], [Da2].

1. Stark's general conjecture

1.1. Let K be an algebraic number field. We denote by $\mathbf{a}_K$ the set of all infinite places of K. We regard $\mathbf{a}_K$ as a subset of J_K by choosing, for an imaginary place, a representative from two complex conjugate isomorphisms of K into $\mathbf{C}$. The analytic class number formula gives

$$\lim_{s\to 1}(s-1)\zeta_K(s) = \frac{2^{r_1(K)}(2\pi)^{r_2(K)}h_K R_K}{w_K|D_K|^{1/2}}. \tag{1.1}$$

Using the functional equation, we see that (1.1) is equivalent to

$$\zeta_K(s) = -\frac{h_K R_K}{w_K} s^{r_1(K)+r_2(K)-1} + O(s^{r_1(K)+r_2(K)}), \qquad s \to 0. \tag{1.2}$$

Let k be an algebraic number field and K be a Galois extension of k of finite degree. We put $G = \mathrm{Gal}(K/k)$; G acts on $\mathbf{a}_K$ on the right by $\mathbf{a}_K \ni w \longrightarrow \sigma^{-1}w \in \mathbf{a}_K$. Let Y be the free abelian group generated by $\mathbf{a}_K$. We put

$$X = \{\sum_{w\in\mathbf{a}_K} n_w w \in Y \mid \sum_{w\in\mathbf{a}_K} n_w = 0\}.$$

We define a mapping λ of E_K into $X \otimes_{\mathbf{Z}} \mathbf{R}$ by

$$\lambda(\epsilon) = \sum_{w\in\mathbf{a}_K} \log|\epsilon|_w \cdot w. \tag{1.3}$$

Here $|\ |_w$ is the usual (resp. the square of the usual) absolute value if w is real (resp. imaginary). Since G leaves real places (resp. imaginary places) stable, we see easily that λ is a G-homomorphism. Taking $\otimes_{\mathbf{Z}}\mathbf{C}$, we obtain a G-homomorphism

$$\lambda_{\mathbf{C}} : E_K \otimes_{\mathbf{Z}} \mathbf{C} \longrightarrow X \otimes_{\mathbf{Z}} \mathbf{C}. \tag{1.4}$$

By Kronecker's theorem, we have $\ker(\lambda) = W_K$; hence $\lambda_{\mathbf{C}}$ is injective. Comparing dimensions of both sides of (1.2) using Dirichlet's unit theorem, we see that $\lambda_{\mathbf{C}}$ is an isomorphism. Therefore the representations of G on $E_K \otimes_{\mathbf{Z}} \mathbf{Q}$ and on $X \otimes_{\mathbf{Z}} \mathbf{Q}$ have the same character. Hence there exists a G-isomorphism

$$f : X \otimes_{\mathbf{Z}} \mathbf{Q} \cong E_K \otimes_{\mathbf{Z}} \mathbf{Q}. \tag{1.5}$$

Let

$$f_{\mathbf{C}} : X \otimes_{\mathbf{Z}} \mathbf{C} \cong E_K \otimes_{\mathbf{Z}} \mathbf{C} \tag{1.6}$$

be the G-isomorphism obtained from (1.5) by taking $\otimes_{\mathbf{Q}}\mathbf{C}$. Let V be a finite dimensional vector space over $\mathbf{C}$ and let $\rho : G \longrightarrow GL(V)$ be a representation of G on V. Let χ be the character of ρ. We set

$$L(s,\chi) = c(\chi)s^{r(\chi)} + O(s^{r(\chi)+1}) \qquad \text{for} \quad s \to 0, \quad c(\chi) \neq 0. \tag{1.7}$$

Let V^* be the dual vector space of V; G acts on V^* according to the contragredient representation to ρ. Then we have $\dim_{\mathbf{C}} \mathrm{Hom}_G(V^*, X \otimes_{\mathbf{Z}} \mathbf{C}) = r(\chi)$. Combining (1.4) with (1.6), we have a G-isomorphism

$$\lambda_{\mathbf{C}} \circ f_{\mathbf{C}} : X \otimes_{\mathbf{Z}} \mathbf{C} \cong X \otimes_{\mathbf{Z}} \mathbf{C},$$

which yields an isomorphism

$$(\lambda \circ f)_V : \mathrm{Hom}_G(V^*, X \otimes_{\mathbf{Z}} \mathbf{C}) \cong \mathrm{Hom}_G(V^*, X \otimes_{\mathbf{Z}} \mathbf{C})$$

defined by $(\lambda \circ f)_V(\varphi) = \lambda_{\mathbf{C}} \circ f_{\mathbf{C}} \circ \varphi$. Taking a basis of $\mathrm{Hom}_G(V^*, X \otimes_{\mathbf{Z}} \mathbf{C})$, we define

$$R(\chi, f) = \det((\lambda \circ f)_V). \tag{1.8}$$

We see that $R(\chi, f)$ depends only on f. Now Stark's general conjecture states:

CONJECTURE 1.1([T], I, 5.1). *Let $c(\chi)$ be the leading coefficient of the Taylor expansion of the Artin L-function $L(s,\chi)$ at $s = 0$ as in (1.7) and let $R(\chi, f)$ be the generalized regulator defined by (1.8). Then, for every $\sigma \in Aut(\mathbf{C})$, we have*

$$\left(\frac{c(\chi)}{R(\chi, f)}\right)^{\sigma} = \frac{c(\chi^{\sigma})}{R(\chi^{\sigma}, f)}.$$

It is shown in [T], I, §6 that if Conjecture 1.1 holds for one choice of f, then it holds for all the other choices.

Suppose that $k = K$. We can choose generators $\{\epsilon_w\}_{w \in \mathbf{a}_K}$ of E_K so that the relations satisfied by ϵ_w are $(\prod_{w \in \mathbf{a}_K} \epsilon_w)^n = 1$, $n \in \mathbf{Z}$. Then the mapping

$$X \ni \sum_{w \in \mathbf{a}_K} n_w w \longrightarrow \prod_{w \in \mathbf{a}_K} \epsilon_w^{n_w} \in E_K$$

can be extended to the isomorphism $f : X \otimes_{\mathbf{Z}} \mathbf{Q} \cong E_K \otimes_{\mathbf{Z}} \mathbf{Q}$. Then we find easily that $R(1, f) = \pm R_K$. Therefore Conjecture 1.1 is compatible with (1.2). We also see that $R(\chi, f)$ satisfies

$$R(\chi_1 + \chi_2, f) = R(\chi_1, f)R(\chi_2, f), \tag{1.9}$$

$$R(\mathrm{Ind}_H^G \eta, f) = R(\eta, f), \tag{1.10}$$

where H is a subgroup of G and η is a character of H. Thus Conjecture 1.1 is compatible with the formalism of the Artin L-functions.

1.2. We are going to write "the generalized regulator" $R(\chi, f)$ explicitly when $\dim \chi = 1$. Put

$$n = [k : \mathbf{Q}], \quad r_1 = r_1(k), \quad r_2 = r_2(k).$$

We have $r_1 + 2r_2 = n$. Let $\rho \in \mathrm{Gal}(\overline{\mathbf{Q}}/\mathbf{Q})$ be the complex conjugation. We can choose a complete set of representatives $\{\beta_i\}_{1 \leqq i \leqq n}$ of $\mathrm{Gal}(\overline{\mathbf{Q}}/k)\backslash\mathrm{Gal}(\overline{\mathbf{Q}}/\mathbf{Q})$ so that the following conditions are satisfied.

(i) k^{β_i} is real for $1 \leqq i \leqq r_1$.

(ii) k^{β_i} is complex and $\beta_{i+r_2} = \beta_i \rho$ for $r_1 + 1 \leqq i \leqq r_1 + r_2$.

(iii) K^{β_i} is real for $1 \leqq i \leqq r_1'$ and is complex for $r_1' + 1 \leqq i \leqq r_1$.

Here r_1' is the number of real infinite places of k which do not ramify in K. For $r_1' < i \leqq r_1$, put $\tilde{\sigma}_i = \beta_i \rho \beta_i^{-1} \in \mathrm{Gal}(\overline{\mathbf{Q}}/\mathbf{Q})$, $\sigma_i = \tilde{\sigma}_i | K$. Since $\tilde{\sigma}_i | k = \mathrm{id}$, we see that $\sigma_i \in G$ and that the order of σ_i is 2. For other i, $1 \leqq i \leqq n$, we put $\sigma_i = 1$ and let G_i be a complete set of representatives of $\langle \sigma_i \rangle \backslash G$.

LEMMA 1.2. *There exist n units $\epsilon_1, \epsilon_2, \ldots, \epsilon_n \in E_K$ such that $\epsilon_i^{\sigma_i} = \epsilon_i$, $1 \leqq i \leqq n$, $\epsilon_{i+r_2} = \epsilon_i$, $r_1 + 1 \leqq i \leqq r_1 + r_2$ and that the units*

$$\{\epsilon_i^\sigma \mid 1 \leqq i \leqq r_1 + r_2,\ \sigma \in G_i\}$$

satisfy the relation

$$\prod_{i=1}^{n} \prod_{\sigma \in G} \epsilon_i^\sigma = \Big(\prod_{1 \leqq i \leqq r_1'} \prod_{\sigma \in G_i} \epsilon_i^\sigma\Big)\Big(\prod_{r_1'+1 \leqq i \leqq r_1+r_2} \prod_{\sigma \in G_i} \epsilon_i^\sigma\Big)^2 = 1,$$

and any relation is of the form of an integral power of this relation.

PROOF. The idea of the proof is to use the isomorphism (1.5) and to employ suitable elements in the simpler G-module $X \otimes_{\mathbf{Z}} \mathbf{Q}$.

We have $|\mathbf{a}_K| = nr_1' + n(r_1 - r_1')/2 + nr_2$. Put $\delta = (\sum_{w \in \mathbf{a}_K} w)/|\mathbf{a}_K|$. Then $\delta \in Y \otimes_{\mathbf{Z}} \mathbf{Q}$ and $\delta^\sigma = \delta$ for every $\sigma \in G$. Regarding $\beta_i | K$ as an element of $\mathbf{a}_K$, we define $u_i \in X \otimes_{\mathbf{Z}} \mathbf{Q}$, $1 \leqq i \leqq n$ by

$$\begin{aligned} u_i &= \beta_i | K - \delta, \qquad 1 \leqq i \leqq r_1', \\ u_i &= \frac{1}{2}(\beta_i | K - \delta), \qquad r_1' + 1 \leqq i \leqq r_1 + r_2, \\ u_{i+r_2} &= u_i, \qquad r_1 + 1 \leqq i \leqq r_1 + r_2. \end{aligned}$$

Then we find

$$\begin{aligned} \sum_{i=1}^{n} \sum_{\sigma \in G} u_i^\sigma &= \sum_{1 \leqq i \leqq r_1'} \sum_{\sigma \in G_i} u_i^\sigma + \sum_{r_1'+1 \leqq i \leqq r_1+r_2} \sum_{\sigma \in G_i} 2u_i^\sigma \\ &= \sum_{w \in \mathbf{a}_K} w - \{nr_1' + \frac{n}{2}(r_1 - r_1') + nr_2\}\delta = 0. \end{aligned} \tag{1.11}$$

Since $\beta_i | K$ and $\beta_i \rho | K$ define the same place of K, we have $u_i^{\sigma_i} = u_i$, $1 \leqq i \leqq n$. Since $\{\sigma\beta_i | K, 1 \leqq i \leqq r_1 + r_2, \sigma \in G_i\}$ generates a subgroup of Y of rank $|\mathbf{a}_K|$, $\{u_i^\sigma \mid 1 \leqq i \leqq r_1 + r_2, \sigma \in G_i\}$ generates a subgroup of $X \otimes_{\mathbf{Z}} \mathbf{Q}$ of rank $|\mathbf{a}_K| - 1$. We can find a positive integer m so that $f(mu_i) \in E_K$, $1 \leqq i \leqq n$. By (1.5) and (1.11), we have $\prod_{i=1}^{n}\prod_{\sigma\in G} f(mu_i)^\sigma = 1$ in $E_K \otimes_{\mathbf{Z}} \mathbf{Q}$. Put $\epsilon_i = f(|W_K|mu_i)$, $1 \leqq i \leqq n$. Then we have $\prod_{i=1}^{n}\prod_{\sigma\in G}\epsilon_i^\sigma = 1$ and $\{\epsilon_i\}_{1\leqq i\leqq n}$ clearly satisfies all the required conditions. This completes the proof.

Lemma 1.2 is a version of Artin's unit theorem given in Stark [St2], III, p. 68. By considering ranks, we see that $\{\epsilon_i^\sigma \mid 1 \leqq i \leqq r_1 + r_2,\ \sigma \in G_i\}$ generates a subgroup of E_K of finite index.

We put $\mathbf{a}_k = \{v_1, v_2, \dots, v_{r_1+r_2}\}$. We may assume that $v_i = \beta_i|k$, $1 \leqq i \leqq r_1 + r_2$. Then we have

$$\mathbf{a}_K = \{\tau^{-1}v_i \mid 1 \leqq i \leqq r_1 + r_2,\ \tau \in G_i\},$$

where $\tau^{-1}v_i$ means $\tau^{-1}\beta_i|K$. Suppose that $\epsilon_1, \epsilon_2, \dots, \epsilon_{r_1+r_2}$ satisfy the conditions of Lemma 1.2. Then the mapping

$$X \ni \sum_{i=1}^{r_1+r_2}\sum_{\tau\in G_i} n_{i,\tau}\tau^{-1}v_i \longrightarrow \prod_{i=1}^{r_1'}\prod_{\tau\in G_i}(\epsilon_i^\tau)^{n_{i,\tau}} \prod_{i=r_1'+1}^{r_1+r_2}\prod_{\tau\in G_i}(\epsilon_i^\tau)^{2n_{i,\tau}} \in E_K$$

is injective. Hence extending $\mathbf{Q}$-linearly, we obtain an explicit isomorphism $f : X \otimes_{\mathbf{Z}} \mathbf{Q} \longrightarrow E_K \otimes_{\mathbf{Z}} \mathbf{Q}$ of G-modules.

Let χ be a one dimensional representation of G. Let $\widetilde{\chi}$ be the character of $k_A^\times$ which corresponds to χ by class field theory. We have $L(s,\chi) = L(s,\widetilde{\chi})$ and $\widetilde{\chi}_{v_i}$ is unramified if and only if $\chi(\sigma_i) = 1$. Changing the numbering of β_i if necessary, we may assume that $\widetilde{\chi}_{v_i}$ is unramified for $1 \leqq i \leqq r_1''$, is ramified for $r_1'' < i \leqq r_1$, where $r_1' \leqq r_1'' \leqq r_1$. Then the gamma factor to go with $L(s,\chi)$ is

$$\Gamma(\frac{s}{2})^{r_1''}\Gamma(\frac{s+1}{2})^{r_1-r_1''}\Gamma(s)^{r_2}.$$

Hence, by the functional equation, we have

$$r(\chi) = \begin{cases} r_1'' + r_2 & \text{if } \chi \neq 1, \\ r_1'' + r_2 - 1 & \text{if } \chi = 1. \end{cases}$$

PROPOSITION 1.3. *Let the notation be the same as above. Assume* $\dim\chi = 1$, $\chi \neq 1$. *For* i, j *such that* $1 \leqq i \leqq r_1''$ *or* $r_1 + 1 \leqq i \leqq r_1 + r_2$, $1 \leqq j \leqq r_1''$ *or* $r_1 + 1 \leqq j \leqq r_1 + r_2$, *we put*

$$r_{ij} = \sum_{\tau\in G}\chi(\tau)\log|\epsilon_i^{\tau v_j}|.$$

We regard (r_{ij}) *as an* $(r_1''+r_2)\times(r_1''+r_2)$*-matrix. Then we have*

$$R(\chi, f) = 2^{2r_2+r_1''-r_1'}\det(r_{ij}).$$

PROOF. For $1 \leqq i \leqq r_1''$ or for $r_1 + 1 \leqq i \leqq r_1 + r_2$, we put

$$E_i = \sum_{\tau\in G_i}\chi(\tau)\tau^{-1}v_i \in X \otimes_{\mathbf{Z}} \mathbf{C}. \tag{1.12}$$

Then $\{E_i\}$ gives a basis of the χ^{-1}-isotypic component of $X \otimes_{\mathbf{Z}} \mathbf{C}$. For i in the same range as above, we put

$$A_i = \sum_{\tau \in G} \sum_{j=1}^{r_1'} \sum_{\mu \in G} \chi(\tau) \log |\epsilon_i^{\tau\mu^{-1}v_j}| \mu^{-1} v_j,$$

$$B_i = \sum_{\tau \in G} \sum_{j=r_1'+1}^{r_1} \sum_{\mu \in G_j} \chi(\tau) \log |\epsilon_i^{\tau\mu^{-1}v_j}| \mu^{-1} v_j,$$

$$C_i = \sum_{\tau \in G} \sum_{j=r_1+1}^{r_1+r_2} \sum_{\mu \in G} \chi(\tau) \log |\epsilon_i^{\tau\mu^{-1}v_j}| \mu^{-1} v_j,$$

where $|\ |$ denotes the usual absolute value. Since $\sigma_j v_j = v_j \rho$, we see that B_i is well defined. We see immediately that

$$A_i = \sum_{j=1}^{r_1'} r_{ij} E_j, \qquad C_i = \sum_{j=r_1+1}^{r_1+r_2} r_{ij} E_j.$$

We have

$$\begin{aligned} B_i &= \sum_{j=r_1'+1}^{r_1} \sum_{\tau \in G} \sum_{\mu \in G_j} \chi(\tau\mu^{-1}) \log |\epsilon_i^{\tau\mu^{-1}v_j}| \chi(\mu) \mu^{-1} v_j \\ &= \sum_{j=r_1'+1}^{r_1} \Big(\sum_{\tau \in G} \chi(\tau) \log |\epsilon_i^{\tau v_j}|\Big) E_j. \end{aligned}$$

If $r_1'' < j \leqq r_1$, we have $\sum_{\tau \in G} \chi(\tau) \log |\epsilon_i^{\tau v_j}| = 0$, since $|\epsilon_i^{\tau\sigma_j v_j}| = |\epsilon_i^{\tau v_j}|$, $\chi(\sigma_j) = -1$. Hence we get

$$B_i = \sum_{j=r_1'+1}^{r_1''} r_{ij} E_j.$$

For $1 \leqq i \leqq r_1''$, we have

$$\begin{aligned} \lambda_{\mathbf{C}}(f_{\mathbf{C}}(E_i)) &= \sum_{\tau \in G_i} \chi(\tau) \sum_{w \in \mathbf{a}_K} \log |\epsilon_i^{\tau}|_w w \\ &= \sum_{\tau \in G} \sum_{j=1}^{r_1'} \sum_{\mu \in G} \chi(\tau) \log |\epsilon_i^{\tau\mu^{-1}v_j}| \mu^{-1} v_j \\ &\quad + \sum_{\tau \in G} \sum_{j=r_1'+1}^{r_1} \sum_{\mu \in G_j} \chi(\tau) 2 \log |\epsilon_i^{\tau\mu^{-1}v_j}| \mu^{-1} v_j \\ &\quad + \sum_{\tau \in G} \sum_{j=r_1+1}^{r_2} \sum_{\mu \in G} \chi(\tau) 2 \log |\epsilon_i^{\tau\mu^{-1}v_j}| \mu^{-1} v_j \\ &= A_i + 2B_i + 2C_i \\ &= \sum_{j=1}^{r_1'} r_{ij} E_j + 2 \sum_{j=r_1'+1}^{r_1''} r_{ij} E_j + 2 \sum_{j=r_1+1}^{r_1+r_2} r_{ij} E_j. \end{aligned}$$

For $r_1+1 \leqq i \leqq r_1+r_2$, we have

$$\lambda_{\mathbf{C}}(f_{\mathbf{C}}(E_i)) = \lambda(\prod_{\tau\in G}(\epsilon_i^\tau)^{2\chi(\tau)}) = \sum_{\tau\in G}\sum_{w\in\mathbf{a}_K} 2\chi(\tau)\log|\epsilon_i^\tau|_w w = 2A_i + 4B_i + 4C_i.$$

Let $\{e\}$ be a basis of V^*. Define $E_i^* \in \mathrm{Hom}_G(V^*, X\otimes_{\mathbf{Z}}\mathbf{C})$ by $E_i^*(e) = E_i$. Then we obtain

$$(\lambda\circ f)_V(E_i^*) = \sum_{j=1}^{r_1'} r_{ij}E_j^* + 2\sum_{j=r_1'+1}^{r_1''} r_{ij}E_j^* + 2\sum_{j=r_1+1}^{r_1+r_2} r_{ij}E_j^* \qquad \text{for} \quad 1\leqq i\leqq r_1'',$$

$$(\lambda\circ f)_V(E_i^*) = \sum_{j=1}^{r_1'} 2r_{ij}E_j^* + 4\sum_{j=r_1'+1}^{r_1''} r_{ij}E_j^* + 4\sum_{j=r_1+1}^{r_1+r_2} r_{ij}E_j^* \qquad \text{for} \quad r_1+1\leqq i\leqq r_1+r_2.$$

Using the basis $\{E_i^*\}$ of $\mathrm{Hom}_G(V^*, X\otimes_{\mathbf{Z}}\mathbf{C})$, we obtain

$$R(\chi, f) = \det\begin{pmatrix} r_{ij} & 2r_{ij} \\ 2r_{ij} & 4r_{ij}\end{pmatrix},$$

where we split rows (resp. columns) by (r_1'', r_2) (resp. $(r_1', r_2+r_1''-r_1')$). Hence the assertion follows.

When $\chi = 1$, we see $R(\chi, f) \equiv R_k \mod \mathbf{Q}^\times$, modifying the proof of Proposition 1.3. By (1.9) and (1.10), we have

$$R_K \equiv R_k \prod_{\chi\in\hat{G},\chi\neq 1} R(\chi, f)^{\dim\chi} \mod \mathbf{Q}^\times.$$

Thus Conjecture 1.1 can be regarded as a factorization of the quotient of the regulators R_K/R_k. Such an idea to factorize transcendental invariants seems to be very fruitful in number theory. We will meet one more example of factorization ideas in Chapter III.

2. Transition to a more precise conjecture

2.1. We retain the notation of the previous section. Thus χ is a non-trivial one dimensional character of $G = \mathrm{Gal}(K/k)$. We are going to consider the case $r(\chi) = 1$ where we can bring the general conjecture to a more precise form.

Since $r(\chi) = 1$, we have $c(\chi) = L'(0,\chi)$. We are going to write down $L'(0,\chi)$ as explicitly as possible assuming Conjecture 1.1. For this purpose, we may assume that χ is a faithful character of G. Since $r(\chi) = r_1''+r_2 = 1$, we have either $r_1'' = 1$, $r_2 = 0$ or $r_1'' = 0$, $r_2 = 1$. Since χ is faithful, we have $r_1' = r_1''$. Therefore only two cases can occur.

Case I. k is totally real, χ and K are unramified at one infinite place of k and ramify at all the other infinite places.

Case II. k has one and only one imaginary place. In K, all the real places of k ramify; χ ramifies at every real place of k.

In this section, we will consider only Case I. We write k as F to make our notation consistent with that in later chapters. Let $\infty_i = v_i$, $1\leqq i\leqq n$. We may assume that K is unramified at ∞_1 and that ∞_1 corresponds to the identical embedding of F into $\mathbf{R}$. By Lemma 1.2, there exist units $\epsilon_1, \epsilon_2, \ldots, \epsilon_n$ such that

$\epsilon_i^{\sigma_i} = \epsilon_i$, $2 \leqq i \leqq n$ and that $\{\epsilon_i^\tau \mid 1 \leqq i \leqq n,\ \tau \in G_i\}$ satisfy essentially only one relation

$$\prod_{i=1}^{n} \prod_{\tau \in G} \epsilon_i^\tau = 1.$$

By Proposition 1.3, we have

$$R(\chi, f) = \sum_{\tau \in G} \chi(\tau) \log |\epsilon_1^\tau|. \tag{2.1}$$

PROPOSITION 2.1. *Let F be a totally real algebraic number field of degree n. Let K be an abelian extension of F of finite degree such that ∞_1 is unramified in K and that $\infty_2, \ldots, \infty_n$ are ramified in K. We assume Conjecture* 1.1. *Then there exist a positive integer m and $\epsilon \in E_K$ such that $\prod_{\tau \in G} \epsilon^\tau = 1$ and that*

$$L'(0, \chi) = \frac{1}{m} \sum_{\tau \in G} \chi(\tau) \log |\epsilon^\tau| \tag{2.2}$$

for all non-trivial characters χ of $G = \mathrm{Gal}(K/F)$. If $n \geqq 3$, (2.2) *holds also for $\chi = 1$.*

PROOF. Take $\epsilon_1, \epsilon_2, \ldots, \epsilon_n \in E_K$ as above. Then (2.1) holds for a character $\chi \neq 1$ of G which is ramified at $\infty_2, \ldots, \infty_n$. (We do not have to assume that χ is faithful.) Hence Conjecture 1.1 implies that

$$L'(0, \chi) = A(\chi) \sum_{\tau \in G} \chi(\tau) \log |\epsilon_1^\tau| \qquad \text{if } \chi \in \widehat{G} \text{ ramifies at } \infty_2, \cdots, \infty_n, \tag{2.3}$$

$$A(\chi)^\alpha = A(\chi^\alpha) \qquad \text{for every} \quad \alpha \in \mathrm{Aut}(\mathbf{C}). \tag{2.4}$$

We have $A(\chi) \in \mathbf{Q}(\chi)$ from (2.4), where $\mathbf{Q}(\chi)$ denotes the field generated over $\mathbf{Q}$ by the values of χ. For $\chi \in \widehat{G}$ which is unramified at ∞_i for some $2 \leqq i \leqq n$, we have $L'(0, \chi) = 0$ except for $n \leqq 2$, $\chi = 1$. We put $A(\chi) = 0$ for such a χ. Then (2.4) holds for all $\chi \in \widehat{G}$ and (2.3) holds except for the case $n \leqq 2$, $\chi = 1$.

For $\sigma \in G$, put

$$b(\sigma) = \frac{1}{|G|} \sum_{\chi \in \hat{G}} A(\chi) \chi(\sigma).$$

Then, for $\alpha \in \mathrm{Aut}(\mathbf{C})$, we have

$$b(\sigma)^\alpha = \frac{1}{|G|} \sum_{\chi \in \hat{G}} A(\chi)^\alpha \chi^\alpha(\sigma) = \frac{1}{|G|} \sum_{\chi \in \hat{G}} A(\chi^\alpha) \chi^\alpha(\sigma) = b(\sigma).$$

Hence $b(\sigma) \in \mathbf{Q}$ and

$$\sum_{\sigma \in G} b(\sigma) \chi(\sigma)^{-1} = A(\chi) \qquad \text{for all} \quad \chi \in \widehat{G}.$$

Therefore we can find a positive integer m and a $\mathbf{Z}$-valued function $a(\sigma)$ on G such that

$$\sum_{\sigma \in G} a(\sigma) \chi(\sigma)^{-1} = m A(\chi) \qquad \text{for all} \quad \chi \in \widehat{G}.$$

Put

$$\epsilon = \prod_{\sigma \in G} (\epsilon_1)^{a(\sigma)\sigma} \in E_K.$$

Then, for $\chi \in \hat{G}$, we have

$$\begin{aligned}\sum_{\tau\in G}\chi(\tau)\log|\epsilon^\tau| &= \sum_{\tau\in G}\chi(\tau)\sum_{\sigma\in G}a(\sigma)\log|\epsilon_1^{\sigma\tau}| \\ &= \sum_{\tau\in G}\sum_{\sigma\in G}a(\sigma)\chi(\sigma)^{-1}\chi(\sigma\tau)\log|\epsilon_1^{\sigma\tau}| = mA(\chi)\sum_{\tau\in G}\chi(\tau)\log|\epsilon_1^\tau|.\end{aligned}$$

Since (2.3) holds except for the case $n \leqq 2$, $\chi = 1$, we get

$$L'(0,\chi) = \frac{1}{m}\sum_{\tau\in G}\chi(\tau)\log|\epsilon^\tau|$$

if $\chi \neq 1$ or $n \geqq 3$. Since $A(1) = 0$, we find easily that $\prod_{\tau\in G}\epsilon^\tau = 1$. Therefore (2.3) holds for all χ if $n \geqq 3$. This completes the proof.

2.2. Let F and K be as in Proposition 2.1. Let $\mathfrak{f}\infty_2\cdots\infty_n$ be the conductor of K as a class field over F, where $\mathfrak{f}$ is an integral ideal of F. For $\sigma \in G$, define a partial zeta function attached to σ by

$$\zeta(s,\sigma) = \sum_{\mathfrak{A},\left(\frac{K/F}{\mathfrak{A}}\right)=\sigma} N(\mathfrak{A})^{-s}, \tag{2.5}$$

where $\mathfrak{A}$ extends over all integral ideals relatively prime to $\mathfrak{f}$. Let χ_* be the character of the ideal class group modulo $\mathfrak{f}\infty_2\cdots\infty_n$ obtained from χ. We have $\chi_*(\mathfrak{A}) = \chi(\left(\frac{K/F}{\mathfrak{A}}\right))$ for a fractional ideal $\mathfrak{A}$ relatively prime to $\mathfrak{f}$. The L-function $L(s,\chi_*)$ attached to the character χ_* is given by

$$L(s,\chi_*) = \sum_{\sigma\in G}\chi(\sigma)\zeta(s,\sigma). \tag{2.6}$$

In general, χ_* may not be primitive. Let $\mathfrak{f}_\chi$ be the finite part of the conductor of χ_*. We have

$$L(s,\chi_*) = \prod_{\mathfrak{p}|\mathfrak{f},\mathfrak{p}\nmid\mathfrak{f}_\chi}(1-\chi(\mathfrak{p})N(\mathfrak{p})^{-s})L(s,\chi). \tag{2.7}$$

Since χ is unramified at ∞_1, we have $L(0,\chi) = 0$ except for the case $n = 1$, $\chi = 1$. We get

$$L'(0,\chi_*) = \prod_{\mathfrak{p}|\mathfrak{f},\mathfrak{p}\nmid\mathfrak{f}_\chi}(1-\chi(\mathfrak{p}))L'(0,\chi) \quad \text{if} \quad \chi\neq 1 \quad \text{or} \quad n \geqq 2. \tag{2.8}$$

PROPOSITION 2.2. *Let the notation and assumptions be the same as in Proposition* 2.1. *We assume Conjecture* 1.1 *and assume* $n \geqq 2$. *Then there exist a positive integer* e *and* $\epsilon \in E_K$ *such that*

$$\exp(e\zeta'(0,\tau)) = \epsilon^\tau \qquad \textit{for every} \quad \tau \in G. \tag{2.9}$$

PROOF. Put $B(\chi) = \prod_{\mathfrak{p}|\mathfrak{f}, \mathfrak{p}\nmid\mathfrak{f}_\chi}(1 - \chi(\mathfrak{p}))$ for $\chi \in \widehat{G}$. Since $B(\chi)^\alpha = B(\chi^\alpha)$ for $\alpha \in \mathrm{Aut}(\mathbf{C})$, we can find a positive integer m' and a $\mathbf{Z}$-valued function $a(\sigma)$ on G such that

$$\sum_{\sigma\in G} a(\sigma)\chi(\sigma)^{-1} = m'B(\chi) \qquad \text{for all} \quad \chi \in \widehat{G}.$$

Let $\epsilon \in E_K$ be as in Proposition 2.1 and put

$$\epsilon_* = \prod_{\sigma\in G} \epsilon^{a(\sigma)\sigma} \in E_K.$$

Then we have

$$\sum_{\tau\in G} \chi(\tau)\log|\epsilon_*^\tau| = m'B(\chi)\sum_{\tau\in G}\chi(\tau)\log|\epsilon^\tau|.$$

Assume $n \geqq 2$. By Proposition 2.1 and (2.8), we obtain

$$L'(0,\chi_*) = \frac{1}{mm'}\sum_{\tau\in G}\chi(\tau)\log|\epsilon_*^\tau|, \qquad \chi \in \widehat{G} \tag{2.10}$$

except for the case $n = 2$, $\chi = 1$. We have $\prod_{\tau\in G}\epsilon_*^\tau = 1$ since $\prod_{\tau\in G}\epsilon^\tau = 1$. Hence the right-hand side of (2.10) is 0 if $\chi = 1$. If $\chi = 1$, $n = 2$, $\mathfrak{f} \neq (1)$, we have $L'(0,\chi_*) = 0$. Therefore we get

$$\zeta'(0,\tau) = \frac{1}{mm'}\log|\epsilon_*^\tau| \qquad \text{for every} \quad \tau \in G \tag{2.11}$$

if $n \geqq 3$ or $n = 2$, $\mathfrak{f} \neq (1)$. There remains the case $n = 2$, $\chi = 1$, $\mathfrak{f} = (1)$. But since there does not exist any abelian extension of F of conductor $(1)\infty_2$, this case cannot occur.

Finally we note that $\epsilon_*^\tau \in \mathbf{R}$ since ∞_1 is unramified in K. Replacing ϵ_* by ϵ_*^2 and mm' by $2mm'$ if necessary, we can drop the symbol $|\ \ |$ in (2.11). This completes the proof.

According to results of Stark and Shintani, it is very plausible that we can take $e = 2$ in (2.9).

2.3. The following observation (discovered independently by Stark and Shintani) is the key point to check (2.9) numerically. For $2 \leqq i \leqq n$, let D_i be the subgroup of the character group of G such that $\chi \in D_i$ if and only if χ is unramified at ∞_i, i.e., $\chi(\sigma_i) = 1$. Assume $n \geqq 3$ or $n = 2$, $\mathfrak{f} \neq (1)$. Then, since

$$\sum_{\tau\in G}\chi(\tau)\zeta'(0,\tau) = 0 \qquad \text{for} \quad \chi \in D_i,$$

we have (cf. III, Lemma 5.1)

$$\zeta'(0,\tau) + \zeta'(0,\tau\sigma_i) = 0 \qquad \text{for every} \quad \tau \in G. \tag{2.12}$$

Suppose that (2.9) holds. We must have

$$\epsilon^\tau\epsilon^{\tau\sigma_i} = 1 \qquad \text{for every} \quad \tau \in G,\ 2 \leqq i \leqq n. \tag{2.13}$$

For an isomorphism β_i as in §1, we see that $\beta_i^{-1}\sigma_i\beta_i$ gives the complex conjugation ρ on K^{β_i}. Hence we get

$$|\epsilon^{\tau\beta_i}|^2 = \epsilon^{\tau\beta_i}\epsilon^{\tau\beta_i\rho} = 1 \qquad \text{for every} \quad \tau \in G,\ 2 \leqq i \leqq n. \tag{2.14}$$

Computing $\zeta'(0,\tau)$ numerically, we can determine the symmetric polynomials of ϵ^τ (or of $\epsilon^\tau+(\epsilon^\tau)^{-1}$) with sufficient accuracy. These polynomials must have coefficients in $\mathfrak{O}_F$, and the i-th conjugates of the coefficients can be bounded, for $i \geqq 2$, using (2.14). Hence, practically, we can identify these coefficients with algebraic integers of F.

3. Shintani's formulas for the partial zeta function

3.1. Let F be a totally real algebraic number field of degree n. Let $\mathfrak{f}$ be an integral ideal of F and $C_\mathfrak{f}$ denote the ideal class group defined modulo $\mathfrak{f}\infty_1\infty_2\cdots\infty_n$. We do not assume that the conductor of $C_\mathfrak{f}$ is exactly $\mathfrak{f}\infty_1\infty_2\cdots\infty_n$. For $c \in C_\mathfrak{f}$, let

$$\zeta_F(s,c)=\sum_{\mathfrak{A}\in c} N(\mathfrak{A})^{-s}, \qquad \Re(s)>1 \tag{3.1}$$

be the partial zeta function of the class c, where $\mathfrak{A}$ extends over integral ideals. Since $\zeta_F(s,c)$ can be written as a linear combination of L-functions attached to characters of $C_\mathfrak{f}$, we see that $\zeta_F(s,c)$ is a meromorphic function on the whole plane, holomorphic except for the simple pole at $s=1$. In this section, applying results in I, §2, we will derive explicit formulas for $\zeta_F(0,c)$ and $\zeta'_F(0,c)$. Here, for later applications, it is important to notice the difference that we now consider the class modulo $\mathfrak{f}\infty_1\infty_2\cdots\infty_n$, whereas $\zeta(s,\sigma)$ in the previous section is the sum of the partial zeta functions attached to classes modulo $\mathfrak{f}\infty_2\cdots\infty_n$.

Let $J_F=\{\sigma_1,\sigma_2,\dots,\sigma_n\}$. We put $x^{(i)}=x^{\sigma_i}$ for $x\in F$ and embed F into $\mathbf{R}^n$ by

$$F\ni x \longrightarrow (x^{(1)},x^{(2)},\dots,x^{(n)})\in \mathbf{R}^n.$$

Then E_F^+ is a discrete subgroup of $\mathbf{R}_+^n$. For r linearly independent vectors v_1, v_2, $\dots$, $v_r\in\mathbf{R}^n$, put

$$C(v_1,v_2,\dots,v_r)=\left\{\sum_{i=1}^r t_iv_i \mid t_1,t_2,\dots,t_r>0\right\}$$

and call $C(v_1,v_2,\dots,v_r)$ *an r-dimensional open simplicial cone with basis* v_1, v_2, $\dots$, v_r.

LEMMA 3.1 (SHINTANI [SHI1], PROPOSITION 4). *There exist finitely many open simplicial cones C_j, $j\in J$ with basis from $\mathfrak{O}_F$ such that*

$$\mathbf{R}_+^n=\sqcup_{\epsilon\in E_F^+}\,\epsilon\,(\sqcup_{j\in J}C_j).$$

In other words, a fundamental domain for $E_F^+\backslash\mathbf{R}_+^n$ can be taken as the disjoint union $\sqcup_{j\in J}C_j$. We call this decomposition *a cone decomposition of* $\mathbf{R}_+^n$. In chapter IV, we will investigate explicit constructions of $\sqcup_{j\in J}C_j$.

Let $r(j)$ be the dimension of C_j and put

$$C_j=C(v_{j1},v_{j2},\dots,v_{jr(j)}), \qquad v_{j1},v_{j2},\dots,v_{jr(j)}\in\mathfrak{O}_F.$$

Of course, v_{ji} is totally positive. For $z\in C_j$, we put

$$z=x_1(z)v_{j1}+x_2(z)v_{j2}+\dots+x_{r(j)}(z)v_{jr(j)},$$
$${}^tx(z)=(x_1(z),x_2(z),\dots,x_{r(j)}(z))\in\mathbf{R}^{r(j)}.$$

For a fractional ideal $\mathfrak{a}$ of F, we put

$$R(C_j, \mathfrak{a}) = \{z \in \mathfrak{a} \cap C_j \mid 0 < x_1(z), x_2(z), \dots, x_{r(j)}(z) \leqq 1, {}^t x(z) \in \mathbf{Q}^{r(j)}\}. \tag{3.2}$$

Let h_0 be the class number of F in the narrow sense and let $\mathfrak{a}_1, \mathfrak{a}_2, \dots, \mathfrak{a}_{h_0}$ be integral ideals which represent narrow ideal classes. For $c \in C_{\mathfrak{f}}$, take $\mathfrak{a}_\mu$ so that c and $\mathfrak{a}_\mu \mathfrak{f}$ belong to the same narrow ideal class and put

$$R(C_j, c) = \{z \in R(C_j, (\mathfrak{a}_\mu \mathfrak{f})^{-1}) \mid (z)\mathfrak{a}_\mu \mathfrak{f} = c \text{ in } C_{\mathfrak{f}}\}. \tag{3.3}$$

LEMMA 3.2.

$$\begin{aligned}&\{z \in (\mathfrak{a}_\mu \mathfrak{f})^{-1} \cap C_j, (z)\mathfrak{a}_\mu \mathfrak{f} = c \text{ in } C_{\mathfrak{f}}\}\\ &= \sqcup_{z \in R(C_j, c)} \{z + x_1 v_{j1} + x_2 v_{j2} + \dots + x_{r(j)} v_{jr(j)} \mid 0 \leqq x_1, x_2, \dots, x_{r(j)} \in \mathbf{Z}\}.\end{aligned}$$

PROOF. We first note that if $z \in F \cap C_j$, then $x_1(z), x_2(z), \dots, x_{r(j)}(z) \in \mathbf{Q}$. In fact, take $w_1, \dots, w_{n-r(j)}$ so that $\{v_1, \dots, v_{r(j)}, w_1, \dots, w_{n-r(j)}\}$ is a basis of F over $\mathbf{Q}$. Let $V_j = \mathbf{Q}v_1 + \dots + \mathbf{Q}_{r(j)}$. Then $z \in F \cap C_j$ implies $z \in F \cap V_j \otimes_{\mathbf{Q}} \mathbf{R}$. We have $z = \sum_{i=1}^{r(j)} a_i v_i + \sum_{i=1}^{n-r(j)} b_i w_i$ with a_i, $b_i \in \mathbf{Q}$ since $z \in F$ and $b_i = 0$, $1 \leqq i \leqq n - r(j)$ since $z \in V_j \otimes_{\mathbf{Q}} \mathbf{R}$.

Suppose that z belongs to the left-hand side. Then we have

$$z = x_1(z)v_{j1} + x_2(z)v_{j2} + \dots + x_{r(j)}(z)v_{jr(j)}, \qquad 0 < x_i(z) \in \mathbf{Q}.$$

Take $y_i \in \mathbf{Q}$ so that

$$x_i(z) \equiv y_i \mod \mathbf{Z}, \quad 0 < y_i \leqq 1, \quad 1 \leqq i \leqq r(j)$$

and put

$$w = y_1 v_{j1} + y_2 v_{j2} + \dots + y_{r(j)} v_{jr(j)}.$$

Let us show that $w \in R(C_j, c)$. Since $z \gg 0$ and $w \gg 0$, it suffices to show that

$$(w)\mathfrak{a}_\mu \mathfrak{f} \equiv (z)\mathfrak{a}_\mu \mathfrak{f} \mod^\times \mathfrak{f}.$$

We have $w - z \in \mathfrak{O}_F$ since $v_{j1}, v_{j2}, \dots, v_{jr(j)} \in \mathfrak{O}_F$. Since $(z)\mathfrak{a}_\mu \mathfrak{f}$ is relatively prime to $\mathfrak{f}$, $(w)\mathfrak{a}_\mu \mathfrak{f}$ is also relatively prime to $\mathfrak{f}$. Hence it suffices to show $(w - z)\mathfrak{a}_\mu \mathfrak{f} \subset \mathfrak{f}$. This is clear since $\mathfrak{a}_\mu$ is an integral ideal. Hence the left-hand side is contained in the right-hand side. The opposite inclusion can be shown similarly. The disjointness of the union of the right-hand side is obvious. This completes the proof.

3.2. By definition, we have

$$\zeta_F(s, c) = \sum_{(z),\, z \in (\mathfrak{a}_\mu \mathfrak{f})^{-1},\, (z)\mathfrak{a}_\mu \mathfrak{f} = c \text{ in } C_{\mathfrak{f}}} N((z)\mathfrak{a}_\mu \mathfrak{f})^{-s}.$$

Applying Lemma 3.1, we get

$$\zeta_F(s, c) = N(\mathfrak{a}_\mu \mathfrak{f})^{-s} \sum_{j \in J} \sum_{z \in (\mathfrak{a}_\mu \mathfrak{f})^{-1} \cap C_j, (z)\mathfrak{a}_\mu \mathfrak{f} = c \text{ in } C_{\mathfrak{f}}} N(z)^{-s}. \tag{3.4}$$

By Lemma 3.2, we have

$$\begin{aligned}
&\sum_{z\in(\mathfrak{a}_\mu\mathfrak{f})^{-1}\cap C_j,(z)\mathfrak{a}_\mu\mathfrak{f}=c \text{ in } C_\mathfrak{f}} N(z)^{-s} \\
&= \sum_{z\in R(C_j,c)} \sum_{x_1,\dots,x_{r(j)}=0}^{\infty} \Big(\prod_{i=1}^{n}(z^{(i)}+x_1v_{j1}^{(i)}+\cdots+x_{r(j)}v_{jr(j)}^{(i)})\Big)^{-s} \\
&= \sum_{z\in R(C_j,c)} \sum_{x_1,\dots,x_{r(j)}=0}^{\infty} \prod_{i=1}^{n}\Big\{\sum_{l=1}^{r(j)} v_{jl}^{(i)}(x_l+x_l(z))\Big\}^{-s}.
\end{aligned}$$

Define an $n\times r(j)$-matrix A_j by $A_j=(v_{jl}^{(i)})$ (the (i,l)-component of A_j is $v_{jl}^{(i)}$). Then we have

$$\zeta_F(s,c)=N(\mathfrak{a}_\mu\mathfrak{f})^{-s}\sum_{j\in J}\sum_{z\in R(C_j,c)}\zeta(s,A_j,x(z)). \tag{3.5}$$

We gave expressions for $\zeta(0,A_j,x(z))$ and $\zeta'(0,A_j,x(z))$ in I, §2. By I, Theorems 2.1 and 2.2, we obtain:

THEOREM 3.3.

$$\begin{aligned}
&\zeta_F(0,c) \\
&=\sum_{j\in J}\frac{(-1)^{r(j)}}{n}\sum_{z\in R(C_j,c)}\sum_{l}\Big[\prod_{m=1}^{r(j)}\Big\{\frac{B_{l_m}(x_m(z))}{l_m!}\Big\}\mathrm{Tr}_{F/\mathbf{Q}}\Big\{\prod_{m=1}^{r(j)}(v_{jm}^{l_m-1})\Big\}\Big],
\end{aligned} \tag{3.6}$$

$$\begin{aligned}
\zeta_F'(0,c) &= \sum_{k=1}^{n}\sum_{j\in J}\sum_{z\in R(C_j,c)}\log\frac{\Gamma_{r(j)}(z^{(k)},A_j^{(k)})}{\rho_{r(j)}(A_j^{(k)})} \\
&\quad+\sum_{j\in J}\frac{(-1)^{r(j)}}{n}\Big[\sum_{l}C_l(A_j)\sum_{z\in R(C_j,c)}\prod_{m=1}^{r(j)}\Big\{\frac{B_{l_m}(x_m(z))}{l_m!}\Big\}\Big] \\
&\quad-\log N(\mathfrak{a}_\mu\mathfrak{f})\zeta_F(0,c).
\end{aligned} \tag{3.7}$$

Here in $\sum_l$, $l=(l_1,l_2,\dots,l_{r(j)})$ extends over all $r(j)$-tuples of nonnegative integers such that $l_1+l_2+\cdots+l_{r(j)}=r(j)$; $A_j^{(k)}$ denotes the k-th row of A_j.

4. An example

We illustrate the Stark-Shintani conjecture in the form of §2 by an example. We will discuss more examples in the relation to absolute CM-periods in chapter III.

4.1. Let $F=\mathbf{Q}(\sqrt{5})$. For $x\in F$, let x' denote the conjugate of x. We have $(19)=\mathfrak{p}_{19}\mathfrak{p}_{19}'$ in F where $\mathfrak{p}_{19}$ (resp. $\mathfrak{p}_{19}'$) is the prime ideal given by $\mathfrak{p}_{19}=(8+3\sqrt{5})$ (resp. $\mathfrak{p}_{19}'=(8-3\sqrt{5})$). Put $\mathfrak{f}=(4)\mathfrak{p}_{19}$ and let $C_\mathfrak{f}$ be the ideal class group modulo $\mathfrak{f}\infty_1\infty_2$. Let ϵ_0 be the fundamental unit of F and put $\epsilon=\epsilon_0^2$. We have $\epsilon_0=\dfrac{1+\sqrt{5}}{2}$, $\epsilon=\dfrac{3+\sqrt{5}}{2}$. Set

$$E_{\mathfrak{f}\infty_1\infty_2}=\{\eta\in E_F \mid \eta\gg 0,\ \eta\equiv 1 \mod \mathfrak{f}\}.$$

Then we have

$$|C_{\mathfrak{f}}| = \frac{\varphi(\mathfrak{f}) \cdot 2^2}{[E_F : E_{\mathfrak{f}\infty_1\infty_2}]},$$

where φ denotes the Euler function. We have $\varphi(\mathfrak{f}) = (2^2 - 1)2^2(19 - 1)$ and $E_{\mathfrak{f}\infty_1\infty_2} = \langle \epsilon^9 \rangle$. Hence $|C_{\mathfrak{f}}| = 24$. Define α, β, $\gamma \in C_{\mathfrak{f}}$ by

$$\alpha = \text{the class of } (\frac{49+\sqrt{5}}{2}), \qquad \beta = \text{the class of } (39), \qquad \gamma = \text{the class of } (69).$$

We see immediately that $\alpha^6 = \beta^2 = \gamma^2 = 1$. In fact, put $\alpha_0 = \dfrac{49+\sqrt{5}}{2}$, $\beta_0 = 39$, $\gamma_0 = 69$. Then we have $\alpha_0 \equiv 1 \mod \mathfrak{p}_{19}$, $\alpha_0^6 \equiv 1 \mod 4$. Hence $\alpha^6 = 1$. We get $\gamma^2 = 1$ from $\epsilon^{-3}\gamma_0^2 \equiv 1 \mod \mathfrak{f}$. Let us show that the orders of α, β, γ are 6, 2, 2 respectively and that

$$C_{\mathfrak{f}} = \langle\alpha\rangle \oplus \langle\beta\rangle \oplus \langle\gamma\rangle \cong \mathbf{Z}/6\mathbf{Z} \oplus \mathbf{Z}/2\mathbf{Z} \oplus \mathbf{Z}/2\mathbf{Z}. \tag{4.1}$$

For this purpose, it suffices to show that if $\alpha^a\beta^b\gamma^c = 1$, then $6|a$, $2|b$, $2|c$. This relation is equivalent to

$$\pm\epsilon_0^i\alpha_0^a\beta_0^b\gamma_0^c \equiv 1 \mod^{\times} \mathfrak{f}\infty_1\infty_2, \tag{4.2}$$

for some $i \in \mathbf{Z}$. Since the left-hand side of (4.2) must be totally positive, the sign $\pm$ is $+$ and i is even. Put $i = 2j$. We have

$$\epsilon^j\alpha_0^a\beta_0^b\gamma_0^c \equiv 1 \mod \mathfrak{f} \tag{4.3}$$

from (4.2). Considering (4.3) mod $\mathfrak{p}_{19}$, we get $\epsilon^j 12^c \equiv 1 \mod \mathfrak{p}_{19}$. Since $(\dfrac{\epsilon}{\mathfrak{p}_{19}}) = 1$ and $(\dfrac{12}{\mathfrak{p}_{19}}) = (\dfrac{12}{19}) = -1$, we have $2|c$. Now we may assume $c = 0$ and we have (4.2), (4.3) with $c = 0$. Then from $\epsilon^j \equiv 1 \mod \mathfrak{p}_{19}$, we get $9|j$. Hence $\epsilon^j \equiv 1 \mod \mathfrak{f}$. Since $\alpha_0 \equiv \epsilon_0 \mod (4)$, (4.3) implies

$$(-1)^b\epsilon_0^a \equiv 1 \mod (4). \tag{4.4}$$

By $\epsilon_0^3 \equiv 2+\sqrt{5}$, $\epsilon_0^4 = \dfrac{7+3\sqrt{5}}{2}$, $\epsilon_0^5 = \dfrac{11+5\sqrt{5}}{2}$, $\epsilon_0^6 = 9+4\sqrt{5}$, we see immediately that $6|a$, $2|b$. Thus we have established (4.1).

Let $D = \widehat{C}_{\mathfrak{f}}$ be the character group of $C_{\mathfrak{f}}$. Define ω, ψ, $\eta \in D$ by

$$\begin{aligned}
&\omega(\alpha) = e^{2\pi i/6}, && \omega(\beta) = \omega(\gamma) = 1,\\
&\psi(\alpha) = \psi(\gamma) = 1, && \psi(\beta) = -1,\\
&\eta(\alpha) = \eta(\beta) = 1, && \eta(\gamma) = -1.
\end{aligned}$$

For $\chi \in D$, let $\mathfrak{f}(\chi)$ denote the conductor of χ. We have

$$\mathfrak{f}(\omega) = (4)\mathfrak{p}_{19}\infty_1\infty_2, \qquad \mathfrak{f}(\psi) = (4)\infty_1, \qquad \mathfrak{f}(\eta) = \mathfrak{p}_{19}\infty_2. \tag{4.5}$$

Let us verify (4.5). We see that $\chi \in D$ can be written in the form

$$\chi((x)) = \chi_4(x \mod 4)\chi_{19}(x \mod \mathfrak{p}_{19})(\text{sgn } x)^{\delta_1}(\text{sgn } x')^{\delta_2} \tag{4.6}$$

for $x \in \mathfrak{O}_F$, $((x), \mathfrak{f}) = 1$. Here χ_4 (resp. χ_{19}) denotes a character of $(\mathfrak{O}_F/(4))^{\times}$ (resp. $(\mathfrak{O}_f/\mathfrak{p}_{19})^{\times}$), $\delta_i = 0$ or 1. We have

$$\chi(\alpha) = \chi_4(\epsilon_0 \mod 4), \quad \chi(\beta) = \chi_4(-1 \mod 4), \quad \chi(\gamma) = \chi_{19}(12 \mod \mathfrak{p}_{19}). \tag{4.7}$$

We note that $\pm\epsilon_0^j \bmod 4$, $0 \leqq j \leqq 5$ exhaust $(\mathfrak{O}_F/(4))^\times$ and χ_4 is determined completely by (4.7). Since $\chi((-1)) = 1$, we have

$$1 = \chi_4(-1 \mod 4)\chi_{19}(-1 \mod \mathfrak{p}_{19})(-1)^{\delta_1+\delta_2}.$$

Noting $12^3 \equiv -1 \bmod 19$, we get

$$(4.8) \qquad (-1)^{\delta_1+\delta_2} = \chi(\beta)\chi(\gamma)^3.$$

By $\chi((\epsilon_0)) = 1$, we have

$$1 = \chi_4(\epsilon_0 \mod 4)\chi_{19}(\epsilon_0 \mod \mathfrak{p}_{19})(-1)^{\delta_2},$$

$$(4.9) \qquad (-1)^{\delta_2} = \chi(\alpha)\chi_{19}(\epsilon_0 \mod \mathfrak{p}_{19}).$$

Since $\epsilon_0^3 \equiv 12 \mod \mathfrak{p}_{19}$, we have

$$(4.10) \qquad \chi_{19}(\epsilon_0 \mod \mathfrak{p}_{19})^3 = \chi(\gamma).$$

Let $\chi = \omega$. By (4.9) and (4.10), we see that $\omega_{19}(\epsilon_0 \bmod \mathfrak{p}_{19}) = -e^{-2\pi i/6} = e^{2\pi i/3}$ and $\delta_2 = 1$. By (4.8), we get $\delta_1 = 1$ and $\mathfrak{f}(\omega) = (4)\mathfrak{p}_{19}\infty_1\infty_2$. Let $\chi = \psi$. By (4.9) and (4.10), we get $\psi_{19}(\epsilon_0 \bmod \mathfrak{p}_{19}) = 1$ and $\delta_2 = 0$. By (4.8), we get $\delta_1 = 1$ and $\mathfrak{f}(\psi) = (4)\infty_1$, since $\epsilon_0 \bmod \mathfrak{p}_{19}$ generates $(\mathfrak{O}_F/\mathfrak{p}_{19})^\times$. Let $\chi = \eta$. By (4.9) and (4.10), we get $\chi_{19}(\epsilon_0 \bmod \mathfrak{p}_{19}) = -1$ and $\delta_2 = 1$. By (4.8), we have $\delta_1 = 0$, $\mathfrak{f}(\eta) = \mathfrak{p}_{19}\infty_2$. Thus we have verified (4.5).

4.2. Let D_0, D_1 and D_2 be the subgroups of D consisting of all characters whose conductors divide $(4)\mathfrak{p}_{19}$, $(4)\mathfrak{p}_{19}\infty_1$, $(4)\mathfrak{p}_{19}\infty_2$ respectively. We see that $|D_0|$ is equal to the order of the ideal class group modulo $(4)\mathfrak{p}_{19}$. We easily get $|D_0| = 6$. Hence we have

$$D_0 = \{1, \omega\psi\eta, \omega^2, \omega^3\psi\eta, \omega^4, \omega^5\psi\eta\}.$$

Now it is clear that

$$D_1 = D_0 \sqcup \psi D_0, \qquad D_2 = D_0 \sqcup \eta D_0.$$

For $0 \leqq i \leqq 2$, let C_i be the annihilator of D_i. We have

$$C_0 = \{1, \alpha^3\beta, \alpha^3\gamma, \beta\gamma\}, \qquad C_1 = \{1, \alpha^3\gamma\}, \qquad C_2 = \{1, \alpha^3\beta\}.$$

Since

$$\sum_{c \in C_\mathfrak{f}} \chi(c)\zeta_F'(0, c) = 0 \qquad \text{for every} \quad \chi \in D_0,$$

we have (cf. III, Lemma 5.1)

$$(4.11) \qquad \sum_{a \in C_0} \zeta_F'(0, ca) = 0 \qquad \text{for every} \quad c \in C_\mathfrak{f}.$$

Now let M (resp. K) be the maximal ray class field over F modulo $(4)\mathfrak{p}_{19}\infty_1\infty_2$ (resp. $(4)\mathfrak{p}_{19}\infty_2$). We see that the conductors of M and K coincide with these moduli respectively. Let $C_\mathfrak{f}'$ be the ideal class group modulo $(4)\mathfrak{p}_{19}\infty_2$; the conductor of $C_\mathfrak{f}'$ is the same. We have $\mathrm{Gal}(M/F) \cong C_\mathfrak{f}$ and $\mathrm{Gal}(M/K)$ corresponds to the subgroup $\{1, \alpha^3\beta\}$. Let $\varphi : C_\mathfrak{f} \longrightarrow C_\mathfrak{f}'$ be the canonical map. Then $\ker\varphi = \{1, \alpha^3\beta\}$. We clearly have

$$(4.12) \qquad \zeta_F(s, c) + \zeta_F(s, c\alpha^3\beta) = \zeta_F(s, \varphi(c)) \qquad \text{for} \quad c \in C_\mathfrak{f}.$$

Here $\zeta_F(s, \varphi(c))$ is defined to be $\zeta(s, \sigma)$ in (2.5) taking $\sigma \in \mathrm{Gal}(K/F)$ which corresponds to $\varphi(c)$ under the Artin map. If (2.9) holds with $e = 2$, then $\exp(2(\zeta_F'(0, c) + \zeta_F'(0, c\alpha^3\beta)))$ is a unit of K for every $c \in C_{\mathfrak{f}}$. Put

$$v(c) = \exp(2(\zeta_F'(0, c) + \zeta_F'(0, c\alpha^3\beta))) + \exp(-2(\zeta_F'(0, c) + \zeta_F'(0, c\alpha^3\beta))), \quad c \in C_{\mathfrak{f}}.$$

By (4.11), we have

$$v(c) = v(ca) \qquad \text{for every} \quad a \in C_0.$$

Taking representatives $\{\alpha^{i-1} \mid 1 \leqq i \leqq 6\}$ for $C_{\mathfrak{f}}/C_0$, put $t_i = v(\alpha^{i-1})$, $1 \leqq i \leqq 6$. A numerical computation gives

$$\begin{aligned}\prod_{i=1}^{6}(X - t_i) = X^6 &- 56.8328157299\ldots X^5 + 922.2480373537\ldots X^4 \\ &- 6571.4920628796\ldots X^3 + 23200.5747692351\ldots X^2 \\ &- 39782.8246598606\ldots X + 26351.9125234287\ldots.\end{aligned}$$

Using (2.14), we can identify the coefficients of this polynomial with integers of $\mathbf{Q}(\sqrt{5})$. For example, let $\theta = 56.83281\ldots$. Suppose that $\theta = \dfrac{a + b\sqrt{5}}{2}$, a, $b \in \mathbf{Z}$, $a \equiv b \mod 2$. By (2.14), we have $|\theta'| \leqq \binom{6}{1} \cdot 2 = 12$. Hence we get $44 < b\sqrt{5} = \theta - \theta' < 69$, $20 \leqq b \leqq 30$. Searching b for which the decimal part of $b\sqrt{5}/2$ (resp. $(1 + b\sqrt{5})/2$) if b is even (resp. odd) matches well with that of θ, we find $b = 24$. Thus we get the identification $\theta = 30 + 12\sqrt{5}$. In this way, we obtain a very plausible equality

$$\begin{aligned}\prod_{i=1}^{6}(X - t_i) = X^6 - (30 + 12\sqrt{5})X^5 + \frac{921 + 413\sqrt{5}}{2}X^4 - (3280 + 1472\sqrt{5})X^3 \\ + \frac{23193 + 10379\sqrt{5}}{2}X^2 - (19893 + 8895\sqrt{5})X + (13177 + 5892\sqrt{5}).\end{aligned}$$

Now let L be the maximal ray class field over F of conductor $(4)\mathfrak{p}_{19}\infty_1$ and let $C_{\mathfrak{f}}''$ be the ideal class group modulo $(4)\mathfrak{p}_{19}\infty_1$. We have $\mathrm{Gal}(L/F) \cong C_{\mathfrak{f}}''$. Suppose that (2.9) holds with $e = 2$. Since $\zeta_F(s, c) = \zeta_F(s, c^\sigma)$, $c \in C_{\mathfrak{f}}''$ for the non-trivial automorphism σ of F over $\mathbf{Q}$, we see that $\exp(2(\zeta_F'(0, c)))$ is a unit of $L^\sigma =$ the maximal ray class field over F of conductor $(4)\mathfrak{p}_{19}'\infty_2$. We put

$$s_i = \exp(2(\zeta_F'(0, \alpha^{i-1}) + \zeta_F'(0, \alpha^{i-1}\alpha^3\gamma))) + \exp(-2(\zeta_F'(0, \alpha^{i-1}) + \zeta_F'(0, \alpha^{i-1}\alpha^3\gamma)))$$

for $1 \leqq i \leqq 6$. Similarly to the above, we obtain a very plausible equality

$$\begin{aligned}\prod_{i=1}^{6}(X - s_i) = X^6 - (56 + 24\sqrt{5})X^5 + (1856 + 832\sqrt{5})X^4 - (24208 + 10832\sqrt{5})X^3 \\ + (141248 + 63168\sqrt{5})X^2 - (367232 + 164224\sqrt{5})X + (345472 + 154496\sqrt{5}).\end{aligned}$$

Exercises

1. For an integer $n \geqq 3$, put $\zeta_n = e^{2\pi i/n}$. Let $\mathbf{Q}(\zeta_n)^+$ denote the maximal real subfield of $\mathbf{Q}(\zeta_n)$.
   (a) For an integer a, $(a, n) = 1$, put
$$E_a^{(n)} = \frac{\zeta_{2n}^a - \zeta_{2n}^{-a}}{\zeta_{2n} - \zeta_{2n}^{-1}} = \frac{\sin(a\pi/n)}{\sin(\pi/n)}.$$
   Show that $E_a^{(n)}$ is a unit of $\mathbf{Q}(\zeta_n)^+$.
   (b) For an integer t, $(t, n) = 1$, let $\sigma(t) \in \mathrm{Gal}(\mathbf{Q}(\zeta_n)/\mathbf{Q})$ be the automorphism determined by $\zeta_n^{\sigma(t)} = \zeta_n^t$. Show that
$$(E_a^{(n)})^{\sigma(t)} = \begin{cases} \frac{\sin(at\pi/n)}{\sin(t\pi/n)} & \text{if } t \text{ is odd}, \\ (-1)^{a-1}\frac{\sin(at\pi/n)}{\sin(t\pi/n)} & \text{if } t \text{ is even}. \end{cases}$$
   (c) For a character χ of $\mathrm{Gal}(\mathbf{Q}(\zeta_n)^+/\mathbf{Q})$, let χ^* denote the corresponding primitive Dirichlet character; the conductor of χ^* divides n and χ^* is even. Show that
$$\sum_{\tau \in \mathrm{Gal}(\mathbf{Q}(\zeta_n)^+/\mathbf{Q})} \chi(\tau) \log|(E_a^{(n)})^\tau| = (\chi^*(a)^{-1} - 1) \sum_{c=1}^{[n/2]} \chi^*(c) \log\left|\sin\frac{c\pi}{n}\right|.$$
   (d) Put
$$E = \prod_{a=1,(a,n)=1}^{[n/2]} E_a^{(n)}.$$
   We assume $\chi \neq 1$. Show that
$$\sum_{\tau \in \mathrm{Gal}(\mathbf{Q}(\zeta_n)^+/\mathbf{Q})} \chi(\tau) \log|E^\tau| = -\frac{\varphi(n)}{2} \sum_{c=1}^{[n/2]} \chi^*(c) \log(\sin\frac{c\pi}{n})$$
   (e) Let K be a totally real abelian extension of $\mathbf{Q}$ with conductor (n), $n > 0$. Put $G = \mathrm{Gal}(K/\mathbf{Q})$, $\epsilon = N_{\mathbf{Q}(\zeta_n)^+/K}(E)$. For a character χ of G, use the same letter χ for the character of $\mathrm{Gal}(\mathbf{Q}(\zeta_n)^+/\mathbf{Q})$ obtained from χ by the canonical homomorphism $\mathrm{Gal}(\mathbf{Q}(\zeta_n)^+/\mathbf{Q}) \longrightarrow G$. We assume $\chi \neq 1$. Show that
$$\sum_{\tau \in G} \chi(\tau) \log|\epsilon^\tau| = -\frac{\varphi(n)}{2} \sum_{c=1}^{[n/2]} \chi^*(c) \log(\sin\frac{c\pi}{n})$$
   (f) In the setting of (e), show that
$$L'(0,\chi) = -\sum_{c=1}^{[n/2]} \chi^*(c) \log(\sin\frac{c\pi}{n}) = \frac{2}{\varphi(n)} \sum_{\tau \in G} \chi(\tau) \log|\epsilon^\tau|$$
   using III, (2.11). Compare with Proposition 2.1.
2. In the notation of Exercise 1, let $3 \leqq f$ be a divisor of n. Show that
$$N_{\mathbf{Q}(\zeta_n)^+/\mathbf{Q}(\zeta_f)^+}(E_a^{(n)}) = \pm E_a^{(f)}.$$

(Use the formula

$$\sin mx = 2^{m-1} \prod_{k=0}^{m-1} \sin(x + \frac{k\pi}{m}), \qquad 1 \leqq m \in \mathbf{Z}$$

which can be proven easily.)

3. Let F be a real quadratic field. Show that the order of the ideal class group of F modulo $(1)\infty_2$ is h_F. Show that there does not exist a class field over F of conductor $(1)\infty_2$.
4. Let F be a totally real algebraic number field of degree $n \geqq 2$. Let K be a quadratic extension of F of conductor $\mathfrak{f}\infty_2 \cdots \infty_n$. Put $G = \mathrm{Gal}(K/F)$ and for $\sigma \in G$, define $\zeta(s, \sigma)$ by (2.5).
   (a) Put $G = \{\sigma_1, \sigma_2\}$, $\sigma_1 = 1$. Show that

$$\zeta(s, \sigma_1) + \zeta(s, \sigma_2) = \prod_{\mathfrak{p} | \mathfrak{f}} (1 - N(\mathfrak{p})^{-s}) \zeta_F(s),$$

$$\zeta(s, \sigma_1) - \zeta(s, \sigma_2) = L(s, \omega)$$

   where ω denotes the Hecke character of $F_A^\times$ which corresponds to the quadratic extension K/F.
   (b) Using (1.2), show that

$$L'(0, \omega) = \frac{R_K}{R_F} \cdot \frac{h_K}{h_F}.$$

   (c) Put $U = \{x \in E_K \mid N_{K/F}(x) = 1\}$. Show that $U/\{\pm 1\} \cong \mathbf{Z}$ and that

$$\frac{R_K}{R_F} = \frac{1}{2} [E_F : N_{K/F}(E_K)] \log u$$

   for a generator u of $U/\{\pm 1\}$. (This is a special case of a theorem of Shimura. The reader may consult [Sh15].)
   (d) Show that

$$\exp(\frac{4}{[E_F : N_{K/F}(E_K)]} \zeta'(0, \sigma_1)) = u^{h_K/h_F},$$

$$\exp(\frac{4}{[E_F : N_{K/F}(E_K)]} \zeta'(0, \sigma_2)) = u^{-h_K/h_F}.$$

   (e) Show that h_F divides h_K.
   (f) Show that $-1 \notin N_{K/F}(E_K)$, $[E_F : N_{K/F}(E_K)] = 2^i$, $1 \leqq i \leqq n$. Compare with Proposition 2.2.
5. Let $F = \mathbf{Q}(\sqrt{5})$. Put $(11) = \mathfrak{p}\mathfrak{p}'$ in $\mathfrak{O}_F$, $\mathfrak{p} = (4 - \sqrt{5})$.
   (a) Show that $|C_\mathfrak{p}| = 2$ and that the conductor of $C_\mathfrak{p}$ (the ideal class group modulo $\mathfrak{p}\infty_1\infty_2$) is $\mathfrak{p}\infty_2$.
   (b) Show that $C_\mathfrak{p} = \{c_1, c_2\}$, $c_1 =$ the class of (1), $c_2 =$ the class of (2).
   (c) Define cones by $C_1 = C(1, \epsilon)$, $C_2 = C(1)$, $\epsilon = \frac{3+\sqrt{5}}{2}$. Show that

$$R(C_1, c_1) = \left\{ \frac{1+2\epsilon}{11}, \frac{9+7\epsilon}{11}, \frac{4+8\epsilon}{11}, \frac{3+6\epsilon}{11}, \frac{5+10\epsilon}{11} \right\},$$

$$R(C_1, c_2) = \left\{ \frac{2+4\epsilon}{11}, \frac{7+3\epsilon}{11}, \frac{8+5\epsilon}{11}, \frac{6+\epsilon}{11}, \frac{10+9\epsilon}{11} \right\},$$

and that $R(C_2, c_i) = \emptyset$, $i = 1, 2$.

(d) Using (3.7), show that

$$\begin{aligned} \zeta_F'(0, c_1) &= 0.383598609125659\ldots, \\ \exp(2\zeta_F'(0, c_1)) &= 2.153721375541767\ldots, \\ \exp(2\zeta_F'(0, c_1)) + \exp(2\zeta_F'(0, c_2)) &= 2.618033988749897\ldots. \end{aligned}$$

(The reader may consult III, §4.1.)

(e) Let K be the maximal ray class field of F of conductor $\mathfrak{p}\infty_2$. Put $u_1 = \exp(2\zeta_F'(0, c_1))$. Take $u \in U$ as in Exercise 4. Put $m = [E_F : N_{K/F}(E_K)]/2$. By Exercise 4, (d), (f), we have $u_1 = u^{mh_K}$ and $m = 1$ or 2. Put $\alpha = u_1 + u_1^{-1} \in \mathfrak{O}_F$. Show that $|\alpha'| \leqq 2$ for the conjugate α' of α. Using the numerical values $\alpha = 2.61808\ldots$, $u_1 = 2.15372\ldots$ in (d), show that

$$\alpha = \frac{3+\sqrt{5}}{2}, \qquad u_1 = \frac{1}{2}\Big(\frac{3+\sqrt{5}}{2} + \sqrt{\frac{3\sqrt{5}-1}{2}}\Big), \qquad K = F\Big(\sqrt{\frac{3\sqrt{5}-1}{2}}\Big).$$

(f) Put $\beta = u + u^{-1} \in \mathfrak{O}_F$. Show that $|\beta'| \leqq 2$. Using $2 < \beta \leqq u_1 + u_1^{-1}$, show that $mh_K = 1$, hence $m = 1$, $h_K = 1$.

(g) Verify that $K = F(\sqrt{\frac{3\sqrt{5}-1}{2}})$ is the maximal ray class field of F of conductor $\mathfrak{p}\infty_2$, by examining the relative different.

(h) Verify that $-\epsilon_0 \in N_{K/F}(E_K)$, $[E_F : N_{K/F}(E_K)] = 2$, where $\epsilon_0 = \frac{1+\sqrt{5}}{2}$ is the fundamental unit of F.

6. Let $F = \mathbf{Q}(\sqrt{5})$. Put $\mathfrak{p} = (14 - 3\sqrt{5})$, $N(\mathfrak{p}) = 151$. Let K be the quadratic extension of F of conductor $\mathfrak{p}\infty_2$. In a similar way to Exercises 4, 5, show that

$$\exp(2\zeta'(0, \sigma_1)) = \frac{1}{2}(64 + 29\sqrt{5} + \sqrt{8297 + 3712\sqrt{5}}),$$

$$K = \mathbf{Q}(\sqrt{8297 + 3712\sqrt{5}}), \qquad h_K = 1, \quad [E_F : N_{K/F}(E_K)] = 2.$$

CHAPTER III

ABSOLUTE CM-PERIODS

In §1, we will review basic properties of Shimura's period symbol $p_K(\xi, \eta)$, where K is a CM-field and ξ, η are $\mathbf{Z}$-linear combinations of isomorphisms of K into $\mathbf{C}$. In §2, we will study $p_K(\xi, \eta)$ when K is abelian over $\mathbf{Q}$. We will prove that $p_K(\xi, \eta)$ can be expressed using the gamma function. This result was predicted by Shimura [Sh7] and proved by Anderson [An]. Our proof given here, which uses the period symbol p_K extensively, is somewhat simpler. In §3, we first formulate a conjecture on the logarithmic derivative of Artin's L-function at $s = 0$, when the representation of Galois group is odd. This conjecture, first given in Colmez [Co], is a natural extension of the result in §2. Next we proceed to define another period symbol $g_K(\mathrm{id}, \tau)$ using the multiple gamma function. We conjecture that $p_K(\mathrm{id}, \tau) \sim g_K(\mathrm{id}, \tau)$ when K is abelian over a totally real number field F and $\tau \in \mathrm{Gal}(K/F)$. The absolute period symbol g_K was introduced by the author in [Y4]. In §4, we will discuss numerical examples, which support this conjecture, extensively. The symbol g_K is defined using certain class invariants $X(c)$. In §5, we will study properties of $X(c)$ when F is a real quadratic field. By this study, we can construct a coherent picture in which the period g_K and the Stark-Shintani unit fit together harmoniously. In §6, we will illustrate our results in §5 by examples.

1. Shimura's period symbol p_K

1.1. Let K be a CM-field of degree $2n$. We denote by J_K the set of all isomorphisms of K into $\mathbf{C}$ and by I_K the free abelian group generated by J_K. We denote by ρ the complex conjugation. We call $\Phi \in I_K$ *a CM-type of* K if $\Phi = \sum_{i=1}^{n} \sigma_i \in I_K$ with $\sigma_i \in J_K$ and if $\Phi + \Phi\rho$ is the sum of all elements of J_K. We call a couple (A, θ) formed by an abelian variety A over $\mathbf{C}$ of dimension n and an isomorphism θ of K into $\mathrm{End}(A) \otimes \mathbf{Q}$ to be of type (K, Φ) if the representation of K on the space of holomorphic differential 1-forms on A, which is obtained through θ, is equivalent to Φ. For a given CM-type of K, there exists a couple (A, θ) of type (K, Φ) defined over $\overline{\mathbf{Q}}$. By definition, we see that for every $\sigma \in \Phi$, there is a non-zero holomorphic differential 1-form ω_σ on A such that ω_σ is multiplied by a^σ by the action of $a \in \theta^{-1}(\theta(K) \cap \mathrm{End}(A))$ and that ω_σ is defined over $\overline{\mathbf{Q}}$.

LEMMA 1.1. *Let $\sigma \in \Phi$. There exists a constant $p_K(\sigma, \Phi) \in \mathbf{C}^\times$ such that*

$$(1.1) \qquad \int_c \omega_\sigma \sim \pi p_K(\sigma, \Phi) \qquad \textit{for every} \quad c \in H_1(A, \mathbf{Z}).$$

Furtheremore $p_K(\sigma, \Phi)$ mod $\overline{\mathbf{Q}}^\times$ *does not depend on the choice of (A, θ) and ω_σ.*

PROOF. Let $\mathfrak{D}(A)$ denote the space of holomorphic differential 1-forms on A. Let θ^* (resp. θ_*) denote the action of K on $\mathfrak{D}(A)$ (resp. $H_1(A, \mathbf{Q})$) obtained from θ. When regarded as a K-module by θ_*, $H_1(A, \mathbf{Q})$ is a one dimensional vector space over K. Let $\langle\ ,\ \rangle$ denote the pairing between $\mathfrak{D}(A)$ and $H_1(A, \mathbf{Q})$ which is the linear extension of

$$\langle \omega, c\rangle = \int_c \omega, \qquad \omega \in \mathfrak{D}(A), \quad c \in H_1(A, \mathbf{Z}),$$

when c is represented by a positive 1-cycle. Since

$$\int_c \omega \circ p = \int_{p(c)} \omega$$

for a continuous mapping $p : A \longrightarrow A$, we have

$$(1.2) \qquad \langle \theta^*(a)\omega, c\rangle = \langle \omega, \theta_*(a)c\rangle, \quad \omega \in \mathfrak{D}(A), \quad c \in H_1(A, \mathbf{Q}), \quad a \in K.$$

Now take $c \in H_1(A, \mathbf{Z})$ so that $\langle \omega_\sigma, c\rangle \neq 0$. Then we have $\langle \omega_\sigma, \theta_*(a)c\rangle \sim \langle \omega_\sigma, c\rangle$ for $a \in K$ by (1.2). Since $H_1(A, \mathbf{Q}) = \{\theta_*(a)c \mid a \in K\}$, (1.1) holds taking $\pi p_K(\sigma, \Phi) = \langle \omega_\sigma, c\rangle$. Clearly $p_K(\sigma, \Phi) \bmod \overline{\mathbf{Q}}^\times$ does not depend on the choice of ω_σ since ω_σ is uniquely determined up to the multiplication by an element in $\overline{\mathbf{Q}}^\times$.

Let (A', θ') be another couple of type (K, Φ) defined over $\overline{\mathbf{Q}}$. Then there exists an isogeny φ of A' to A defined over $\overline{\mathbf{Q}}$ such that $\varphi \circ \theta' = \theta \circ \varphi$. We have

$$\langle \omega, \varphi(c)\rangle = \langle \varphi^*(\omega), c\rangle, \qquad \omega \in \mathfrak{D}(A), \quad c \in H_1(A', \mathbf{Z}),$$

where $\varphi^*(\omega)$ denotes the pullback of ω by φ. If ω_σ is as in our lemma, we have $\theta^{*\prime}(a)\varphi^*(\omega_\sigma) = \varphi^*(\theta^*(a)\omega_\sigma) = a^\sigma \varphi^*(\omega_\sigma)$ and $\varphi^*(\omega_\sigma)$ is rational over $\overline{\mathbf{Q}}$. Hence $p_K(\sigma, \Phi)$ does not depend on the choice of (A, θ). This completes the proof.

1.2. The following theorem, which gives a factorization of the geometric period $p_K(\sigma, \Phi)$, is considerably deeper. The crucial step is to prove various monomial algebraic relations satisfied by $p_K(\sigma, \Phi)$'s, which is done in Shimura [Sh7]. The present formulation was first given in [Sh8].

THEOREM 1.2([SH12], THEOREM 32.5). *For every CM-field K, there exists a map $p_K : I_K \times I_K \longrightarrow \mathbf{C}^\times$ with the following properties.*

(1) $p_K(\sigma, \Phi)$ is defined by (1.1) if Φ is a CM-type of K and $\sigma \in \Phi$.

(2) $p_K(\sigma_1 + \sigma_2, \tau) \sim p_K(\sigma_1, \tau)p_K(\sigma_2, \tau)$, $p_K(\sigma, \tau_1 + \tau_2) \sim p_K(\sigma, \tau_1)p_K(\sigma, \tau_2)$ for every σ, σ_1, σ_2, τ, τ_1, $\tau_2 \in I_K$.

(3) $p_K(\xi\rho, \eta) \sim p_K(\xi, \eta\rho) \sim p_K(\xi, \eta)^{-1}$ for every ξ, $\eta \in I_K$.

(4) $p_K(\xi, \mathrm{Res}_{L/K}(\zeta)) \sim p_L(\mathrm{Inf}_{L/K}(\xi), \zeta)$ if $\xi \in I_K$, $\zeta \in I_L$ and $K \subset L$, L is a CM-field.

(5) $p_K(\mathrm{Res}_{L/K}(\zeta), \xi) \sim p_L(\zeta, \mathrm{Inf}_{L/K}(\xi))$ if $\xi \in I_K$, $\zeta \in I_L$ and $K \subset L$, L is a CM-field.

(6) $p_{K'}(\gamma\xi, \gamma\eta) \sim p_K(\xi, \eta)$ if γ is an isomorphism of K' onto K.

EXAMPLE 1.3. We consider the elliptic curve E defined by the equation $y^2 = 1 - x^4$. The mapping $\iota : (x, y) \longrightarrow (\sqrt{-1}x, y)$ defines an automorphism of E. We can extend the mapping $\sqrt{-1} \longrightarrow \iota$ to the isomorphism $\theta : K \cong \mathrm{End}(E) \otimes \mathbf{Q}$, where

$K = \mathbf{Q}(\sqrt{-1})$. Then (E, θ) is of CM-type $(K, \{\text{id}\})$. Since dx/y is a $\mathbf{Q}$-rational holomorphic differential 1-form, we get

$$\pi p_K(\text{id}, \text{id}) \sim \int_0^1 \frac{dx}{\sqrt{1-x^4}}.$$

Recall the well known formula for the beta function $B(p,q)$ (cf. [WW], p. 253):

$$B(p,q) = \Gamma(p)\Gamma(q)/\Gamma(p+q),$$

$$(1.3) \qquad \int_0^1 x^{p-1}(1-x)^{q-1}dx = B(p,q), \qquad \Re(p) > 0, \quad \Re(q) > 0.$$

We get

$$\int_0^1 \frac{dx}{\sqrt{1-x^4}} = \frac{1}{4}B(\frac{1}{4}, \frac{1}{2}) = \frac{\sqrt{\pi}}{4}\frac{\Gamma(1/4)}{\Gamma(3/4)}.$$

Hence we obtain

$$\sqrt{\pi} p_{\mathbf{Q}(\sqrt{-1})}(\text{id}, \text{id}) \sim \frac{\Gamma(1/4)}{\Gamma(3/4)}.$$

More generally, let K be an imaginary quadratic field of discriminant $-d$ and let χ be the Dirichlet character which corresponds to the quadratic extension $K/\mathbf{Q}$. Then the Chowla-Selberg formula ([SC], §12) states

$$(1.4) \qquad \pi p_K(\text{id}, \text{id})^2 \sim \prod_{a=1}^{d-1} \Gamma(\frac{a}{d})^{w\chi(a)/2h},$$

where w is the number of roots of unity contained in K and h is the class number of K. Except for the cases $K = \mathbf{Q}(\sqrt{-1})$ and $K = \mathbf{Q}(\sqrt{-3})$, there is no known way to derive (1.4) from the direct evaluation of an elliptic integral. We will give a geometric proof of (1.4) in the next section.

1.3. Shimura's period symbol p_K plays a fundamental role in the analysis of the behavior of arithmetic automorphic forms at CM-points. Let us state a result in the simplest case.

Let F be a totally real algebraic number field of degree n. Let $J_F = \{\tau_1, \tau_2, \ldots, \tau_n\}$ and we embed $SL(2, F)$ into $SL(2, \mathbf{R})^n$ by

$$\begin{pmatrix} a & b \\ c & d \end{pmatrix} \longrightarrow (\begin{pmatrix} a^{\tau_1} & b^{\tau_1} \\ c^{\tau_1} & d^{\tau_1} \end{pmatrix}, \begin{pmatrix} a^{\tau_2} & b^{\tau_2} \\ c^{\tau_2} & d^{\tau_2} \end{pmatrix}, \ldots, \begin{pmatrix} a^{\tau_n} & b^{\tau_n} \\ c^{\tau_n} & d^{\tau_n} \end{pmatrix}).$$

Let Γ be a congruence subgroup of $SL(2, \mathfrak{O}_F)$ and $k = (k_1, k_2, \ldots, k_n) \in \mathbf{Z}^n$. A holomorphic function f on $\mathfrak{H}^n$ is called a Hilbert modular form of weight k if the transformation law

$$(1.5) \qquad f(\begin{pmatrix} a & b \\ c & d \end{pmatrix} z) = f(z) \prod_{i=1}^{n} (c^{\tau_i} z_i + d^{\tau_i})^{k_i}, \qquad z \in \mathfrak{H}^n$$

is satisfied for every $\begin{pmatrix} a & b \\ c & d \end{pmatrix} \in \Gamma$ and if f satisfies the usual condition at the cusps when $F = \mathbf{Q}$. Here we write $z = (z_1, z_2, \ldots, z_n)$. Let $G_k(\Gamma)$ be the space of all such forms. Every $f \in G_k(\Gamma)$ has the Fourier expansion

$$f(z) = \sum_{\alpha \in L} a_f(\alpha) \exp(2\pi\sqrt{-1}\sum_{i=1}^{n} \alpha^{\tau_i} z_i), \qquad z = (z_1, z_2, \ldots, z_n) \in \mathfrak{H}^n, \tag{1.6}$$

where L is a lattice in F determined by Γ; $a_f(\alpha) = 0$ if $\alpha^{\tau_i} < 0$ for some τ_i. Let $G_k(\Gamma, \overline{\mathbf{Q}})$ be the set of all $f \in G_k(\Gamma)$ which satisfy $a_f(\alpha) \in \overline{\mathbf{Q}}$ for every $\alpha \in L$. Put

$$G_k = \bigcup_{\Gamma} G_k(\Gamma), \qquad G_k(\overline{\mathbf{Q}}) = \bigcup_{\Gamma} G_k(\Gamma, \overline{\mathbf{Q}})$$

where Γ extends over all congruence subgroups of $SL(2, \mathfrak{O}_F)$. Let M_k (resp. $M_k(\overline{\mathbf{Q}})$) be the space of all functions f/g, $f \in G_l$, $0 \neq g \in G_m$ (resp. $f \in G_l(\overline{\mathbf{Q}})$, $0 \neq g \in G_m(\overline{\mathbf{Q}})$), $l - m = k$. We have (cf. [Sh6], §1)

$$G_k = G_k(\overline{\mathbf{Q}}) \otimes_{\overline{\mathbf{Q}}} \mathbf{C}, \qquad M_k = M_k(\overline{\mathbf{Q}}) \otimes_{\overline{\mathbf{Q}}} \mathbf{C}.$$

THEOREM 1.4 ([SH12], THEOREM 32.2). *Let K be a CM-field such that $[K : F] = 2$. Take $z \in K - F$. Let $\Phi = \{\sigma_1, \sigma_2, \ldots, \sigma_n\}$ be the CM-type of K such that*

$$\Im(z^{\sigma_i}) > 0, \qquad \sigma_i | F = \tau_i, \quad 1 \leqq i \leqq n.$$

If $f \in M_k(\overline{\mathbf{Q}})$ is finite at $(z^{\sigma_1}, z^{\sigma_2}, \ldots, z^{\sigma_n}) \in \mathfrak{H}^n$, then we have

$$f(z^{\sigma_1}, z^{\sigma_2}, \ldots, z^{\sigma_n}) \sim p_K(\sum_{i=1}^{n} k_i \sigma_i, \Phi).$$

Let K be a CM-field, Φ be a CM-type of K and λ be a character of $I_{\mathfrak{f}}(K)$ (i.e., a homomorphism of $I_{\mathfrak{f}}(K)$ into the group $\{z \in \mathbf{C} \mid |z| = 1\}$) such that

$$\lambda((\alpha)) = \prod_{\sigma \in \Phi} (\alpha^{\sigma\rho}/|\alpha^{\sigma}|)^{t_\sigma} \qquad \text{if} \quad \alpha \equiv 1 \mod^{\times} \mathfrak{f}, \tag{1.7}$$

where t_σ, $\sigma \in \Phi$ are non-negative integers. In Hecke's terminology, λ is called *a Grössencharacter* of $I_{\mathfrak{f}}(K)$. Let $L(s, \lambda)$ denote the L-function attached to λ. We write $L_K(s, \lambda)$ for $L(s, \lambda)$ when we emphasize the dependence on K.

When t_σ mod 2, $\sigma \in \Phi$ are independent of σ, λ is called a Grössencharacter of A_0-type in Weil's old terminology. Now λ (or rather λ_0 given below) is often called an algebraic Hecke character. We assume that t_σ mod 2, $\sigma \in \Phi$ are independent of σ. Put

$$t_0 = \max_{\sigma \in \Phi} t_\sigma, \tag{1.8}$$

$$\lambda_0(\mathfrak{a}) = \lambda(\mathfrak{a}) N(\mathfrak{a})^{t_0/2}, \qquad \mathfrak{a} \in I_{\mathfrak{f}}(K). \tag{1.9}$$

Then we have $L(s, \lambda) = L(s + t_0/2, \lambda_0)$ and

$$\lambda_0((\alpha)) = \prod_{\sigma \in \Phi} (\alpha^{\sigma\rho})^{(t_0 + t_\sigma)/2} (\alpha^{\sigma})^{(t_0 - t_\sigma)/2} \qquad \text{if} \quad \alpha \equiv 1 \mod^{\times} \mathfrak{f}. \tag{1.10}$$

By (1.10), we see easily that $\lambda_0(\mathfrak{a})$ lies in an algebraic number field of finite degree when $\mathfrak{a}$ runs over $I_{\mathfrak{f}}(K)$. For $\sigma \in \mathrm{Aut}(\mathbf{C})$, we put

$$\lambda_0^{\sigma}(\mathfrak{a}) = \lambda_0(\mathfrak{a})^{\sigma}, \qquad \lambda^{\sigma}(\mathfrak{a}) = \lambda_0^{\sigma}(\mathfrak{a}) N(\mathfrak{a})^{-t_0/2}, \qquad \mathfrak{a} \in I_{\mathfrak{f}}(K).$$

Thus $\lambda^\sigma(\mathfrak{a}) = \lambda(\mathfrak{a})^\sigma$ when t_0 is even, but not necessary so when t_0 is odd.

The following theorem is obtained by Shimura by applying Theorem 1.2 to Eisenstein series. (Actually the notion of near holomorphy is needed to handle all cases.)

THEOREM 1.5. *Let m be an integer such that $m \equiv t_\sigma \bmod 2$ and $-t_\sigma < m \leqq t_\sigma$ for every $\sigma \in \Phi$. Then we have*

$$L(m/2, \lambda) \sim \pi^{e/2} p_K(\sum_{\sigma\in\Phi} t_\sigma \cdot \sigma,\ \Phi),$$

where $e = m[F : \mathbf{Q}] + \sum_{\sigma\in\Phi} t_\sigma$, F being the maximal real subfield of K.

For a proof, see [Sh3], Theorem 2, [Sh8], Theorem 1.1; see also [Sh6], §5 and [Sh12], Theorem 32.12.

2. The case of abelian fields

2.1. Let $n \geqq 3$ be an integer. We set $\zeta_n = e^{2\pi i/n}$ and put $K = \mathbf{Q}(\zeta_n)$. This notation will be retained until the end of subsection 2.3. For $a \in (\mathbf{Z}/n\mathbf{Z})^\times$, let $\sigma(a) \in \mathrm{Gal}(K/\mathbf{Q})$ denote the automorphism given by $\zeta_n^{\sigma(a)} = \zeta_n^a$. We may regard $\sigma(a) \in J_K$.

We consider the Fermat curve

$$F_n : x^n + y^n = 1.$$

We can regard F_n as a projective smooth curve defined over $\mathbf{Q}$. The genus of F_n is $(n-1)(n-2)/2$. For a triplet of integers r, s, t such that

$$0 < r, s, t < n, \qquad r + s + t = n, \tag{2.1}$$

we consider a differential form

$$\eta_{r,s,t} = x^{r-1} y^{s-n} dx$$

on F_n. Then $\{\eta_{r,s,t}, \mid 0 < r, s, t < n,\ r+s+t = n\}$ makes a basis of the space of differential forms of the first kind on F_n. Rohrlich showed that (cf. the Theorem of the appendix of [G])

$$\int_\gamma \eta_{r,s,t} \sim B(\frac{r}{n}, \frac{s}{n}) \qquad \text{for every} \quad \gamma \in H_1(F_n, \mathbf{Z}), \tag{2.2}$$

where B denotes the beta function. For $a \in \mathbf{Z}$, let $\langle\langle a\rangle\rangle$ denote the integer such that

$$0 \leqq \langle\langle a\rangle\rangle < n, \qquad \langle\langle a\rangle\rangle \equiv a \mod n.$$

For $a \in \mathbf{Z}/n\mathbf{Z}$, by abuse of notation, we set $\langle\langle a\rangle\rangle = \langle\langle \tilde{a}\rangle\rangle$ taking $\tilde{a} \in \mathbf{Z}$ such that $\tilde{a} \bmod n = a$. For integers r, s, t satisfying (2.1), we set

$$H_{r,s,t} = \{a \in (\mathbf{Z}/n\mathbf{Z})^\times \mid \langle\langle ar\rangle\rangle + \langle\langle as\rangle\rangle + \langle\langle at\rangle\rangle = n\}.$$

Then we have $|H_{r,s,t}| = \varphi(n)/2$. Assume that $(r, s, t, n) = 1$. Then there exists an abelian variety $A_{r,s,t}$ defined over $\mathbf{Q}$ which is a factor of the Jacobian variety J_n of F_n such that $\dim A_{r,s,t} = \varphi(n)/2$ and that $\mathbf{Z}[\zeta_n] \subseteq \mathrm{End}(A_{r,s,t})$. The differential forms $\eta_{\langle\langle ar\rangle\rangle,\langle\langle as\rangle\rangle,\langle\langle at\rangle\rangle}$, $a \in H_{r,s,t}$ correspond to the basis of $\overline{\mathbf{Q}}$-rational holomorphic differential 1-forms on $A_{r,s,t}$. We have an isomorphism $\theta : K \longrightarrow \mathrm{End}(A)\otimes\mathbf{Q}$. Then $(A_{r,s,t},\theta)$ is of type $(K,\Phi_{r,s,t})$ where

$$\Phi_{r,s,t} = \{\sigma(a) \mid a \in H_{r,s,t}\}.$$

We refer the reader to Gross [G], §2 for the facts concerning the Fermat curve stated above. Now by (2.2) and Theorem 1.2, (1), (2), we obtain

$$\begin{aligned} \pi p_K(\sigma(a),\Phi_{r,s,t}) &\sim \pi \prod_{u\in\Phi_{r,s,t}} p_K(\sigma(a),\sigma(u)) \\ &\sim B(\frac{\langle\langle ar\rangle\rangle}{n},\frac{\langle\langle as\rangle\rangle}{n}) \qquad \text{for every} \quad a \in H_{r,s,t}. \end{aligned} \tag{2.3}$$

We are going to write (2.3) in a more convenient form for calculation. Take $a \in \mathbf{Z}$ so that $1 \leqq a \leqq n-2$, and set $r=1$, $s=a$, $t=n-(a+1)$. For $x \in \mathbf{R}$, let $\langle x\rangle$ denote the fractional part of x, i.e., $0 \leqq \langle x\rangle < 1$, $x-\langle x\rangle \in \mathbf{Z}$, and let $[x] = x - \langle x\rangle$ denote the integer part of x. We can show easily that, for $u \in (\mathbf{Z}/n\mathbf{Z})^\times$, $\langle\frac{au}{n}\rangle + \langle\frac{u}{n}\rangle < 1$ if and only if $\langle\langle au\rangle\rangle + \langle\langle u\rangle\rangle + \langle\langle n-(a+1)u\rangle\rangle = n$. Here we set $\langle\frac{u}{n}\rangle = \langle\frac{\tilde{u}}{n}\rangle$ taking $\tilde{u} \in \mathbf{Z}$ so that $\tilde{u} \mod n = u$. Put

$$T_a = \{t \in (\mathbf{Z}/n\mathbf{Z})^\times \mid \langle\frac{at}{n}\rangle + \langle\frac{t}{n}\rangle < 1\}, \qquad \Phi_a = \{\sigma(t) \mid t \in T_a\}.$$

We have

$$H_{1,a,n-(a+1)} = T_a, \qquad \Phi_{1,a,n-(a+1)} = \Phi_a.$$

Hence (2.3) can be written as[1]

$$\prod_{t\in T_a} p_K(\sigma(1),\sigma(t)) \sim \pi^{-1}B(\frac{1}{n},\frac{a}{n}). \tag{2.4}$$

For $t \in (\mathbf{Z}/n\mathbf{Z})^\times$, put

$$\epsilon_{at} = \begin{cases} 1 & \text{if} \quad t \in T_a, \\ -1 & \text{if} \quad t \notin T_a. \end{cases}$$

Since $p_K(\sigma(1),\sigma(-t)) \sim p_K(\sigma(1),\sigma(t))^{-1}$ by Theorem 1.2, (3), and noting that

$$\sigma(t) \in \Phi_a \Longleftrightarrow \sigma(n-t) \notin \Phi_a,$$

we obtain

$$\prod_{t=1,(t,n)=1}^{[(n-1)/2]} p_K(\sigma(1),\sigma(t))^{\epsilon_{at}} \sim \pi^{-1}\frac{\Gamma(\frac{a}{n})\Gamma(\frac{1}{n})}{\Gamma(\frac{a+1}{n})} \tag{2.5}$$

by (2.4). We note that (2.5) holds for every a, $1 \leqq a \leqq n-2$.

---

[1] When n is an odd prime, (2.4) follows immediately from a formula of Weil [We3], p. 815.

2.2. We are going to relate CM-periods to derivatives of Dirichlet's L-functions at $s = 0$. For this purpose, we first consider Hurwitz-Lerch's zeta function. For $0 < x \leqq 1$, set

$$\zeta(s, x) = \sum_{m=0}^{\infty} (x + m)^{-s}, \qquad \Re(s) > 1.$$

Then $\zeta(s, x)$ can be meromorphically continued to the whole s-plane, holomorphic except for the simple pole at $s = 1$. For $c \in \mathbf{Z}$, $1 \leqq c \leqq n - 1$, we set

$$\zeta_{\mathbf{Q}}(s, c) = \sum_{m=1,\, m \equiv c \mod n}^{\infty} m^{-s}.$$

Then we have

$$\zeta_{\mathbf{Q}}(s, c) = n^{-s} \zeta(s, \frac{c}{n}). \tag{2.6}$$

By I, (1.16), (1.17), we get

$$\zeta_{\mathbf{Q}}(0, c) = \frac{1}{2} - \frac{c}{n}, \tag{2.7}$$

$$\zeta'_{\mathbf{Q}}(0, c) = \log \Gamma(\frac{c}{n}) - \frac{1}{2} \log(2\pi) - \log n \cdot \zeta_{\mathbf{Q}}(0, c). \tag{2.8}$$

Let η be a Dirichlet character whose conductor divides n. We do not assume η is primitive. We call η *odd* (resp. *even*) if $\eta(-1) = -1$ (resp. $\eta(-1) = 1$). Let

$$L(s, \eta) = \sum_{m=1}^{\infty} \eta(m) m^{-s}$$

be the Dirichlet L-function attached to η. Since

$$L(s, \eta) = \sum_{c=1}^{n-1} \eta(c) \zeta_{\mathbf{Q}}(s, c), \tag{2.9}$$

we obtain

$$L(0, \eta) = -\frac{1}{n} \sum_{c=1}^{n-1} \eta(c) c \qquad \text{if} \quad \eta \neq 1, \tag{2.10}$$

$$L'(0, \eta) = \sum_{c=1}^{n-1} \eta(c) \log \Gamma(\frac{c}{n}) - \log n \cdot L(0, \eta) \qquad \text{if} \quad \eta \neq 1. \tag{2.11}$$

Now let η be a primitive Dirichlet character of conductor f. We assume that f divides n. Let η_* be the Dirichlet character to the modulus n obtained from η. Since

$$\eta_*(m) = \begin{cases} \eta(m) & \text{if} \quad (m, n) = 1, \\ 0 & \text{if} \quad (m, n) > 1, \end{cases}$$

we have

$$L(s, \eta_*) = \prod_{p \mid \frac{n}{f}} (1 - \eta(p) p^{-s}) \times L(s, \eta). \tag{2.12}$$

It is well known that $L(0, \eta) \neq 0$ if η is odd and primitive.

2.3. Let $\widehat{G}$ denote the set of all primitive Dirichlet characters whose conductors divide n. Let $\widehat{G}_-$ (resp. $\widehat{G}_+$) denote the subset of $\widehat{G}$ consisting of all odd (resp. even) Dirichlet characters in $\widehat{G}$.

LEMMA 2.1. *For $1 \leqq a \leqq n-2$ and $t \in (\mathbf{Z}/n\mathbf{Z})^\times$, we have*

$$\epsilon_{at} = -(2\langle \frac{t}{n} \rangle - 1) - (2\langle \frac{at}{n} \rangle - 1) + (2\langle \frac{(a+1)t}{n} \rangle - 1).$$

PROOF. Put $\mu_{at} = \langle \frac{t}{n} \rangle + \langle \frac{at}{n} \rangle - \langle \frac{(a+1)t}{n} \rangle$. It suffices to show $\epsilon_{at} = 1 - 2\mu_{at}$. Since $\mu_{at} \in \mathbf{Z}$, $0 \leqq \langle \frac{t}{n} \rangle, \langle \frac{at}{n} \rangle, \langle \frac{(a+1)t}{n} \rangle < 1$, we have $\mu_{at} = 0$ or 1. If $\langle \frac{t}{n} \rangle + \langle \frac{at}{n} \rangle < 1$, then $\mu_{at} < 1$; hence $\mu_{at} = 0$. If $\langle \frac{t}{n} \rangle + \langle \frac{at}{n} \rangle > 1$, then $\mu_{at} > 0$; hence $\mu_{at} = 1$. This proves our Lemma.

LEMMA 2.2. *Let $\eta \in \widehat{G}_-$ and let f be the conductor of η. For $1 \leqq a \leqq n-1$, put*

$$\frac{a}{n} = \frac{b}{m}, \qquad (m, b) = 1, \qquad 1 \leqq b \leqq m-1.$$

Then we have

$$\sum_{t=1,(t,n)=1}^{[(n-1)/2]} (2\langle \frac{at}{n} \rangle - 1)\eta(t)$$

$$= \begin{cases} -\frac{\varphi(n)}{\varphi(m)} \eta(b)^{-1} \prod_{p \mid \frac{m}{f}} (1 - \eta(p)) L(0, \eta) & \text{if } f \mid m, \\ 0 \quad \text{if } f \nmid m. \end{cases}$$

PROOF. Let S denote the sum in question. Since η is odd, we have

$$S = \sum_{t=1,(t,n)=1}^{n-1} (\langle \frac{at}{n} \rangle - \frac{1}{2})\eta(t) = \sum_{t=1,(t,n)=1}^{n-1} \langle \frac{at}{n} \rangle \eta(t) = \sum_{t=1,(t,n)=1}^{n-1} \langle \frac{bt}{m} \rangle \eta(t).$$

Put $\langle\langle bt \rangle\rangle_m = m\langle \frac{bt}{m} \rangle$, $n = md$. Then we have

$$S = \frac{1}{m} \sum_{t=1,(t,n)=1}^{n-1} \langle\langle bt \rangle\rangle_m \eta_*(t) = \frac{1}{m} \sum_{k=0}^{d-1} \sum_{u=1}^{m-1} \langle\langle bu \rangle\rangle_m \eta_*(mk+u).$$

We have $\eta_*(mk+u) = 0$ if $(u, m) > 1$. Assume $(u, m) = 1$. Take $u' \in (\mathbf{Z}/n\mathbf{Z})^\times$ so that $u' \equiv u \mod m$. Then, regarding η_* also as a character of $(\mathbf{Z}/n\mathbf{Z})^\times$, we have

$$\sum_{k=0}^{d-1} \eta_*(mk+u) = \sum_{v \in (\mathbf{Z}/n\mathbf{Z})^\times, v \equiv u \mod m} \eta_*(v)$$

$$= \sum_{v \in (\mathbf{Z}/n\mathbf{Z})^\times, v \equiv u' \mod m} \eta_*(u')\eta_*(u'^{-1}v) = \eta_*(u') \sum_{v \in (\mathbf{Z}/n\mathbf{Z})^\times, v \equiv 1 \mod m} \eta_*(v).$$

Set $Y_m = \{v \in (\mathbf{Z}/n\mathbf{Z})^\times \mid v \equiv 1 \mod m\}$. Then η_* is trivial on Y_m if and only if f divides m. If $f \mid m$, then $\eta_*(u') = \eta(u)$. Therefore we get

$$\sum_{k=0}^{d-1} \eta_*(mk+u) = \begin{cases} \frac{\varphi(n)}{\varphi(m)}\eta(u) & \text{if } f \mid m, \\ 0 & \text{if } f \nmid m. \end{cases}$$

We have shown that $S = 0$ if $f \nmid m$. Assume $f \mid m$. Then we have

$$S = \frac{1}{m} \sum_{u=1,(u,m)=1}^{m-1} \langle\langle bu \rangle\rangle_m \cdot \frac{\varphi(n)}{\varphi(m)}\eta(u) = \frac{\varphi(n)}{\varphi(m)} \cdot \frac{1}{m} \sum_{v=1,(v,m)=1}^{m-1} v\eta(b)^{-1}\eta(v)$$
$$= \frac{\varphi(n)}{\varphi(m)} \cdot \eta(b)^{-1} \cdot \frac{1}{m} \sum_{v=1,(v,m)=1}^{m-1} \eta(v)v.$$

By (2.10) and (2.12), we obtain

$$\frac{1}{m} \sum_{v=1,(v,m)=1}^{m-1} \eta(v)v = -\prod_{p \mid \frac{m}{f}} (1-\eta(p))L(0,\eta).$$

This completes the proof.

LEMMA 2.3. *For $1 \leqq a \leqq n-2$, regard ϵ_{at} as a function of t, $1 \leqq t \leqq [(n-1)/2]$, $(t,n) = 1$. Then the dimension of the vector space V spanned by ϵ_{at} over $\mathbf{C}$ is equal to $\varphi(n)/2$, where $\varphi(n) = |(\mathbf{Z}/n\mathbf{Z})^\times|$. In other words, the rank of $(n-2) \times (\varphi(n)/2)$-matrix (ϵ_{at}) is $\varphi(n)/2$.*

PROOF. For $1 \leqq a \leqq n-1$, we set

$$f_a(t) = 2\langle\frac{at}{n}\rangle - 1, \qquad 1 \leqq t \leqq [(n-1)/2], \quad (t,n) = 1.$$

We see easily that

$$f_a + f_{n-a} = 0, \qquad 1 \leqq a \leqq n-1.$$

By Lemma 2.1, we have

$$\epsilon_{at} = -f_1(t) - f_a(t) + f_{a+1}(t), \qquad 1 \leqq a \leqq n-2.$$

Hence we have

$$V \ni 2f_1 - f_2, f_1 + f_2 - f_3, f_1 + f_3 - f_4, \dots, f_1 + f_{n-2} - f_{n-1}.$$

Adding successive terms, we get

$$V \ni 2f_1 - f_2, 3f_1 - f_3, 4f_1 - f_4, \dots, (n-1)f_1 - f_{n-1}.$$

Since $f_{n-1} = -f_1$, we get $V \ni f_1$. Therefore we have $V \ni f_1, f_2, f_3, \dots, f_{n-1}$. It suffices to show that $\dim\langle f_1, f_2, \dots, f_{n-1}\rangle_{\mathbf{C}} = \varphi(n)/2$.

Now we enlarge the domain of definition of f_a: we set

$$f_a(t) = 2\langle\frac{at}{n}\rangle - 1, \qquad 1 \leqq t \leqq n-1, \quad (t,n) = 1$$

and we regard f_a as a function on $(\mathbf{Z}/n\mathbf{Z})^\times$. Since $f_a(n-t) = -f_a(t)$, f_a is an odd function. Therefore it suffices to show that the space of odd functions on $(\mathbf{Z}/n\mathbf{Z})^\times$ is spanned by f_a, $1 \leqq a \leqq n-1$. For this purpose, set

$$W = \{g \mid g \text{ is a complex valued odd function on } (\mathbf{Z}/n\mathbf{Z})^\times \text{ such that } \sum_{t\in(\mathbf{Z}/n\mathbf{Z})^\times} g(t)f_a(t) = 0 \quad \text{for every} \quad f_a,\ 1 \leqq a \leqq n-1\}.$$

It is sufficient to show that $W = \{0\}$. Since $\sum_{t\in(\mathbf{Z}/n\mathbf{Z})^\times} g(t) = 0$ if g is odd, we have

$$W = \{g \mid g \text{ is a complex valued odd function on } (\mathbf{Z}/n\mathbf{Z})^\times \text{ such that } \sum_{t\in(\mathbf{Z}/n\mathbf{Z})^\times} g(t)\langle\frac{at}{n}\rangle = 0 \quad \text{for every} \quad a,\ 1 \leqq a \leqq n-1\}.$$

If $g \in W$, $c \in (\mathbf{Z}/n\mathbf{Z})^\times$, we see easily that $g(ct) \in W$. Therefore $(\mathbf{Z}/n\mathbf{Z})^\times$ acts on W. Assume $W \neq \{0\}$. Then we have a representation of $(\mathbf{Z}/n\mathbf{Z})^\times$ on W which splits into a direct sum of one dimensional representations. Let ψ be a one dimensional constituent and let $g \in W$ be a non-zero function which transforms according to ψ. Since

$$g(ct) = \psi(c)g(t), \qquad c, t \in (\mathbf{Z}/n\mathbf{Z})^\times,$$

g is a constant multiple of ψ. Hence we may assume that $\psi \in W$. Since ψ must be odd, we have $\psi = \eta_*$ with some $\eta \in \widehat{G}_-$. Let f be the conductor of η and put $a = n/f$. We have $1 \leqq a < n$, $a/n = 1/f$. By Lemma 2.2, we have

$$\sum_{t\in(\mathbf{Z}/n\mathbf{Z})^\times} f_a(t)\eta_*(t) = 2\sum_{t=1,(t,n)=1}^{[(n-1)/2]} (2\langle\frac{at}{n}\rangle - 1)\eta(t) = -\frac{2\varphi(n)}{\varphi(f)} \cdot L(0,\eta) \neq 0.$$

This is a contradiction and completes the proof.

For $\eta \in \widehat{G}$, let M_η be the field generated over $\mathbf{Q}$ by $\eta(m)$, $m \in \mathbf{Z}$. We see easily that $M_\eta = \mathbf{Q}(\eta_*(m) \mid m \in \mathbf{Z})$. Let J_η be the set of all isomorphisms of M_η into $\mathbf{C}$. Then $\{\eta^\sigma \mid \sigma \in J_\eta\}$ is the set of all conjugates of η over $\mathbf{Q}$.

LEMMA 2.4. *Let $\eta \in \widehat{G}_-$ and f be the conductor of η. For every $a \in M_\eta$, we have*

$$\prod_{c=1}^{f-1} \Gamma(\frac{c}{f})^{\sum_\sigma a^\sigma \prod_p (1-\eta^\sigma(p))\eta^\sigma(c)} \sim \prod_{c=1,(c,n)=1}^{n-1} \Gamma(\frac{c}{n})^{\sum_\sigma a^\sigma\eta^\sigma(c)},$$

where σ extends over J_η and p runs over all prime divisors of n/f.

PROOF. Take any $a \in M_\eta$. By (2.11), we have

$$\prod_\sigma \exp(a^\sigma L'(0,\eta_*^\sigma)) = n^{-\sum_\sigma a^\sigma L(0,\eta_*^\sigma)} \prod_{c=1}^{n-1} \Gamma(\frac{c}{n})^{\sum_\sigma a^\sigma \eta_*^\sigma(c)},$$

where σ extends over J_η. Since

$$\sum_{\sigma\in J_\eta} a^\sigma L(0,\eta_*^\sigma) = \sum_{\sigma\in J_\eta} (aL(0,\eta_*))^\sigma \in \mathbf{Q},$$

we obtain

$$\prod_{\sigma\in J_\eta} \exp(a^\sigma L'(0,\eta_*^\sigma)) \sim \prod_{c=1,(c,n)=1}^{n-1} \Gamma(\frac{c}{n})^{\sum_{\sigma\in J_\eta} a^\sigma \eta_*^\sigma(c)}. \tag{2.13}$$

By (2.11), we similarly obtain

$$\prod_{\sigma\in J_\eta} \exp(b^\sigma L'(0,\eta^\sigma)) \sim \prod_{c=1}^{f-1} \Gamma(\frac{c}{f})^{\sum_{\sigma\in J_\eta} b^\sigma \eta^\sigma(c)} \qquad \text{for every} \quad b\in M_\eta. \tag{2.14}$$

By (2.12), we have

$$\prod_{\sigma\in J_\eta} \exp(a^\sigma L'(0,\eta_*^\sigma)) = \prod_{\sigma\in J_\eta} \exp(a^\sigma \prod_{p\mid \frac{n}{f}} (1-\eta^\sigma(p)) L'(0,\eta^\sigma)). \tag{2.15}$$

Now, taking $b = a\prod_{p\mid\frac{n}{f}}(1-\eta(p))$ in (2.14), the assertion follows immediately from (2.13) $\sim$ (2.15).

THEOREM 2.5. *We have*

$$p_K(\sigma(1),\sigma(t)) \sim \pi^{-\delta_{1t}/2} \prod_{\eta\in\hat{G}_-} \prod_{c=1}^{n-1} \Gamma(\frac{c}{n})^{\eta(tc)/(L(0,\eta)\varphi(n))} \tag{2.16}$$

for $1 \leqq t \leqq [(n-1)/2]$, $(t,n)=1$. *Here* δ *denotes Kronecker's delta, i.e.,* $\delta_{11}=1$, $\delta_{1t}=0$ *if* $t\neq 1$.

PROOF. Set

$$p'_K(\sigma(1),\sigma(t)) = \pi^{-\delta_{1t}/2} \prod_{\eta\in\hat{G}_-} \prod_{c=1}^{n-1} \Gamma(\frac{c}{n})^{\eta(tc)/(L(0,\eta)\varphi(n))}.$$

We will first show that $p'_K(\sigma(1),\sigma(t))$ satisfies (2.5), i.e.,

$$\prod_{t=1,(t,n)=1}^{[(n-1)/2]} p'_K(\sigma(1),\sigma(t))^{\epsilon_{at}} \sim \pi^{-1}\frac{\Gamma(\frac{a}{n})\Gamma(\frac{1}{n})}{\Gamma(\frac{a+1}{n})} \tag{2.17}$$

for $1\leqq a\leqq n-2$. Take any $a\in\mathbf{Z}$, $1\leqq a\leqq n-1$ and put

$$\frac{a}{n} = \frac{b}{m}, \qquad (b,m)=1, \qquad 1\leqq b\leqq m-1.$$

For $\eta\in\widehat{G}$, let f_η denote the conductor of η. By (2.11), if $\eta\neq 1$, we have

$$L'(0,\eta) = \sum_{c=1}^{n-1} \eta(c)\log\Gamma(\frac{c}{n}) - \log n\cdot L(0,\eta) = \sum_{c=1}^{f_\eta-1} \eta(c)\log\Gamma(\frac{c}{f_\eta}) - \log f_\eta \cdot L(0,\eta).$$

Hence we have

$$\prod_{c=1}^{n-1} \Gamma(\frac{c}{n})^{\sum_\sigma \eta^\sigma(c)/(L(0,\eta^\sigma)\varphi(n))} \sim \prod_{c=1}^{f_\eta-1} \Gamma(\frac{c}{f_\eta})^{\sum_\sigma \eta^\sigma(c)/(L(0,\eta^\sigma)\varphi(n))} \qquad \text{if} \quad \eta\neq 1,$$

where σ extends over J_η. Therefore we have

$$p'_K(\sigma(1),\sigma(t)) \sim \pi^{-\delta_{1t}/2} \prod_{\eta\in\hat{G}_-, f_\eta\nmid m} \prod_{c=1}^{n-1} \Gamma(\frac{c}{n})^{\eta(tc)/(L(0,\eta)\varphi(n))}$$
$$\times \prod_{\eta\in\hat{G}_-, f_\eta|m} \prod_{c=1}^{f_\eta-1} \Gamma(\frac{c}{f_\eta})^{\eta(tc)/(L(0,\eta)\varphi(n))}.$$

By Lemma 2.2, we obtain

$$\prod_{t=1,(t,n)=1}^{[(n-1)/2]} p'_K(\sigma(1),\sigma(t))^{2\langle at/n\rangle-1}$$
$$\sim \pi^{(1-2a/n)/2} \prod_{\eta\in\hat{G}_-, f_\eta|m} \prod_{c=1}^{f_\eta-1} \Gamma(\frac{c}{f_\eta})^{-\eta(b)^{-1}\eta(c)\prod_p(1-\eta(p))/\varphi(m)},$$

where p extends over all prime divisors of m/f_η. By Lemma 2.4, for $\eta \in \widehat{G}_-$, we have

$$\prod_{\sigma\in J_\eta} \prod_{c=1}^{f-1} \Gamma(\frac{c}{f})^{-\eta^\sigma(b)^{-1}\eta^\sigma(c)\prod_p(1-\eta^\sigma(p))/\varphi(m)}$$
$$\sim \prod_{\sigma\in J_\eta} \prod_{c=1,(c,m)=1}^{m-1} \Gamma(\frac{c}{m})^{-\eta^\sigma(b)^{-1}\eta^\sigma(c)/\varphi(m)},$$

where f is the conductor of η and $f \mid m$ is assumed; p extends over all prime divisors of m/f. Therefore we obtain

$$\prod_{t=1,(t,n)=1}^{[(n-1)/2]} p'_K(\sigma(1),\sigma(t))^{2\langle at/n\rangle-1}$$
$$\sim\pi^{(1-2a/n)/2} \prod_{c=1,(c,m)=1}^{m-1} \Gamma(\frac{c}{m})^{-\sum_\eta \eta(b)^{-1}\eta(c)/\varphi(m)},$$

where η extends over Dirichlet characters in $\widehat{G}_-$ such that $f_\eta \mid m$. Since

$$\sum_{\eta\in\hat{G}_-, f_\eta|m} \eta(b)^{-1}\eta(c) = \begin{cases} \varphi(m)/2 & \text{if} \quad c\equiv b \mod m, \\ -\varphi(m)/2 & \text{if} \quad c\equiv -b \mod m, \\ 0 & \text{otherwise,} \end{cases}$$

we have

$$\prod_{t=1,(t,n)=1}^{[(n-1)/2]} p'_K(\sigma(1),\sigma(t))^{2\langle at/n\rangle-1} \sim \pi^{(1-2a/n)/2}\Gamma(\frac{b}{m})^{-1/2}\Gamma(\frac{m-b}{m})^{1/2}$$
$$\sim \pi^{(1-2a/n)/2}\pi^{1/2}\Gamma(\frac{b}{m})^{-1} = \pi^{1-a/n}\Gamma(\frac{a}{n})^{-1}.$$

By Lemma 2.1, we have

$$\prod_{t=1,(t,n)=1}^{[(n-1)/2]} p'_K(\sigma(1),\sigma(t))^{\epsilon_{at}} \sim \pi^{1/n-1}\Gamma(\frac{1}{n})\pi^{a/n-1}\Gamma(\frac{a}{n})\pi^{1-(a+1)/n}\Gamma(\frac{a+1}{n})^{-1}$$

$$= \pi^{-1}\frac{\Gamma(\frac{1}{n})\Gamma(\frac{a}{n})}{\Gamma(\frac{a+1}{n})}$$

Thus we have established (2.17). By (2.5), we get

$$\begin{aligned} &\prod_{t=1,(t,n)=1}^{[(n-1)/2]} p_K(\sigma(1),\sigma(t))^{\epsilon_{at}} \\ \sim &\prod_{t=1,(t,n)=1}^{[(n-1)/2]} p'_K(\sigma(1),\sigma(t))^{\epsilon_{at}}, \qquad 1 \leqq a \leqq n-2. \end{aligned} \tag{2.18}$$

By Lemma 2.3, for every t_0, $1 \leqq t_0 \leqq [(n-1)/2]$, $(t_0,n)=1$, there exist integers u_a, $1 \leqq a \leqq n-2$ and a positive integer m such that $\sum_{a=1}^{n-2} \epsilon_{at}u_a = m\delta_{t_0t}$. Taking u_a-th power of (2.18) and making a product over a, we obtain $p'_K(\sigma(1),\sigma(t_0))^m \sim p_K(\sigma(1),\sigma(t_0))^m$. This completes the proof.

2.4. Let K be an algebraic number field which is abelian over $\mathbf{Q}$. Then K is a CM-field if K is totally imaginary. Set $G = \mathrm{Gal}(K/\mathbf{Q})$ and let $\widehat{G}$ be the set of all irreducible characters of G. Let $\widehat{G}_+$ (resp. $\widehat{G}_-$) be the subset of $\widehat{G}$ consisting of characters η such that $\eta(\rho)=1$ (resp. $\eta(\rho)=-1$). When $K=\mathbf{Q}(\zeta_n)$, this notation is consistent with the previous one if we identify $\eta \in \widehat{G}$ with the corresponding primitive Dirichlet character.

THEOREM 2.6. *Let K be a totally imaginary algebraic number field which is abelian over $\mathbf{Q}$. Then we have*

$$p_K(\mathrm{id},\sigma) \sim \pi^{-\mu(\sigma)/2} \prod_{\eta\in\hat{G}_-} \exp(\frac{\eta(\sigma)}{[K:\mathbf{Q}]}\frac{L'(0,\eta)}{L(0,\eta)}), \qquad \sigma \in G \tag{2.19}$$

where

$$\mu(\sigma) = \begin{cases} 1 & \text{if } \sigma = 1, \\ -1 & \text{if } \sigma = \rho, \\ 0 & \text{if } \sigma \neq 1,\rho. \end{cases}$$

PROOF. If $K=\mathbf{Q}(\zeta_n)$, the assertion follows from (2.11), Theorem 2.5 and Theorem 1.2, (3). Now let K be a totally imaginary subfield of $\mathbf{Q}(\zeta_n)$. We set

$$L = \mathbf{Q}(\zeta_n), \qquad \widetilde{G} = \mathrm{Gal}(L/\mathbf{Q}), \qquad H = \mathrm{Gal}(L/K).$$

By Theorem 1.2, (5), we have

$$p_K(\mathrm{id},\sigma) \sim \prod_{\tau\in H} p_L(\mathrm{id},\tau\widetilde{\sigma})$$

where $\widetilde{\sigma} \in \widetilde{G}$ denotes an element such that $\widetilde{\sigma}|K = \sigma$. Since (2.19) holds for p_L, we obtain

$$p_K(\mathrm{id}, \sigma) \sim \prod_{\tau \in H} \pi^{-\mu(\tau\tilde{\sigma})/2} \prod_{\eta \in \hat{\tilde{G}}_-} \exp\Big(\frac{\eta(\tau\widetilde{\sigma})}{[L:\mathbf{Q}]} \cdot \frac{L'(0,\eta)}{L(0,\eta)}\Big)$$

$$= \pi^{-\sum_{\tau \in H} \mu(\tau\tilde{\sigma})/2} \prod_{\eta \in \hat{\tilde{G}}_-} \exp\Big(\frac{\sum_{\tau \in H} \eta(\tau\widetilde{\sigma})}{[L:\mathbf{Q}]} \cdot \frac{L'(0,\eta)}{L(0,\eta)}\Big).$$

We see easily that $\sum_{\tau \in H} \mu(\tau\widetilde{\sigma}) = \mu(\sigma)$. It is clear that $\sum_{\tau \in H} \eta(\tau\widetilde{\sigma}) = 0$ if $\eta|H$ is non-trivial; if $\eta|H$ is trivial, then η can be identified with an element of $\widehat{G}_-$ and we have $\sum_{\tau \in H} \eta(\tau\widetilde{\sigma}) = |H|\,\eta(\sigma)$. Hence the assertion follows.

Let K be an imaginary quadratic field. Let d, χ, h, w be the same as in Example 1.3. By Theorem 2.6 and (2.11), we have

$$\pi p_K(\mathrm{id}, \mathrm{id})^2 \sim \exp\Big(\frac{L'(0,\chi)}{L(0,\chi)}\Big) = \frac{1}{d} \prod_{c=1}^{d-1} \Gamma\Big(\frac{c}{d}\Big)^{\chi(c)/L(0,\chi)}.$$

Since $\zeta_K(s) = \zeta(s)L(s,\chi)$, $\zeta(0) = -1/2$, we obtain $L(0,\chi) = 2h/w$ from II, (1.2). Therefore we can derive the Chowla-Selberg formula (1.4) from Theorem 2.6.

3. Conjectures on absolute CM-periods

In this section, we will state a conjecture which expresses period symbol p_K using division values of the multiple gamma function. A refinement of this conjecture will be given and we will examine it in detail by numerical examples in the next section. Let K be a finite Galois extension of an algebraic number field k. Let η be a finite dimensional representation of $\mathrm{Gal}(K/k)$ and χ_η be the character of η. In this section, we abbreviate the Artin L-function $L(s, \chi_\eta, K/k)$ attached to η to $L(s,\eta)$ for simplicity.

3.1. Let K be a CM-field. We first assume that K is normal over $\mathbf{Q}$. Put $G = \mathrm{Gal}(K/\mathbf{Q})$ and let $\rho \in G$ be the complex conjugation. A representation η of G is called *odd* (resp. *even*) if $\eta(\rho) = -\mathrm{id}$ (resp. $\eta(\rho) = \mathrm{id}$). Let $\widehat{G}$ (resp. $\widehat{G}_-$, resp. $\widehat{G}_+$) denote the set of all equivalence classes of irreducible (resp. irreducible odd, resp. irreducible even) representations of G. We note that $L(s,\eta) \neq 0$ if $\eta \in \widehat{G}_-$. Since ρ belongs to the center of G, we have $\widehat{G} = \widehat{G}_+ \sqcup \widehat{G}_-$.

For a conjugacy class c of G, put $c\rho = \{g\rho \mid g \in c\}$. Then $c\rho$ is also a conjugacy class of G.

LEMMA 3.1. *Let h be the number of conjugacy classes of G. Let h_- (resp. h_+) be the number of conjugacy classes c of G such that $c \neq c\rho$ (resp. $c = c\rho$). Then we have*

$$|\widehat{G}_-| = \frac{1}{2}h_-, \qquad |\widehat{G}_+| = h_+ + \frac{1}{2}h_-.$$

PROOF. Put $\overline{G} = G/\langle\rho\rangle$ and let $\varphi : G \longrightarrow \overline{G}$ be the canonical map. Clearly $|\widehat{G}_+|$ is equal to the number of conjugacy classes of $\overline{G}$. We see that φ induces a surjection from conjugacy classes of G to those of $\overline{G}$ and that $\varphi(c_1) = \varphi(c_2)$ if and only if $c_1 = c_2$ or $c_1 = c_2\rho$, where c_1 and c_2 are conjugacy classes of G. Hence we get $|\widehat{G}_+| = h_+ + \frac{1}{2}h_-$. Since $|\widehat{G}| = |\widehat{G}_-| + |\widehat{G}_+| = h_+ + h_-$, we obtain $|\widehat{G}_-| = \frac{1}{2}h_-$. This completes the proof.

First we naturally conjecture the following generalization of Theorem 2.6.

CONJECTURE 3.2. *Let K be a CM-field normal over $\mathbf{Q}$. Let $G = \mathrm{Gal}(K/\mathbf{Q})$ and let $\widehat{G}_-$ be the set of all equivalence classes of irreducible odd representations of G. Let c be a conjugacy class of G. Then*

$$\prod_{\sigma\in c} p_K(\mathrm{id},\sigma) \sim \pi^{-\mu(c)/2} \prod_{\eta\in\hat{G}_-} \exp\Big(\frac{|c|\chi_\eta(c)}{[K:\mathbf{Q}]}\frac{L'(0,\eta)}{L(0,\eta)}\Big),$$

where χ_η is the character of η and

$$\mu(c) = \begin{cases} 1 & \text{if} \quad c = \{1\}, \\ -1 & \text{if} \quad c = \{\rho\}, \\ 0 & \text{if} \quad c \neq \{1\}, \{\rho\}. \end{cases}$$

This conjecture is essentially equivalent to a conjecture of Colmez [Co]; since the period symbol p_K is not employed, the conjecture takes more complicated form in [Co].

If we use Shintani's formulas (II, §3), Conjecture 3.2 gives an expression of $\prod_{\sigma\in c} p_K(\mathrm{id},\sigma) \mod \overline{\mathbf{Q}}^\times$ in terms of the multiple gamma function. However this is not sufficient in general to give $p_K(\mathrm{id},\sigma) \mod \overline{\mathbf{Q}}^\times$ itself by the multiple gamma function.

3.2. We are going to deepen Conjecture 3.2 so that $p_K(\mathrm{id},\sigma) \mod \overline{\mathbf{Q}}^\times$ can be described by the multiple gamma function. Let F be a totally real algebraic number field of degree n. Let K be a Galois extension of F of finite degree. We assume that K is a CM-field and put $G = \mathrm{Gal}(K/F)$. Then the complex conjugation ρ belongs to the center of G.

PROPOSITION 3.3. *Let $\widehat{G}_-$ be the set of equivalence classes of all irreducible odd representations of G. For $\tau \in G$, let $\mu(\tau) = 1$ (resp. -1) if $\tau = 1$ (resp. $\tau = \rho$) and $\mu(\tau) = 0$ otherwise. If Conjecture 3.2 holds, then*

$$\prod_{\sigma\in J_K} p_K(\sigma,\tau\sigma) \sim \pi^{-[K:\mathbf{Q}]\mu(\tau)/2} \prod_{\omega\in\hat{G}_-} \exp\Big(\chi_\omega(\tau)\frac{L'(0,\omega)}{L(0,\omega)}\Big) \qquad \text{for} \quad \tau\in G. \tag{3.1}$$

Here χ_ω denotes the character of ω.

If $F = \mathbf{Q}$, then taking the $2n/|c|$-th power of the formula of Conjecture 3.2, we obtain (3.1) using Theorem 1.2, (6), where c is the conjugacy class of G which contains τ. Hence Conjecture 3.2 is the special case $F = \mathbf{Q}$ of (3.1). We gave a proof of Proposition 3.3 in [Y4], §3. Here we will give a slightly different proof. The next two lemmas will also be used in §3.8.

LEMMA 3.4. *Let F and F_0 be totally real algebraic number fields and let K be a CM-field. We assume that $F_0 \subset F \subset K$ and that K is normal over F_0. We put*

$$G = \mathrm{Gal}(K/F_0), \qquad H = \mathrm{Gal}(K/F).$$

Then we have

$$\sum_{\omega\in\hat{H}_-}\chi_\omega(\tau)\frac{L'(0,\omega)}{L(0,\omega)}=\sum_{\eta\in\hat{G}_-}\chi_\eta(\tau)\frac{L'(0,\eta)}{L(0,\eta)}\qquad \textit{for}\quad \tau\in H.$$

PROOF. For $\eta\in\widehat{G}_-$, put

$$\eta|H\cong\oplus_{\omega\in\hat{H}_-}a_{\omega,\eta}\omega.$$

Then, by the Frobenius reciprocity, we get

$$\mathrm{Ind}_H^G\omega\cong\oplus_{\eta\in\hat{G}_-}a_{\omega,\eta}\eta,\qquad \frac{L'(0,\omega)}{L(0,\omega)}=\sum_{\eta\in\hat{G}_-}a_{\omega,\eta}\frac{L'(0,\eta)}{L(0,\eta)}.$$

We have

$$\sum_{\omega\in\hat{H}_-}\chi_\omega(\tau)\frac{L'(0,\omega)}{L(0,\omega)}=\sum_{\omega\in\hat{H}_-}\sum_{\eta\in\hat{G}_-}\chi_\omega(\tau)a_{\omega,\eta}\frac{L'(0,\eta)}{L(0,\eta)}=\sum_{\eta\in\hat{G}_-}\chi_\eta(\tau)\frac{L'(0,\eta)}{L(0,\eta)}$$

and the assertion follows.

LEMMA 3.5. *Let K and L be CM-fields and let F be a totally real algebraic number field. We assume that $F\subset K\subset L$ and that both of K and L are normal over F. We put*

$$\mathfrak{G}=\mathrm{Gal}(L/F),\qquad H=\mathrm{Gal}(L/K),\qquad G=\mathrm{Gal}(K/F).$$

Then we have

$$\sum_{\omega\in\hat{G}_-}\chi_\omega(\tau)\frac{L'(0,\omega)}{L(0,\omega)}=\frac{1}{|H|}\sum_{\eta\in\hat{\mathfrak{G}}_-}\sum_{h\in H}\chi_\eta(h\tilde{\tau})\frac{L'(0,\eta)}{L(0,\eta)}$$

for $\tau\in G$. Here $\tilde{\tau}$ is any element of $\mathfrak{G}$ which satisfies $\tilde{\tau}|K=\tau$.

PROOF. Take $\eta\in\widehat{\mathfrak{G}}_-$. If $\eta|H$ does not contain the trivial representation, then $\sum_{h\in H}\chi_\eta(h\tilde{\tau})=0$. Suppose that $\eta|H$ contains the trivial representation. Then η is a constituent of $\mathrm{Ind}_H^G 1$. Hence η factors through the map $\mathfrak{G}\to\mathfrak{G}/H$ and can be identified with a representation of G. We get $\sum_{h\in H}\chi_\eta(h\tilde{\tau})=|H|\chi_\eta(\tilde{\tau})$. Hence the assertion follows.

PROOF OF PROPOSITION 3.3. First we assume that K is normal over $\mathbf{Q}$. Put $\widetilde{G}=\mathrm{Gal}(K/\mathbf{Q})$. By Theorem 1.2, (6), we have

$$\prod_{\sigma\in J_K}p_K(\sigma,\tau\sigma)\sim\prod_{\sigma\in c}p_K(\mathrm{id},\sigma)^{[K:\mathbf{Q}]/|c|},$$

where c is the conjugacy class of $\widetilde{G}$ which contains τ. By Lemma 3.4, we have

$$\sum_{\omega\in\hat{G}_-}\chi_\omega(\tau)\frac{L'(0,\omega)}{L(0,\omega)}=\sum_{\eta\in(\hat{\tilde{G}})_-}\chi_\eta(\tau)\frac{L'(0,\eta)}{L(0,\eta)}. \tag{3.2}$$

Hence the assertion holds in this case.

In the general case, let L be the normal closure of K over $\mathbf{Q}$. Put $\mathfrak{G} = \mathrm{Gal}(L/F)$, $H = \mathrm{Gal}(L/K)$, $\widetilde{G} = \mathrm{Gal}(L/\mathbf{Q})$. For $\sigma \in J_K$, take $\widetilde{\sigma} \in \widetilde{G}$ so that $\widetilde{\sigma}|K = \sigma$. Take $\widetilde{\tau} \in \mathfrak{G}$ so that $\widetilde{\tau} \mod H = \tau$. By Theorem 1.2, we have

$$\begin{aligned}
&\prod_{\sigma\in J_K} p_K(\sigma,\tau\sigma) \sim \prod_{\sigma\in J_K} p_L(\mathrm{Inf}_{L/K}(\sigma),\widetilde{\tau}\widetilde{\sigma}) \sim \prod_{\sigma\in J_K} p_L(\sum_{h\in H} h\widetilde{\sigma},\widetilde{\tau}\widetilde{\sigma}) \\
&\sim \prod_{\sigma\in J_K}\prod_{h\in H} p_L(\mathrm{id},\widetilde{\sigma}^{-1}h^{-1}\widetilde{\tau}\widetilde{\sigma}) \sim \prod_{\sigma\in\tilde{G}}\prod_{h\in H} p_L(\mathrm{id},\sigma^{-1}h^{-1}\widetilde{\tau}\sigma)^{1/|H|} \\
&\sim \prod_{\sigma\in\tilde{G}}\prod_{h\in H} p_L(\sigma,h\widetilde{\tau}\sigma)^{1/|H|} \sim \prod_{\sigma\in J_L}\prod_{h\in H} p_L(\sigma,h\widetilde{\tau}\sigma)^{1/|H|}.
\end{aligned}$$

Since we have shown Proposition 3.3 when K is normal over $\mathbf{Q}$, we get

$$\prod_{\sigma\in J_K} p_K(\sigma,\tau\sigma) \sim \pi^{-[L:\mathbf{Q}]\mu(\tau)/2|H|} \prod_{\omega\in\hat{\mathfrak{G}}_-} \exp(\frac{1}{|H|}\sum_{h\in H}\chi_\omega(h\widetilde{\tau})\frac{L'(0,\omega)}{L(0,\omega)}).$$

By Lemma 3.5, the right-hand side equals

$$\pi^{-[K:\mathbf{Q}]\mu(\tau)/2} \prod_{\omega\in\hat{G}_-} \exp(\chi_\omega(\tau)\frac{L'(0,\omega)}{L(0,\omega)})$$

and we complete the proof.

COROLLARY 3.6. *Let $J_F = \{\sigma_1, \sigma_2, \dots, \sigma_n\}$ and use the same letter σ_i for any extension σ_i to J_K. We assume that K is abelian over F. Then we have*

$$\prod_{i=1}^{n} p_K(\sigma_i,\tau\sigma_i) \sim \pi^{-n\mu(\tau)/2} \prod_{\omega\in\widehat{G}_-} \exp(\frac{\omega(\tau)}{|G|}\frac{L'(0,\omega)}{L(0,\omega)}). \tag{3.3}$$

PROOF. If we change σ_i to $g\sigma_i$ with $g \in G$, we have

$$p_K(g\sigma_i,\tau g\sigma_i) = p_K(g\sigma_i, g\tau\sigma_i) \sim p_K(\sigma_i,\tau\sigma_i)$$

by Theorem 1.2, (6), since G is abelian. Thus $p_K(\sigma_i,\tau\sigma_i)$ does not depend on the choice of an extension of σ_i to J_K. Then we see that

$$\prod_{\sigma\in J_K} p_K(\sigma,\tau\sigma) \sim (\prod_{i=1}^{n} p_K(\sigma_i,\tau\sigma_i))^{|G|}.$$

Now (3.3) follows from (3.1) noting $[K:\mathbf{Q}]/|G| = [F:\mathbf{Q}] = n$.

For the sake of completeness, let us show that (3.3) implies Conjecture 3.2.

PROPOSITION 3.7. *Assume that (3.3) holds for any totally real algebraic number field F and any abelian extension K of F which is a CM-field. Then Conjecture 3.2 holds.*

PROOF. It suffices to show (3.1) for a CM-field K which is normal over F. Put $G = \mathrm{Gal}(K/F)$ and let $\tau \in G$. Put $G_0 = \langle \tau, \rho \rangle$ and let F_0 be the fixed field of G_0. Then F_0 is totally real and K is abelian over F_0. Put $m = [F_0 : \mathbf{Q}]$, $J_{F_0} = \{\sigma_1, \sigma_2, \ldots, \sigma_m\}$. We use the same letter σ_i for any extension of σ_i to J_K. Then, by our assumption, we get

$$\prod_{\sigma \in J_K} p_K(\sigma, \tau\sigma) \sim (\prod_{i=1}^{m} p_K(\sigma_i, \tau\sigma_i))^{|G_0|} \sim \pi^{-m|G_0|\mu(\tau)/2} \prod_{\omega \in (\hat{G}_0)_-} \exp(\omega(\tau)\frac{L'(0,\omega)}{L(0,\omega)}).$$

Since $m|G_0| = [K : \mathbf{Q}]$, it suffices to show that

$$\sum_{\omega \in (\hat{G}_0)_-} \omega(\tau)\frac{L'(0,\omega)}{L(0,\omega)} = \sum_{\eta \in \hat{G}_-} \chi_\eta(\tau)\frac{L'(0,\eta)}{L(0,\eta)}. \tag{3.4}$$

Since this is a special case of Lemma 3.4, we complete the proof.

3.3. Now let F be a totally real algebraic number field of degree n and K be an abelian extension of F of finite degree. We assume that K is a CM-field. Let $\tilde{\mathfrak{f}}\infty_1\infty_2\cdots\infty_n$ be the conductor of K as a class field over F, $\tilde{\mathfrak{f}}$ being an integral ideal of F. For an integral ideal $\mathfrak{f}$ of F, let $C_\mathfrak{f}$ denote the ideal class group modulo $\mathfrak{f}\infty_1\infty_2\cdots\infty_n$. Set $G = \mathrm{Gal}(K/F)$. For $\omega \in \widehat{G}_-$, we regard ω as a character of $C_{\tilde{\mathfrak{f}}}$ by the Artin map and let $\mathfrak{f}_\omega\infty_1\infty_2\cdots\infty_n$ denote the conductor of ω. For an integral ideal $\mathfrak{f}|\tilde{\mathfrak{f}}$, we set $(\widehat{G}_-)_\mathfrak{f} = \{\omega \in \widehat{G}_- \mid \mathfrak{f}_\omega = \mathfrak{f}\}$. Then we have

$$L'(0,\omega) = \sum_{c \in C_{\mathfrak{f}_\omega}} \omega(c)\zeta_F'(0,c), \qquad \omega \in (\widehat{G})_-,$$

$$\sum_{\omega \in \hat{G}_-} \frac{\omega(\tau)}{L(0,\omega)} L'(0,\omega) = \sum_{\mathfrak{f}|\tilde{\mathfrak{f}}} \sum_{\omega \in (\hat{G}_-)_\mathfrak{f}} \frac{\omega(\tau)}{L(0,\omega)} \sum_{c \in C_\mathfrak{f}} \omega(c)\zeta_F'(0,c).$$

Hence (3.3) can be written as

$$\begin{aligned} &\prod_{i=1}^{n} p_{K^{\sigma_i}}(\mathrm{id}, \sigma_i^{-1}\tau\sigma_i) \\ &\sim \pi^{-n\mu(\tau)/2} \exp(\frac{1}{|G|}\sum_{\mathfrak{f}|\tilde{\mathfrak{f}}} \sum_{\omega \in (\hat{G}_-)_\mathfrak{f}} \frac{\omega(\tau)}{L(0,\omega)} \sum_{c \in C_\mathfrak{f}} \omega(c)\zeta_F'(0,c)), \end{aligned} \tag{3.5}$$

noting $p_K(\sigma_i, \tau\sigma_i) \sim p_{K^{\sigma_i}}(\mathrm{id}, \sigma_i^{-1}\tau\sigma_i)$ by Theorem 1.2, (6).

We are going to factorize the right-hand side of (3.5) in accordance with the factorization of the left-hand side, using Shintani's formula II, (3.7). Let the notation be the same as in II, §3. We choose $\sigma_1 = \mathrm{id}$. Let $c \in C_\mathfrak{f}$. We put

$$G(c) = \sum_{j \in J} \sum_{z \in R(C_j, c)} \log \frac{\Gamma_{r(j)}(z^{(1)}, A_j^{(1)})}{\rho_{r(j)}(A_j^{(1)})}. \tag{3.6}$$

Undoutedly this should be the main term of the factorization; it turns out, as we will show later, that it is best to define

$$
\begin{aligned}
V(c) = \sum_{j\in J}(-1)^{r(j)}\sum_{l}\Big[\frac{1}{n}\sum_{k=2}^{n}(C_{l,1,k}(A_j)+C_{l,k,1}(A_j)) \\
-\frac{1}{n^2}\sum_{1\leqq i,k\leqq n, i\neq k} C_{l,i,k}(A_j)\Big]\times \sum_{z\in R(C_j,c)}\prod_{m=1}^{r(j)}\frac{B_{l_m}(x_m(z))}{l_m!},
\end{aligned}
\tag{3.7}
$$

$$
W(c) = -\frac{1}{n}\log N(\mathfrak{a}_\mu\mathfrak{f})\sum_{j\in J}(-1)^{r(j)}\sum_{l}\sum_{z\in R(C_j,c)}\prod_{m=1}^{r(j)}\frac{B_{l_m}(x_m(z))}{l_m!}(v_{jm}^{(1)})^{l_m-1},
\tag{3.8}
$$

$$
X(c) = G(c)+V(c)+W(c),
\tag{3.9}
$$

$$
g_K(\mathrm{id},\tau) = \pi^{-\mu(\tau)/2}\exp\Big(\frac{1}{|G|}\sum_{\mathfrak{f}|\tilde{\mathfrak{f}}}\sum_{\omega\in(\hat{G}_-)_\mathfrak{f}}\frac{\omega(\tau)}{L(0,\omega)}\sum_{c\in C_\mathfrak{f}}\omega(c)X(c)\Big).
\tag{3.10}
$$

Here $C_{l,i,k}(A_j)$ is defined by I, (2.13). The quantity $X(c)$ will play a fundamental role in our succeeding investigations; $X(c)$ depends on the choice of a cone decomposition $\{C_j\}_{j\in J}$ and the choice of representatives $\{\mathfrak{a}_\mu\}$ of the narrow ideal classes. When it is necessary to emphasize the dependence on these data, we write $X(c)$ as $X(c;\{C_j\}_{j\in J},\mathfrak{a}_\mu)$; $g_K(\mathrm{id},\tau)$ will also be denoted by $g_K(\mathrm{id},\tau;\{C_j\}_{j\in J},\{\mathfrak{a}_\mu\})$. We will often abbreviate $\{C_j\}_{j\in J}$ to $\{C_j\}$. We call g_K *the absolute period symbol.* Let us first verify that this definition gives a factorization of the right-hand side of (3.5).

REMARK. The definition of $V(c)$ does not depend on the numbering of $\sigma_2,\ldots,\sigma_n$.

LEMMA 3.8. *We have*

$$
\sum_{i=1}^{n} X(c^{\sigma_i};\{C_j^{\sigma_i}\}_{j\in J},\mathfrak{a}_\mu^{\sigma_i}) = \zeta_F'(0,c) \qquad \textit{for} \quad c\in C_\mathfrak{f},
\tag{3.11}
$$

$$
\begin{aligned}
&\prod_{i=1}^{n} g_{K^{\sigma_i}}(\mathrm{id},\sigma_i^{-1}\tau\sigma_i;\{C_j^{\sigma_i}\}_{j\in J},\{\mathfrak{a}_\mu^{\sigma_i}\}) \\
=&\pi^{-n\mu(\tau)/2}\exp\Big(\frac{1}{|G|}\sum_{\mathfrak{f}|\tilde{\mathfrak{f}}}\sum_{\omega\in(\hat{G}_-)_\mathfrak{f}}\frac{\omega(\tau)}{L(0,\omega)}\sum_{c\in C_\mathfrak{f}}\omega(c)\zeta_F'(0,c)\Big).
\end{aligned}
\tag{3.12}
$$

PROOF. Put $G_i = \mathrm{Gal}(K^{\sigma_i}/F^{\sigma_i})$. Then we have $G_i = \sigma_i^{-1}G\sigma_i$. By definition, we have

$$
\begin{aligned}
&g_{K^{\sigma_i}}(\mathrm{id},\sigma_i^{-1}\tau\sigma_i;\{C_j^{\sigma_i}\},\{\mathfrak{a}_\mu^{\sigma_i}\}) \\
=&\pi^{-\mu(\sigma_i^{-1}\tau\sigma_i)/2}\exp\Big(\frac{1}{|G|}\sum_{\mathfrak{f}^{\sigma_i}|\tilde{\mathfrak{f}}^{\sigma_i}}\sum_{\omega\in(\hat{G}_{i,-})_{\mathfrak{f}^{\sigma_i}}}\frac{\omega(\sigma_i^{-1}\tau\sigma_i)}{L(0,\omega)}\sum_{c\in C_{\mathfrak{f}^{\sigma_i}}}\omega(c)X(c;\{C_j^{\sigma_i}\},\mathfrak{a}_\mu^{\sigma_i})\Big).
\end{aligned}
$$

For $\omega \in \widehat{G}_-$, define $\omega^{(i)} \in (\widehat{G}_i)_-$ by $\omega^{(i)}(x) = \omega(\sigma_i x \sigma_i^{-1})$, $x \in G_i$. Then $(\widehat{G}_-)_\mathfrak{f} \ni \omega \longrightarrow \omega^{(i)} \in (\widehat{G}_{i,-})_{\mathfrak{f}^{\sigma_i}}$ is a one to one correspondence and

$$\omega^{(i)}(c^{\sigma_i}) = \omega(c), \quad c \in C_\mathfrak{f}, \qquad L(s, \omega^{(i)}) = L(s, \omega).$$

Here $(\widehat{G}_{i,-})_{\mathfrak{f}^{\sigma_i}} = ((\widehat{G}_i)_-)_{\mathfrak{f}^{\sigma_i}}$. Hence, putting $\omega = \omega^{(i)}$, $c = c^{\sigma_i}$ in the above formula, we obtain

$$\begin{aligned} &g_{K^{\sigma_i}}(\mathrm{id}, \sigma_i^{-1}\tau\sigma_i; \{C_j^{\sigma_i}\}, \{\mathfrak{a}_\mu^{\sigma_i}\}) \\ &= \pi^{-\mu(\tau)/2} \exp\Big(\frac{1}{|G|} \sum_{\mathfrak{f}|\tilde{\mathfrak{f}}} \sum_{\omega \in (\hat{G}_{i,-})_\mathfrak{f}} \frac{\omega(\tau)}{L(0,\omega)} \sum_{c \in C_\mathfrak{f}} \omega(c) X(c^{\sigma_i}; \{C_j^{\sigma_i}\}, \mathfrak{a}_\mu^{\sigma_i})\Big). \end{aligned} \tag{3.13}$$

Comparing with (3.10), we see that (3.12) follows from (3.11).

Our task is to prove (3.11). We note that $C_j^{\sigma_i} = C(v_{j1}^{\sigma_i}, \cdots, v_{jr(j)}^{\sigma_i})$ and the mapping $R(C_j, c) \ni z \longrightarrow z^{\sigma_i} \in R(C_j^{\sigma_i}, c^{\sigma_i})$ is a bijection such that $x_m(z^{\sigma_i}) = x_m(z)$. Hence

$$\begin{aligned} &W(c^{\sigma_i}; \{C_j^{\sigma_i}\}, \mathfrak{a}_\mu^{\sigma_i})) \\ &= -\frac{1}{n} \log N(\mathfrak{a}_\mu^{\sigma_i}\mathfrak{f}^{\sigma_i}) \sum_{j \in J} (-1)^{r(j)} \sum_l \sum_{z \in R(C_j^{\sigma_i}, c^{\sigma_i})} \prod_{m=1}^{r(j)} \frac{B_{l_m}(x_m(z))}{l_m!} (v_{jm}^{\sigma_i})^{l_m - 1} \\ &= -\frac{1}{n} \log N(\mathfrak{a}_\mu\mathfrak{f}) \sum_{j \in J} (-1)^{r(j)} \sum_l \sum_{z \in R(C_j, c)} \prod_{m=1}^{r(j)} \frac{B_{l_m}(x_m(z))}{l_m!} (v_{jm}^{\sigma_i})^{l_m - 1}. \end{aligned}$$

Comparing with II, (3.6), we obtain

$$\sum_{i=1}^n W(c^{\sigma_i}; \{C_j^{\sigma_i}\}, \mathfrak{a}_\mu^{\sigma_i}) = -\log N(\mathfrak{a}_\mu\mathfrak{f}) \zeta_F(0, c). \tag{3.14}$$

Therefore, in view of II, (3.7), we can reduce (3.11) to

$$\begin{aligned} &\sum_{i=1}^n V(c^{\sigma_i}; \{C_j^{\sigma_i}\}, \mathfrak{a}_\mu^{\sigma_i}) \\ &= \sum_{j \in J} \frac{(-1)^{r(j)}}{n} \sum_l C_l(A_j) \sum_{z \in R(C_j, c)} \prod_{m=1}^{r(j)} \left\{ \frac{B_{l_m}(x_m(z))}{l_m!} \right\}. \end{aligned} \tag{3.15}$$

Let $A_{j,i}$ be the $n \times r(j)$-matrix determined by the cone $C_j^{\sigma_i}$. Since $J_{F^{\sigma_i}} = \{\tilde{\sigma}_i^{-1}\sigma_l\tilde{\sigma}_i \mid 1 \leqq l \leqq n\}$, the (l,m)-component of $A_{j,i}$ is $v_{jm}^{\sigma_l\tilde{\sigma}_i}$, when we extend σ_i to an element of $\mathrm{Gal}(\overline{\mathbf{Q}}/\mathbf{Q})$. Thus $A_{j,i}$ depends on the choice of $\tilde{\sigma}_i$, except for its first row, but this does not affect the definition of $V(c^{\sigma_i}; \{C_j^{\sigma_i}\}, \mathfrak{a}_\mu^{\sigma_i}))$ (cf. Remark preceding Lemma 3.8). Since

$$V(c^{\sigma_i};\{C_j^{\sigma_i}\},\mathfrak{a}_\mu^{\sigma_i})=\sum_{j\in J}(-1)^{r(j)}\sum_l$$
$$\left[\frac{1}{n}\sum_{k=2}^{n}(C_{l,1,k}(A_{j,i})+C_{l,k,1}(A_{j,i}))-\frac{1}{n^2}\sum_{1\leqq i,k\leqq n,i\neq k}C_{l,i,k}(A_{j,i})\right]$$
$$\times\sum_{z\in R(C_j,c)}\prod_{m=1}^{r(j)}\frac{B_{l_m}(x_m(z))}{l_m!},$$

(3.15) follows if we can show that

$$C_l(A_j)=\sum_{i=1}^{n}\left[\sum_{k=2}^{n}C_{l,1,k}(A_{j,i})+C_{l,k,1}(A_{j,i})-\frac{1}{n}\sum_{1\leqq p,k\leqq n,p\neq k}C_{l,p,k}(A_{j,i})\right],\tag{3.16}$$

for $j\in J$ and for an arbitrary $r(j)$-tuple of non-negative integers $l=(l_1,l_2,\cdots,l_{r(j)})$ such that $l_1+l_2+\cdots+l_{r(j)}=r(j)$. Since rows of $A_{j,i}$ coincide with rows of A_j up to a permutation, we have

$$\sum_{1\leqq p,k\leqq n,p\neq k}C_{l,p,k}(A_{j,i})=\sum_{1\leqq p,k\leqq n,p\neq k}C_{l,p,k}(A_j)=C_l(A_j)$$

for every i. We also have

$$\sum_{k=2}^{n}C_{l,1,k}(A_{j,i})+C_{l,k,1}(A_{j,i})=\sum_{k=1,k\neq i}^{n}C_{l,i,k}(A_j)+C_{l,k,i}(A_j),$$
$$\sum_{i=1}^{n}\sum_{k=2}^{n}C_{l,1,k}(A_{j,i})+C_{l,k,1}(A_{j,i})=2C_l(A_j).$$

Hence (3.16) holds. This completes the proof.

Now the factorization being established, we make:

CONJECTURE 3.9. *Let K be a CM-field which is an abelian extension of a totally real algebraic number field F. For $\tau\in\mathrm{Gal}(K/F)$, define $g_K(\mathrm{id},\tau)$ by (3.6) $\sim$ (3.10). Then for $\tau\in\mathrm{Gal}(K/F)$, we have $p_K(\mathrm{id},\tau)\sim g_K(\mathrm{id},\tau)$.*

Apparently Conjecture 3.9 may seem too special to express $p_K(\sigma,\tau)$ for any CM-field and for any σ, $\tau\in I_K$ in terms of the multiple gamma function. However this is not so. In fact, let L be the normal closure of K over $\mathbf{Q}$. Take $\tilde\sigma\in I_L$ so that $\mathrm{Res}_{L/K}(\tilde\sigma)=\sigma$. By Theorem 1.2, (5), we have $p_K(\sigma,\tau)\sim p_L(\tilde\sigma,\mathrm{Inf}_{L/K}(\tau))$. Put $\tilde\sigma=\sum_{i=1}^k a_i\sigma_i$, $\mathrm{Inf}_{L/K}(\tau)=\sum_{j=1}^l b_j\tau_j$ with σ_i, $\tau_j\in J_L$, a_i, $b_j\in\mathbf{Z}$. Then, by Theorem 1.2, (6), we get

$$p_K(\sigma,\tau)\sim\prod_{i=1}^{k}\prod_{j=1}^{l}p_L(\mathrm{id},\sigma_i^{-1}\tau_j)^{a_ib_j}.$$

Put $\alpha=\sigma_i^{-1}\tau_j$. We may regard α as an element of $\mathrm{Gal}(L/\mathbf{Q})$. Let F_α be the fixed field of $\langle\alpha,\rho\rangle$. Then F_α is totally real and L is abelian over F_α. This reduces to the situation of Conjecture 3.9 and every $p_L(\mathrm{id}, \sigma_i^{-1}\tau_j) \mod \overline{\mathbf{Q}}^\times$ can be expressed in terms of the multiple gamma function.

3.4. Next we are going to refine Conjecture 3.9. Let K be a CM-field which is abelian over a totally real algebraic number field F. Let $\mathfrak{q}$ be an integral ideal of K. Suppose that there are given q characters λ_i, $1 \leqq i \leqq q$ of $I_{\mathfrak{q}}(K)$ such that

$$\lambda_i((\alpha)) = \prod_{\sigma \in \Phi_i} (\alpha^{\sigma\rho}/|\alpha^\sigma|)^{t_\sigma^{(i)}}, \qquad \alpha \equiv 1 \mod^\times \mathfrak{q}, \tag{3.17}$$

where Φ_i is a CM-type of K and $t_\sigma^{(i)}$ are positive integers. We assume that there exists an integer m which satisfies

$$m \equiv t_\sigma^{(i)} \mod 2, \qquad -t_\sigma^{(i)} < m \leqq t_\sigma^{(i)}$$

for every i and every $\sigma \in \Phi_i$. Then, by Theorem 1.5, we have

$$L(m/2, \lambda_i) \sim \pi^{e_i/2} p_K(\sum_{\sigma \in \Phi_i} t_\sigma^{(i)} \cdot \sigma,\ \Phi_i), \qquad e_i = m[F : \mathbf{Q}] + \sum_{\sigma \in \Phi_i} t_\sigma^{(i)}. \tag{3.18}$$

Let L be the normal closure of K over $\mathbf{Q}$ and S be a CM-type of L. We put $\eta_i = \mathrm{Inf}_{L/K}(\sum_{\sigma \in \Phi_i} t_\sigma^{(i)} \cdot \sigma)$ and write $\eta_i = \sum_{\alpha \in J_L} a_\alpha \cdot \alpha$, $a_\alpha \in \mathbf{Z}$. We set $\eta_i^{-1} = \sum_{\alpha \in J_L} a_\alpha \cdot \alpha^{-1} \in J_L$. Take $\widetilde{\Phi}_i \in I_L$ so that $\mathrm{Res}_{L/K}(\widetilde{\Phi}_i) = \Phi_i$ and set $\eta_i^{-1}\widetilde{\Phi}_i = \sum_{\sigma \in \widetilde{\Phi}_i} \eta_i^{-1}\sigma \in I_L$. Since J_L can be identified with $\mathrm{Gal}(L/\mathbf{Q})$, these notations are legitimate. Using the CM-type S, we can write

$$\eta_i^{-1}\widetilde{\Phi}_i = \sum_{\sigma \in S} l_\sigma^{(i)}\sigma + m_\sigma^{(i)}\sigma\rho.$$

Then, by Theorem 1.2, we have

$$p_K(\sum_{\sigma \in \Phi_i} t_\sigma^{(i)}\sigma, \Phi_i) \sim p_L(\eta_i, \widetilde{\Phi}_i) \sim p_L(\mathrm{id}, \eta_i^{-1}\widetilde{\Phi}_i) \sim p_L(\mathrm{id}, \sum_{\sigma \in S}(l_\sigma^{(i)} - m_\sigma^{(i)})\sigma).$$

Let $\epsilon_1, \epsilon_2, \ldots, \epsilon_q$ be integers. We have

$$\prod_{i=1}^q p_K(\sum_{\sigma \in \Phi_i} t_\sigma^{(i)} \cdot \sigma, \Phi_i)^{\epsilon_i} \sim p_L(\mathrm{id}, \sum_{\sigma \in S} n_\sigma \cdot \sigma), \qquad n_\sigma = \sum_{i=1}^q \epsilon_i(l_\sigma^{(i)} - m_\sigma^{(i)}).$$

Let $\tau_1, \tau_2, \ldots, \tau_r \in G = \mathrm{Gal}(K/F)$, $\sigma_1, \sigma_2, \ldots, \sigma_r \in J_K$, $a_1, a_2, \ldots, a_r \in \mathbf{Z}$ and put

$$\sum_{i=1}^r a_i \,\mathrm{Inf}_{L/K^{\sigma_i}}(\sigma_i^{-1}\tau_i\sigma_i) = \sum_{\sigma \in S} l'_\sigma\sigma + m'_\sigma\sigma\rho, \qquad n'_\sigma = l'_\sigma - m'_\sigma.$$

Then we have

$$\prod_{i=1}^r p_K(\sigma_i, \tau_i\sigma_i)^{a_i} \sim \prod_{i=1}^r p_{K^{\sigma_i}}(\mathrm{id}, \sigma_i^{-1}\tau_i\sigma_i)^{a_i} \sim p_L(\mathrm{id}, \sum_{\sigma \in S} n'_\sigma\sigma).$$

We assume that

$$\sum_{\sigma \in S} n_\sigma\sigma = \sum_{\sigma \in S} n'_\sigma\sigma \tag{3.19}$$

holds. If $L(m/2, \lambda_i) \neq 0$ for $\epsilon_i < 0$, we obtain

$$(3.20)\qquad \prod_{i=1}^{q} L(m/2, \lambda_i)^{\epsilon_i} \sim \pi^A \prod_{i=1}^{r} p_{K^{\sigma_i}}(\mathrm{id}, \sigma_i^{-1}\tau_i\sigma_i)^{a_i}, \qquad A = \sum_{i=1}^{q} \frac{\epsilon_i e_i}{2}.$$

Now a refinement of Conjecture 3.9 is given by

CONJECTURE 3.10. *Let the assumptions be the same as in Conjecture 3.7. Let λ_i, $1 \leqq i \leqq q$ be characters of $I_{\mathfrak{q}}(K)$ which satisfy (3.17) and (3.20), where $\mathfrak{q}$ is an integral ideal of K. Assume that $L(m/2, \lambda_i) \neq 0$ if $\epsilon_i < 0$. Then for every $\sigma \in \mathrm{Gal}(\overline{\mathbf{Q}}/\mathbf{Q})$, there exists a root of unity $\zeta(\sigma)$ such that*

$$\begin{aligned}&\Big(\frac{\prod_{i=1}^{q} L(m/2, \lambda_i)^{\epsilon_i}}{\pi^A \prod_{i=1}^{r} g_{K^{\sigma_i}}(\mathrm{id}, \sigma_i^{-1}\tau_i\sigma_i; \{C_j^{\sigma_i}\}_{j\in J}, \{\mathfrak{a}_\mu^{\sigma_i}\})^{a_i}}\Big)^{\sigma}\\ &= \zeta(\sigma)\cdot \frac{\prod_{i=1}^{q} L(m/2, \lambda_i^{\sigma})^{\epsilon_i}}{\pi^A \prod_{i=1}^{r} g_{K^{\sigma_i\sigma}}(\mathrm{id}, \sigma^{-1}\sigma_i^{-1}\tau_i\sigma_i\sigma; \{C_j^{\sigma_i\sigma}\}_{j\in J}, \{\mathfrak{a}_\mu^{\sigma_i\sigma}\})^{a_i}}.\end{aligned}$$

Here $g_{K^{\sigma_i\sigma}}(\mathrm{id}, \sigma^{-1}\sigma_i^{-1}\tau_i\sigma_i\sigma; \{C_j^{\sigma_i\sigma}\}_{j\in J}, \{\mathfrak{a}_\mu^{\sigma_i\sigma}\})$ is defined regarding $K^{\sigma_i\sigma}$ as an abelian extension of $F^{\sigma_i\sigma}$ and $\sigma^{-1}\sigma_i^{-1}\tau_i\sigma_i\sigma \in \mathrm{Gal}(K^{\sigma_i\sigma}/F^{\sigma_i\sigma})$.

REMARK. We see that $g_{K^{\sigma_i\sigma}}(\mathrm{id}, \sigma^{-1}\sigma_i^{-1}\tau_i\sigma_i\sigma; \{C_j^{\sigma_i\sigma}\}, \{\mathfrak{a}_\mu^{\sigma_i\sigma}\})$ depends only on $\sigma|F^{\sigma_i}$ when $\{C_j\}$ and $\{\mathfrak{a}_\mu\}$ are fixed. The conjecture above was stated in [Y4] as Conjecture B when $r = 1$, $\sigma_1 = \mathrm{id}$. The general case can be derived from this special case without difficulty.

The conjectured mapping $\sigma \longrightarrow \zeta$ from $\mathrm{Gal}(\overline{\mathbf{Q}}/\mathbf{Q})$ to the group formed by the roots of unity will be called *the reciprocity map.*

REMARK. The validity of Conjecture 3.10 does not depend on the choice of $\{\lambda_i\}$ which satisfies (3.20). This follows from a result of Blasius [Bl] and can be shown in the same way as in [Y3], §6.

REMARK 3.11. When all of ϵ_i, a_i are positive, it may be possible to strengthen Conjecture 3.10 by controlling the denominator of

$$\frac{\prod_{i=1}^{q} L(m/2, \lambda_i)^{\epsilon_i}}{\pi^A (\prod_{i=1}^{r} g_{K^{\sigma_i}}(\mathrm{id}, \sigma_i^{-1}\tau_i\sigma_i; \{C_j^{\sigma_i}\}_{j\in J}, \{\mathfrak{a}_\mu^{\sigma_i}\})^{a_i})^e}.$$

See examples in the next section and Remark 4.5.

EXAMPLE 3.12. Let us give an example for (3.20). Let K be a CM-field. Put $2l = [K : \mathbf{Q}]$ and let $\Phi = \{\sigma_1, \sigma_2, \ldots, \sigma_l\}$ be a CM-type of K. Let F_0 be the maximal real subfield of K. It is well known that $[E_K : E_{F_0}W_K] = 1$ or 2, where W_K is the group formed by the roots of unity contained in K. Put $w = |W_K|$. By a theorem of Chevalley [Che], we can find an integral ideal $\mathfrak{q}$ of K such that $\{\epsilon \in E_K \mid \epsilon \equiv 1 \mod \mathfrak{q}\} \subset E_K^{2w}$. We have $E_K^{2w} \subset E_{F_0}^{w} \subset E_{F_0}^{+}$. Let $\delta(i)$, $1 \leqq i \leqq l$ be 0 or 1. Then $\{\sigma_1\rho^{\delta(1)}, \sigma_2\rho^{\delta(2)}, \ldots \sigma_l\rho^{\delta(l)}\}$ is also a CM-type of K. For positive integers a_1, a_2, $\ldots$, a_l, there exists a character $\lambda_{a_1,a_2,\ldots,a_l}^{\delta(1),\delta(2),\ldots,\delta(l)}$ of $I_{\mathfrak{q}}(K)$ such that

$$(3.21)\qquad \lambda_{a_1,a_2,\ldots,a_l}^{\delta(1),\delta(2),\ldots,\delta(l)}((\alpha)) = \prod_{i=1}^{l} \Big(\frac{\alpha^{\sigma_i\rho^{1+\delta(i)}}}{|\alpha^{\sigma_i\rho^{\delta(i)}}|}\Big)^{a_i}, \qquad \alpha \equiv 1 \mod^{\times} \mathfrak{q},$$

since $\prod_{i=1}^{l}(\frac{\epsilon^{\sigma_i\rho^{1+\delta(i)}}}{|\epsilon^{\sigma_i\rho^{\delta(i)}}|})^{a_i}=1$ for $\epsilon\in E_K$ satisfying $\epsilon\equiv 1 \mod \mathfrak{f}$. Assume that a_1, a_2, $\dots$, a_l are even and $\sigma_1=\mathrm{id}$. Then, by Theorem 1.5, we have

$$\frac{L(1,\lambda^{0,\delta(2),\dots,\delta(l)}_{a_1+2,a_2,\dots,a_l})}{L(1,\lambda^{0,\delta(2),\dots,\delta(l)}_{a_1,a_2,\dots,a_l})}\sim \pi p_K(2\cdot\mathrm{id},\,\mathrm{id}+\sigma_2\rho^{\delta(2)}+\cdots+\sigma_l\rho^{\delta(l)}). \tag{3.22}$$

Thus we have

$$\frac{L(1,\lambda^{0,0,\dots,0}_{a_1+2,a_2,\dots,a_l})}{L(1,\lambda^{0,0,\dots,0}_{a_1,a_2,\dots,a_l})}\frac{L(1,\lambda^{0,1,\dots,1}_{a_1+2,a_2,\dots,a_l})}{L(1,\lambda^{0,1,\dots,1}_{a_1,a_2,\dots,a_l})}\sim\pi^2 p_K(\mathrm{id},\mathrm{id})^4, \tag{3.23}$$

using Theorem 1.2. Similarly we can obtain an expression for $p_K(\mathrm{id},\sigma_2)$. For example,

$$\frac{L(1,\lambda^{0,0,\dots,0}_{a_1+2,a_2,\dots,a_l})}{L(1,\lambda^{0,0,\dots,0}_{a_1,a_2,\dots,a_l})}\frac{L(1,\lambda^{0,0,1,\dots,1}_{a_1+2,a_2,\dots,a_l})}{L(1,\lambda^{0,0,1,\dots,1}_{a_1,a_2,\dots,a_l})}\sim\pi^2 p_K(\mathrm{id},\mathrm{id}+\sigma_2)^4$$

holds and taking the quotient by the left-hand side of (3.23), we obtain an expression for $p_K(\mathrm{id},\sigma_2)^4$.

3.5. Next we will investigate the dependence of $g_K(\mathrm{id},\tau;\{C_j\},\{\mathfrak{a}_\mu\})$ on $\{C_j\}$ and $\{\mathfrak{a}_\mu\}$. The consideration given below will show that Conjectures 3.9 and 3.10 are consistently formulated so that the validity of them does not depend on the choice of auxiliary data $\{C_j\}$ and $\{\mathfrak{a}_\mu\}$. For this purpose, the following lemma is crucial.

Let E be a subgroup of finite index of E_F^+. Let

$$\mathbf{R}^n_+=\sqcup_{\epsilon\in E}\ \epsilon\ (\sqcup_{j\in J}C_j)$$

be a cone decomposition of $\mathbf{R}^n_+$, where C_j is an open simplicial cone with basis in $\mathfrak{O}_F$. Let $\{C'_l\}_{l\in L}$ be a finite family of open simplicial cones with basis in $\mathfrak{O}_F$. We call $\{C'_l\}_{l\in L}$ *a refinement* of $\{C_j\}_{j\in J}$ if every C'_l is contained in some C_j and if

$$\mathbf{R}^n_+=\sqcup_{\epsilon\in E}\ \epsilon\ (\sqcup_{l\in L}C'_l) \tag{3.24}$$

holds. We see that (3.24) holds if and only if $C_j=\sqcup_{l\in L,C'_l\subset C_j}C'_l$ for every j.

LEMMA 3.13. *Let $\{C_j\}_{j\in J}$ and $\{D_i\}_{i\in I}$ be two finite families of open simplicial cones with basis in $\mathfrak{O}_F$ such that*

$$\mathbf{R}^n_+=\sqcup_{\epsilon\in E}\ \epsilon\ (\sqcup_{j\in J}C_j),\qquad \mathbf{R}^n_+=\sqcup_{\epsilon\in E}\ \epsilon\ (\sqcup_{i\in I}D_i).$$

Then there exist a finite family $\{C'_l\}_{l\in L}$ of open simplicial cones with basis in $\mathfrak{O}_F$ and a family of units $\{\epsilon_l\}_{l\in L}$, $\epsilon_l\in E$ such that $\{C'_l\}_{l\in L}$ is a refinement of $\{C_j\}_{j\in J}$ and that $\{\epsilon_l C'_l\}_{l\in L}$ is a refinement of $\{D_i\}_{i\in I}$.

PROOF. We have

$$C_j=\sqcup_{i\in I}\sqcup_{\epsilon\in E,C_j\cap\epsilon D_i\neq\emptyset}(C_j\cap\epsilon D_i),$$
$$D_i=\sqcup_{j\in J}\sqcup_{\epsilon\in E,D_i\cap\epsilon^{-1}C_j\neq\emptyset}(D_i\cap\epsilon^{-1}C_j).$$

We put $H=\{(x_1,x_2,\dots,x_n)\in\mathbf{R}^n_+\mid x_1x_2\cdots x_n=1\}$. Then E is a discrete subgroup of H such that H/E is compact. Let $C=C(v_1,v_2,\dots,v_r)$ be an r-dimensional open simplicial cone in $\mathbf{R}^n_+$. Put $C^1=C\cap H$. Then the closure $\overline{C^1}$ of

C^1 is a compact subset of H. In fact, put $v_k = (v_{k1}, v_{k2}, \dots, v_{kn}) \in \mathbf{R}_+^n$, $1 \leqq k \leqq r$. For $x = \sum_{k=1}^r t_k v_k \in \overline{C^1}$, we have $t_k^n \prod_{i=1}^n v_{ki} \leqq \prod_{i=1}^n (\sum_{k=1}^r t_k v_{ki}) = 1$. Hence $\overline{C^1}$ is bounded and closed; therefore it is compact.

Now fix $i \in I$, $j \in J$. If $C_j \cap \epsilon D_i \neq \emptyset$, $\epsilon \in E$, then $C_j^1 \cap \epsilon D_i^1 \neq \emptyset$. We have $\epsilon \in (\overline{D_i^1})^{-1}\overline{C_j^1} \cap E$, which is a finite set, since E is discrete and $(\overline{D_i^1})^{-1}\overline{C_j^1}$ is compact. Therefore there are only finitely many $\epsilon \in E$ such that $C_j \cap \epsilon D_i \neq \emptyset$. Let E_{ij} be the set of such units. By Shintani [Sh1], Corollary to Lemma 2, we have

$$C_j \cap \epsilon D_i = \sqcup_{p \in L(\epsilon,i,j)} C'_p$$

where $L(\epsilon, i, j)$ is a finite set and C'_p is an open simplicial cone with basis in $\mathfrak{O}_F$. We obtain

$$C_j = \sqcup_{i \in I, E_{ij} \neq \emptyset} \sqcup_{\epsilon \in E_{ij}} \sqcup_{p \in L(\epsilon,i,j)} C'_p,$$
$$D_i = \sqcup_{j \in J, E_{ij} \neq \emptyset} \sqcup_{\epsilon \in E_{ij}} \sqcup_{p \in L(\epsilon,i,j)} \epsilon^{-1} C'_p.$$

Now the assertion follows by taking $L = \cup_{i \in I} \cup_{j \in J} \cup_{\epsilon \in E_{ij}} L(\epsilon, i, j)$.

Fix $c \in C_\mathfrak{f}$ and take $\mathfrak{a}_\mu$ so that the narrow ideal class of c is the same as that of $\mathfrak{a}_\mu \mathfrak{f}$. Fix $j \in J$. For $1 \leqq i \leqq n$, put

$$\zeta^{(i)}(s, C_j, \mathfrak{a}_\mu) = \sum_{z \in (\mathfrak{a}_\mu \mathfrak{f})^{-1} \cap C_j, (z)\mathfrak{a}_\mu \mathfrak{f} \equiv c} (z^{(i)})^{-s}. \tag{3.25}$$

For $1 \leqq i, k \leqq n$, $i \neq k$, put

$$\zeta^{(i,k)}(s, C_j, \mathfrak{a}_\mu) = \sum_{z \in (\mathfrak{a}_\mu \mathfrak{f})^{-1} \cap C_j, (z)\mathfrak{a}_\mu \mathfrak{f} \equiv c} (z^{(i)} z^{(k)})^{-s}. \tag{3.26}$$

In (3.25) and (3.26), $\equiv$ means the equality in $C_\mathfrak{f}$. Hereafter in this section, $\equiv$ will always have this meaning. By II, Lemma 3.2, we have

$$\zeta^{(i)}(s, C_j, \mathfrak{a}_\mu) = \sum_{z \in R(C_j, c)} \zeta(s, A_j^{(i)}, x(z^{(i)})),$$
$$\zeta^{(i,k)}(s, C_j, \mathfrak{a}_\mu) = \sum_{z \in R(C_j, c)} \zeta(s, A_j^{(i,k)}, (x(z^{(i)}), x(z^{(k)}))),$$

where $A_j^{(i,k)}$ is the $2 \times r(j)$-matrix whose first row is $A_j^{(i)}$ and the second row is $A_j^{(k)}$. By I, Theorems 2.1 and 2.2, we obtain:

$$\zeta^{(i,k)}(0, C_j, \mathfrak{a}_\mu) = \frac{1}{2}(\zeta^{(i)}(0, C_j, \mathfrak{a}_\mu) + \zeta^{(k)}(0, C_j, \mathfrak{a}_\mu)), \tag{3.27}$$

$$G(c) = \sum_{j \in J} \frac{d}{ds} \zeta^{(1)}(s, C_j, \mathfrak{a}_\mu)\Big|_{s=0}, \tag{3.28}$$

$$V(c) = \sum_{j \in J} \Big(\frac{2}{n} \sum_{k=2}^{n} v_{1,k}^{(j)} - \frac{1}{n^2} \sum_{1 \leqq i,k \leqq n, i \neq k} v_{i,k}^{(j)}\Big), \tag{3.29}$$

where

$$(3.30)\quad v_{i,k}^{(j)} = \frac{d}{ds}\zeta^{(i,k)}(s, C_j, \mathfrak{a}_\mu)\Big|_{s=0} - \frac{d}{ds}\zeta^{(i)}(s, C_j, \mathfrak{a}_\mu)\Big|_{s=0} - \frac{d}{ds}\zeta^{(k)}(s, C_j, \mathfrak{a}_\mu)\Big|_{s=0},$$

$$(3.31)\qquad W(c) = -\frac{1}{n}\log N(\mathfrak{a}_\mu \mathfrak{f}) \sum_{j\in J} \zeta^{(1)}(0, C_j, \mathfrak{a}_\mu).$$

3.6. Let us first examine the dependence of $X(c; \{C_j\}, \mathfrak{a}_\mu)$ on $\{C_j\}_{j\in J}$. By Lemma 3.13, it suffices to consider the following two operations.

(I) Replace $\{C_j\}_{j\in J}$ by a refinement $\{C'_l\}_{l\in L}$ of $\{C_j\}_{j\in J}$.

(II) Replace a member C_j of $\{C_j\}_{j\in J}$ by ϵC_j with $\epsilon \in E_F^+$.

The formulas (3.28) $\sim$ (3.31) show that $G(c)$, $V(c)$ and $W(c)$ do not change under (I). Let us consider the operation (II). Clearly we have

$$(3.32)\qquad \zeta^{(i)}(s, \epsilon C_j, \mathfrak{a}_\mu) = (\epsilon^{(i)})^{-s} \zeta^{(i)}(s, C_j, \mathfrak{a}_\mu),$$

$$(3.33)\qquad \zeta^{(i,k)}(s, \epsilon C_j, \mathfrak{a}_\mu) = (\epsilon^{(i)}\epsilon^{(k)})^{-s} \zeta^{(i,k)}(s, C_j, \mathfrak{a}_\mu).$$

Hence $G(c)$ changes to $G(c) - \log \epsilon^{(1)}\zeta^{(1)}(0, C_j, \mathfrak{a}_\mu)$ and $W(c)$ does not change. When i is fixed, using (3.27), we see that $\sum_{1\leqq k\leqq n, k\neq i} v_{ik}^{(j)}$ changes to

$$\begin{aligned}
&\sum_{1\leqq k\leqq n, k\neq i} v_{ik}^{(j)} + \frac{1}{2}\sum_{1\leqq k\leqq n, k\neq i} (\log \epsilon^{(i)}\zeta^{(i)}(0, C_j, \mathfrak{a}_\mu) + \log \epsilon^{(k)}\zeta^{(k)}(0, C_j, \mathfrak{a}_\mu) \\
&\qquad\qquad - \log \epsilon^{(i)}\zeta^{(k)}(0, C_j, \mathfrak{a}_\mu) - \log \epsilon^{(k)}\zeta^{(i)}(0, C_j, \mathfrak{a}_\mu)) \\
&= \sum_{1\leqq k\leqq n, k\neq i} v_{ik}^{(j)} + \frac{1}{2}\Bigg[n \log \epsilon^{(i)}\zeta^{(i)}(0, C_j, \mathfrak{a}_\mu) + \sum_{k=1}^{n} \log \epsilon^{(k)}\zeta^{(k)}(0, C_j, \mathfrak{a}_\mu) \\
&\qquad\qquad - \log \epsilon^{(i)} \sum_{k=1}^{n} \zeta^{(k)}(0, C_j, \mathfrak{a}_\mu)\Bigg].
\end{aligned}$$

Hence $V(c)$ changes to

$$\begin{aligned}
&V(c) + \log \epsilon^{(1)}\zeta^{(1)}(0, C_j, \mathfrak{a}_\mu) + \frac{1}{n}\sum_{k=1}^{n} \log \epsilon^{(k)}\zeta^{(k)}(0, C_j, \mathfrak{a}_\mu) \\
&- \frac{1}{n}\log \epsilon^{(1)} \sum_{k=1}^{n} \zeta^{(k)}(0, C_j, \mathfrak{a}_\mu) - \frac{1}{2n}\sum_{i=1}^{n}\Bigg[\log \epsilon^{(i)}\zeta^{(i)}(0, C_j, \mathfrak{a}_\mu) \\
&+ \frac{1}{n}\sum_{k=1}^{n} \log \epsilon^{(k)}\zeta^{(k)}(0, C_j, \mathfrak{a}_\mu) - \frac{1}{n}\log \epsilon^{(i)} \sum_{k=1}^{n} \zeta^{(k)}(0, C_j, \mathfrak{a}_\mu)\Bigg] \\
&= V(c) + \log \epsilon^{(1)}\zeta^{(1)}(0, C_j, \mathfrak{a}_\mu) - \frac{1}{n}\log \epsilon^{(1)} \sum_{k=1}^{n} \zeta^{(k)}(0, C_j, \mathfrak{a}_\mu).
\end{aligned}$$

We summalize this result as

$$(3.34)\qquad \begin{aligned}
&V(c; \{C_{i, i\neq j}, \epsilon C_j\}, \mathfrak{a}_\mu) \\
&= V(c; \{C_j\}, \mathfrak{a}_\mu) + \log \epsilon^{(1)}\zeta^{(1)}(0, C_j, \mathfrak{a}_\mu) - \frac{1}{n}\log \epsilon^{(1)} \sum_{k=1}^{n} \zeta^{(k)}(0, C_j, \mathfrak{a}_\mu),
\end{aligned}$$

$$X(c;\{C_{i,i\neq j},\epsilon C_j\},\mathfrak{a}_\mu)=X(c;\{C_j\},\mathfrak{a}_\mu)-\frac{1}{n}\log\epsilon^{(1)}\sum_{k=1}^{n}\zeta^{(k)}(0,C_j,\mathfrak{a}_\mu). \tag{3.35}$$

Put

$$\zeta(s,C_j,\mathfrak{a}_\mu)=\sum_{z\in(\mathfrak{a}_\mu\mathfrak{f})^{-1}\cap C_j,(z)\mathfrak{a}_\mu\mathfrak{f}\equiv c}N(z)^{-s}. \tag{3.36}$$

By II, Lemma 3.2, we have

$$\zeta(s,C_j,\mathfrak{a}_\mu)=\sum_{z\in R(C_j,c)}\zeta(s,A_j,x(z)).$$

Then by I, Theorem 2.1, we see that

$$\zeta(0,C_j,\mathfrak{a}_\mu)=\frac{1}{n}\sum_{k=1}^{n}\zeta^{(k)}(s,C_j,\mathfrak{a}_\mu)\in\mathbf{Q}. \tag{3.37}$$

By (3.10) and (3.35), we obtain

$$g_K(\mathrm{id},\tau;\{C_{i,i\neq j},\epsilon C_j\},\mathfrak{a}_\mu)=(\epsilon^{(1)})^{-a}g_K(\mathrm{id},\tau;\{C_j\},\mathfrak{a}_\mu), \tag{3.38}$$

$$a=\frac{1}{|G|}\sum_{\mathfrak{f}|\tilde{\mathfrak{f}}}\sum_{\omega\in(\hat{G}_-)_{\mathfrak{f}}}\frac{\omega(\tau)}{L(0,\omega)}\sum_{c\in C_{\mathfrak{f}}}\omega(c)\zeta(0,C_j,\mathfrak{a}_\mu). \tag{3.39}$$

For $\omega\in\widehat{G}$ and $\sigma\in\mathrm{Aut}(\mathbf{C})$, put $\omega^\sigma(g)=\omega(g)^\sigma$, $g\in G$. If $\omega\in(\widehat{G}_-)_{\mathfrak{f}}$, then $\omega^\sigma\in(\widehat{G}_-)_{\mathfrak{f}}$ and $L(0,\omega)^\sigma=L(0,\omega^\sigma)$. Hence we see that $a\in\mathbf{Q}$. Therefore the validity of Conjecture 3.9 does not depend on the choice of a cone decomposition. By (3.13) and (3.35), we obtain

$$\begin{aligned}&g_{K^\sigma}(\mathrm{id},\sigma^{-1}\tau\sigma;\{C^\sigma_{i,i\neq j},(\epsilon C_j)^\sigma\},\{\mathfrak{a}^\sigma_\mu\})\\&=(\epsilon^\sigma)^{-a}g_{K^\sigma}(\mathrm{id},\sigma^{-1}\tau\sigma;\{C^\sigma_j\},\{\mathfrak{a}^\sigma_\mu\}),\end{aligned} \tag{3.40}$$

since $\zeta(0,C^\sigma_j,\mathfrak{a}^\sigma_\mu)=\zeta(0,C_j,\mathfrak{a}_\mu)$. By (3.38) and (3.40), we see that the validity of Conjecture 3.10 does not depend on the choice of a cone decomposition. When $r=1$, $\sigma_1=\mathrm{id}$, $\tau_1=\tau$, the reciprocity map $\sigma\longrightarrow\zeta(\sigma)$ changes to

$$\sigma\longrightarrow\zeta(\sigma)\cdot\frac{(\epsilon^{aa_1})^\sigma}{(\epsilon^\sigma)^{aa_1}},\qquad\sigma\in\mathrm{Gal}(\overline{\mathbf{Q}}/\mathbf{Q}). \tag{3.41}$$

3.7. Now suppose that a representative $\mathfrak{a}_\mu$ for a narrow ideal class is replaced by $(\alpha)\mathfrak{a}_\mu$ with $0\ll\alpha\in F$. Since we have examined the change of a cone decomposition already, it suffices (to establish the consistency of Conjectures) to consider the simultaneous transformation

(III) $\mathfrak{a}_\mu\longrightarrow(\alpha)\mathfrak{a}_\mu$, $\quad\{C_j\}_{j\in J}\longrightarrow\{\alpha^{-1}C_j\}_{j\in J}$.

We note that the cone $\alpha^{-1}C_j$ has a basis in $\mathfrak{O}_F$. First suppose that c and $\mathfrak{a}_\mu\mathfrak{f}$ belong to the same narrow ideal class. Since the mapping

$$\begin{aligned}&\{z\mid z\in((\alpha)\mathfrak{a}_\mu\mathfrak{f})^{-1}\cap\alpha^{-1}C_j,(z)(\alpha)\mathfrak{a}_\mu\mathfrak{f}\equiv c\}\ni z\\&\longrightarrow\alpha z\in\{z\mid z\in(\mathfrak{a}_\mu\mathfrak{f})^{-1}\cap C_j,(z)\mathfrak{a}_\mu\mathfrak{f}\equiv c\}\end{aligned}$$

is a bijection, we have

$$\zeta^{(i)}(s,\alpha^{-1}C_j,(\alpha)\mathfrak{a}_\mu)=(\alpha^{(i)})^s\zeta^{(i)}(s,C_j,(\alpha)\mathfrak{a}_\mu), \tag{3.42}$$

$$\zeta^{(i,k)}(s,\alpha^{-1}C_j,(\alpha)\mathfrak{a}_\mu)=(\alpha^{(i)}\alpha^{(k)})^s\zeta^{(i,k)}(s,C_j,(\alpha)\mathfrak{a}_\mu). \tag{3.43}$$

Hence we see that $G(c)$ changes to

$$G(c)+\sum_{j\in J}\log\alpha^{(1)}\zeta^{(1)}(0,C_j,\mathfrak{a}_\mu)$$

and $W(c)$ changes to

$$W(c)-\frac{1}{n}\sum_{j\in J}\log N(\alpha)\zeta^{(1)}(0,C_j,\mathfrak{a}_\mu).$$

When i is fixed, using (3.27), we see that $\sum_{1\leqq k\leqq n,k\neq i}v_{ik}^{(j)}$ changes to

$$\sum_{1\leqq k\leqq n,k\neq i}v_{ik}^{(j)}+\frac{1}{2}\Bigg[-n\log\alpha^{(i)}\zeta^{(i)}(0,C_j,\mathfrak{a}_\mu)-\sum_{k=1}^{n}\log\alpha^{(k)}\zeta^{(k)}(0,C_j,\mathfrak{a}_\mu)$$
$$+\log\alpha^{(i)}\sum_{k=1}^{n}\zeta^{(k)}(0,C_j,\mathfrak{a}_\mu)+\log N(\alpha)\zeta^{(i)}(0,C_j,\mathfrak{a}_\mu)\Bigg].$$

Hence we obtain

$$\begin{aligned}&V(c;\{\alpha^{-1}C_j\},(\alpha)\mathfrak{a}_\mu)=V(c;\{C_j\},(\alpha)\mathfrak{a}_\mu)\\&-\sum_{j\in J}\Bigg[(\log\alpha^{(1)}-\frac{1}{n}\log N(\alpha))(\zeta^{(1)}(0,C_j,\mathfrak{a}_\mu)-\frac{1}{n}\sum_{k=1}^{n}\zeta^{(k)}(0,C_j,\mathfrak{a}_\mu)\Bigg].\end{aligned} \tag{3.44}$$

Since $\zeta_F(0,c)=\frac{1}{n}\sum_{j\in J}\sum_{k=1}^{n}\zeta^{(k)}(0,C_j,\mathfrak{a}_\mu)$, we get

$$X(c;\{\alpha^{-1}C_j\},(\alpha)\mathfrak{a}_\mu)=X(c;\{C_j\},\mathfrak{a}_\mu)+(\log\alpha^{(1)}-\frac{1}{n}\log N(\alpha))\zeta_F(0,c). \tag{3.45}$$

For $c\in C_\mathfrak{f}$, we write $c\sim\mathfrak{a}_\nu$ (resp. $c\nsim\mathfrak{a}_\nu$) if c and $\mathfrak{a}_\nu\mathfrak{f}$ belong (resp. do not belong) to the same narrow ideal class. For $c\sim\mathfrak{a}_\mu$, (3.45) describes the change of $X(c)$ under the transformation (III). For a class $c\sim\mathfrak{a}_\nu$, $\nu\neq\mu$, we see that we can write

$$X(c;\{\alpha^{-1}C_j\},\mathfrak{a}_\nu)=X(c;\{C_j\},\mathfrak{a}_\nu)+\sum_{i=1}^{n-1}b_i(c)\log\epsilon_i^{(1)} \tag{3.46}$$

with $b_i(c)\in\mathbf{Q}$, $\epsilon_i\in E_F^+$, by Lemma 3.13 and (3.35), and also that

$$X(c^\sigma;\{(\alpha^\sigma)^{-1}C_j^\sigma\},\mathfrak{a}_\nu^\sigma)=X(c^\sigma;\{C_j^\sigma\},\mathfrak{a}_\nu^\sigma)+\sum_{i=1}^{n-1}b_i(c)\log\epsilon_i^\sigma \tag{3.47}$$

for $\sigma\in\mathrm{Gal}(\overline{\mathbf{Q}}/\mathbf{Q})$. Here in (3.46) and (3.47), $\epsilon_1,\epsilon_2,\ldots,\epsilon_{n-1}$ denote generators of E_F^+. Put

$$b_i=\frac{1}{|G|}\sum_{\mathfrak{f}|\tilde{\mathfrak{f}}}\sum_{\omega\in(\hat{G}_-)_\mathfrak{f}}\frac{\omega(\tau)}{L(0,\omega)}\sum_{c\in C_\mathfrak{f},c\nsim\mathfrak{a}_\mu}\omega(c)b_i(c) \tag{3.48}$$

for $1\leqq i\leqq n-1$,

$$b=\frac{1}{|G|}\sum_{\mathfrak{f}|\tilde{\mathfrak{f}}}\sum_{\omega\in(\hat{G}_-)_\mathfrak{f}}\frac{\omega(\tau)}{L(0,\omega)}\sum_{c\in C_\mathfrak{f},c\sim\mathfrak{a}_\mu}\omega(c)\zeta_F(0,c). \tag{3.49}$$

By the same argument below (3.39), we see that b_i, $b \in \mathbf{Q}$. By (3.10), (3.45) and (3.46), we obtain

$$\begin{aligned} &g_K(\mathrm{id}, \tau; \{\alpha^{-1}C_j\}, \{(\alpha)\mathfrak{a}_\mu, \mathfrak{a}_{\nu,\nu\neq\mu}\}) \\ &= \prod_{i=1}^{n-1} (\epsilon_i^{(1)})^{b_i} (\alpha^{(1)} N(\alpha)^{-1/n})^b g_K(\mathrm{id}, \tau; \{C_j\}, \{\mathfrak{a}_\mu\}). \end{aligned} \tag{3.50}$$

Therefore the validity of Conjecture 3.9 does not depend on the choice of representatives $\{\mathfrak{a}_\mu\}$. By (3.13), (3.45) and (3.47), we obtain

$$\begin{aligned} &g_{K^\sigma}(\mathrm{id}, \sigma^{-1}\tau\sigma; \{(\alpha^\sigma)^{-1}C_j^\sigma\}, \{(\alpha^\sigma)\mathfrak{a}_\mu^\sigma, \mathfrak{a}_{\nu^\sigma,\nu\neq\mu}\}) \\ &= \prod_{i=1}^{n-1} (\epsilon_i^\sigma)^{b_i} (\alpha^\sigma N(\alpha)^{-1/n})^b g_{K^\sigma}(\mathrm{id}, \sigma^{-1}\tau\sigma; \{C_j^\sigma\}, \{\mathfrak{a}_\mu^\sigma\}). \end{aligned} \tag{3.51}$$

By (3.50) and (3.51), we see that the validity of Conjecture 3.10 does not depend on the choice of representatives $\{\mathfrak{a}_\nu\}$. When $r = 1$, $\sigma_1 = \mathrm{id}$, $\tau_1 = \tau$, the reciprocity map changes to

$$\sigma \longrightarrow \zeta(\sigma) \cdot \frac{\prod_{i=1}^{n-1} (\epsilon_i^\sigma)^{b_i a_1}}{\prod_{i=1}^{n-1} (\epsilon_i^{b_i a_1})^\sigma} \cdot \frac{(\alpha^\sigma N(\alpha)^{-1/n})^{b a_1}}{((\alpha N(\alpha)^{-1/n})^{b a_1})^\sigma}. \tag{3.52}$$

3.8. Strictly speaking, $g_K(\mathrm{id}, \tau; \{C_j\}, \{\mathfrak{a}_\mu\})$ depends also on F. Focusing our attention on the dependence of F, let us write this quantity as $g_{K/F}(\mathrm{id}, \tau; \{C_j\}, \{\mathfrak{a}_\mu\})$ or as $g_{K/F}(\mathrm{id}, \tau)$ when no confusion is likely. Now it can happen that a CM-field K is abelian over two distinct totally real fields F_1 and F_2. Then Conjecture 3.9 implies that

$$g_{K/F_1}(\mathrm{id}, \tau) \sim g_{K/F_2}(\mathrm{id}, \tau) \qquad \text{if} \quad \tau \in \mathrm{Gal}(K/F_1) \cap \mathrm{Gal}(K/F_2). \tag{G1}$$

In this subsection, we will investigate (G1) and the problem to define $g_K(\xi, \eta)$ for a CM-field K and ξ, $\eta \in I_K$.

First we consider (G1).

LEMMA 3.14. *Let F be a totally real algebraic number field. We put $[F : \mathbf{Q}] = n$, $J_F = \{\sigma_1, \sigma_2, \dots, \sigma_n\}$. Let K be a CM-field which is abelian over F. Let L be the normal closure of K over $\mathbf{Q}$. We put*

$$\mathfrak{G} = \mathrm{Gal}(L/\mathbf{Q}), \qquad G = \mathrm{Gal}(K/F), \qquad H = \mathrm{Gal}(L/K).$$

Then we have

$$\begin{aligned} &\prod_{i=1}^{n} g_{K^{\sigma_i}/F^{\sigma_i}}(\mathrm{id}, \sigma_i^{-1}\tau\sigma_i; \{C_j^{\sigma_i}\}, \{\mathfrak{a}_\mu^{\sigma_i}\}) \\ &= \pi^{-n\mu(\tau)/2} \exp\Big(\frac{n}{|\mathfrak{G}|} \sum_{\eta \in \hat{\mathfrak{G}}_-} \sum_{h \in H} \chi_\eta(h\tilde{\tau}) \frac{L'(0,\eta)}{L(0,\eta)}\Big) \qquad \text{for} \quad \tau \in G. \end{aligned} \tag{3.53}$$

Here $\tilde{\tau}$ is any element of $\mathrm{Gal}(L/F)$ *which satisfies $\tilde{\tau}|K = \tau$.*

PROOF. Let P denote the left-hand side of (3.53). By Lemma 3.8, we have

$$P=\pi^{-n\mu(\tau)/2}\exp(\frac{1}{|G|}\sum_{\omega\in\hat{G}_-}\omega(\tau)\frac{L'(0,\omega)}{L(0,\omega)}).$$

By Lemmas 3.5 and 3.4, we get

$$\begin{aligned}P&=\pi^{-n\mu(\tau)/2}\exp(\frac{1}{|G||H|}\sum_{\eta\in\widehat{\mathrm{Gal}(L/F)}_-}\sum_{h\in H}\chi_\eta(h\tilde{\tau})\frac{L'(0,\eta)}{L(0,\eta)})\\&=\pi^{-n\mu(\tau)/2}\exp(\frac{1}{|G||H|}\sum_{\eta\in\hat{\mathfrak{G}}_-}\sum_{h\in H}\chi_\eta(h\tilde{\tau})\frac{L'(0,\eta)}{L(0,\eta)}).\end{aligned}$$

Hence the assertion follows.

The right-hand side of (3.53) is the n-th power of the quantity which depends only on K and τ. Therefore we obtain:

COROLLARY 3.15. *Let F_1 and F_2 be totally real algebraic number fields and K be a CM-field. We assume that K is abelian over F_1 and also over F_2. Put $n_l=[F_l:\mathbf{Q}]$, $J_{F_l}=\{\sigma_1^{(l)},\sigma_2^{(l)},\dots,\sigma_{n_l}^{(l)}\}$, $l=1,2$. Let $\{C_j\}$ and $\{\mathfrak{a}_\mu\}$ (resp. $\{D_k\}$ and $\{\mathfrak{b}_\nu\}$) be data for a cone decomposition and representatives of narrow ideal classes of F_1 (resp. F_2). Then we have*

$$\begin{aligned}&(\prod_{i=1}^{n_1}g_{K^{\sigma_i^{(1)}}/F_1^{\sigma_i^{(1)}}}(\mathrm{id},(\sigma_i^{(1)})^{-1}\tau\sigma_i^{(1)};\{C_j^{\sigma_i^{(1)}}\},\{\mathfrak{a}_\mu^{\sigma_i^{(1)}}\})^{n_2}\\&=(\prod_{i=1}^{n_2}g_{K^{\sigma_i^{(2)}}/F_2^{\sigma_i^{(2)}}}(\mathrm{id},(\sigma_i^{(2)})^{-1}\tau\sigma_i^{(2)};\{D_k^{\sigma_i^{(2)}}\},\{\mathfrak{b}_\nu^{\sigma_i^{(2)}}\})^{n_1}\end{aligned}$$

for $\tau\in\mathrm{Gal}(K/F_1)\cap\mathrm{Gal}(K/F_2)=\mathrm{Gal}(K/(F_1\vee F_2))$.

Clearly if (G1) holds for the case $F_1\subset F_2$, it holds in general.

PROPOSITION 3.16. *Let F_1 and F_2 be totally real algebraic number fields and K be a CM-field. We assume that $F_1\subset F_2\subset K$ and that K is abelian over F_1. Put $n=[F_1:\mathbf{Q}]$, $J_{F_1}=\{\sigma_1,\sigma_2,\dots,\sigma_n\}$. Then we have*

$$\prod_{i=1}^{n}g_{K^{\sigma_i}/F_1^{\sigma_i}}(\mathrm{id},\sigma_i^{-1}\tau\sigma_i)\sim\prod_{i=1}^{n}g_{K^{\sigma_i}/F_2^{\sigma_i}}(\mathrm{id},\sigma_i^{-1}\tau\sigma_i),\qquad\tau\in\mathrm{Gal}(K/F_2).$$

This proposition follows from Corollary 3.15 since $g_K(\mathrm{id},\tau;\{C_j\},\{\mathfrak{a}_\mu\})$ considered mod $\overline{\mathbf{Q}}^\times$ does not depend on $\{C_j\}$ and $\{\mathfrak{a}_\mu\}$.

COROLLARY 3.17. *In Proposition 3.16, we assume that K, F_1 and F_2 are normal over $\mathbf{Q}$. Then, if $\tau\in\mathrm{Gal}(K/F_2)$ belongs to the center of $\mathrm{Gal}(K/\mathbf{Q})$, we have*

$$g_{K/F_1}(\mathrm{id},\tau)\sim g_{K/F_2}(\mathrm{id},\tau).$$

In particular, (G1) holds when K is abelian over $\mathbf{Q}$. Using Corollary 3.17, we can prove (G1) in a more general situation.

LEMMA 3.18. *Let F_1, F_2 and K be fields in a fixed algebraically closed field such that K is a Galois extension of finite degree over both of F_1 and F_2. We assume that K is a separable extension of finite degree over the prime field. Then K is a Galois extension of $F_1 \cap F_2$.*

PROOF. Let L be the normal closure of K over $F_1 \cap F_2$. Put

$$G = \mathrm{Gal}(L/F_1 \cap F_2), \qquad H = \mathrm{Gal}(L/K), \qquad H_i = \mathrm{Gal}(L/F_i), \quad i = 1, 2.$$

Then H_i normalizes H for $i = 1$, 2 and G is generated by H_1 and H_2. Therefore G normalizes H and this implies that K is normal over $F_1 \cap F_2$. This completes the proof.

PROPOSITION 3.19. *Let F_1 and F_2 be totally real algebraic number fields and K be a CM-field. We assume that K is abelian over both of F_1 and F_2. We further assume that F_1 and F_2 are normal over $\mathbf{Q}$ and that $F_1 \cap F_2 = \mathbf{Q}$. Then we have*

$$g_{K/F_1}(\mathrm{id}, \tau) \sim g_{K/F_2}(\mathrm{id}, \tau) \qquad \textit{for} \quad \tau \in \mathrm{Gal}(K/F_1) \cap \mathrm{Gal}(K/F_2).$$

PROOF. By Lemma 3.18, K is normal over $\mathbf{Q}$. Put $G = \mathrm{Gal}(K/\mathbf{Q})$, $H_i = \mathrm{Gal}(K/F_i)$, $i = 1$, 2. We apply Corollary 3.17 for pairs $(F_1, F_1 \vee F_2)$, $(F_2, F_1 \vee F_2)$. It suffices to show that $H_1 \cap H_2 = \mathrm{Gal}(K/F_1 \vee F_2)$ is contained in the center of G. This is clear since $G = H_1 H_2$ and H_i is abelian, $i = 1$, 2. This completes the proof.

Next we are going to consider the problem to define the symbol $g_K(\xi, \eta)$, ξ, $\eta \in I_K$, well defined modulo $\overline{\mathbf{Q}}^\times$, which satisfies the functoriality property of the same type as for $p_K(\xi, \eta)$. We assume (G1). First for a CM-field K which is abelian over a totally real algebraic number field F, we define

$$g_K(\mathrm{id}, \tau) = g_{K/F}(\mathrm{id}, \tau; \{C_j\}, \{\mathfrak{a}_\mu\}), \qquad \tau \in \mathrm{Gal}(K/F). \tag{3.54}$$

When F is fixed, $g_K(\mathrm{id}, \tau) \bmod \overline{\mathbf{Q}}^\times$ is well defined, and by the assumption (G1), its class mod $\overline{\mathbf{Q}}^\times$ does not depend on F. For $\xi = \sum_{i=1}^m a_i \tau_i \in I_K$, $a_i \in \mathbf{Z}$, $\tau_i \in J_K$, we define

$$g_K(\mathrm{id}, \sum_{i=1}^m a_i \tau_i) = \prod_{i=1}^m g_K(\mathrm{id}, \tau_i)^{a_i}. \tag{3.55}$$

Let L be the normal closure of K over $\mathbf{Q}$. We identify J_L with $\mathrm{Gal}(L/\mathbf{Q})$. For $\tau \in J_L$, $g_L(\mathrm{id}, \tau)$ can be defined, since $\tau \in \mathrm{Gal}(L/F)$ when we take F as the fixed field of $\langle \rho, \tau \rangle$. Hence $g_L(\mathrm{id}, \xi)$ is defined for $\xi \in I_L$ by (3.55). Let σ, $\tau \in J_K$. Take $\tilde{\sigma} \in J_L$ so that $\tilde{\sigma}|K = \sigma$. We define

$$g_K(\sigma, \tau) = g_L(\mathrm{id}, \tilde{\sigma}^{-1} \mathrm{Inf}_{L/K}(\tau)). \tag{3.56}$$

To assure the compatibility with the definition (3.54), we assume:

(G2) Let K be a CM-field which is abelian over a totally real algebraic number field F. Let L be a CM-field which contains K and which is normal over $\mathbf{Q}$. Then
$$g_K(\mathrm{id}, \tau) \sim g_L(\mathrm{id}, \mathrm{Inf}_{L/K}(\tau)) \qquad \text{for} \quad \tau \in \mathrm{Gal}(K/F).$$

Since $\tilde{\sigma}^{-1}\mathrm{Inf}_{L/K}(\tau)$ does not depend on the choice of $\tilde{\sigma}$, $g_K(\sigma,\tau) \mod \overline{\mathbf{Q}}^\times$ is well defined by (3.56). For $\xi = \sum_{i=1}^{l} a_i\sigma_i \in I_K$, $\eta = \sum_{j=1}^{m} b_j\tau_j \in I_K$, a_i, $b_j \in \mathbf{Z}$, σ_i, $\tau_j \in J_K$, we define

$$g_K(\xi,\eta) = \prod_{i=1}^{l}\prod_{j=1}^{m} g_K(\sigma_i,\tau_j)^{a_ib_j}. \tag{3.57}$$

Now it is clear that $g_K(\xi,\eta) \mod \overline{\mathbf{Q}}^\times$ is well defined under the assumptions (G1) and (G2).

Under these assumptions, we are going to show that the symbol g_K satisfies similar properties to those are described by (2) $\sim$ (6) of Theorem 1.2 due to Shimura. The property (2) is obviously satisfied. Let K and K' be CM-fields and γ be an isomorphism of K' onto K. Let us prove that

$$g_{K'}(\gamma\xi,\gamma\eta) \sim g_K(\xi,\eta), \qquad \xi,\eta \in I_K. \tag{3.58}$$

We denote by L the same normal closure of K and K' over $\mathbf{Q}$. We use the same letter γ for an element of $\mathrm{Gal}(L/\mathbf{Q})$ whose restrictions to K' gives γ. It suffices to prove (3.58) when $\xi = \sigma$, $\eta = \tau$, σ, $\tau \in J_K$. Take $\tilde{\sigma} \in J_L$ so that $\tilde{\sigma}|K = \sigma$. Then $\gamma\tilde{\sigma}|K' = \gamma\sigma$. Since

$$\mathrm{Inf}_{L/K'}(\gamma\tau) = \gamma\mathrm{Inf}_{L/K}(\tau),$$

by (3.56), we have

$$g_{K'}(\gamma\sigma,\gamma\tau) = g_L(\mathrm{id},(\gamma\tilde{\sigma})^{-1}\mathrm{Inf}_{L/K'}(\gamma\tau)) = g_L(\mathrm{id},\tilde{\sigma}^{-1}\mathrm{Inf}_{L/K}(\tau)) = g_K(\sigma,\tau).$$

Hence (3.58) is proved. By the definition (3.10), we have

$$g_{K/F}(\mathrm{id},\tau;\{C_j\},\{\mathfrak{a}_\mu\})g_{K/F}(\mathrm{id},\tau\rho;\{C_j\},\{\mathfrak{a}_\mu\}) = 1, \qquad \tau \in \mathrm{Gal}(K/F), \tag{3.59}$$

noting $\mu(\tau)+\mu(\tau\rho) = 0$. It follows that

$$g_K(\xi\rho,\eta) \sim g_K(\xi,\eta\rho) \sim g_K(\xi,\eta)^{-1} \qquad \text{for} \quad \xi,\eta \in I_K. \tag{3.60}$$

Let K and L be CM-fields such that $K \subset L$. Let us show that

$$g_K(\xi,\mathrm{Res}_{L/K}(\zeta)) \sim g_L(\mathrm{Inf}_{L/K}(\xi),\zeta), \qquad \xi \in I_K, \zeta \in I_L, \tag{3.61}$$

$$g_K(\mathrm{Res}_{L/K}(\zeta),\xi) \sim g_L(\zeta,\mathrm{Inf}_{L/K}(\xi)), \qquad \xi \in I_K, \zeta \in I_L \tag{3.62}$$

assuming (G2). Let K' and L' be the normal closures over $\mathbf{Q}$ of K and L respectively. It suffices to prove (3.61) and (3.62) for $\xi = \sigma \in J_K$, $\zeta = \tau \in J_L$. Take $\tilde{\sigma} \in J_{K'}$ (resp. $\tilde{\tau} \in J_{L'}$) so that $\tilde{\sigma}|K = \sigma$ (resp. $\tilde{\tau}|L = \tau$). By the definition (3.56), we have

$$g_K(\mathrm{Res}_{L/K}(\tau),\sigma) = g_{K'}(\mathrm{id},(\tilde{\tau}|K')^{-1}\mathrm{Inf}_{K'/K}(\sigma)),$$
$$g_L(\tau,\mathrm{Inf}_{L/K}(\sigma)) = g_{L'}(\mathrm{id},\tilde{\tau}^{-1}\mathrm{Inf}_{L'/K}(\sigma)).$$

Since

$$\mathrm{Inf}_{L'/K'}((\tilde{\tau}|K')^{-1}\mathrm{Inf}_{K'/K}(\sigma)) = \tilde{\tau}^{-1}\mathrm{Inf}_{L'/K}(\sigma),$$

(3.62) follows from (G2) applied for $K = K'$, $L = L'$. Take $\tilde{\tilde{\sigma}} \in J_{L'}$ so that $\tilde{\tilde{\sigma}}|K' = \tilde{\sigma}$. Put $\mathrm{Inf}_{L/K} = \sigma_1 + \sigma_2 + \cdots + \sigma_l$, $\sigma_i \in J_L$ and take $\tilde{\sigma}_i \in J_{L'}$ so that $\tilde{\sigma}_i|L = \sigma_i$. By the definition (3.56) and (G2), we have

$$\begin{aligned} g_K(\sigma, \mathrm{Res}_{L/K}(\tau)) &= g_{K'}(\mathrm{id}, \tilde{\sigma}^{-1}\mathrm{Inf}_{K'/K}(\mathrm{Res}_{L/K}(\tau))) \\ &= g_{L'}(\mathrm{id}, \tilde{\tilde{\sigma}}^{-1}\mathrm{Inf}_{L'/K}(\mathrm{Res}_{L/K}(\tau))), \\ g_L(\mathrm{Inf}_{L/K}(\sigma), \tau) &= g_{L'}(\mathrm{id}, \sum_{i=1}^{l} \tilde{\sigma}_i^{-1}\mathrm{Inf}_{L'/L}(\tau)). \end{aligned}$$

Therefore it suffices to show

$$\tilde{\tilde{\sigma}}^{-1}\mathrm{Inf}_{L'/K}(\mathrm{Res}_{L/K}(\tau)) = \sum_{i=1}^{l} \tilde{\sigma}_i^{-1}\mathrm{Inf}_{L'/L}(\tau). \tag{3.63}$$

Put

$$G = \mathrm{Gal}(L'/K), \qquad H = \mathrm{Gal}(L'/L).$$

We have

$$\mathrm{Inf}_{L'/K}(\mathrm{Res}_{L/K}(\tau)) = \sum_{g \in G} g\tilde{\tau}, \qquad \mathrm{Inf}_{L'/L}(\tau) = \sum_{h \in H} h\tilde{\tau},$$

$$\tilde{\sigma}_1 + \tilde{\sigma}_2 + \cdots + \tilde{\sigma}_l = \sum_{g \in H\backslash G} g\tilde{\tilde{\sigma}}.$$

Hence we get

$$\begin{aligned} \sum_{i=1}^{l} \tilde{\sigma}_i^{-1}\mathrm{Inf}_{L'/L}(\tau) &= \sum_{g \in H\backslash G} (g\tilde{\tilde{\sigma}})^{-1} \sum_{h \in H} h\tilde{\tau} = \sum_{g \in H\backslash G} \sum_{h \in H} \tilde{\tilde{\sigma}}^{-1} g^{-1} h\tilde{\tau} \\ &= \sum_{g \in G} \tilde{\tilde{\sigma}}^{-1} g\tilde{\tau} = \tilde{\tilde{\sigma}}^{-1}\mathrm{Inf}_{L'/K}(\mathrm{Res}_{L/K}(\tau)). \end{aligned}$$

This proves (3.61) assuming (G2). The next proposition gives a partial answer to (G2).

PROPOSITION 3.20. *Let L be a CM-field which is abelian over a totally real algebraic number field F. Let K be a CM-field such that $F \subset K \subset L$. Then we have*

$$g_{K/F}(\mathrm{id}, \tau; \{C_j\}, \{\mathfrak{a}_\mu\}) = g_{L/F}(\mathrm{id}, \mathrm{Inf}_{L/K}(\tau); \{C_j\}, \{\mathfrak{a}_\mu\}) \qquad \textit{for} \quad \tau \in \mathrm{Gal}(K/F).$$

PROOF. Put

$$\tilde{G} = \mathrm{Gal}(L/F), \qquad G = \mathrm{Gal}(K/F), \qquad H = \mathrm{Gal}(L/K).$$

Take $\tilde{\tau} \in \tilde{G}$ so that $\tau = \tilde{\tau}|K$. Then $\mathrm{Inf}_{L/K}(\tau) = \sum_{h \in H} h\tilde{\tau}$. Let $\tilde{\mathfrak{h}}$ (resp. $\tilde{\mathfrak{f}}$) be the finite part of the conductor of L (resp. K) as a class field over F. By definition, we have

$$\begin{aligned} &g_{L/F}(\mathrm{id}, \mathrm{Inf}_{L/K}(\tau); \{C_j\}, \{\mathfrak{a}_\mu\}) \\ &= \pi^{-\mu(\tau)/2} \exp\Big(\frac{1}{|G|} \sum_{\mathfrak{h}|\tilde{\mathfrak{h}}} \sum_{\omega \in (\hat{\tilde{G}}_-)_{\mathfrak{h}}} \frac{\sum_{h \in H} \omega(h\tilde{\tau})}{L(0,\omega)} \sum_{c \in C_{\mathfrak{h}}} \omega(c) X(c)\Big). \end{aligned}$$

Here we abbreviated $X(c;\{C_j\},\mathfrak{a}_\mu)$ to $X(c)$. Take $\omega \in (\hat{\tilde{G}}_-)_{\mathfrak{h}}$. If $\omega|H$ is non-trivial, then we have $\sum_{h\in H}\omega(h\tilde{\tau})=0$. If $\omega|H$ is trivial, we have $\sum_{h\in H}\omega(h\tilde{\tau})=|H|\omega(\tilde{\tau})$. In this case, ω can be identified with a character of G. Therefore we obtain

$$\begin{aligned}&g_{L/F}(\mathrm{id},\mathrm{Inf}_{L/K}(\tau);\{C_j\},\{\mathfrak{a}_\mu\})\\ &=\pi^{-\mu(\tau)/2}\exp(\frac{|H|}{|G|}\sum_{\mathfrak{f}|\tilde{\mathfrak{f}}}\sum_{\omega\in(\hat{G}_-)_{\mathfrak{f}}}\frac{\omega(\tau)}{L(0,\omega)}\sum_{c\in C_{\mathfrak{f}}}\omega(c)X(c))=g_{K/F}(\mathrm{id},\tau;\{C_j\},\{\mathfrak{a}_\mu\}).\end{aligned}$$

This completes the proof.

4. Numerical examples

4.1. First we consider the case where F is a real quadratic field. For $z\in F$, we write $z^{(1)}=z$, $z^{(2)}=z'$. By ϵ, we denote the generator of E_F^+ such that $\epsilon>1$. We can obtain a cone decomposition of $\mathbf{R}_+^2$ taking $C_1=C(1,\epsilon)$, $C_2=C(1)$. For $c\in C_{\mathfrak{f}}$, we put

$$R(\epsilon,c)=\{z=x+y\epsilon\in(\mathfrak{a}_\mu\mathfrak{f})^{-1}\mid x,y\in\mathbf{Q},\ 0<x\leqq 1,\ 0\leqq y<1,\ (z)\mathfrak{a}_\mu\mathfrak{f}=c\text{ in }C_{\mathfrak{f}}\}.$$

We have

$$\begin{aligned}G(c)&=\sum_{z\in R(C_1,c)}\log\frac{\Gamma_2(z,(1,\epsilon))}{\rho_2((1,\epsilon))}+\sum_{z\in R(C_2,c)}\log\frac{\Gamma_1(z,1)}{\rho_1(1)}\\ &=\sum_{z=x+y\epsilon\in R(\epsilon,c),0<y<1}\log\frac{\Gamma_2(z,(1,\epsilon))}{\rho_2((1,\epsilon))}+\sum_{z=x+y\epsilon\in R(\epsilon,c),y=0}\log\frac{\Gamma_2(z+\epsilon,(1,\epsilon))}{\rho_2((1,\epsilon))}\\ &\quad+\sum_{z=x+y\epsilon\in R(\epsilon,c),y=0}\log\frac{\Gamma_1(z,1)}{\rho_1(1)}.\end{aligned}$$

The difference equation I, (1.13) gives

$$\log\frac{\Gamma_2(z+\epsilon,(1,\epsilon))}{\rho_2((1,\epsilon))}-\log\frac{\Gamma_2(z,(1,\epsilon))}{\rho_2((1,\epsilon))}=-\log\frac{\Gamma_1(z,1)}{\rho_1(1)}.$$

Hence we have

$$G(c)=\sum_{z\in R(\epsilon,c)}\log\frac{\Gamma_2(z,(1,\epsilon))}{\rho_2((1,\epsilon))}. \tag{4.1}$$

Let σ be the non-trivial element of $\mathrm{Gal}(F/\mathbf{Q})$. Since $R(C_j,c)\ni z\longrightarrow z^\sigma\in R(C_j^\sigma,c^\sigma)$ is a bijection, we have

$$G(c^\sigma)=G(c^\sigma;\{C_j^\sigma\},\mathfrak{a}_\mu^\sigma)=\sum_{z\in R(\epsilon,c)}\log\frac{\Gamma_2(z',(1,\epsilon'))}{\rho_2((1,\epsilon'))}, \tag{4.2}$$

by the same calculation as above. Using $C_1(A_2)=0$, $B_2(x)=B_2(1-x)$ and

$$C_{(2,0)}(\begin{pmatrix}1&\epsilon\\1&\epsilon'\end{pmatrix})=2(\epsilon'-\epsilon)\log\epsilon,\quad C_{(1,1)}(\begin{pmatrix}1&\epsilon\\1&\epsilon'\end{pmatrix})=0,\quad C_{(0,2)}(\begin{pmatrix}1&\epsilon\\1&\epsilon'\end{pmatrix})=0,$$

we obtain

$$V(c)=\frac{1}{4}(\epsilon'-\epsilon)\log\epsilon\sum_{z=x+y\epsilon\in R(\epsilon,c)}B_2(x). \tag{4.3}$$

By a similar calculation, we get

$$V(c^\sigma) = V(c^\sigma; \{C_j^\sigma\}, \mathfrak{a}_\mu^\sigma) = V(c). \tag{4.4}$$

By definition, I, Theorem 2.1 and II, Lemma 3.2, we have

$$\begin{aligned}
W(c)/\log N(\mathfrak{a}_\mu \mathfrak{f}) &= -\frac{1}{2}(\zeta^{(1)}(0, C_1, \mathfrak{a}_\mu) + \zeta^{(1)}(0, C_2, \mathfrak{a}_\mu)) \\
= -\frac{1}{2}\Bigg[&\sum_{z \in R(C_1, c)} (\frac{1}{2}B_2(x_1(z))\epsilon' + B_1(x_1(z))B_1(x_2(z)) + \frac{1}{2}B_2(x_2(z))\epsilon) \\
&- \sum_{z \in R(C_2, c)} B_1(x_1(z))\Bigg] \\
= -\frac{1}{4}\Bigg[&\sum_{z = x + y\epsilon \in R(\epsilon, c), 0 < y < 1} (\epsilon' B_2(x) + 2B_1(x)B_1(y) + \epsilon B_2(y)) \\
&+ \sum_{z = x + y\epsilon \in R(\epsilon, c), y = 0} (\epsilon' B_2(x) + 2B_1(x)B_1(1) + \epsilon B_2(1)) \\
&- \sum_{z = x + y\epsilon \in R(\epsilon, c), y = 0} 2B_1(x)\Bigg] \\
= -\frac{1}{4} &\sum_{z = x + y\epsilon \in R(\epsilon, c)} (\epsilon' B_2(x) + \epsilon B_2(y) + 2B_1(x)B_1(y)).
\end{aligned}$$

Thus we get

$$W(c) = -\frac{1}{4}\log N(\mathfrak{a}_\mu \mathfrak{f}) \sum_{z = x + y\epsilon \in R(\epsilon, c)} (\epsilon' B_2(x) + \epsilon B_2(y) + 2B_1(x)B_1(y)). \tag{4.5}$$

Similarly we obtain

$$\begin{aligned}
W(c^\sigma) &= W(c^\sigma; \{C_j^\sigma\}, \mathfrak{a}_\mu^\sigma) \\
&= -\frac{1}{4}\log N(\mathfrak{a}_\mu \mathfrak{f}) \sum_{z = x + y\epsilon \in R(\epsilon, c)} (\epsilon B_2(x) + \epsilon' B_2(y) + 2B_1(x)B_1(y)).
\end{aligned} \tag{4.6}$$

Adding these terms, we have

$$\begin{aligned}
X(c) = \sum_{z = x + y\epsilon \in R(\epsilon, c)} \Bigg[&\log \frac{\Gamma_2(z, (1, \epsilon))}{\rho_2((1, \epsilon))} + \frac{1}{4}(\epsilon' - \epsilon)\log \epsilon B_2(x) \\
&- \frac{1}{4}\log N(\mathfrak{a}_\mu \mathfrak{f})(\epsilon B_2(x) + \epsilon' B_2(y) + 2B_1(x)B_1(y))\Bigg],
\end{aligned} \tag{4.7}$$

$$\begin{aligned}
X(c^\sigma) &= X(c^\sigma; \{C_j^\sigma\}, \mathfrak{a}_\mu^\sigma) \\
&= \sum_{z = x + y\epsilon \in R(\epsilon, c)} \Bigg[\log \frac{\Gamma_2(z', (1, \epsilon'))}{\rho_2((1, \epsilon'))} + \frac{1}{4}(\epsilon' - \epsilon)\log \epsilon B_2(x) \\
&\qquad - \frac{1}{4}\log N(\mathfrak{a}_\mu \mathfrak{f})(\epsilon' B_2(x) + \epsilon B_2(y) + 2B_1(x)B_1(y))\Bigg].
\end{aligned} \tag{4.8}$$

By (3.14) and (3.11) (or directly from II, (3.6), (3.7)), we obtain

$$\zeta_F(0,c) = \sum_{z=x+y\epsilon\in R(\epsilon,c)} \left\{\frac{\epsilon+\epsilon'}{4}(B_2(x)+B_2(y)) + B_1(x)B_1(y)\right\}, \tag{4.9}$$

$$\begin{aligned}\zeta_F'(0,c) = &\sum_{z=x+y\epsilon\in R(\epsilon,c)} \left\{\log\frac{\Gamma_2(z,(1,\epsilon))\Gamma_2(z',(1,\epsilon'))}{\rho_2((1,\epsilon))\rho_2((1,\epsilon'))}\right. \\ &\left.+\frac{\epsilon'-\epsilon}{2}\log\epsilon\, B_2(x)\right\} - \log N(\mathfrak{a}_\mu\mathfrak{f})\zeta_F(0,c).\end{aligned} \tag{4.10}$$

Let K be a totally imaginary quadratic extension of F and let $\mathfrak{f}\infty_1\infty_2$ be the conductor of K as a class field over F. Let χ be the Hecke character of $F_A^\times$ which corresponds to K. Then (3.10) and (3.13) can be written as

$$g_K(\mathrm{id},\mathrm{id}) = \pi^{-1/2}\exp(\frac{1}{2L(0,\chi)}\sum_{c\in C_\mathfrak{f}}\chi(c)X(c)), \tag{4.11}$$

$$\begin{aligned}g_{K^\sigma}(\mathrm{id},\mathrm{id}) &= g_{K^\sigma}(\mathrm{id},\mathrm{id};\{C_j^\sigma\},\{\mathfrak{a}_\mu^\sigma\}) \\ &= \pi^{-1/2}\exp(\frac{1}{2L(0,\chi)}\sum_{c\in C_\mathfrak{f}}\chi(c)X(c^\sigma)).\end{aligned} \tag{4.12}$$

By (3.12), we have

$$g_K(\mathrm{id},\mathrm{id})g_{K^\sigma}(\mathrm{id},\mathrm{id}) = \pi^{-1}\exp(\frac{L'(0,\chi)}{2L(0,\chi)}). \tag{4.13}$$

Let $F=\mathbf{Q}(\sqrt{d})$, $0<d\in\mathbf{Q}$ and $x+y\sqrt{d}$, x, $y\in\mathbf{Q}$ be a totally positive element. Set $\xi=\sqrt{x+y\sqrt{d}}\,i$, $\xi'=\sqrt{x-y\sqrt{d}}\,i$, $K=\mathbf{Q}(\xi)$. Then K is a CM-field. Define $\sigma\in J_K$ by $\xi^\sigma=\xi'$. Assume $y\neq 0$. Then $\{\mathrm{id},\sigma\}$ is a CM-type of K. Let $\mathfrak{q}$ be an integral ideal of K. For positive even integers a and b, let $\lambda_{a,b}^{(1)}$ and $\lambda_{a,b}^{(2)}$ be characters of $I_\mathfrak{q}(K)$ such that

$$\lambda_{a,b}^{(1)}((\alpha)) = (\frac{\alpha^\rho}{|\alpha|})^a(\frac{\alpha^{\sigma\rho}}{|\alpha^\sigma|})^b, \qquad \alpha\equiv 1 \mod^\times\mathfrak{q},$$

$$\lambda_{a,b}^{(2)}((\alpha)) = (\frac{\alpha^\rho}{|\alpha|})^a(\frac{\alpha^{\sigma}}{|\alpha^{\sigma\rho}|})^b, \qquad \alpha\equiv 1 \mod^\times\mathfrak{q}.$$

By Theorem 1.5, we have

$$L(1,\lambda_{a,b}^{(1)}) \sim \pi^{(4+a+b)/2}p_K(a\cdot\mathrm{id}+b\cdot\sigma,\mathrm{id}+\sigma),$$

$$L(1,\lambda_{a,b}^{(2)}) \sim \pi^{(4+a+b)/2}p_K(a\cdot\mathrm{id}+b\cdot\sigma\rho,\mathrm{id}+\sigma\rho).$$

By Theorem 1.2, we obtain

$$L(1,\lambda_{a,b}^{(1)})L(1,\lambda_{a,b}^{(2)}) \sim \pi^{4+a+b}p_K(\mathrm{id},\mathrm{id})^{2a}p_K(\sigma,\sigma)^{2b}. \tag{4.14}$$

In this case, Conjecture 3.10 states that

$$\left(\frac{L(1,\lambda_{a,b}^{(1)})L(1,\lambda_{a,b}^{(2)})}{\pi^{4+a+b}g_K(\mathrm{id},\mathrm{id})^{2a}g_{K^\sigma}(\mathrm{id},\mathrm{id})^{2b}}\right)^\alpha = \zeta\cdot\frac{L(1,(\lambda_{a,b}^{(1)})^\alpha)L(1,(\lambda_{a,b}^{(2)})^\alpha)}{\pi^{4+a+b}g_{K^\alpha}(\mathrm{id},\mathrm{id})^{2a}g_{K^{\sigma\alpha}}(\mathrm{id},\mathrm{id})^{2b}} \tag{4.15}$$

for every $\alpha \in \mathrm{Gal}(\overline{\mathbf{Q}}/\mathbf{Q})$ with a root of unity ζ. Put

$$Q = \frac{L(1,\lambda_{2,2}^{(1)})L(1,\lambda_{2,2}^{(2)})}{\pi^8 g_K(\mathrm{id},\mathrm{id})^4 g_{K^\sigma}(\mathrm{id},\mathrm{id})^4}, \tag{4.16}$$

$$Q_1 = \frac{L(1,\lambda_{4,2}^{(1)})L(1,\lambda_{4,2}^{(2)})}{\pi^{10} g_K(\mathrm{id},\mathrm{id})^8 g_{K^\sigma}(\mathrm{id},\mathrm{id})^4}, \qquad Q_2 = \frac{L(1,\lambda_{2,4}^{(1)})L(1,\lambda_{2,4}^{(2)})}{\pi^{10} g_K(\mathrm{id},\mathrm{id})^4 g_{K^\sigma}(\mathrm{id},\mathrm{id})^8}. \tag{4.17}$$

Assume that the conductors of $\lambda_{a,b}^{(i)}$ are (1) and $h_K = 1$. Then we see that $L(1,(\lambda_{a,b}^{(i)})^\rho) = L(1,\lambda_{a,b}^{(i)})$. Using this fact, we see easily that (4.15) can be written as

$$Q^\alpha = \zeta Q, \tag{4.18}$$

$$\begin{cases} (Q_1)^\alpha = \zeta_1\cdot Q_1, & (Q_2)^\alpha = \zeta_2\cdot Q_2, \quad \text{if} \quad \alpha|F = \mathrm{id}, \\ (Q_1)^\alpha = \zeta_1\cdot Q_2, & (Q_2)^\alpha = \zeta_2\cdot Q_1, \quad \text{if} \quad \alpha|F \neq \mathrm{id}, \end{cases} \tag{4.19}$$

with roots of unity ζ, ζ_1 and ζ_2 which depend on α.

EXAMPLE 4.1. Let $F = \mathbf{Q}(\sqrt{29})$. Then $h_F = 1$ and $\epsilon_0 = \dfrac{5+\sqrt{29}}{2}$ is the fundamental unit of F. We have $\epsilon = \epsilon_0^2 = \dfrac{27+5\sqrt{29}}{2}$. Let $\mathfrak{p} = (\dfrac{9+\sqrt{29}}{2})$. Then $N(\mathfrak{p}) = 13$, $C_\mathfrak{p} = \{c_1, c_2\}$, $c_1 =$ the class of (1), $c_2 =$ the class of (2). Put $K = \mathbf{Q}(\sqrt{\dfrac{9+\sqrt{29}}{2}}i)$. We see that

$$\mathfrak{O}_K = \mathfrak{O}_F\cdot\frac{1}{2}(\sqrt{\frac{9+\sqrt{29}}{2}}i + \frac{3+\sqrt{29}}{2}) \oplus \mathfrak{O}_F$$

and that K is the maximal ray class field of conductor $\mathfrak{p}\infty_1\infty_2$. We have $L(0,\chi) = 2$. Hence $h_K = 1$. By numerical computation, we find

$$Q = \frac{2^4\cdot 3}{29^2}, \qquad Q_1 Q_2 = \frac{2^2\cdot 1889}{29^4}.$$

We also find

$$\epsilon^{-9/13}Q_1 + \epsilon^{9/13}Q_2 = \frac{5870}{13\cdot 29^2}.$$

Solving the quadratic equation and comparing with numerical values to determine the signature of the square root, we find

$$Q_1 = \epsilon^{9/13}\cdot\frac{2935+503\sqrt{29}}{13\cdot 29^2}, \qquad Q_2 = \epsilon^{-9/13}\cdot\frac{2935-503\sqrt{29}}{13\cdot 29^2}.$$

The numerical values coincide to the 50th decimal place. We have $|R(\epsilon, c_i)| = 30$, $i = 1, 2$ and

$$V(c_1) = -\frac{1}{4\cdot 13}\sqrt{29}\log\epsilon, \qquad V(c_2) = \frac{3}{4\cdot 13}\sqrt{29}\log\epsilon,$$
$$V(c_1) - V(c_2) = -\frac{1}{13}\sqrt{29}\log\epsilon.$$

We have the congruence[2]

$$9 - \sqrt{29} \equiv 0 \mod \mathfrak{p}',$$

where $\mathfrak{p}' = (\frac{9-\sqrt{29}}{2})$.

Still taking $F = \mathbf{Q}(\sqrt{29})$ as the basic field, we made experiments for CM-extensions of F of prime (relative) discriminant. Let us describe them briefly.

Let $\mathfrak{p} = (13 + 2\sqrt{29})$, $K = \mathbf{Q}(\sqrt{13 + 2\sqrt{29}}\,i)$. Then $N(\mathfrak{p}) = 53$ and K is the maximal ray class field of conductor $\mathfrak{p}\infty_1\infty_2$. We have $h_K = 1$. A numerical computation suggests

$$Q = \frac{2^4\cdot 1181}{29^2}, \qquad Q_1Q_2 = \frac{2^2\cdot 5^2\cdot 7^2\cdot 227\cdot 317371}{29^4},$$

$$Q_1 = \epsilon^{19/53}\cdot\frac{304304579 + 56204573\sqrt{29}}{53\cdot 29^2},$$
$$Q_2 = \epsilon^{-19/53}\cdot\frac{304304579 - 56204573\sqrt{29}}{53\cdot 29^2}.$$

We have $C_{\mathfrak{p}} = \{c_1, c_2\}$, $c_1 =$ the class of (1),

$$V(c_1) - V(c_2) = -\frac{7}{53}\sqrt{29}\log\epsilon,$$

$$19 - 7\sqrt{29} \equiv 0 \mod \mathfrak{p}', \qquad \mathfrak{p}' = (13 - 2\sqrt{29}).$$

The examples presented below concern the case where $h_K > 1$. Then there are h_K choices for $\lambda_{a,b}^{(1)}$ of conductor (1). We put

$$Q = \frac{\prod_\omega L(1,\lambda_{2,2}^{(1)}\omega)L(1,\lambda_{2,2}^{(2)}\omega)}{(\pi^8 g_K(\mathrm{id},\mathrm{id})^4 g_{K^\sigma}(\mathrm{id},\mathrm{id})^4)^{h_K}}, \tag{4.16$'$}$$

$$(4.17') \qquad Q_1 = \frac{\prod_\omega L(1,\lambda_{4,2}^{(1)}\omega)L(1,\lambda_{4,2}^{(2)}\omega)}{(\pi^{10} g_K(\mathrm{id},\mathrm{id})^8 g_{K^\sigma}(\mathrm{id},\mathrm{id})^4)^{h_K}}, \quad Q_2 = \frac{\prod_\omega L(1,\lambda_{2,4}^{(1)}\omega)L(1,\lambda_{2,4}^{(2)}\omega)}{(\pi^{10} g_K(\mathrm{id},\mathrm{id})^4 g_{K^\sigma}(\mathrm{id},\mathrm{id})^8)^{h_K}},$$

where we choose $\lambda_{a,b}^{(i)}$ of conductor (1) arbitrarily and ω extends over all the characters of the ideal class group of K. Then (4.15) yields (4.18) and (4.19).

[2]The meaning of this type of congruence, which holds in many examples, is somewhat mysterious, though the congruence can be appreciated as quite natural.

Let $\mathfrak{p} = (17 + 2\sqrt{29})$, $K = \mathbf{Q}(\sqrt{17+2\sqrt{29}}\,i)$. Then $N(\mathfrak{p}) = 173$ and K is the maximal ray class field of conductor $\mathfrak{p}\infty_1\infty_2$. We have $h_K = 3$. A numerical computation suggests

$$Q = \frac{2^{14} \cdot 19177633}{29^5}, \qquad Q_1Q_2 = \frac{2^{16} \cdot 11^2 \cdot 13 \cdot 223962373 \cdot 89559567206967409}{29^{11}},$$

$$\begin{aligned} Q_1 &= \epsilon^{72/173} \cdot 2^7 \\ &\times \frac{518652963873357398048336 45 + 963114357665080525951940 3\sqrt{29}}{173^3 \cdot 29^6}, \\ Q_2 &= \epsilon^{-72/173} \cdot 2^7 \\ &\times \frac{518652963873357398048336 45 - 963114357665080525951940 3\sqrt{29}}{173^3 \cdot 29^6}. \end{aligned}$$

We have $C_{\mathfrak{p}} = \{c_1, c_2\}$, $c_1 =$ the class of (1),

$$V(c_1) - V(c_2) = -\frac{39}{173}\sqrt{29}\log\epsilon,$$

$$72 - 39\sqrt{29} \equiv 0 \mod \mathfrak{p}', \qquad \mathfrak{p}' = (17 - 2\sqrt{29}).$$

Let $\mathfrak{p} = (\frac{25+\sqrt{29}}{2})$, $K = \mathbf{Q}(\sqrt{\frac{25+\sqrt{29}}{2}}\,i)$. Then $N(\mathfrak{p}) = 149$ and K is the maximal ray class field of conductor $\mathfrak{p}\infty_1\infty_2$. We have $h_K = 5$. A numerical computation suggests

$$Q = \frac{2^{20} \cdot 7 \cdot 19 \cdot 460589}{29^8},$$

$$Q_1Q_2 = \frac{2^{26} \cdot 29863 \cdot 251336777809681 \cdot 5075977770734034851}{29^{20}},$$

$$Q_1 = \epsilon^{130/149} \cdot 2^{12} \cdot \frac{A + B\sqrt{29}}{149^5 \cdot 29^{10}}, \qquad Q_2 = \epsilon^{-130/149} \cdot 2^{12} \cdot \frac{A - B\sqrt{29}}{149^5 \cdot 29^{10}},$$

where

$$\begin{aligned} A &= 4932947005844846344067203947829705 09, \\ B &= 9160252624582871538572648333571181 9. \end{aligned}$$

We have $C_{\mathfrak{p}} = \{c_1, c_2\}$, $c_1 =$ the class of (1),

$$V(c_1) - V(c_2) = -\frac{35}{149}\sqrt{29}\log\epsilon,$$

$$130 - 35\sqrt{29} \equiv 0 \mod \mathfrak{p}', \qquad \mathfrak{p}' = (\frac{25 - \sqrt{29}}{2}).$$

In this Example 4.1, it is remarkable that we can controle the denominators; it is conjecturable that

$$(29)^{2h_K} Q \in \mathbf{Z}, \qquad (29)^{4h_K} Q_1Q_2 \in \mathbf{Z}$$

and that

$$\epsilon^a (N(\mathfrak{p})29^2)^{h_K} Q_1 \in \mathfrak{O}_F$$

with $a \in N(\mathfrak{p})^{-1}\mathbf{Z}$, for a CM-extension K over F such that $D(K/F) = \mathfrak{p}$ is a prime ideal of degree 1.

EXAMPLE 4.2. We consider the example of II, §4 from the point of view of §3. We use the same notation as there. Thus $F = \mathbf{Q}(\sqrt{5})$, $\mathfrak{p}_{19} = (8+3\sqrt{5})$, $\mathfrak{f} = (4)\mathfrak{p}_{19}$, M is the maximal ray class field of conductor $\mathfrak{f}\infty_1\infty_2$ over F. Now let $K = \mathbf{Q}(\sqrt{(8+3\sqrt{5})\,i})$. Since the relative different of $\sqrt{(8+3\sqrt{5})\,i}$ is $2\sqrt{(8+3\sqrt{5})\,i}$, only (2) and $\mathfrak{p}_{19}$ can ramify in K/F. It is clear that $\mathfrak{p}_{19}$ ramifies but we can show that (2) also ramifies (cf. [Y1], §5). Hence we obtain

$$\mathfrak{O}_K = \mathfrak{O}_F \cdot \sqrt{(8+3\sqrt{5})\,i} \oplus \mathfrak{O}_F, \qquad D(K/F) = (4)\mathfrak{p}_{19}, \qquad K \subset M.$$

Let $(2) = \mathfrak{P}_2^2$ in K. We see easily that $\mathfrak{P}_2$ is not principal and that

$$\mathfrak{P}_2 = (1 + \sqrt{(8+3\sqrt{5})\,i})\mathfrak{O}_F \oplus \mathfrak{O}_F.$$

Let χ be the non-trivial character of $\mathrm{Gal}(K/F)$ and we regard χ also as a character of $C_\mathfrak{f}$. The conductor $\mathfrak{f}(\chi)$ of χ is $\mathfrak{f}(\chi) = (4)\mathfrak{p}_{19}\infty_1\infty_2$ by the conductor discriminant theorem. In view of II, (4.5), we have $\chi = \psi\eta$. Calculating $\zeta_F(0, c)$, $c \in C_\mathfrak{f}$, using (4.9), we obtain $L(0, \chi) = 4$. Hence we get $h_K = 2$ (cf. [Y3], §3).

We have $\lambda_{a,b}^{(1)}(\mathfrak{P}_2) = \pm 1$ and there are two choices of such characters. Define Q, Q_1 and Q_2 by (4.16) and (4.17). If $\lambda_{a,b}^{(i)}(\mathfrak{P}_2)$ are chosen simultaneously as $+1$ or -1, then (4.18) and (4.19) follow from (4.15).

Choose $\lambda_{a,b}^{(i)}$ so that $\lambda_{a,b}^{(i)}(\mathfrak{P}_2) = 1$ for all a, b and i. Then we find

$$Q = \frac{2^3}{5}, \qquad Q_1Q_2 = \frac{2^6 \cdot 3089}{5^4},$$

$$Q_1 = \epsilon^{1/2} \cdot 2^2 \cdot \frac{2879 + 875\sqrt{5}}{5^2 \cdot 19}, \qquad Q_2 = \epsilon^{-1/2} \cdot 2^2 \cdot \frac{2879 - 875\sqrt{5}}{5^2 \cdot 19}.$$

Next choose $\lambda_{a,b}^{(i)}$ so that $\lambda_{a,b}^{(i)}(\mathfrak{P}_2) = -1$ for all a, b and i. Then we find

$$Q = \frac{2^3 \cdot \sqrt{5}}{5^2}, \qquad Q_1Q_2 = \frac{2^6 \cdot 13^2}{5^4}.$$

$$Q_1 = \epsilon^{1/2} \cdot 2^2 \cdot \frac{533 + 325\sqrt{5}}{5^2 \cdot 19}, \qquad Q_2 = \epsilon^{-1/2} \cdot 2^2 \cdot \frac{-533 + 325\sqrt{5}}{5^2 \cdot 19}.$$

The numerical values coincide to the 45th decimal place. We have

$$\frac{1}{2} \sum_{c \in C_\mathfrak{f}} \chi(c)V(c) = -\frac{1}{4}\sqrt{5}\log\epsilon.$$

The phenomenon concerning the exponent of ϵ is rather different from those observed in Example 4.1 and in Examples $1 \sim 7$ of [Y4].

Let D_0 be the subgroup of the character group of $C_\mathfrak{f}$ consisting of those characters which are unramified at ∞_1 and ∞_2, as in II, §4. Experimentally we observed that

$$\sum_{c \in C_\mathfrak{f}} \chi(c)X(c) = 0 \qquad \text{if} \quad \chi \in \psi D_0, \tag{4.20}$$

$$\sum_{c \in C_\mathfrak{f}} \chi(c)X(c^\sigma) = 0 \qquad \text{if} \quad \chi \in \eta D_0. \tag{4.21}$$

In other words, $\sum_{c\in C_\mathfrak{f}}\chi(c)X(c)=0$ (resp. $\sum_{c\in C_\mathfrak{f}}\chi(c)X(c^\sigma)=0$) if χ is ramified at ∞_1 and unramified at ∞_2 (resp. unramified at ∞_1 and ramified at ∞_2). This shows that the invariant $X(c)$ has still more intriguing properties, a part of which will be investigated in the next section. We will prove (4.20) and (4.21) in §6.1.

Let ξ, ξ' and $K=\mathbf{Q}(\xi)$ be the same as in the beginning of this subsection. We assume that K is not normal over $\mathbf{Q}$. Let L be the normal closure of K over $\mathbf{Q}$ and put $G=\mathrm{Gal}(K/\mathbf{Q})$. We have $L=\mathbf{Q}(\xi,\xi')$. Define σ, $\tau\in G$ by

$$\sigma:(\xi,\xi')\longrightarrow(\xi',-\xi),\qquad \tau:(\xi,\xi')\longrightarrow(\xi',\xi).$$

Then G is generated by σ, τ and they satisfy the relations

$$\sigma^4=\tau^2=1,\qquad \tau\sigma=\sigma^3\tau.$$

Hence G is isomorphic to the dihedral group of order 8 (cf. [Sh12], p. 65). We have $\sigma^2=\rho$. Put $d'=x^2-y^2d$, $F_*=\mathbf{Q}(\sqrt{dd'})$. Then $\mathrm{Gal}(L/F_*)$ is the cyclic group of order 4 generated by σ. By Theorems 1.2 and 1.5, we have

$$L(1,\lambda^{(1)}_{a,b})L(1,\lambda^{(2)}_{b,a})\sim\pi^{4+a+b}p_L(\mathrm{id},\mathrm{id})^{2a+2b}p_L(\mathrm{id},\sigma)^{2a-2b}. \tag{4.22}$$

We put

$$Q_1=\frac{L(1,\lambda^{(1)}_{4,2})L(1,\lambda^{(2)}_{2,4})}{\pi^{10}g_L(\mathrm{id},\mathrm{id})^{12}g_L(\mathrm{id},\sigma)^4}. \tag{4.23}$$

Here we define $g_L(\mathrm{id},\mathrm{id})$ and $g_L(\mathrm{id},\sigma)$ regarding L as an abelian extension of F_*. Since $\lambda^{(i)}_{a,b}$ is not a Hecke character of L, we cannot apply Conjecture 3.10 directly to Q_1. However this difficulty can be overcome by the following device. We regard $\lambda^{(i)}_{a,b}$ as a Hecke character of $K_A^\times$ and let ω be the Hecke character of $K_A^\times$ which corresponds to the quadratic extension L/K. Then $\lambda^{(i)}_{a,b}\circ N_{L/K}$ is a Hecke character of $L_A^\times$ and we have

$$L(s,\lambda^{(i)}_{a,b}\circ N_{L/K})=L(s,\lambda^{(i)}_{a,b})L(s,\lambda^{(i)}_{a,b}\omega).$$

Put

$$R_1=\frac{L(1,\lambda^{(1)}_{4,2})L(1,\lambda^{(1)}_{4,2}\omega)L(1,\lambda^{(2)}_{2,4})L(1,\lambda^{(2)}_{2,4}\omega)}{(\pi^{10}g_L(\mathrm{id},\mathrm{id})^{12}g_L(\mathrm{id},\sigma)^4)^2}.$$

Conjecture 3.10 can be applied to R_1. It states that

$$R_1^\alpha=\zeta\cdot\frac{L(1,(\lambda^{(1)}_{4,2})^\alpha)L(1,(\lambda^{(1)}_{4,2}\omega)^\alpha)L(1,(\lambda^{(2)}_{2,4})^\alpha)L(1,(\lambda^{(2)}_{2,4}\omega)^\alpha)}{(\pi^{10}g_L(\mathrm{id},\mathrm{id})^{12}g_{L^\alpha}(\mathrm{id},\alpha^{-1}\sigma\alpha)^4)^2} \tag{4.24}$$

for $\alpha\in\mathrm{Gal}(\overline{\mathbf{Q}}/\mathbf{Q})$ with a root of unity ζ. Here we understand

$$g_{L^\alpha}(\mathrm{id},\beta)=g_{L^\alpha/F_*^\alpha}(\mathrm{id},\beta;\{C_j^\alpha\},\{\mathfrak{a}_\mu^\alpha\})$$

for $\beta=\mathrm{id}$, $\alpha^{-1}\sigma\alpha$. By [Y3], Proposition 7.3, we have

$$\left(\frac{L(1,\lambda^{(1)}_{4,2}\omega)L(1,\lambda^{(2)}_{2,4}\omega)}{L(1,\lambda^{(1)}_{4,2})L(1,\lambda^{(2)}_{2,4})}\right)^\alpha=\pm\frac{L(1,(\lambda^{(1)}_{4,2}\omega)^\alpha)L(1,(\lambda^{(1)}_{2,4}\omega)^\alpha)}{L(1,(\lambda^{(1)}_{4,2})^\alpha)L(1,(\lambda^{(2)}_{2,4})^\alpha)}$$

for $\alpha \in \mathrm{Gal}(\overline{\mathbf{Q}}/\mathbf{Q})$. Since

$$R_1 = Q_1^2 \cdot \frac{L(1, \lambda_{4,2}^{(1)}\omega)L(1, \lambda_{2,4}^{(2)}\omega)}{L(1, \lambda_{4,2}^{(1)})L(1, \lambda_{2,4}^{(2)})},$$

we obtain, by (4.24), that

$$Q_1^\alpha = \zeta \cdot \frac{L(1, (\lambda_{4,2}^{(1)})^\alpha)L(1, (\lambda_{2,4}^{(2)})^\alpha)}{\pi^{10} g_{L^\alpha}(\mathrm{id}, \mathrm{id})^{12} g_{L^\alpha}(\mathrm{id}, \alpha^{-1}\sigma\alpha)^4} \tag{4.25}$$

for $\alpha \in \mathrm{Gal}(\overline{\mathbf{Q}}/\mathbf{Q})$ with a root of unity ζ.

EXAMPLE 4.3. We take $F = \mathbf{Q}(\sqrt{29})$, $K = \mathbf{Q}(\sqrt{\frac{9+\sqrt{29}}{2}}\, i)$. This is the same situation as the first case of Example 4.1. We have $F_* = \mathbf{Q}(\sqrt{377})$. The fundamental unit ϵ_0 of F_* is $233 + 12\sqrt{377}$, which is totally positive. Set $\epsilon = \epsilon_0$. We have $h_{F_*} = 2$, $h_0 = 4$. It is not difficult to see that L is the maximal ray class field of conductor $(1)\infty_1\infty_2$ of F_*. Set

$$\mathfrak{p}_2 = 2\mathbf{Z} + \frac{1+\sqrt{377}}{2}\mathbf{Z}, \qquad \mathfrak{p}_2' = 2\mathbf{Z} + \frac{1-\sqrt{377}}{2}\mathbf{Z}.$$

Then we have $(2) = \mathfrak{p}_2\mathfrak{p}_2'$, $(\frac{L/F_*}{\mathfrak{p}_2}) = \sigma$. As the representatives of the narrow ideal class group of F_*, we take

$$\mathfrak{a}_1 = (1), \qquad \mathfrak{a}_2 = \mathfrak{p}_2, \qquad \mathfrak{a}_3 = \mathfrak{p}_2^2 = (\frac{19-\sqrt{377}}{2}), \qquad \mathfrak{a}_4 = \mathfrak{p}_2'.$$

We take c_i as the class of $\mathfrak{a}_i$. We have

$$\zeta_{F_*}(0, c_1) = 1, \qquad \zeta_{F_*}(0, c_2) = 0, \qquad \zeta_{F_*}(0, c_3) = -1, \qquad \zeta_{F_*}(0, c_4) = 0.$$

By definition, we have

$$\begin{aligned} g_L(\mathrm{id}, \mathrm{id}) &= \pi^{-1/2} \exp(\frac{1}{4}(X(c_1) - X(c_3))), \\ g_L(\mathrm{id}, \mathrm{id}) &= \exp(\frac{1}{4}(X(c_4) - X(c_2))), \\ g_{L^\alpha}(\mathrm{id}, \mathrm{id}) &= \pi^{-1/2} \exp(\frac{1}{4}(X(c_1') - X(c_3'))) \end{aligned}$$

if $\alpha | F_*$ is non-trivial. Here $X(c_i')$ is defined by (4.8) with c_i' in place c^σ. (Thus $R(\epsilon, c)$ should be read as $R(\epsilon, c_i)$ in (4.8). Of course, we now take $F = F_*$ in (4.8).) Define Q_1 by (4.23) and put

$$Q_2 = \frac{L(1, \lambda_{2,4}^{(1)})L(1, \lambda_{4,2}^{(2)})}{\pi^{10} g_{L'}(\mathrm{id}, \mathrm{id})^{12} g_L(\mathrm{id}, \sigma)^{-4}}. \tag{4.26}$$

Here $g_{L'}(\mathrm{id}, \mathrm{id}) = \pi^{-1/2} \exp(\frac{1}{4}(X(c_1') - X(c_3')))$. We use the same letter σ, τ for their extensions to elements of $\mathrm{Gal}(\overline{\mathbf{Q}}/\mathbf{Q})$. Then (4.25) and a similar formula for Q_2 imply that

$$Q_1^\sigma = \zeta_1 Q_1, \qquad Q_2^\sigma = \zeta_2 Q_2, \tag{4.27}$$

$$Q_1^\tau = \zeta_1 Q_2, \qquad Q_2^\tau = \zeta_2 Q_1 \tag{4.28}$$

with roots of unity ζ_1 and ζ_2. First we find that

$$Q_1Q_2 = \frac{2^2 \cdot 1889}{29^4}.$$

This is consistent with the first case of Example 4.1. To see this, note that we have

$$g_L(\mathrm{id}, \mathrm{id})g_{L'}(\mathrm{id}, \mathrm{id}) = \pi^{-1}\exp(\frac{L'(0,\omega)}{2L(0,\omega)})$$

for a generator ω of the character group of $\mathrm{Gal}(L/F_*)$ by (3.12) and by $L(s,\omega) = L(s,\chi)$ (cf. (4.13)). Next we find that

$$(\epsilon^{-3/32}Q_1)^2 + (\epsilon^{3/32}Q_2)^2 = \frac{3 \cdot 53 \cdot 2543 \cdot 21841 \cdot 200381}{2^2 \cdot 29^4},$$

$$\epsilon^{-3/32}Q_1 + \epsilon^{3/32}Q_2 = \frac{5 \cdot 7^3 \cdot 6803 \cdot \sqrt{13}}{2 \cdot 29^2}.$$

We finally get

$$Q_1 = \epsilon^{3/32}\frac{11667145\sqrt{13} + 7811551\sqrt{29}}{2^2 \cdot 29^2},$$

$$Q_2 = \epsilon^{-3/32}\frac{11667145\sqrt{13} - 7811551\sqrt{29}}{2^2 \cdot 29^2}.$$

The numerical values coincide to the 41st decimal place. These values are consistent with (4.27) and (4.28) since

$$\sqrt{29}^{\,\sigma} = -\sqrt{29}, \qquad \sqrt{13}^{\,\sigma} = -\sqrt{13}, \qquad \sqrt{29}^{\,\tau} = -\sqrt{29}, \qquad \sqrt{13}^{\,\tau} = \sqrt{13}.$$

4.2. In this subsection, we will give an example for the case where F is a totally real cubic field.

EXAMPLE 4.4. Let F be a totally real cubic field and let K be a totally imaginary quadratic extension of F. Let $J_F = \{\sigma_1, \sigma_2, \sigma_3\}$ with $\sigma_1 = \mathrm{id}$ and extend each σ_i to an element of J_K. Let $\mathfrak{q}$ be an integral ideal of K. For positive even integers a, b and c, let $\lambda^{(i)}_{a,b,c}$, $1 \leqq i \leqq 4$ be characters of $I_{\mathfrak{q}}(K)$ such that

$$\lambda^{(1)}_{a,b,c}((\alpha)) = (\frac{\alpha^\rho}{|\alpha|})^a(\frac{\alpha^{\sigma_2\rho}}{|\alpha^{\sigma_2}|})^b(\frac{\alpha^{\sigma_3\rho}}{|\alpha^{\sigma_3}|})^c,$$

$$\lambda^{(2)}_{a,b,c}((\alpha)) = (\frac{\alpha^\rho}{|\alpha|})^a(\frac{\alpha^{\sigma_2\rho}}{|\alpha^{\sigma_2}|})^b(\frac{\alpha^{\sigma_3}}{|\alpha^{\sigma_3\rho}|})^c,$$

$$\lambda^{(3)}_{a,b,c}((\alpha)) = (\frac{\alpha^\rho}{|\alpha|})^a(\frac{\alpha^{\sigma_2}}{|\alpha^{\sigma_2\rho}|})^b(\frac{\alpha^{\sigma_3\rho}}{|\alpha^{\sigma_3}|})^c,$$

$$\lambda^{(4)}_{a,b,c}((\alpha)) = (\frac{\alpha^\rho}{|\alpha|})^a(\frac{\alpha^{\sigma_2}}{|\alpha^{\sigma_2\rho}|})^b(\frac{\alpha^{\sigma_3}}{|\alpha^{\sigma_3\rho}|})^c.$$

for $\alpha \equiv 1 \mod{}^\times \mathfrak{q}$. By Theorems 1.2 and 1.5, we see that

$$\prod_{i=1}^{4}\frac{L(1,\lambda^{(i)}_{4,2,2})}{L(1,\lambda^{(i)}_{2,2,2})} \sim \pi^4 p_K(\mathrm{id},\mathrm{id})^8$$

holds (independently of the extensions σ_i). Conjecture 3.10 states that

$$(4.29)\qquad \begin{aligned}&\Big(\frac{\prod_{i=1}^4 L(1,\lambda_{4,2,2}^{(i)})L(1,\lambda_{2,2,2}^{(i)})^{-1}}{\pi^4 g_K(\mathrm{id},\mathrm{id})^8}\Big)^\alpha\\ &=\zeta\cdot\frac{\prod_{i=1}^4 L(1,(\lambda_{4,2,2}^{(i)})^\alpha)L(1,(\lambda_{2,2,2}^{(i)})^\alpha)^{-1}}{\pi^4 g_{K^\alpha}(\mathrm{id},\mathrm{id})^8}\end{aligned}$$

for every $\alpha\in\mathrm{Gal}(\overline{\mathbf{Q}}/\mathbf{Q})$ with a root of unity ζ.

Let $f(X)=X^3-4X-1$; $f(X)$ is irreducible over $\mathbf{Q}$ and has three real roots $\rho=2.11490\ldots$, $\rho'=-0.25410\ldots$, $\rho''=-1.86080\ldots$. We set $F=\mathbf{Q}(\rho)$. Then F is a totally real cubic field non-normal over $\mathbf{Q}$. Define σ_1, σ_2, $\sigma_3\in J_F$ by $\sigma_1=\mathrm{id}$, $\rho^{\sigma_2}=\rho'$, $\rho^{\sigma_3}=\rho''$. The discriminant of F is 229 and we see easily that $F\vee\mathbf{Q}(\sqrt{229})$, which is the normal closure of F over $\mathbf{Q}$, is the absolute class field of $\mathbf{Q}(\sqrt{229})$. A basis of $\mathfrak{O}_F$ and generators of E_F are given in Billevič [Bi]; it reads as (see also Pohst and Zassenhaus [PZ], p. 468)

$$\mathfrak{O}_F=\mathbf{Z}+\mathbf{Z}\rho+\mathbf{Z}\rho^2,\qquad E_F=\langle\, -1,\rho^2+2\rho,2\rho^2+4\rho+1\,\rangle.$$

Put $\eta_0=2\rho^2+4\rho+1$. Since ρ is a unit of F and $\rho^{-1}\eta_0=\rho^2+2\rho$, we have

$$E_F=\langle\, -1,\rho,\eta_0\,\rangle,\qquad E_F^+=\langle\,\rho^2,\eta_0\,\rangle=\langle\,\epsilon,\eta\,\rangle,$$

where $\epsilon=\rho^2$, $\eta=\rho^{-2}\eta_0=\rho+2$. Using Minkowski's bound, we see easily that the class number of F is 1. The class number of F in the narrow sense is 2 since $[E_F:E_F^+]=4$. Put $\alpha=1+\rho$. Then $N(\alpha)=-2$. As representatives of the narrow ideal classes of F, we can take $\mathfrak{a}_1=(1)$, $\mathfrak{a}_2=(\alpha)$. By Thomas-Vasquez [TV], Theorem 1 (cf. chapter IV, §4.2), we have $\mathbf{R}_+^3=\sqcup_{u\in E_F^+}u(\sqcup_{j=1}^6 C_j)$ with

$$\begin{aligned}&C_1=C(1,\epsilon,\epsilon\eta),\qquad C_2=C(1,\eta,\eta\epsilon),\qquad C_3=C(1,\epsilon),\\ &C_4=C(1,\eta),\qquad C_5=C(1,\epsilon\eta),\qquad C_6=C(1).\end{aligned}$$

We put $\delta=4-\rho$, $K=F(\sqrt{\delta}\,i)$. We see that δ is totally positive and $N(\delta)=47$. We have $(47)=\mathfrak{p}\mathfrak{q}$ in F with prime ideals $\mathfrak{p}$ and $\mathfrak{q}$ such that $N(\mathfrak{p})=47$, $N(\mathfrak{q})=47^2$. Hence $\mathfrak{p}=(4-\rho)$. Since $\dfrac{\sqrt{\delta}+\rho^2}{2}$ is integral over $\mathfrak{O}_F$, we see that $\mathfrak{p}$ is the only prime ideal which ramifies in K/F and that $\mathfrak{O}_K=\mathfrak{O}_F\cdot\dfrac{\sqrt{\delta}+\rho^2}{2}\oplus\mathfrak{O}_F$. We have $C_{\mathfrak{p}}=\{c_1,c_2,c_3,c_4\}$ where $c_1=$ the class of (1), $c_2=$ the class of (5), $c_3=$ the class of (α), $c_4=$ the class of (5α). We see that

$$\begin{aligned}&|R(C_1,c_1)|=|R(C_1,c_2)|=|R(C_2,c_3)|=|R(C_2,c_4)|=92,\\ &|R(C_1,c_3)|=|R(C_1,c_4)|=184,\qquad |R(C_2,c_1)|=|R(C_2,c_2)|=46\end{aligned}$$

and $R(C_j,c_i)=\emptyset$, $1\leqq i\leqq 4$ if $j\geqq 3$. Using II, (3.6), we obtain

$$\zeta_F(0,c_1)=2,\qquad \zeta_F(0,c_2)=-2,\qquad \zeta_F(0,c_3)=2,\qquad \zeta_F(0,c_4)=-2.$$

Let χ be the Hecke character of $F_A^\times$ which corresponds to the quadratic extension K/F. We have $L(0,\chi)=8$. Hence we get $h_K=2$. By somewhat laborious computations[3], we obtain

$$(4.30)\qquad \begin{aligned} &W(c_1) = -W(c_2) = W(c_3) = -W(c_4) = -\frac{2}{3}\log 47,\\ &\frac{1}{2}\sum_{i=1}^{4}\chi(c_i)W(c_i) = -\frac{4}{3}\log 47, \end{aligned}$$

$$(4.31)\qquad \begin{aligned} V(c_1) = -V(c_2) =& \frac{1}{2^2\cdot 3^2\cdot 47^2}\{(77224 + 9231\rho - 28959\rho^2)\log\epsilon\\ &+ (-234744 - 44799\rho + 88029\rho^2)\log\eta\},\\ V(c_3) = -V(c_4) =& \frac{1}{2^2\cdot 3^2\cdot 47^2}\{(8872 - 16833\rho - 3327\rho^2)\log\epsilon\\ &+ (-108600 - 20823\rho + 40725\rho^2)\log\eta\},\\ \frac{1}{2}\sum_{i=1}^{4}\chi(c_i)V(c_i) =& \frac{1}{2^2\cdot 3^2\cdot 47^2}\{(86096 - 7602\rho - 32286\rho^2)\log\epsilon\\ &+ (-343344 - 65622\rho + 128754\rho^2)\log\eta\}. \end{aligned}$$

By definition, we have

$$g_{K^\sigma}(\mathrm{id},\mathrm{id}) = \pi^{-1/2}\exp(\frac{1}{16}(X(c_1^\sigma) - X(c_2^\sigma) + X(c_3^\sigma) - X(c_4^\sigma))), \qquad \sigma \in \mathrm{Gal}(\overline{\mathbf{Q}}/\mathbf{Q}).$$

Here $g_{K^\sigma}(\mathrm{id},\mathrm{id}) = g_{K^\sigma}(\mathrm{id},\mathrm{id};\{C_j^\sigma\}_{1\leqq j\leqq 6},\{\mathfrak{a}_\mu^\sigma\}_{1\leqq\mu\leqq 2})$. We extend σ_1, σ_2, $\sigma_3 \in \mathrm{Gal}(F/\mathbf{Q})$ to σ_1, σ_2, $\sigma_3 \in J_K$ by $\sigma_1 = \mathrm{id}$,

$$(\sqrt{\delta}\,i)^{\sigma_2} = \sqrt{\delta^{(2)}}\,i, \qquad (\sqrt{\delta}\,i)^{\sigma_3} = \sqrt{\delta^{(3)}}\,i.$$

We have the decomposition into prime ideals in F:

$$(2) = \mathfrak{p}_2\mathfrak{q}_2, \qquad N(\mathfrak{p}_2) = 2, \qquad N(\mathfrak{q}_2) = 4, \qquad \mathfrak{p}_2 = (\alpha).$$

Then $\mathfrak{p}_2$ decomposes completely in K/F so that $\mathfrak{p}_2 = \mathfrak{P}_2\mathfrak{P}_2'$. We can verify easily that $\mathfrak{P}_2$ is not a principal ideal. We can choose $\mathfrak{P}_2$ so that $\dfrac{\sqrt{\delta}\,i + \rho^2}{2} \equiv 0 \mod \mathfrak{P}_2$. Then we have

$$\mathfrak{P}_2 = \frac{\sqrt{\delta}\,i + \rho^2}{2}\mathfrak{O}_F \oplus (1+\rho)\mathfrak{O}_F.$$

We also find $\mathfrak{P}_2^2 = (\beta)$, $\beta = \dfrac{\sqrt{\delta}\,i + \rho^2}{2} - 2$. We take $\mathfrak{q} = (1)$. Then for every positive even integers a, b, c and $1 \leqq i \leqq 4$, there exist two choices of $\lambda_{a,b,c}^{(i)}$. They are determined by choosing a square root of $\lambda_{a,b,c}^{(i)}((\beta))$ as the value of $\lambda_{a,b,c}^{(i)}(\mathfrak{P}_2)$.

[3] The computation is especially laborious for $V(c_i)$

Let ω be the non-trivial character of the ideal class group of K. We put

$$Q_1 = \frac{\prod_{i=1}^4 L(1,\lambda^{(i)}_{4,2,2})L(1,\lambda^{(i)}_{4,2,2}\omega)L(1,\lambda^{(i)}_{2,2,2})^{-1}L(1,\lambda^{(i)}_{2,2,2}\omega)^{-1}}{\pi^8 g_K(\mathrm{id},\mathrm{id})^{16}},$$

$$Q_2 = \frac{\prod_{i=1}^4 L(1,\lambda^{(i)}_{2,4,2})L(1,\lambda^{(i)}_{2,4,2}\omega)L(1,\lambda^{(i)}_{2,2,2})^{-1}L(1,\lambda^{(i)}_{2,2,2}\omega)^{-1}}{\pi^8 g_{K^{\sigma_2}}(\mathrm{id},\mathrm{id})^{16}},$$

$$Q_3 = \frac{\prod_{i=1}^4 L(1,\lambda^{(i)}_{2,2,4})L(1,\lambda^{(i)}_{2,2,4}\omega)L(1,\lambda^{(i)}_{2,2,2})^{-1}L(1,\lambda^{(i)}_{2,2,2}\omega)^{-1}}{\pi^8 g_{K^{\sigma_3}}(\mathrm{id},\mathrm{id})^{16}}.$$

Then Q_1, Q_2 and Q_3 are defined independently of the choice of $\lambda^{(i)}_{a,b,c}$.

Let L be the normal closure of F over $\mathbf{Q}$. Let τ be the generator of $\mathrm{Gal}(L/F)$ and define $\sigma \in \mathrm{Gal}(L/\mathbf{Q}(\sqrt{229}))$ by $\rho^\sigma = \rho'$. Then $\mathrm{Gal}(L/\mathbf{Q}) \cong S_3$ is generated by σ and τ which are subject to the relations $\sigma^3 = 1$, $\tau^2 = 1$, $\tau\sigma\tau = \sigma^2$. For $\alpha \in \mathrm{Gal}(\overline{\mathbf{Q}}/\mathbf{Q})$, by (4.29), we obtain

(4.32)
$$\begin{array}{llll} Q_1^\alpha = \zeta_1 Q_2, & Q_2^\alpha = \zeta_2 Q_3, & Q_3^\alpha = \zeta_3 Q_1 & \text{if } \alpha|L = \sigma, \\ Q_1^\alpha = \zeta_1 Q_1, & Q_2^\alpha = \zeta_2 Q_3, & Q_3^\alpha = \zeta_3 Q_2 & \text{if } \alpha|L = \tau. \end{array}$$

Here ζ_1, ζ_2 and ζ_3 are roots of unity which depend on α. The behavior of Q_i for general α up to the multiplication by the roots of unity can be obtained by (4.29). If we assume $Q_1 \in \overline{\mathbf{Q}}$ and (4.29), we can show easily that $Q_1^e \in F$ for some positive integer e.

For the numerical values, we obtain

$$Q_1 = 1317208.89315730055380323815\ldots, \qquad Q_2 = 0.00003809580444796224\ldots,$$
$$Q_3 = 5.34268242096579435122\ldots.$$

It is not difficult to find

(4.33)
$$Q_1Q_2Q_3 = \frac{3^3 \cdot l}{2^{17}\cdot 229^4}, \qquad l = 3579144678850673.$$

Our problem is to identify Q_i , $1 \leqq i \leqq 3$ with algebraic numbers. Since $l = 59873247^2 - 229 \cdot 157228^2$, the prime l decomposes completely in F. Hence there exists $\gamma \in \mathfrak{O}_F$ such that $N(\gamma) = l$. We are going to apply the same idea as in [Y4], Example 11. For this purpose, we need explicit values of γ. We have[4]

(4.34)
$$\begin{array}{ll} N(\gamma_1) = l, & \gamma_1 = 167810 + 707786\rho - 362803\rho^2 \\ N(\gamma_2) = l, & \gamma_2 = -(831882 + 504799\rho + 30312\rho^2) \\ N(\gamma_3) = l, & \gamma_3 = 196001 + 174022\rho - 135476\rho^2 \end{array}$$

and $\gamma_1\gamma_2\gamma_3 = l$. We have

$$2 = (1+\rho)(3+\rho-\rho^2), \qquad 47 = (4-\rho)(12+4\rho+\rho^2), \qquad 229 = (4-3\rho^2)^2(16-3\rho^2)$$

and respective numbers on the right-hand side generate prime ideals of F. We suppose that $Q_1^3 \times$ (fractional powers of ϵ and η) $\in F$ and, in view of (4.33), that

[4]The author thanks Professor Yuji Kida for suggesting a method to find such a γ starting from the factorization of $X^3 - 4X - 1 \mod l$.

Q_1^3 is of the form

$$Q_1^3 = 47^8 \cdot \epsilon^a \eta^b \cdot (3\gamma)^3 \cdot \frac{1}{2^s(1+\rho)^{51-3s}} \cdot \frac{1}{229^c(4-3\rho^2)^{12-3c}} \cdot \frac{1}{47^d(4-\rho)^{24-3d}}. \tag{4.35}$$

Here a, $b \in (2^6 \cdot 3^6 \cdot 47^6)^{-1}\mathbf{Z}$, s, c, d are integers and $\gamma = \pm\gamma_i$ for some i. Considering the action of σ, Q_2^3 should be of the form

$$\begin{aligned} Q_2^3 =& 47^8 \cdot (\epsilon^{(2)})^a (\eta^{(2)})^b \cdot (3\gamma^{(2)})^3 \\ &\cdot \frac{1}{2^s(1+\rho^{(2)})^{51-3s}} \cdot \frac{1}{229^c(4-3(\rho^{(2)})^2)^{12-3c}} \cdot \frac{1}{47^d(4-\rho^{(2)})^{24-3d}}. \end{aligned} \tag{4.36}$$

When s, c, d and γ are given, we can solve (4.35) and (4.36) if the right-hand sides are positive and obtain a and b. We search s, c, d and γ so that $2^6 \cdot 3^6 \cdot 47^6 a$ is close to an integer for the solution (a, b). This procedure works well for $s = 14$, $c = d = 0$, $\gamma = -\gamma_2$ and we find

$$a = \frac{229523}{2 \cdot 3 \cdot 47^2}, \qquad b = \frac{6687}{2 \cdot 3 \cdot 47^2}.$$

Thus we obtain identifications

$$\begin{aligned} Q_1^3 =& \epsilon^{229523/2\cdot 3\cdot 47^2} \eta^{6687/2\cdot 3\cdot 47^2} \frac{1}{2^{14}(1+\rho)^9} \cdot 3^3 \cdot \frac{1}{(4-3\rho^2)^{12}} \cdot \frac{47^8}{(4-\rho)^{24}} \\ &\cdot (831882 + 504799\rho + 30312\rho^2)^3. \end{aligned}$$

This can be written as

$$\begin{aligned} Q_1 =& \epsilon^{30713/2\cdot 3^2\cdot 47^2} \eta^{6687/2\cdot 3^2\cdot 47^2} \frac{\sqrt[3]{2}}{2^5(1+\rho)^3} \cdot 3 \cdot \frac{1}{(4-3\rho^2)^4} \cdot \frac{47^{8/3}}{(4-\rho)^8} \\ &\cdot (183006652 + 818452906\rho + 386627159\rho^2). \end{aligned}$$

Taking "conjugates", we obtain

$$\begin{aligned} Q_2 =& (\epsilon^{(2)})^{30713/2\cdot 3^2\cdot 47^2} (\eta^{(2)})^{6687/2\cdot 3^2\cdot 47^2} \frac{\sqrt[3]{2}}{2^5(1+\rho^{(2)})^3} \cdot 3 \cdot \frac{1}{(4-3(\rho^{(2)})^2)^4} \\ &\cdot \frac{47^{8/3}}{(4-\rho^{(2)})^8} \cdot (183006652 + 818452906\rho^{(2)} + 386627159(\rho^{(2)})^2), \\ Q_3 =& (\epsilon^{(3)})^{30713/2\cdot 3^2\cdot 47^2} (\eta^{(3)})^{6687/2\cdot 3^2\cdot 47^2} \frac{\sqrt[3]{2}}{2^5(1+\rho^{(3)})^3} \cdot 3 \cdot \frac{1}{(4-3(\rho^{(3)})^2)^4} \\ &\cdot \frac{47^{8/3}}{(4-\rho^{(3)})^8} \cdot (183006652 + 818452906\rho^{(3)} + 386627159(\rho^{(3)})^2). \end{aligned}$$

The numerical values coincide with these algebraic numbers to the 35th decimal place. Concerning the exponents of ϵ and η, we find (cf. (4.31))

$$\begin{aligned} &2 \cdot 30713 + (86096 - 7602\rho - 32286\rho^2) \equiv 0 \mod \mathfrak{q}^2, \\ &2 \cdot 6687 + (-343344 - 65622\rho + 128754\rho^2) \equiv 0 \mod \mathfrak{q}^2. \end{aligned}$$

Here $(47) = \mathfrak{pq}$ in F, $\mathfrak{p} = (4-\rho)$, $N(\mathfrak{p}) = 47$, $\mathfrak{q} = (12 + 4\rho + \rho^2)$, $N(\mathfrak{q}) = 47^2$. This phenomenon is quite similar to Example 11 of [Y4].

Put

$$Q = \frac{\prod_{i=1}^4 L(1, \lambda_{2,2,2}^{(i)}) L(1, \lambda_{2,2,2}^{(i)}\omega)}{\pi^{48} g_K(\mathrm{id}, \mathrm{id})^{16} g_{K^{\sigma_2}}(\mathrm{id}, \mathrm{id})^{16} g_{K^{\sigma_3}}(\mathrm{id}, \mathrm{id})^{16}}.$$

Then we find numerically that

$$Q = \frac{2^{24}}{3 \cdot 229^6}, \qquad (QQ_1)(QQ_2)(QQ_3) = \frac{2^{55} \cdot l}{229^{22}}$$

and that $47^6 \cdot 229^8 \cdot QQ_i$, $i = 1, 2, 3$ are integral over $\mathbf{Z}$.

REMARK 4.5. Let the notation be the same as in Remark 3.11 and assume that all of ϵ_i, a_i are positive. Put

$$Q = \frac{\prod_{i=1}^{q} L(m/2, \lambda_i)^{\epsilon_i}}{\pi^A \prod_{i=1}^{r} g_{K^{\sigma_i}}(\mathrm{id}, \sigma_i^{-1}\tau_i\sigma_i; \{C_j^{\sigma_i}\}, \{\mathfrak{a}_\mu^{\sigma_i}\})^{a_i}}$$

The examples discussed in this section suggest that the denominator of Q can be taken as the product of powers of $N(\mathfrak{f})$, $|D_K|$ and $N(\mathfrak{a}_\mu)$, $1 \leqq \mu \leqq h_0$. Though this suggestion is consistent with (3.38) and (3.50), it is not true in general. See Example 2 of [Y3], p. 238 and the example in the footnote of [Y3], p. 239. We may expect that the other factors of the denominator is "small" and possibly related to the constant terms of Eisenstein series used to compute $L(m/2, \lambda_i)$. For the actual computation of $L(m/2, \lambda_i)$, we refer the reader to [Y3], §3, §4. We used UBASIC created by Yuji Kida for numerical computations.

5. Further investigations on the invariant $X(c)$

The purpose of this section is to show that the invariant $X(c)$ introduced in §3 has still more interesting and intricate properties.

5.1. Let us begin with general preliminaries fixing notations. Let F be a totally real number field of degree n and $\mathfrak{f}$ be an integral ideal of F. Let D denote the character group of $C_\mathfrak{f}$. For $1 \leqq i \leqq n$, we put

$$D_i = \{\chi \in D \mid \chi \text{ is unramified at } \infty_i\}.$$

The mapping $D \ni \chi \longrightarrow \chi_{\infty_i}(-1)$ gives an injection of D/D_i into $\{1, -1\}$. Hence we have $D/D_i \cong \{1\}$ or $\mathbf{Z}/2\mathbf{Z}$. Let S_i be the annihilator of D_i and s_i be the generator of S_i. Then we see easily that

$$\chi(s_i) = \chi_{\infty_i}(-1) \qquad \text{for} \quad \chi \in D, \quad 1 \leqq i \leqq n. \tag{5.1}$$

Since (5.1) characterizes s_i uniquely, we have $s_i =$ the class of (ν_i), taking $\nu_i \in \mathfrak{O}_F$ so that $\nu_i \equiv 1 \mod \mathfrak{f}$, $\nu_i^{(i)} < 0$, $\nu_i^{(j)} > 0$, $j \neq i$.

LEMMA 5.1. *Let G be a finite abelian group and $\widehat{G}$ be the character group of G. Let K be a subgroup of $\widehat{G}$ and H be the annihilator of K. Let f be a complex valued function on G. Then*

$$\sum_{g \in G} \chi(g) f(g) = 0 \qquad \textit{for every} \quad \chi \in K$$

holds if and only if

$$\sum_{h \in H} f(gh) = 0 \qquad \textit{for every} \quad g \in G$$

holds.

PROOF. Define a function φ on G/H by

$$\varphi(gH) = \sum_{h\in H} f(gh), \quad g \in G.$$

Since K can be identified with the character group of G/H canonically,

$$\sum_{x\in G/H} \psi(x)\varphi(x) = 0 \qquad \text{for every character } \psi \text{ of } G/H$$

is equivalent to $\varphi = 0$. Hence the assertion follows.

LEMMA 5.2. *Let $c \in C_{\mathfrak{f}}$. If $n \geqq 2$ or $\mathfrak{f} \neq (1)$, we have*

$$\zeta_F(0, c) + \zeta_F(0, cs_i) = 0 \qquad \textit{for} \quad 1 \leqq i \leqq n. \tag{5.2}$$

If $n \geqq 3$ or $\mathfrak{f} \neq (1)$, we have

$$\zeta_F'(0, c) + \zeta_F'(0, cs_i) + \zeta_F'(0, cs_j) + \zeta_F'(0, cs_is_j) = 0 \quad \textit{for} \quad 1 \leqq i, j \leqq n,\ i \neq j. \tag{5.3}$$

PROOF. Let $\chi \in D$ and $\mathfrak{f}_\chi$ be the finite part of the conductor of χ. We have

$$\sum_{c\in C_{\mathfrak{f}}} \chi(c)\zeta_F(s, c) = \prod_{\mathfrak{p}|\mathfrak{f}, \mathfrak{p}\nmid\mathfrak{f}_\chi} (1 - \chi^*(\mathfrak{p})N(\mathfrak{p})^{-s})L(s, \chi^*), \tag{5.4}$$

where χ^* denotes the primitive character associated to χ. For $\chi \in D_i$, we have $L(0, \chi^*) = 0$ if $\chi \neq 1$ or $n \geqq 2$. Hence, excluding the case $F = \mathbf{Q}$ and $\mathfrak{f} = (1)$, we get

$$\sum_{c\in C_{\mathfrak{f}}} \chi(c)\zeta_F(0, c) = 0 \qquad \text{for every} \quad \chi \in D_i. \tag{5.5}$$

If $D_i \neq D$, we have $s_i \neq 1$ and (5.2) follows from (5.5) using Lemma 5.1. If $D_i = D$, we have $s_i = 1$. By (5.5), we have $\zeta_F(0, c) = 0$ for every $c \in C_{\mathfrak{f}}$. Hence (5.2) holds.

Assume $i \neq j$ and $\chi \in D_i \cap D_j$. Then $n \geqq 2$. If $\chi \neq 1$ or $n \geqq 3$, we have $L(0, \chi^*) = L'(0, \chi^*) = 0$. Therefore, excluding the case $n = 2$ and $\mathfrak{f} = (1)$, we get

$$\sum_{c\in C_{\mathfrak{f}}} \chi(c)\zeta_F'(0, c) = 0 \qquad \text{for every} \quad \chi \in D_i \cap D_j. \tag{5.6}$$

If $[D : D_i] = 2$, $[D : D_j] = 2$ and $D_i \neq D_j$, then the annihilator of $D_i \cap D_j$ is $\{1, s_i, s_j, s_is_j\}$ and (5.3) follows from (5.6) using Lemma 5.1. The other cases can be treated similarly. This completes the proof.

We define a subgroup S_0 of $C_{\mathfrak{f}}$ by $S_0 = \langle s_i \mid 1 \leqq i \leqq n\rangle$. Let M be the maximal ray class field modulo $\mathfrak{f}\infty_1 \cdots \infty_n$ over F. We do not assume that the conductor of M over F is equal to $\mathfrak{f}\infty_1 \cdots \infty_n$. We have $\mathrm{Gal}(M/F) \cong C_{\mathfrak{f}}$.

LEMMA 5.3. *Let F_0 be the subfield of M which corresponds to S_0. Then F_0 is the maximum totally real subfield contained in M.*

PROOF. Let K be a subfield of M and X be the subgroup of $C_{\mathfrak{f}}$ which corresponds to K. Then K is totally real if and only if every character of $\mathrm{Gal}(K/F)$, regarded as a character of $C_{\mathfrak{f}}$ using the canonical map $\mathrm{Gal}(M/F) \longrightarrow \mathrm{Gal}(K/F)$ and the Artin map, is unramified at every infinite place of F. In other words, K is totally real if and only if every character χ of $C_{\mathfrak{f}}$ which is trivial on X does not ramify at all infinite places of F. By (5.1), this condition is equivalent to $S_0 \subset X$. Therefore that K is totally real is equivalent to that $F_0 \supset K$. This completes the proof.

Let $D_0 = \cap_{i=1}^{n} D_i$. The character group of $\mathrm{Gal}(F_0/F)$ can be identified with D_0 canonically. We define a subgroup S' of S_0 by

$$S' = \{s_{i_1} s_{i_2} \cdots s_{i_t} \mid 1 \leqq i_1 < i_2 < \ldots < i_t \leqq n,\ t \text{ is even}\}.$$

LEMMA 5.4. *Let K be the subfield of M which corresponds to S'. If D does not contain a character which ramifies at every infinite place, then M has no CM-subfields. If D contains a character which ramifies at every infinite place, then K is the maximum CM-subfield of M.*

PROOF. Suppose that M contains a CM-field K_1 and let F_1 be its maximal real subfield. Let η be the non-trivial character of $\mathrm{Gal}(K_1/F_1)$. We can find a character χ of $\mathrm{Gal}(K_1/F)$ such that $\chi|\mathrm{Gal}(K_1/F) = \eta$. If χ does not ramify at an infinite place ∞_i, then all characters of $\mathrm{Gal}(K_1/F)$ (regarded as characters of $C_{\mathfrak{f}}$) do not ramify at ∞_i. This contradicts the fact that K_1 is totally imaginary. Hence the first assertion follows.

Next assume that $\chi \in D$ ramifies at every infinite place of F. By (5.1), we have $\chi(s_i) = -1$, $1 \leqq i \leqq n$. Since $\chi|S' = 1$, we get $[S_0 : S'] = 2$. Therefore K is a quadratic extension of F_0. We can regard χ as a character of $\mathrm{Gal}(K/F)$. Since χ ramifies at every infinite place of F, K must be totally imaginary. Hence K is a CM-field. Let K' be a CM-subfield of M and let F^* be the maximal real subfield of $K \vee K'$. By Lemma 5.3, we have $F^* = F_0$. Since $[K \vee K' : F^*] = 2$, we get $K \vee K' = K$, hence $K' \subset K$. This completes the proof.

5.2. We are going to investigate relations between the invariants $X(c)$ when we change the modulus $\mathfrak{f}$. Let $\mathfrak{q}$ be a prime ideal of F and let

$$\varphi : C_{\mathfrak{f}\mathfrak{q}} \longrightarrow C_{\mathfrak{f}}$$

be the canonical map. Take $c \in C_{\mathfrak{f}}$ and put $\varphi^{-1}(c) = \{c_1, c_2, \ldots, c_t\}$. First let us consider the relation between the partial zeta functions. An integral ideal $\mathfrak{a}$ of F belongs to the class c_i for some i, $1 \leqq i \leqq t$ if and only if $(\mathfrak{a}, \mathfrak{f}\mathfrak{q}) = 1$ and the class of $\mathfrak{a}$ in $C_{\mathfrak{f}}$ is c. Therefore we obtain

$$\zeta_F(s, c) = \sum_{l=1}^{t} \zeta_F(s, c_l) + \sum_{\mathfrak{a} \in c,\ \mathfrak{a}:\text{integral}, \mathfrak{q}|\mathfrak{a}} N(\mathfrak{a})^{-s}. \tag{5.7}$$

Hence

$$\zeta_F(s, c) = \sum_{l=1}^{t} \zeta_F(s, c_l) \qquad \text{if} \quad \mathfrak{q} \mid \mathfrak{f}. \tag{5.8}$$

Assume that $\mathfrak{q}$ does not divide $\mathfrak{f}$ and let q be the class of $\mathfrak{q}$ in $C_{\mathfrak{f}}$. We have

$$\sum_{\mathfrak{a}\in c,\ \mathfrak{a}:\text{integral},\mathfrak{q}|\mathfrak{a}} N(\mathfrak{a})^{-s} = \sum_{\mathfrak{b}:\text{integral},\ \mathfrak{b}\in q^{-1}c} N(\mathfrak{q}\mathfrak{b})^{-s} = N(\mathfrak{q})^{-s}\zeta_F(s, q^{-1}c).$$

Hence, in this case, we get

$$\zeta_F(s,c) = \sum_{l=1}^{t} \zeta_F(s,c_l) + N(\mathfrak{q})^{-s}\zeta_F(s,q^{-1}c) \qquad \text{if} \quad \mathfrak{q}\nmid\mathfrak{f}. \tag{5.9}$$

Now we are going to consider $X(c)$. If $\mathfrak{q}\nmid\mathfrak{f}$, let q be the class of $\mathfrak{q}$ in $C_{\mathfrak{f}}$. We may assume that $\mathfrak{a}_\mu$ is of the form $\mathfrak{a}_\mu = \mathfrak{b}_\mu\mathfrak{q}$ with an integral ideal $\mathfrak{b}_\mu$. Choose μ so that $\mathfrak{a}_\mu\mathfrak{f}$ and c belong to the same narrow ideal class. We have

$$\begin{aligned}&\{z\in(\mathfrak{b}_\mu\mathfrak{q}\mathfrak{f})^{-1}\cap C_j \mid (z)\mathfrak{b}_\mu\mathfrak{q}\mathfrak{f}\equiv c \quad \text{in} \quad C_{\mathfrak{f}}\}\\ &= \sqcup_{l=1}^{t}\{z\in(\mathfrak{b}_\mu\mathfrak{q}\mathfrak{f})^{-1}\cap C_j \mid (z)\mathfrak{b}_\mu\mathfrak{q}\mathfrak{f}\equiv c_l \quad \text{in} \quad C_{\mathfrak{f}\mathfrak{q}}\}\\ &\quad\sqcup\{z\in(\mathfrak{b}_\mu\mathfrak{f})^{-1}\cap C_j \mid (z)\mathfrak{b}_\mu\mathfrak{q}\mathfrak{f}\equiv q^{-1}c \quad \text{in} \quad C_{\mathfrak{f}}\}.\end{aligned}\tag{5.10}$$

Here we understand that the second set of the right-hand side is empty if $\mathfrak{q}\mid\mathfrak{f}$. We write $\zeta^{(i)}(s,C_j,\mathfrak{a}_\mu)$ (resp. $\zeta^{(i,k)}(s,C_j,\mathfrak{a}_\mu)$) defined by (3.25) (resp. (3.26)) as $\zeta_c^{(i)}(s,C_j,\mathfrak{a}_\mu)$ (resp. $\zeta_c^{(i,k)}(s,C_j,\mathfrak{a}_\mu)$) to emphasize the dependence on c. By (5.10), we have

$$\zeta_c^{(i)}(s,C_j,\mathfrak{b}_\mu\mathfrak{q}) = \sum_{l=1}^{t}\zeta_{c_l}^{(i)}(s,C_j,\mathfrak{b}_\mu) + \zeta_{q^{-1}c}^{(i)}(s,C_j,\mathfrak{b}_\mu), \tag{5.11}$$

$$\zeta_c^{(i,k)}(s,C_j,\mathfrak{b}_\mu\mathfrak{q}) = \sum_{l=1}^{t}\zeta_{c_l}^{(i,k)}(s,C_j,\mathfrak{b}_\mu) + \zeta_{q^{-1}c}^{(i,k)}(s,C_j,\mathfrak{b}_\mu). \tag{5.12}$$

Here, in (5.11) and (5.12), the second terms of the right-hand sides are omitted if $\mathfrak{q}\mid\mathfrak{f}$. By (5.11), (5.12) and (3.27) $\sim$ (3.31), we obtain

$$\begin{aligned}X(c;\{C_j\},\mathfrak{b}_\mu\mathfrak{q}) &= \sum_{l=1}^{t} X(c_l;\{C_j\},\mathfrak{b}_\mu) + X(q^{-1}c;\{C_j\},\mathfrak{b}_\mu)\\ &\quad -\frac{1}{n}\log N(\mathfrak{q})\sum_{j\in J}\zeta_{q^{-1}c}^{(1)}(0,C_j,\mathfrak{b}_\mu).\end{aligned}\tag{5.13}$$

Here, if $\mathfrak{q}\mid\mathfrak{f}$, the last two terms of the right-hand side are omitted. In §6 of the next chaper, we will prove $\sum_{j\in J}\zeta_{q^{-1}c}^{(1)}(0,C_j,\mathfrak{b}_\mu) = \zeta_F(0,q^{-1}c)$. Admitting this fact, (5.13) can be written as

$$\begin{aligned}X(c;\{C_j\},\mathfrak{b}_\mu\mathfrak{q}) &= \sum_{l=1}^{t} X(c_l;\{C_j\},\mathfrak{b}_\mu) + X(q^{-1}c;\{C_j\},\mathfrak{b}_\mu)\\ &\quad -\frac{1}{n}\log N(\mathfrak{q})\zeta_F(0,q^{-1}c).\end{aligned}\tag{5.14}$$

This formula is consistent with (3.11), (5.8), (5.9).

5.3. Now let us begin the study of finer properties of $X(c)$. Let F be a real quadratic field. We use the notation of §4. Thus $X(c) = X(c; \{C_j\}, \mathfrak{a}_\mu)$ is defined by (4.7). Let ϵ_0 denote the fundamental unit of F. First we investigate the case where $N(\epsilon_0) = -1$. We have $\epsilon = \epsilon_0^2$.

LEMMA 5.5. *If* $N(\epsilon_0) = -1$, *then we have*

$$C(1, \epsilon_0) = C(1, \epsilon) \sqcup C(\epsilon_0, \epsilon) \sqcup C(\epsilon).$$

PROOF. Though the assertion is obvious by drawing a picture, we give an analytic proof. We note that $\mathrm{Tr}(\epsilon_0) = \epsilon_0 + \epsilon_0' \geqq 1$. We have $x + y\epsilon_0 = x_1 + y_1\epsilon$ where x, y and x_1, y_1 are related by

$$x = x_1 + y_1, \qquad y = (\epsilon_0 + \epsilon_0')y_1, \qquad x_1 = x - \frac{y}{\epsilon_0 + \epsilon_0'}, \qquad y_1 = \frac{y}{\epsilon_0 + \epsilon_0'}.$$

This relation shows that $C(1, \epsilon) \subset C(1, \epsilon_0)$ and

$$C(1, \epsilon) = \{x + y\epsilon_0 \in C(1, \epsilon_0) \mid x,\ y > 0, \quad (\epsilon_0 + \epsilon_0')x > y\}. \tag{5.15}$$

We have $x + y\epsilon_0 = x_1\epsilon_0 + y_1\epsilon$ with

$$x = y_1, \qquad y = x_1 + (\epsilon_0 + \epsilon_0')y_1, \qquad x_1 = y - (\epsilon_0 + \epsilon_0')x, \qquad y_1 = x.$$

Hence we have $C(\epsilon_0, \epsilon) \subset C(1, \epsilon_0)$ and

$$C(\epsilon_0, \epsilon) = \{x + y\epsilon_0 \in C(1, \epsilon_0) \mid x,\ y > 0, \quad (\epsilon_0 + \epsilon_0')x < y\}. \tag{5.16}$$

Noting $\epsilon x = x + (\epsilon_0 + \epsilon_0')x\epsilon_0$, we see that

$$C(\epsilon) = \{x + y\epsilon_0 \in C(1, \epsilon_0) \mid x,\ y > 0, \quad (\epsilon_0 + \epsilon_0')x = y\}. \tag{5.17}$$

Our assertion follows from (5.15), (5.16) and (5.17).

We put

$$C_1 = C(1, \epsilon), \quad C_2 = C(1), \quad C_3 = C(1, \epsilon_0), \quad C_4 = C(\epsilon_0, \epsilon), \quad C_5 = C(\epsilon).$$

Take $c \in C_\mathfrak{f}$ and choose $\mathfrak{a}_\mu$ so that $\mathfrak{a}_\mu\mathfrak{f}$ and c belong to the same narrow ideal class. For $1 \leqq j \leqq 5$, we put

$$L_j(c) = \{z \mid z \in (\mathfrak{a}_\mu\mathfrak{f})^{-1} \cap C_j,\ (z)\mathfrak{a}_\mu\mathfrak{f} \equiv c \ \text{ in } \ C_\mathfrak{f}\}, \tag{5.18}$$

$$\mathcal{Z}_j(s, c) = \sum_{z \in L_j(c)} (z^{(1)})^{-s}. \tag{5.19}$$

Thus $\mathcal{Z}_j(s, c) = \zeta^{(1)}(s, C_j, \mathfrak{a}_\mu)$ for $j = 1, 2$. By (3.28), we have

$$G(c; \{C_j\}, \mathfrak{a}_\mu) = \sum_{j=1}^{2} \left(\frac{d}{ds}\mathcal{Z}_j(s, c)\right)\bigg|_{s=0}. \tag{5.20}$$

By a simple counting argument, we see that the number of $z \in (\mathfrak{a}_\mu\mathfrak{f})^{-1} \cap C_j$ which satisfies $z^{(1)} < t$ is of order $O(t^{\dim C_j})$ for $t \longrightarrow +\infty$. Hence the sum in (5.19) converges absolutely when $\Re(s) > \dim C_j + 1$. By Lemma 5.5, we have

$$\mathcal{Z}_3(s, c) = \mathcal{Z}_1(s, c) + \mathcal{Z}_4(s, c) + \mathcal{Z}_5(s, c). \tag{5.21}$$

Clearly the mappings

$$L_3(c) \ni z \longrightarrow \epsilon_0 z \in L_4(c), \qquad L_2(c) \ni z \longrightarrow \epsilon z \in L_5(c)$$

are bijections. Hence we have

$$\mathcal{Z}_4(s,c) = \epsilon_0^{-s}\mathcal{Z}_3(s,c), \qquad \mathcal{Z}_5(s,c) = \epsilon^{-s}\mathcal{Z}_2(s,c). \tag{5.22}$$

By (5.21) and (5.22), we have

$$\mathcal{Z}_1(s,c) + \epsilon^{-s}\mathcal{Z}_2(s,c) = (1-\epsilon_0^{-s})\mathcal{Z}_3(s,c). \tag{5.23}$$

This formula shows that $\mathcal{Z}_3(s,c)$ can be continued to the whole s-plane as a meromorphic function and $(1-\epsilon_0^{-s})\mathcal{Z}_3(s,c)$ is holomorphic at $s=0$. Differentiating at $s=0$, we obtain

$$G(c;\{C_j\},\mathfrak{a}_\mu) - \log\epsilon\cdot\mathcal{Z}_2(0,c) = \frac{d}{ds}((1-\epsilon_0^{-s})\mathcal{Z}_3(s,c))\Big|_{s=0}. \tag{5.24}$$

Since $N(\epsilon_0) = -1$, the narrow ideal class of s_2 is trivial. Applying (5.24) to the class cs_2 and adding the formulas, we obtain

$$\begin{aligned} &G(c;\{C_j\},\mathfrak{a}_\mu) + G(cs_2;\{C_j\},\mathfrak{a}_\mu)\\ &=\frac{d}{ds}((1-\epsilon_0^{-s})(\mathcal{Z}_3(s,c)) + \mathcal{Z}_3(s,cs_2)))\Big|_{s=0} + \log\epsilon\cdot(\mathcal{Z}_2(0,c)+\mathcal{Z}_2(0,cs_2)). \end{aligned} \tag{5.25}$$

For $1 \leqq j \leqq 5$, we put

$$L_j^*(c) = \{z \mid z\in(\mathfrak{a}_\mu\mathfrak{f})^{-1}\cap C_j,\ (z)\mathfrak{a}_\mu\mathfrak{f}\equiv c \mod^\times \mathfrak{f}\infty_1\}, \tag{5.26}$$

$$\mathcal{Z}_j^*(s,c) = \sum_{z\in L_j^*(c)} (z^{(1)})^{-s}. \tag{5.27}$$

Here $(z)\mathfrak{a}_\mu\mathfrak{f}\equiv c \mod^\times \mathfrak{f}\infty_1$ means that the classes of $(z)\mathfrak{a}_\mu\mathfrak{f}$ and (the image of) c are same in the ideal class group modulo $\mathfrak{f}\infty_1$. Then, if $s_2\neq 1$, we have

$$L_j^*(c) = L_j(c)\sqcup L_j(cs_2), \qquad \mathcal{Z}_j^*(s,c) = \mathcal{Z}_j(s,c)+\mathcal{Z}_j(s,cs_2).$$

Hence (5.25) can be written as

$$G(c;\{C_j\},\mathfrak{a}_\mu)+G(cs_2;\{C_j\},\mathfrak{a}_\mu) = \frac{d}{ds}((1-\epsilon_0^{-s})\mathcal{Z}_3^*(s,c))\Big|_{s=0} + \log\epsilon\cdot\mathcal{Z}_2^*(0,c). \tag{5.28}$$

If $s_2 = 1$, this formula holds multiplying the right-hand side by 2. We put

$$R_3^*(c) = \{z = x+y\epsilon_0\in(\mathfrak{a}_\mu\mathfrak{f})^{-1} \mid x,y\in\mathbf{Q},\ 0<x,\ y\leqq 1,\ (z)\mathfrak{a}_\mu\mathfrak{f}\equiv c \mod^\times\mathfrak{f}\infty_1\}.$$

Then $R_3^*(c)$ is a finite set. By the same proof as that of II, Lemma 3.2, we have

$$L_3^*(c) = \sqcup_{z\in R_3^*(c)}\{z+x+y\epsilon_0 \mid 0\leqq x,\ y\in\mathbf{Z}\}.$$

Therefore we obtain

$$Z_3^*(s,c) = \sum_{z=x+y\epsilon_0\in R_3^*(c)} \zeta(s,(1,\epsilon_0),(x,y)). \tag{5.29}$$

By I, Theorem 2.1, we have

$$Z_3^*(0,c) = \sum_{z=x+y\epsilon_0 \in R_3^*(c)} (\frac{1}{2}\epsilon_0^{-1}B_2(x) + B_1(x)B_1(y) + \frac{1}{2}\epsilon_0 B_2(y)). \tag{5.30}$$

We have $W(c) = -\frac{1}{2}\log N(\mathfrak{a}_\mu\mathfrak{f})\zeta_F(0,c)$. (cf. [Y4], Lemma 4. The corresponding fact for any totally real field will be proved in §6 of the next chapter.) By Lemma 5.2, we have $W(c)+W(cs_2)=0$. By (5.28), (5.30) and (4.3), we obtain

$$\begin{aligned} &X(c;\{C_j\},\mathfrak{a}_\mu) + X(cs_2;\{C_j\},\mathfrak{a}_\mu) \\ &= \log\epsilon_0 \sum_{z=x+y\epsilon_0 \in R_3^*(c)} (\frac{1}{2}\epsilon_0^{-1}B_2(x) + B_1(x)B_1(y) + \frac{1}{2}\epsilon_0 B_2(y)) \\ &+\frac{1}{4}(\epsilon'-\epsilon)\log\epsilon(\sum_{z=x+y\epsilon\in R(\epsilon,c)} B_2(x) + \sum_{z=x+y\epsilon\in R(\epsilon,cs_2)} B_2(x)) \\ &+\log\epsilon\cdot \mathcal{Z}_2^*(0,c) \end{aligned} \tag{5.31}$$

when $s_2 \neq 1$; if $s_2 = 1$, (5.31) holds by multiplying the right-hand side by 2. To simplify this formula, we need some preparations. For a positive integer e, we temporarily put

$$R(e) = \{z = xe + ye\epsilon_0 \in (\mathfrak{a}_\mu\mathfrak{f})^{-1} \mid x,y\in\mathbf{Q},\ 0<x,\ y\leqq 1,\ (z)\mathfrak{a}_\mu\mathfrak{f}\equiv c \mod^\times \mathfrak{f}\infty_1\}.$$

Regarding $\{e, e\epsilon_0\}$ as a basis of $C(1,\epsilon_0)$, we obtain, by I, Theorem 2.1, that

$$Z_3^*(0,c) = \sum_{z=xe+ye\epsilon_0 \in R(e)} (\frac{1}{2}\epsilon_0^{-1}B_2(x) + B_1(x)B_1(y) + \frac{1}{2}\epsilon_0 B_2(y)). \tag{5.32}$$

LEMMA 5.6. *We have*

$$\sum_{z=x+y\epsilon\in R(\epsilon,c)} B_2(x) = \sum_{z=x+y\epsilon\in R(\epsilon,c)} B_2(y),$$
$$\sum_{z=xe+ye\epsilon_0\in R(e)} B_2(x) = \sum_{z=xe+ye\epsilon_0\in R(e)} B_2(y).$$

PROOF. The first equality is proved in [Y4], Lemma 4. The proof of the latter equality is similar. In fact, let $z = xe+ye\epsilon_0 \in R(e)$. Then $\epsilon_0^{-1}z = ye-(\epsilon_0+\epsilon_0')xe+xe\epsilon_0$ satisfies $(\epsilon_0^{-1}z)\mathfrak{a}_\mu\mathfrak{f}\equiv c \mod \mathfrak{f}\infty_1$. Take $x_1\in\mathbf{Q}$ so that $x_1\equiv y-(\epsilon_0+\epsilon_0')x \mod \mathbf{Z}$, $0<x_1\leqq 1$. Put $y_1 = x$, $z_1 = x_1e+y_1e\epsilon_0$. Then $z_1\in R(e)$ and the mapping $z\longrightarrow z_1$ is a bijection. Hence the assertion follows.

LEMMA 5.7. *If $s_2\neq 1$, we have*

$$(\epsilon_0+\epsilon_0')(\sum_{z=x+y\epsilon\in R(\epsilon,c)} B_2(x) + \sum_{z=x+y\epsilon\in R(\epsilon,cs_2)} B_2(x)) = \sum_{z=x+y\epsilon_0\in R_3^*(c)} B_2(x).$$

If $s_2 = 1$, this formula holds multiplying the right-hand side by 2.

PROOF. Comparing the irrational parts of the right-hand side of (5.32) for $R(1) = R_3^*(c)$ and $R(e)$, we obtain

$$\sum_{z=x+y\epsilon_0\in R_3^*(c)} B_2(y) = \sum_{z=xe+ye\epsilon_0\in R(e)} B_2(y) \tag{5.33}$$

in view of Lemma 5.6. Let $e = \epsilon_0 + \epsilon_0'$. Put $R_0(\epsilon, c) = R(\epsilon, c) \cup R(\epsilon, cs_2)$. This union is disjoint if and only if $s_2 \neq 1$. We have

$$\begin{aligned} R_0(\epsilon, c) = \{z = &x + y\epsilon \in (\mathfrak{a}_\mu \mathfrak{f})^{-1} \\ &\mid x, y \in \mathbf{Q},\ 0 < x \leqq 1,\ 0 \leqq y < 1,\ (z)\mathfrak{a}_\mu \mathfrak{f} \equiv c \mod^\times \mathfrak{f}\infty_1\}. \end{aligned}$$

It suffices to show that

$$e \sum_{z=x+y\epsilon\in R_0(\epsilon,c)} B_2(y) = \sum_{z=xe+ye\epsilon_0\in R(e)} B_2(y). \tag{5.34}$$

We define a mapping φ of $R(e)$ into $R_0(\epsilon, c)$ in the following way. Let $z = xe + ye\epsilon_0 \in R(e)$. We have $z = (ex - y) + y\epsilon$. Take $x_1 \in \mathbf{Q}$ so that $x_1 \equiv ex - y \mod \mathbf{Z}$, $0 < x_1 \leqq 1$. Let $y_1 = y$ if $0 < y < 1$ and $y_1 = 0$ if $y = 1$. Put $\varphi(z) = z_1 = x_1 + y_1\epsilon$. Then $\varphi(z) \in R_0(\epsilon, c)$. Let us show that φ is surjective. Take $z = x + y\epsilon \in R_0(\epsilon, c)$. Then $z = (x + y) + ye\epsilon_0$. Take $x_0 \in \mathbf{Q}$ so that $x_0 \equiv (x + y)/e \mod \mathbf{Z}$, $0 < x_0 \leqq 1$. Let $y_0 = y$ if $0 < y < 1$ and $y_0 = 1$ if $y = 0$. Put $z_0 = x_0 e + y_0 e\epsilon_0$. Then $z_0 \in R(e)$. We have $\varphi(z_0) = x_1 + y\epsilon$ and x_1 is determined by

$$x_1 \in \mathbf{Q}, \quad 0 < x_1 \leqq 1, \quad x_1 \equiv ex_0 - y \mod \mathbf{Z}.$$

Since $ex_0 \equiv x + y \mod e\mathbf{Z}$, we have $x_1 \equiv x \mod \mathbf{Z}$. Hence $x_1 = x$, $\varphi(z_0) = z$. This proves that φ is surjective. Since $B_2(0) = B_2(1)$, (5.34) follows if we can show that

$$|\varphi^{-1}(z)| = e \quad \text{for every} \quad z \in R_0(\epsilon_0, c). \tag{5.35}$$

Let $z = x + y\epsilon \in R_0(\epsilon, c)$ and take $z_0 = x_0 e + y_0 e\epsilon_0 \in R(e)$. The conditions for $z_0 \in \varphi^{-1}(z)$ are

$$ex_0 - y_0 \equiv x \mod \mathbf{Z}, \qquad y_0 \equiv y \mod \mathbf{Z}.$$

When z is given, y_0 is uniquely determined and there are e choices for x_0. Hence (5.35) follows. This completes the proof.

By Lemmas 5.6 and 5.7, (5.31) can be written as

$$\begin{aligned} &X(c; \{C_j\}, \mathfrak{a}_\mu) + X(cs_2; \{C_j\}, \mathfrak{a}_\mu) \\ &= \log \epsilon_0 \sum_{z=x+y\epsilon_0\in R_3^*(c)} B_1(x)B_1(y) + \log \epsilon \cdot \mathcal{Z}_2^*(0, c). \end{aligned} \tag{5.36}$$

We put

$$\begin{aligned} R(\epsilon_0, c) = \{&z = x + y\epsilon_0 \in (\mathfrak{a}_\mu \mathfrak{f})^{-1} \mid \\ &x, y \in \mathbf{Q},\ 0 < x \leqq 1,\ 0 \leqq y < 1, (z)\mathfrak{a}_\mu \mathfrak{f} \equiv c \mod^\times \mathfrak{f}\infty_1\}. \end{aligned} \tag{5.37}$$

By II, Theorem 2.1, we obtain

$$\mathcal{Z}_2^*(0, c) = -\sum_{z=x+y\epsilon_0\in R(\epsilon_0,c), y=0} B_1(x). \tag{5.38}$$

We see that

$$\begin{aligned}
&\sum_{z=x+y\epsilon_0\in R_3^*(c)} B_1(x)B_1(y)\\
&= \sum_{z=x+y\epsilon_0\in R(\epsilon_0,c),0<y<1} B_1(x)B_1(y) + \sum_{z=x+y\epsilon_0\in R(\epsilon_0,c),y=0} \frac{1}{2}B_1(x)\\
&= \sum_{z=x+y\epsilon_0\in R(\epsilon_0,c)} B_1(x)B_1(y) + \sum_{z=x+y\epsilon_0\in R(\epsilon_0,c),y=0} B_1(x).
\end{aligned}$$

Thus we obtain the following theorem.

THEOREM 5.8. *Let F be a real quadratic field and ϵ_0 be the fundamental unit of F. We assume that $N(\epsilon_0) = -1$. Let $c \in C_{\mathfrak{f}}$ and define $R(\epsilon_0, c)$ by (5.37). If $s_2 \neq 1$, then we have*

$$\begin{aligned}
&X(c; \{C_j\}, \mathfrak{a}_\mu) + X(cs_2; \{C_j\}, \mathfrak{a}_\mu)\\
&= \log\epsilon_0 \Bigg[\sum_{z=x+y\epsilon_0\in R(\epsilon_0,c)} B_1(x)B_1(y) - \sum_{z=x+y\epsilon_0\in R(\epsilon_0,c),y=0} B_1(x) \Bigg].
\end{aligned}$$

If $s_2 = 1$, this formula holds multiplying the right-hand side by 2.

5.4. Next we will investigate the case where $N(\epsilon_0) = 1$. First we prove the following Lemma which holds without this condition.

LEMMA 5.9. *For any $\alpha \in F$ satisfying $\alpha > 0$, $\alpha' < 0$, we have*

$$C(1, \alpha) = C(1, \epsilon) \sqcup C(\epsilon, \epsilon\alpha) \sqcup C(\epsilon\alpha, \alpha) \sqcup C(\epsilon) \sqcup C(\epsilon\alpha).$$

PROOF. By drawing a picture, we clearly see the assertion. An analytic proof can be given as follows. Suppose that $u = (u_1, u_2) \in \mathbf{R}^2$ and $v = (v_1, v_2) \in \mathbf{R}^2$ are given. We assume that

$$u_1, v_1 > 0, \qquad \frac{v_2}{v_1} < \frac{u_2}{u_1}.$$

Then $(z_1, z_2) \in \mathbf{R}^2$ belongs to $C(u, v)$ if and only if

$$z_1 > 0 \quad \text{and} \quad \frac{v_2}{v_1} < \frac{z_2}{z_1} < \frac{u_2}{u_1}.$$

In fact, examining the condition that the simultaneous equations

$$xu_1 + yv_1 = z_1, \qquad xu_2 + yv_2 = z_2$$

have positive solutions x, y, we see easily this assertion. Hence we have

$$\begin{aligned}
C(1, \alpha) &= \{(z_1, z_2) \in \mathbf{R}^2 \mid z_1 > 0,\ \frac{\alpha'}{\alpha} < \frac{z_2}{z_1} < 1\},\\
C(1, \epsilon) &= \{(z_1, z_2) \in \mathbf{R}^2 \mid z_1 > 0,\ \frac{\epsilon'}{\epsilon} < \frac{z_2}{z_1} < 1\},\\
C(\epsilon, \epsilon\alpha) &= \{(z_1, z_2) \in \mathbf{R}^2 \mid z_1 > 0,\ \frac{\epsilon'\alpha'}{\epsilon\alpha} < \frac{z_2}{z_1} < \frac{\epsilon'}{\epsilon}\},\\
C(\epsilon\alpha, \alpha) &= \{(z_1, z_2) \in \mathbf{R}^2 \mid z_1 > 0,\ \frac{\alpha'}{\alpha} < \frac{z_2}{z_1} < \frac{\epsilon'\alpha'}{\epsilon\alpha}\}.
\end{aligned}$$

Since

$$C(\epsilon) = \{(z_1, z_2) \in \mathbf{R}^2 \mid z_1 > 0, \ \frac{z_2}{z_1} = \frac{\epsilon'}{\epsilon}\},$$
$$C(\epsilon\alpha) = \{(z_1, z_2) \in \mathbf{R}^2 \mid z_1 > 0, \ \frac{z_2}{z_1} = \frac{\epsilon'\alpha'}{\epsilon\alpha}\},$$

the assertion follows.

Taking $\alpha \in \mathfrak{O}_F$ such that $\alpha > 0$, $\alpha' < 0$, we put

$$C_1 = C(1, \epsilon), \quad C_2 = C(1), \quad C_3 = C(1, \alpha), \quad C_4 = C(\epsilon, \epsilon\alpha), \quad C_5 = C(\epsilon\alpha, \alpha),$$
$$C_6 = C(\epsilon), \quad C_7 = C(\epsilon\alpha).$$

Take $c \in C_{\mathfrak{f}}$ and choose $\mathfrak{a}_\mu$ so that $\mathfrak{a}_\mu\mathfrak{f}$ and c belong to the same narrow ideal class. For $j = 1$, 2, put

$$L_j(c) = \{z \mid z \in (\mathfrak{a}_\mu\mathfrak{f})^{-1} \cap C_j, \ (z)\mathfrak{a}_\mu\mathfrak{f} \equiv c \quad \text{in} \quad C_{\mathfrak{f}}\},$$
$$\mathcal{Z}_j(s, c) = \sum_{z \in L_j(c)} (z^{(1)})^{-s}.$$

By (3.28), we have

$$G(c; \{C_j\}, \mathfrak{a}_\mu) = \sum_{j=1}^{2} (\frac{d}{ds}\mathcal{Z}_j(s, c))\Big|_{s=0}. \tag{5.39}$$

We see that $(\alpha)\mathfrak{a}_\mu\mathfrak{f}$ and cs_2 belong to the same narrow ideal class. For $j = 1$, 2, put

$$L_j(cs_2) = \{z \mid z \in ((\alpha)\mathfrak{a}_\mu\mathfrak{f})^{-1} \cap C_j, \ (z)(\alpha)\mathfrak{a}_\mu\mathfrak{f} \equiv cs_2 \quad \text{in} \quad C_{\mathfrak{f}}\},$$
$$\mathcal{Z}_j(s, cs_2) = \sum_{z \in L_j(cs_2)} (z^{(1)})^{-s}.$$

By (3.28), we have

$$G(cs_2; \{C_j\}, (\alpha)\mathfrak{a}_\mu) = \sum_{j=1}^{2} (\frac{d}{ds}\mathcal{Z}_j(s, cs_2))\Big|_{s=0}. \tag{5.40}$$

For $1 \leqq j \leqq 7$, we put

$$L_j^*(c) = \{z \mid z \in (\mathfrak{a}_\mu\mathfrak{f})^{-1} \cap C_j, \ (z)\mathfrak{a}_\mu\mathfrak{f} \equiv c \quad \operatorname{mod}^\times \mathfrak{f}\infty_1\}, \tag{5.41}$$

$$\mathcal{Z}_j^*(s, c) = \sum_{z \in L_j^*(c)} (z^{(1)})^{-s}. \tag{5.42}$$

We see that the series defining $\mathcal{Z}_j^*(s, c)$ converges absolutely when $\Re(s) > \dim C_j + 1$. Hereafter, until the end of Remark 5.13, we assume that $N(\epsilon_0) = 1$. We have $\epsilon = \epsilon_0$. We note that

$$L_j(c) = L_j^*(c) \qquad \text{for} \quad j = 1, 2. \tag{5.43}$$

In fact $L_j(c) \subset L_j^*(c)$ is obvious. Suppose that $z \in L_j^*(c)$. Then the class of $(z)\mathfrak{a}_\mu\mathfrak{f}$ in $C_{\mathfrak{f}}$ is c or cs_2. If the class of $(z)\mathfrak{a}_\mu\mathfrak{f}$ in $C_{\mathfrak{f}}$ is cs_2, then the narrow ideal class defined by s_2 must be 1 since $z \gg 0$. This contradicts the assumption $N(\epsilon_0) = 1$. Similarly we have

$$(5.44)\quad L_j(cs_2) = \{z \mid z \in ((\alpha)\mathfrak{a}_\mu\mathfrak{f})^{-1} \cap C_j,\ (z)(\alpha)\mathfrak{a}_\mu\mathfrak{f} \equiv c \mod^{\times} \mathfrak{f}\infty_1\}, \quad j = 1, 2.$$

By Lemma 5.9, we have

$$(5.45)\qquad \mathcal{Z}_1^*(s,c) = \mathcal{Z}_3^*(s,c) - \sum_{i=4}^{7} \mathcal{Z}_j^*(s,c)$$

for $\Re(s) > 3$. Clearly the mappings

$$L_3^*(c) \ni z \longrightarrow \epsilon z \in L_4^*(c), \qquad L_2^*(c) \ni z \longrightarrow \epsilon z \in L_6^*(c)$$

are bijections. Hence, by (5.43) and (5.45), we obtain

$$(5.46)\quad \mathcal{Z}_1(s,c)+\mathcal{Z}_2(s,c) = (1-\epsilon^{-s})\mathcal{Z}_3^*(s,c)+(1-\epsilon^{-s})\mathcal{Z}_2(s,c)-\mathcal{Z}_5^*(s,c)-\mathcal{Z}_7^*(s,c).$$

In view of (5.44), we see that the mappings

$$L_1(cs_2) \ni z \longrightarrow \alpha z \in L_5^*(c), \qquad L_2(cs_2) \ni z \longrightarrow \epsilon\alpha z \in L_7^*(c)$$

are bijections. By (5.46), we obtain

$$(5.47)\qquad \begin{aligned} &\mathcal{Z}_1(s,c) + \mathcal{Z}_2(s,c) + \alpha^{-s}\mathcal{Z}_1(s,cs_2) + (\epsilon\alpha)^{-s}\mathcal{Z}_2(s,cs_2) \\ &= (1-\epsilon^{-s})\mathcal{Z}_3^*(s,c) + (1-\epsilon^{-s})\mathcal{Z}_2(s,c). \end{aligned}$$

Put

$$R_3^*(c) = \{z = x + y\alpha \in (\mathfrak{a}_\mu\mathfrak{f})^{-1} \mid x, y \in \mathbf{Q},\ 0 < x, y \leqq 1,\ (z)\mathfrak{a}_\mu\mathfrak{f} \equiv c \mod^{\times} \mathfrak{f}\infty_1\}.$$

By the same proof as that of II, Lemma 3.2, we have

$$(5.48)\qquad L_3^*(c) = \sqcup_{z \in R_3^*(c)}\{z + x + y\alpha \mid 0 \leqq x, y \in \mathbf{Z}\}.$$

Hence

$$(5.49)\qquad \mathcal{Z}_3^*(s,c) = \sum_{z = x + y\alpha \in R_3^*(c)} \zeta(s, (1,\alpha), (x,y)).$$

Differentiating (5.47) at $s = 0$, we obtain

$$(5.50)\qquad \begin{aligned} &G(c; \{C_j\}, \mathfrak{a}_\mu) + G(cs_2; \{C_j\}, (\alpha)\mathfrak{a}_\mu) \\ &= \log\alpha \cdot \mathcal{Z}_1(0, cs_2) + (\log\epsilon\alpha)\cdot\mathcal{Z}_2(0,cs_2) + \log\epsilon\cdot\mathcal{Z}_3^*(0,c) + \log\epsilon\cdot\mathcal{Z}_2(0,c). \end{aligned}$$

By [Y4], Lemma 4 and (5.2), we have

$$\begin{aligned} &\mathcal{Z}_1(0, cs_2) + \mathcal{Z}_2(0, cs_2) = \zeta_F(0, cs_2), \\ &W(c) + W(cs_2) = -\frac{1}{2}\log N(\mathfrak{a}_\mu\mathfrak{f})\zeta_F(0,c) - \frac{1}{2}\log N((\alpha)\mathfrak{a}_\mu\mathfrak{f})\zeta_F(0,cs_2) \\ &= -\frac{1}{2}\log|N(\alpha)|\zeta_F(0, cs_2). \end{aligned}$$

By (4.3) and (5.50), we obtain

$$\begin{aligned} &X(c;\{C_j\},\mathfrak{a}_\mu)+X(cs_2;\{C_j\},(\alpha)\mathfrak{a}_\mu)\\ &=(\log\alpha-\frac{1}{2}\log|N(\alpha)|)\zeta_F(0,cs_2)+\log\epsilon\,(\mathcal{Z}_2(0,c)+\mathcal{Z}_2(0,cs_2)+\mathcal{Z}_3^*(0,c))\\ &+\frac{1}{4}(\epsilon'-\epsilon)\log\epsilon\Bigg[\sum_{z=x+y\epsilon\in R(\epsilon,c)}B_2(x)+\sum_{z=x+y\epsilon\in R(\epsilon,cs_2)}B_2(x)\Bigg]. \end{aligned} \tag{5.51}$$

We have

$$\mathcal{Z}_2(0,c)=-\sum_{z=x+y\epsilon\in R(\epsilon,c),y=0}B_1(x),\qquad \mathcal{Z}_2(0,cs_2)=-\sum_{z=x+y\epsilon\in R(\epsilon,cs_2),y=0}B_1(x).$$

By (5.49) and I, Theorem 2.1, we have

$$\mathcal{Z}_3^*(0,c)=\sum_{z=x+y\alpha\in R_3^*(c)}(\frac{1}{2}\alpha^{-1}B_2(x)+B_1(x)B_1(y)+\frac{1}{2}\alpha B_2(y)). \tag{5.52}$$

Inserting these formulas into (5.51), we obtain

$$\begin{aligned} &X(c;\{C_j\},\mathfrak{a}_\mu)+X(cs_2;\{C_j\},(\alpha)\mathfrak{a}_\mu)\\ &=(\log\alpha-\frac{1}{2}\log|N(\alpha)|)\zeta_F(0,cs_2)\\ &-\log\epsilon\Bigg[\sum_{z=x+y\epsilon\in R(\epsilon,c),y=0}B_1(x)+\sum_{z=x+y\epsilon\in R(\epsilon,cs_2),y=0}B_1(x)\Bigg]\\ &+\frac{1}{4}(\epsilon'-\epsilon)\log\epsilon\Bigg[\sum_{z=x+y\epsilon\in R(\epsilon,c)}B_2(x)+\sum_{z=x+y\epsilon\in R(\epsilon,cs_2)}B_2(x)\Bigg]\\ &+\log\epsilon\sum_{z=x+y\alpha\in R_3^*(c)}(\frac{1}{2}\alpha^{-1}B_2(x)+B_1(x)B_1(y)+\frac{1}{2}\alpha B_2(y)). \end{aligned} \tag{5.53}$$

We are going to simplify this formula. For positive integers e_1, e_2, we temporarily put

$$\begin{aligned} R(e_1,e_2)=\{z=&xe_1+ye_2\alpha\in(\mathfrak{a}_\mu\mathfrak{f})^{-1}\\ &\mid x,y\in\mathbf{Q},\ 0<x,y\leqq 1,\ (z)\mathfrak{a}_\mu\mathfrak{f}\equiv c \mod^\times\mathfrak{f}\infty_1\}. \end{aligned}$$

Regarding $\{e_1,e_2\alpha\}$ as a basis of $C(1,\alpha)$, we obtain, by I, Theorem 2.1, that

$$\mathcal{Z}_3^*(0,c)=\sum_{z=xe_1+ye_2\alpha\in R(e_1,e_2)}(\frac{1}{2}\frac{e_1}{e_2\alpha}B_2(x)+B_1(x)B_1(y)+\frac{1}{2}\frac{e_2\alpha}{e_1}B_2(y)). \tag{5.54}$$

By (3.29) and (3.30), we have

$$V(c)=\frac{1}{2}\sum_{j=1}^{2}\Bigg[\frac{d}{ds}\zeta^{(1,2)}(s,C_j,\mathfrak{a}_\mu)\Big|_{s=0}-\frac{d}{ds}\zeta^{(1)}(s,C_j,\mathfrak{a}_\mu)\Big|_{s=0}-\frac{d}{ds}\zeta^{(2)}(s,C_j,\mathfrak{a}_\mu)\Big|_{s=0}\Bigg].$$

By I, Theorem 2.2, we see that the contribution to $V(c)$ from the cone $C_2=C(1)$ is zero. For a positive integer q, we temporarily put

$$S(q,c)=\{z=xq+yq\epsilon\in(\mathfrak{a}_\mu\mathfrak{f})^{-1}\mid x,y\in\mathbf{Q},\ 0<x,y\leqq 1,\ (z)\mathfrak{a}_\mu\mathfrak{f}=c\text{ in }C_\mathfrak{f}\}.$$

Regarding $\{q, q\epsilon\}$ as a basis of $C(1,\epsilon)$, we have

$$V(c) = \frac{1}{4}\sum_{l} C_l(A_1) \sum_{z\in S(q,c)} \prod_{m=1}^{2} \frac{B_{l_m}(x_m(z))}{l_m!}$$

by I, Theorem 2.2. Here $A_1 = \begin{pmatrix} q & q\epsilon \\ q & q\epsilon' \end{pmatrix}$ and $l = (l_1, l_2)$ extends over $\{(2,0),(1,1),(0,2)\}$. Calculating $C_l(A_1)$, we obtain

$$V(c) = \frac{1}{4}(\epsilon' - \epsilon)\log\epsilon \sum_{z=xq+yq\epsilon\in S(q,c)} B_2(x). \tag{5.55}$$

LEMMA 5.10. *We have*

$$\sum_{z=xq+yq\epsilon\in S(q,c)} B_2(x) = \sum_{z=xq+yq\epsilon\in S(q,c)} B_2(y).$$

PROOF. Let $z = xq + yq\epsilon \in S(q,c)$. We have $\epsilon^{-1}z + q\epsilon = \{y + x(\epsilon+\epsilon')\}q + (1-x)q\epsilon$, $(\epsilon^{-1}z + q\epsilon)\mathfrak{a}_\mu\mathfrak{f} \equiv c$ in $C_\mathfrak{f}$. Take $x_1 \in \mathbf{Q}$ so that $x_1 \equiv y + x(\epsilon+\epsilon') \mod \mathbf{Z}$, $0 < x_1 \leqq 1$. Let $y_1 = 1 - x$ if $0 < x < 1$, $y_1 = 1$ if $x = 1$. Put $z_1 = x_1 q + y_1 q\epsilon$, $\varphi(z) = z_1$. Then $z_1 \in S(q,c)$ and φ gives a bijection from $S(q,c)$ to itself. Hence the assertion follows.

LEMMA 5.11. *Let $F = \mathbf{Q}(\sqrt{d})$, $0 < d \in \mathbf{Z}$ with the square free integer d. Let $\alpha = \sqrt{d}$. Then we have*

$$\begin{aligned}&\frac{1}{2}(\epsilon - \epsilon')\Big(\sum_{z=x+y\epsilon\in R(\epsilon,c)} B_2(x) + \sum_{z=x+y\epsilon\in R(\epsilon,cs_2)} B_2(x)\Big)\\ &= \sum_{z=x+y\alpha\in R_3^*(c)} (\alpha^{-1}B_2(x) + \alpha B_2(y)).\end{aligned}$$

PROOF. Comparing the irrational parts of the right-hand side of (5.52) and (5.54), we have

$$\sum_{z=x+y\alpha\in R_3^*(c)} (\alpha^{-1}B_2(x) + \alpha B_2(y)) = \sum_{z=xe_1+ye_2\alpha\in R(e_1,e_2)} \Big(\frac{e_1}{e_2\alpha}B_2(x) + \frac{e_2\alpha}{e_1}B_2(y)\Big)$$

for any $0 < e_1$, $e_2 \in \mathbf{Z}$. By (4.3) and (5.55), we have

$$\sum_{z=x+y\epsilon\in R(\epsilon,c)} B_2(x) = \sum_{z=xq+yq\epsilon\in S(q,c)} B_2(x)$$

for any $0 < q \in \mathbf{Z}$. We put $\epsilon = \dfrac{u+v\sqrt{d}}{2}$, u, $v \in \mathbf{Z}$. We take $e_1 = dv$, $e_2 = v$, $q = 2$. Then, in view of Lemma 5.10, our assertion is reduced to

$$\begin{aligned}&\frac{v}{2}\Big(\sum_{z=2x+2y\epsilon\in S(2,c)} B_2(y) + \sum_{z=2x+2y\epsilon\in S(2,cs_2)} B_2(y)\Big)\\ &= \sum_{z=xdv+yv\alpha\in R(dv,v)} \Big(B_2(x) + \frac{1}{d}B_2(y)\Big).\end{aligned} \tag{5.56}$$

We put

$$\begin{aligned} T(c) = \{ z = 2x + 2y\epsilon \in (\mathfrak{a}_\mu \mathfrak{f})^{-1} \mid \\ x, y \in \mathbf{Q},\ 0 < x \leqq 1/2,\ 0 < y \leqq 1,\ (z)\mathfrak{a}_\mu \mathfrak{f} \equiv c \text{ in } C_\mathfrak{f} \}, \\ T(cs_2) = \{ z \mid z = 2x + 2y\epsilon \in ((\alpha)\mathfrak{a}_\mu \mathfrak{f})^{-1} \mid \\ x, y \in \mathbf{Q},\ 0 < x \leqq 1/2,\ 0 < y \leqq 1,\ (z)(\alpha)\mathfrak{a}_\mu \mathfrak{f} \equiv cs_2 \text{ in } C_\mathfrak{f} \}. \end{aligned}$$

Then we see easily that

$$\begin{aligned} S(2, c) = \{ 2x + 2y\epsilon, 2(x + \frac{1}{2}) + 2y\epsilon \mid 2x + 2y\epsilon \in T(c) \}, \\ S(2, cs_2) = \{ 2x + 2y\epsilon, 2(x + \frac{1}{2}) + 2y\epsilon \mid 2x + 2y\epsilon \in T(cs_2) \}. \end{aligned}$$

Therefore (5.56) is equivalent to

$$(5.57) \qquad \begin{aligned} v(\sum_{z = 2x + 2y\epsilon \in T(c)} B_2(y) + \sum_{z = 2x + 2y\epsilon \in T(cs_2)} B_2(y)) \\ = \sum_{z = xdv + yv\alpha \in R(dv, v)} (B_2(x) + \frac{1}{d} B_2(y)). \end{aligned}$$

First we define a mapping φ from $R(dv, v)$ to $T(c)$. Let $z = xdv + yv\alpha \in R(dv, v)$. We have $z = (xdv - yu) + 2y\epsilon$. Take $x_1 \in \mathbf{Q}$ so that $2x_1 \equiv xdv - yu \mod \mathbf{Z}$, $0 < x_1 \leqq 1/2$ and put $z_1 = 2x_1 + 2y\epsilon$. Then, since $z_1 > 0$ and $z > 0$, we have

$$(z_1)\mathfrak{a}_\mu \mathfrak{f} \equiv (z)\mathfrak{a}_\mu \mathfrak{f} \equiv c \mod^\times \mathfrak{f}\infty_1.$$

If $(z_1)\mathfrak{a}_\mu \mathfrak{f} \equiv cs_2 \mod^\times \mathfrak{f}\infty_1\infty_2$, then the narrow ideal class of s_2 coincides with that of (z_1). This is a contradiction since $z_1 \gg 0$. Therefore $(z_1)\mathfrak{a}_\mu \mathfrak{f} \equiv c \mod^\times \mathfrak{f}\infty_1\infty_2$ holds and $z_1 \in T(c)$. Put $\varphi(z) = z_1$.

Let us show that φ is surjective. Take any $z = 2x + 2y\epsilon \in T(c)$. We have $z = (2x + yu) + yv\alpha$. Take $x_0 \in \mathbf{Q}$ so that $x_0 \equiv \dfrac{2x + yu}{dv} \mod \mathbf{Z}$, $0 < x_0 \leqq 1$ and put $z_0 = x_0 dv + yv\alpha$. Then we have $z_0 \in R(dv, v)$. Put $\varphi(z_0) = 2x_1 + 2y_1\epsilon$. We have $y_1 = y$ and $x_1 \in \mathbf{Q}$ is determined by the conditions

$$2x_1 \equiv x_0 dv - yu \mod \mathbf{Z}, \qquad 0 < x_1 \leqq 1/2.$$

Since $2x_1 \equiv 2x \mod \mathbf{Z}$, $0 < x \leqq 1/2$, we get $x_1 = x$. Hence φ is surjective. Let us show that

$$(5.58) \qquad |\varphi^{-1}(z)| = dv \qquad \text{for every} \quad z \in T(c).$$

Let $z = 2x + 2y\epsilon \in T(c)$ and take $z_0 = x_0 dv + y_0 v\alpha \in R(dv, v)$. The conditions for $z_0 \in \varphi^{-1}(z)$ are

$$x_0 dv - y_0 u \equiv 2x \mod \mathbf{Z}, \qquad y_0 = y.$$

We see easily that there are dv choices of x_0 for which $z_0 \in R(dv, v)$. Hence (5.58) is proved. By (5.58), we obtain

$$(5.59) \qquad dv \sum_{z = 2x + 2y\epsilon \in T(c)} B_2(y) = \sum_{z = xdv + yv\alpha \in R(dv, v)} B_2(y).$$

Next let us define a mapping ψ from $R(dv, v)$ to $T(cs_2)$. Let $z = xdv + yv\alpha \in R(dv, v)$. We have $z/\alpha = (yv - xu) + 2x\epsilon$. Take $x_1 \in \mathbf{Q}$ so that $2x_1 \equiv yv - xu \mod \mathbf{Z}$, $0 < x_1 \leqq 1/2$ and put $z_1 = 2x_1 + 2x\epsilon$. Then we have

$$(z_1)(\alpha)\mathfrak{a}_\mu\mathfrak{f} \equiv (z/\alpha)(\alpha)\mathfrak{a}_\mu\mathfrak{f} \equiv (z)\mathfrak{a}_\mu\mathfrak{f} \mod^\times \mathfrak{f}.$$

Since z_1, z, $\alpha > 0$, we have

$$(z_1)(\alpha)\mathfrak{a}_\mu\mathfrak{f} \equiv (z)\mathfrak{a}_\mu\mathfrak{f} \equiv c \mod^\times \mathfrak{f}\infty_1.$$

If $(z_1)(\alpha)\mathfrak{a}_\mu\mathfrak{f} \equiv c \mod^\times \mathfrak{f}\infty_1\infty_2$, then the narrow ideal class of $(z_1\alpha)$ is trivial. This is a contradiction since $z_1 \gg 0$, $\alpha > 0$, $\alpha' < 0$. Therefore $(z_1)(\alpha)\mathfrak{a}_\mu\mathfrak{f} \equiv cs_2 \mod^\times \mathfrak{f}\infty_1\infty_2$ holds and $z_1 \in T(cs_2)$. Put $\psi(z) = z_1$.

Let us show that ψ is surjective. Take any $z = 2x + 2y\epsilon \in T(cs_2)$. We have $z\alpha = ydv + (2x + yu)\alpha$. Take $y_0 \in \mathbf{Q}$ so that $y_0 \equiv \dfrac{2x + yu}{v} \mod \mathbf{Z}$, $0 < y_0 \leqq 1$ and put $z_0 = ydv + y_0 v\alpha$. Then z_0 satisfies

$$(z_0)\mathfrak{a}_\mu\mathfrak{f} \equiv (z\alpha)\mathfrak{a}_\mu\mathfrak{f} \equiv c \mod^\times \mathfrak{f}\infty_1.$$

Hence $z_0 \in R(dv, v)$. Put $\psi(z_0) = 2x_1 + 2y_1\epsilon$. We have $y_1 = y$ and $x_1 \in \mathbf{Q}$ is determined by the conditions

$$2x_1 \equiv y_0 v - yu \mod \mathbf{Z}, \qquad 0 < x_1 \leqq 1/2.$$

Since $2x_1 \equiv 2x \mod \mathbf{Z}$, $0 < x \leqq 1/2$, we get $x_1 = x$. Hence ψ is surjective. Let us show that

$$|\psi^{-1}(z)| = v \qquad \text{for every} \quad z \in T(cs_2). \tag{5.60}$$

Let $z = 2x + 2y\epsilon \in T(cs_2)$ and take $z_0 = x_0 dv + y_0 v\alpha \in R(dv, v)$. The conditions for $z_0 \in \psi^{-1}(z)$ are

$$y_0 v - x_0 u \equiv 2x \mod \mathbf{Z}, \qquad x_0 = y.$$

We see easily that there are v choices of y_0 for which $z_0 \in R(dv, v)$. Hence (5.60) is proved. Consequently we obtain

$$v \sum_{z=2x+2y\epsilon \in T(cs_2)} B_2(y) = \sum_{z=xdv+yv\alpha \in R(dv,v)} B_2(x). \tag{5.61}$$

By (5.59) and (5.61), we obtain (5.57). This completes the proof.

Using Lemma 5.11, we can simplify (5.53) to

$$\begin{aligned} &X(c; \{C_j\}, \mathfrak{a}_\mu) + X(cs_2; \{C_j\}, (\alpha)\mathfrak{a}_\mu) \\ &= \log \epsilon \sum_{z=x+y\alpha \in R_3^*(c)} B_1(x)B_1(y) \\ &\quad - \log \epsilon \Bigg(\sum_{z=x+y\epsilon \in R(\epsilon,c), y=0} B_1(x) + \sum_{z=x+y\epsilon \in R(\epsilon,cs_2), y=0} B_1(x) \Bigg). \end{aligned} \tag{5.62}$$

We put

$$\begin{aligned} R(\alpha, c) = \{ z = x + y\alpha \in (\mathfrak{a}_\mu\mathfrak{f})^{-1} \mid \\ x, y \in \mathbf{Q},\ 0 < x \leqq 1,\ 0 \leqq y < 1,\ (z)\mathfrak{a}_\mu\mathfrak{f} \equiv c \mod^\times \mathfrak{f}\infty_1 \}. \end{aligned} \tag{5.63}$$

Then we see that

$$\begin{aligned}
&\sum_{z=x+y\alpha\in R_3^*(c)} B_1(x)B_1(y)\\
&= \sum_{z=x+y\alpha\in R(\alpha,c),0<y<1} B_1(x)B_1(y) + \sum_{z=x+y\alpha\in R(\alpha,c),y=0} \frac{1}{2}B_1(x)\\
&= \sum_{z=x+y\alpha\in R(\alpha,c)} B_1(x)B_1(y) + \sum_{z=x+y\alpha\in R(\alpha,c),y=0} B_1(x),
\end{aligned}$$

$$\sum_{z=x+y\alpha\in R(\alpha,c),y=0} B_1(x) = \sum_{z=x+y\epsilon\in R(\epsilon,c),y=0} B_1(x).$$

Thus we obtain the following theorem.

THEOREM 5.12. *Let $F = \mathbf{Q}(\sqrt{d})$ be a real quadratic field, where d is a square free integer. We put $\alpha = \sqrt{d}$. Let $c \in C_{\mathfrak{f}}$ and define $R(\alpha, c)$ by (5.63). If $N(\epsilon_0) = 1$, then we have*

$$\begin{aligned}
&X(c; \{C_j\}, \mathfrak{a}_\mu) + X(cs_2; \{C_j\}, (\alpha)\mathfrak{a}_\mu)\\
&= \log\epsilon\left[\sum_{z=x+y\alpha\in R(\alpha,c)} B_1(x)B_1(y) - \sum_{z=x+y\epsilon\in R(\epsilon,cs_2),y=0} B_1(x)\right].
\end{aligned}$$

REMARK 5.13. If $\mathfrak{f} \neq (1)$, then we have

$$\sum_{z=x+y\epsilon\in R(\epsilon,cs_2),y=0} B_1(x) = \sum_{z=x+y\alpha\in R(\alpha,c),x=1} B_1(y). \tag{5.64}$$

In fact, $1 \notin R(\alpha, c)$ since $\mathfrak{f} \neq (1)$. We can verify that the mapping

$$R(\alpha, c) \ni 1 + y\alpha \longrightarrow y + 0\cdot\epsilon \in R(\epsilon, cs_2)$$

gives a bijection between $\{z \mid z = x + y\alpha \in R(\alpha, c),\ x = 1\}$ and $\{z \mid z = x + y\epsilon \in R(\epsilon, cs_2),\ y = 0\}$.

5.5. Using Theorems 5.8, 5.12 and the Stark-Shintani conjecture, we can clarify the nature of the invariant $X(c)$. By (3.11), we have

$$X(c; \{C_j\}, \mathfrak{a}_\mu) + X(c^\sigma; \{C_j^\sigma\}, \mathfrak{a}_\mu^\sigma) = \zeta_F'(0, c), \tag{5.65}$$

where $\sigma \in \mathrm{Gal}(F/\mathbf{Q})$ denotes the non-trivial automorphism. By (3.35), we obtain

$$X(c^\sigma; \{C_j^\sigma\}, \mathfrak{a}_\mu^\sigma) = X(c^\sigma; \{C_j\}, \mathfrak{a}_\mu^\sigma) + \frac{1}{2}\log\epsilon\sum_{k=1}^{2}\zeta^{(k)}(0, C_1, \mathfrak{a}_\mu^\sigma)$$

since $C_1^\sigma = C(1, \epsilon^\sigma) = \epsilon^{-1}C_1$, $C_2^\sigma = C_2$. Using

$$\zeta_F(0, c) = \zeta_F(0, c^\sigma) = \frac{1}{2}\sum_{j=1}^{2}\sum_{k=1}^{2}\zeta^{(k)}(0, C_j, \mathfrak{a}_\mu^\sigma),$$

we get

$$X(c^\sigma; \{C_j^\sigma\}, \mathfrak{a}_\mu^\sigma) - X(c^\sigma; \{C_j\}, \mathfrak{a}_\mu^\sigma) = \log\epsilon\cdot\zeta_F(0, c) - \frac{1}{2}\log\epsilon\sum_{k=1}^{2}\zeta^{(k)}(0, C_1, \mathfrak{a}_\mu^\sigma).$$

By I, Theorem 2.1, we have

$$\frac{1}{2}\sum_{k=1}^{2}\zeta^{(k)}(0,C_1,\mathfrak{a}_\mu^\sigma) = -\sum_{z\in R(C_1,c)} B_1(x) = \sum_{z=x+y\epsilon\in R(\epsilon,c),y=0} (\frac{1}{2}-x).$$

Thus we obtain

$$\begin{aligned} &X(c^\sigma;\{C_j^\sigma\},\mathfrak{a}_\mu^\sigma) - X(c^\sigma;\{C_j\},\mathfrak{a}_\mu^\sigma) \\ &= \log\epsilon\Big[\zeta_F(0,c) + \sum_{z=x+y\epsilon\in R(\epsilon,c),y=0}(x-\frac{1}{2})\Big]. \end{aligned} \tag{5.66}$$

For $c\in C_\mathfrak{f}$, we put

$$X(c) = X(c;\{C_j\},\mathfrak{a}_\mu),$$
$$X(cs_2) = \begin{cases} X(cs_2;\{C_j\},\mathfrak{a}_\mu) & \text{if} \quad N(\epsilon_0)=-1, \\ X(cs_2;\{C_j\},(\alpha)\mathfrak{a}_\mu) & \text{if} \quad N(\epsilon_0)=1, \end{cases}$$

where $\alpha=\sqrt{d}$ as before. We put

$$X(c)+X(cs_2) = a(c)\log\epsilon, \qquad c\in C_\mathfrak{f} \tag{5.67}$$

where $a(c)\in\mathbf{Q}$ is explicitly given by Theorems 5.8 and 5.12. We also put

$$X(c^\sigma) = X(c^\sigma;\{C_j\},\mathfrak{a}_\mu^\sigma), \qquad c\in C_\mathfrak{f}.$$

By (5.65) and (5.66), we have

$$X(c)+X(c^\sigma) = \zeta_F'(0,c) - \log\epsilon(\zeta_F(0,c)+r(c)), \tag{5.68}$$

where

$$r(c) = \sum_{z=x+y\epsilon\in R(\epsilon,c),y=0}(x-\frac{1}{2}).$$

Hereafter we assume that $\mathfrak{f}\neq(1)$. Then, by Lemma 5.2 and (5.68), we obtain

$$\begin{aligned} &X(c)+X(c^\sigma)+X(cs_1)+X(c^\sigma s_1^\sigma)+X(cs_2)+X(c^\sigma s_2^\sigma)+X(cs_1s_2)+X(c^\sigma s_1^\sigma s_2^\sigma) \\ &= -\log\epsilon\,(r(c)+r(cs_1)+r(cs_2)+r(cs_1s_2)). \end{aligned}$$

Since $c^\sigma\in C_{\mathfrak{f}^\sigma}$ and s_1^σ plays the role of s_2 in $C_{\mathfrak{f}^\sigma}$, we obtain

$$a(c)+a(cs_1)+a(c^\sigma)+a(c^\sigma s_2^\sigma) = -(r(c)+r(cs_1)+r(cs_2)+r(cs_1s_2)) \tag{5.69}$$

by (5.67).

Let us recall that $C_\mathfrak{f}$ is the ideal class group modulo $\mathfrak{f}\infty_1\infty_2$ of a real quadratic filed F. We *do not* assume that the conductor of $C_\mathfrak{f}$ is $\mathfrak{f}\infty_1\infty_2$ but assume that $\mathfrak{f}\neq(1)$. Let M be the maximal ray class field modulo $\mathfrak{f}\infty_1\infty_2$ over F. For $i=1$, 2, let M_i be the subfield of M which corresponds to $\langle s_i\rangle$. By (5.1), we see that ∞_i is unramified in M_i. We assume that:

(A) The conductor of M_1 over F is $\mathfrak{f}\infty_2$.

By the Stark-Shintani conjecture (cf. II, Proposition 2.2; here we take the normalization $e=-2$ as in Tate [T]), there exists a unit $\epsilon(c)$ of M_1 such that

$$\exp(-2(\zeta_F'(0,c)+\zeta_F'(0,cs_1)) = \epsilon(c). \tag{5.70}$$

Furthermore $\epsilon(c)$ should satisfy the reciprocity law

$$\epsilon(c)^{[\frac{M_1/F}{s}]} = \epsilon(cs), \qquad s \in C_{\mathfrak{f}}. \tag{5.71}$$

We note that $\epsilon(c)$ depends only on the class of c modulo $\mathfrak{f}\infty_2$. By (5.68), (5.70) and Lemma 5.2, we get

$$X(c) + X(c^\sigma) + X(cs_1) + X(c^\sigma s_1^\sigma) = -\frac{1}{2}\log\epsilon(c) - (r(c) + r(cs_1))\log\epsilon.$$

By (5.67), we have

$$X(c^\sigma) + X(c^\sigma s_1^\sigma) = a(c^\sigma)\log\epsilon, \qquad X(cs_1) + X(cs_1 s_2) = a(cs_1)\log\epsilon.$$

Hence we obtain

$$X(cs_1) = -X(c) - \frac{1}{2}\log\epsilon(c) - (a(c^\sigma) + r(c) + r(cs_1))\log\epsilon, \tag{5.72}$$

$$X(cs_2) = -X(c) + a(c)\log\epsilon, \tag{5.73}$$

$$X(cs_1 s_2) = X(c) + \frac{1}{2}\log\epsilon(c) + (a(cs_1) + a(c^\sigma) + r(c) + r(cs_1))\log\epsilon. \tag{5.74}$$

Thus the Stark-Shintani conjecture implies that there is only one transcendental invariant among

$$\exp(X(c)), \quad \exp(X(cs_1)), \quad \exp(X(cs_2)), \quad \exp(X(cs_1 s_2)).$$

In other words, $\exp(X(cs))$ for $s = s_1,\ s_2,\ s_1 s_2$ can be written as $\exp(X(c))^{\pm 1}\times$ an algebraic number. Also remarkable here is that $\exp(-2(X(c) + X(cs_1)))$ is of the form $\epsilon^a\epsilon(c)$, $a \in \mathbf{Q}$ where $\epsilon(c)$ is an important and complicated unit whereas $\exp(-2(X(c) + X(cs_2)))$ being of the form ϵ^b, $b \in \mathbf{Q}$ is of quite simple nature. Put

$$b(c) = 2(a(c^\sigma) + r(c) + r(cs_1)).$$

Then we have

$$\exp(-2(X(c) + X(cs_1))) = \epsilon^{b(c)}\epsilon(c).$$

Let $\alpha \in \mathrm{Gal}(\overline{\mathbf{Q}}/F)$ and take $s \in C_{\mathfrak{f}}$ so that $\alpha|M = [\frac{M/F}{s}]$. By (5.71), we have

$$(\exp(-2(X(c) + X(cs_1))))^\alpha = (\epsilon^{b(c)})^\alpha\epsilon^{-b(cs)}\exp(-2(X(cs) + X(cs_1 s))). \tag{5.75}$$

REMARK 5.14. Let us consider the case $F = \mathbf{Q}$. We can take $\mathfrak{a}_1 = (1)$, $C_1 = C(1)$. Let $\mathfrak{f} = (f)$ with $3 \leqq f \in \mathbf{Z}$. Any class $c \in C_{\mathfrak{f}}$ can be represented as the class of (a) with $a \in \mathbf{Z}$, $(a, f) = 1$, $0 < a < f$. We have (cf. (2.6))

$$\zeta^{(1)}(s, C_1, \mathfrak{a}_1) = f^s \sum_{m=1,\ m\equiv a \mod f}^{\infty} m^{-s} = \zeta(s, \frac{a}{f}).$$

We have $V(c) = 0$ and

$$X(c) = G(c) + W(c) = \zeta'(0, \frac{a}{f}) - \log f \cdot \zeta(0, \frac{a}{f})$$

by (3.28) and (3.31). By I, (1.16) and (1.17), we obtain

$$X(c) = \log\Gamma(\frac{a}{f}) - \frac{1}{2}\log(2\pi) - \log f(\frac{1}{2} - \frac{a}{f}). \tag{5.76}$$

We have s_1 = the class of $(1-f)$ and cs_1 is the class of $(f-a)$. Hence we get

$$\exp(X(c)+X(cs_1)) = (2\sin\frac{a\pi}{f})^{-1} \sim 1.$$

On the other hand $\exp(X(c)-X(cs_1))$ is related to CM-periods.

Let us consider "the real case" of the Chowla-Selberg formula. Namely let K be a real quadratic field of discriminant d and let χ be the Dirichlet character which corresponds to the quadratic extension $K/\mathbf{Q}$. By II, (1.2) and $\zeta_K(s)=\zeta(s)L(s,\chi)$, we have

$$L(s,\chi) = (h_K \log \epsilon_0)s + O(s^2), \qquad s \to 0,$$

where ϵ_0 denotes the fundamental unit of K. By (2.11), we have

$$L'(0,\chi) = \sum_{a=1}^{d-1} \chi(a)\log(\Gamma(\frac{a}{d})).$$

Therefore we obtain

$$\prod_{a=1}^{d-1} \Gamma(\frac{a}{d})^{\chi(a)/h_K} = \epsilon_0.$$

This formula is equivalent to Dirichlet's class number formula for real quadratic fields. Our Theorems 5.8 and 5.12 may be regarded as a generalization of these facts.

6. Numerical examples (continued)

6.1. We will illustrate the results in §5 by examples. First let us prove the observations (4.20) and (4.21) using Theorem 5.8.

LEMMA 6.1. *Let F be a real quadratic field such that $N(\epsilon_0)=-1$. For $z=x+y\epsilon_0 \in R(\epsilon_0,c)$, put*

$$z^* = \begin{cases} (1-x)+(1-y)\epsilon_0 & \text{if } 0<x,\, y<1, \\ 1-x & \text{if } 0<x<1,\, y=0, \\ 1+(1-y)\epsilon_0 & \text{if } x=1,\, 0<y<1, \\ 1 & \text{if } x=1,\, y=0. \end{cases}$$

Then $z^* \in R(\epsilon_0,cs_1)$ and the mapping $z \longrightarrow z^*$ is a bijection from $R(\epsilon_0,c)$ to $R(\epsilon_0,cs_1)$.

PROOF. To prove $z^* \in R(\epsilon_0,cs_1)$, it suffices to show

$$(z^*)\mathfrak{a}_\mu\mathfrak{f} \equiv cs_1 \mod^\times \mathfrak{f}\infty_1.$$

Recall that s_1 is the class of (ν_1) where ν_1 is an element of $\mathfrak{O}_F$ such that $\nu_1<0$, $\nu_1^\sigma>0$, $\nu_1\equiv 1 \mod \mathfrak{f}$. Since

$$(z)\mathfrak{a}_\mu\mathfrak{f} \equiv c \mod^\times \mathfrak{f}\infty_1,$$

it suffices to show that

$$(\frac{z^*}{z}) \equiv (\nu_1) \mod^\times \mathfrak{f}\infty_1.$$

Since $z>0$, $z^*>0$, $-\nu_1>0$, it suffices to show

$$\frac{z^*}{z} \equiv -1 \mod^\times \mathfrak{f}. \tag{6.1}$$

Since (6.1) is clear if $\mathfrak{f} = (1)$, we may assume $\mathfrak{f} \neq (1)$. Then $1 \notin R(\epsilon_0, c)$ since $\mathfrak{a}_\mu$ is an integral ideal. We have

$$\frac{z^*}{z} = \begin{cases} \frac{1+\epsilon_0}{z} - 1 & \text{if} \quad 0 < x,\, y < 1, \\ \frac{1}{z} - 1 & \text{if} \quad 0 < x < 1,\, y = 0, \\ \frac{2+\epsilon_0}{z} - 1 & \text{if} \quad x = 1,\, 0 < y < 1. \end{cases}$$

Put $(z)\mathfrak{a}_\mu\mathfrak{f} = \mathfrak{b}$. Then $\mathfrak{b}$ is an integral ideal relatively prime to $\mathfrak{f}$ and $(\frac{1}{z}) = \mathfrak{b}^{-1}\mathfrak{a}_\mu\mathfrak{f}$. Hence we see that (6.1) holds. Put $\varphi_c(z) = z^*$. Then it is immediate to see that $\varphi_{cs_1} \circ \varphi_c$ (resp. $\varphi_c \circ \varphi_{cs_1}$) is the identity map on $R(\epsilon_0, c)$ (resp. $R(\epsilon_0, cs_1)$). This completes the proof.

By Lemma 6.1, we have

$$\begin{aligned} &\sum_{z=x+y\epsilon_0 \in R(\epsilon_0, cs_1)} B_1(x)B_1(y) \\ = &\sum_{z=x+y\epsilon_0 \in R(\epsilon_0,c), 0<x,y<1} B_1(x)B_1(y) + \sum_{z=x+y\epsilon_0 \in R(\epsilon_0,c), 0<x<1, y=0} B_1(1-x)\cdot(-\frac{1}{2}) \\ + &\sum_{z=x+y\epsilon_0 \in R(\epsilon_0,c), x=1, 0<y<1} \frac{1}{2}B_1(1-y) + \sum_{z=x+y\epsilon_0 \in R(\epsilon_0,c), x=1, y=0} B_1(x)B_1(y) \\ = &\sum_{z=x+y\epsilon_0 \in R(\epsilon_0,c)} B_1(x)B_1(y) + \sum_{z=x+y\epsilon_0 \in R(\epsilon_0,c), 0<x<1, y=0} B_1(x) \\ - &\sum_{z=x+y\epsilon_0 \in R(\epsilon_0,c), x=1, 0<y<1} B_1(y). \end{aligned}$$

By a similar argument to the proof of Lemma 6.1, we see that the mapping

$$\begin{aligned} &\{z \mid z = x + y\epsilon_0 \in R(\epsilon_0, c), 0 < x < 1, y = 0\} \ni x \\ &\longrightarrow 1 + x\epsilon_0 \in \{z \mid z = x + y\epsilon_0 \in R(\epsilon_0, c), x = 1, 0 < y < 1\} \end{aligned}$$

is a bijection. Hence we obtain

$$\sum_{z=x+y\epsilon_0 \in R(\epsilon_0, cs_1)} B_1(x)B_1(y) = \sum_{z=x+y\epsilon_0 \in R(\epsilon_0,c)} B_1(x)B_1(y). \tag{6.2}$$

To derive (4.20), let us prove that

$$\{z \mid z = x + y\epsilon_0 \in R(\epsilon_0, c), y = 0\} = \emptyset \quad \text{if} \quad \mathfrak{a}_\mu^\sigma = \mathfrak{a}_\mu \quad \text{and} \quad \mathfrak{f}^\sigma \neq \mathfrak{f}. \tag{6.3}$$

In fact, suppose that $0 < x \leqq 1$, $x \in \mathbf{Q}$ satisfies

$$x \in (\mathfrak{a}_\mu\mathfrak{f})^{-1}, \qquad (x)\mathfrak{a}_\mu\mathfrak{f} \equiv c \mod^\times \mathfrak{f}\infty_1.$$

Then $\mathfrak{b} = (x)\mathfrak{a}_\mu\mathfrak{f}$ is an integral ideal relatively prime to $\mathfrak{f}$. By $(x) = \mathfrak{b}\mathfrak{a}_\mu^{-1}\mathfrak{f}^{-1} = (x^\sigma) = \mathfrak{b}^\sigma(\mathfrak{a}_\mu^\sigma)^{-1}(\mathfrak{f}^\sigma)^{-1}$, we obtain $\mathfrak{b}\mathfrak{f}^\sigma = \mathfrak{b}^\sigma\mathfrak{f}$. Since $(\mathfrak{b}, \mathfrak{f}) = 1$, $(\mathfrak{b}^\sigma, \mathfrak{f}^\sigma) = 1$, we get $\mathfrak{f} = \mathfrak{f}^\sigma$, which contradicts our assumption. By (6.2), (6.3) and Theorem 5.8, we obtain

PROPOSITION 6.2. *Let F be a real quadratic field such that $N(\epsilon_0) = -1$. If $\mathfrak{f}^\sigma \neq \mathfrak{f}$ and $\mathfrak{a}_\mu^\sigma = \mathfrak{a}_\mu$, then we have*

$$X(c; \{C_j\}, \mathfrak{a}_\mu) + X(cs_2; \{C_j\}, \mathfrak{a}_\mu) = X(cs_1; \{C_j\}, \mathfrak{a}_\mu) + X(cs_1 s_2; \{C_j\}, \mathfrak{a}_\mu).$$

From this proposition, (4.20) follows. Considering $C_{\mathfrak{f}^\sigma}$ instead of $C_\mathfrak{f}$, (4.21) also follows.

6.2. We are going to discuss two examples in view of examples in Shintani [Shi2] and Stark [St3].

EXAMPLE 6.3. Let $F = \mathbf{Q}(\sqrt{5})$, $\mathfrak{f} = (4)$. This case was considered by Shintani [Shi2], §3, Example 1. We are going to examine this case in greater detail using the invariant $X(c)$.

The conductor of $C_\mathfrak{f}$ is $\mathfrak{f}\infty_1\infty_2$. Define $c_i \in C_\mathfrak{f}$ by

$$c_1 = \text{the class of } (1), \qquad c_2 = \text{the class of } (3),$$
$$c_3 = \text{the class of } (4+\sqrt{5}), \qquad c_4 = \text{the class of } (6+\sqrt{5}).$$

Then we have

$$C_\mathfrak{f} = \{c_1, c_2, c_3, c_4\} \cong \mathbf{Z}/2\mathbf{Z} \oplus \mathbf{Z}/2\mathbf{Z},$$

which can be seen easily by noting $\epsilon^3 = 9 + 4\sqrt{5} \equiv 1 \mod (4)$. Let D denote the character group of $C_\mathfrak{f}$. Define χ_1, $\chi_2 \in D$ by

$$\chi_1(c_1) = 1, \qquad \chi_1(c_2) = -1, \qquad \chi_1(c_3) = 1, \qquad \chi_1(c_4) = -1,$$
$$\chi_2(c_1) = 1, \qquad \chi_2(c_2) = -1, \qquad \chi_2(c_3) = -1, \qquad \chi_2(c_4) = 1.$$

As in II, §4, we write $\chi \in D$ as

$$\chi((x)) = \chi_4(x \mod 4)(\operatorname{sgn} x)^{\delta_1}(\operatorname{sgn} x')^{\delta_2}, \qquad x \in \mathfrak{O}_F, \quad ((x),(4)) = 1.$$

Then we have

$$\chi_4(-1 \mod 4)(-1)^{\delta_1+\delta_2} = 1, \qquad \chi_4(\epsilon_0 \mod 4)(-1)^{\delta_2} = 1.$$

Since $\epsilon_0^3 = 2 + \sqrt{5} \equiv 6 + \sqrt{5} \mod 4$, we have

$$\chi(c_2)(-1)^{\delta_1+\delta_2} = 1, \qquad \chi(c_4)(-1)^{\delta_2} = 1.$$

Thus, for $\chi = \chi_1$, we get

$$\delta_1 = 0, \quad \delta_2 = 1, \quad \chi_4(\epsilon_0 \mod 4) = -1, \quad \chi_4(-1 \mod 4) = -1,$$

and for $\chi = \chi_2$, we get

$$\delta_1 = 1, \quad \delta_2 = 0, \quad \chi_4(\epsilon_0 \mod 4) = 1, \quad \chi_4(-1 \mod 4) = -1.$$

Hence we obtain

$$\mathfrak{f}(\chi_1) = (4)\infty_2, \qquad \mathfrak{f}(\chi_2) = (4)\infty_1, \qquad s_1 = c_3, \qquad s_2 = c_4.$$

Let M be the maximal ray class field of conductor $(4)\infty_1\infty_2$ of F. We have $[M : F] = |C_\mathfrak{f}| = 4$. Put $K = \mathbf{Q}(\sqrt{5}, \sqrt{-1})$. Then $K \subset M$. Let χ denote the character of $C_\mathfrak{f}$ which corresponds to the extension K/F. Clearly we have $\chi = \chi_1\chi_2$.

Put $K_1 = F(\sqrt{\epsilon_0})$. Since $\mathfrak{d}_{K_1/F} \mid (2)$, we have $D(K_1/F) \mid (4)$. We see that ∞_1 is unramified and ∞_2 is ramified in K_1. Therefore χ_1 is the character which corresponds to the quadratic extension K_1/F and the conductor of K_1 over F is $(4)\infty_2$. Put $K_2 = F(\sqrt{\epsilon_0'})$. Similarly to the above, we see that the conductor of K_2 over F is $(4)\infty_1$ and χ_2 is the corresponding character. By $\zeta_{K_1}(s) = \zeta_F(s)L(s,\chi_1)$ and II, (1.2), we have

$$L'(0,\chi_1) = \frac{2h_{K_1}}{w_{K_1}} \cdot \frac{R_{K_1}}{R_F}.$$

It is clear that $w_{K_1} = 2$. It is known that $h_{K_1} = 1$ and $\{\sqrt{\epsilon_0},\ 1+\sqrt{\epsilon_0}\}$ are generators of a free part of E_{K_1} (cf. Billevič[Bi]). Hence

$$\begin{aligned} R_{K_1} =& \left|\det\begin{pmatrix} \log\sqrt{\epsilon_0} & \log(1+\sqrt{\epsilon_0}) \\ \log\sqrt{\epsilon_0} & \log|(1-\sqrt{\epsilon_0})| \end{pmatrix}\right| = \frac{1}{2}\log\epsilon_0\left|\log\left|\frac{1+\sqrt{\epsilon_0}}{1-\sqrt{\epsilon_0}}\right|\right| \\ =& \frac{1}{2}\log\epsilon_0\log(\epsilon_0(1+\sqrt{\epsilon_0})^2), \\ L'(0,\chi_1) =& \frac{1}{2}\log\epsilon_0 + \log(1+\sqrt{\epsilon_0}). \end{aligned}$$

Since $\chi_2(c) = \chi_1(c^\sigma)$ for $c \in C_{\mathfrak{f}}$, we have

$$L'(0,\chi_1) = L'(0,\chi_2) = \frac{1}{2}\log\epsilon_0 + \log(1+\sqrt{\epsilon_0}). \tag{6.4}$$

Since $L'(0,\chi_j) = \sum_{i=1}^{4}\chi_j(c_i)\zeta_F'(0,c_i)$, $j = 1, 2$, (6.4) yields

$$\zeta_F'(0,c_1) - \zeta_F'(0,c_2) = \frac{1}{2}\log\epsilon_0 + \log(1+\sqrt{\epsilon_0}), \tag{6.5}$$

$$\zeta_F'(0,c_3) = \zeta_F'(0,c_4). \tag{6.6}$$

By Lemma 5.2, we have

$$\zeta_F'(0,c_1) + \zeta_F'(0,c_2) = -2\zeta_F'(0,c_3). \tag{6.7}$$

Let ψ (resp. η) be the Dirichlet character which corresponds to the quadratic extension $\mathbf{Q}(\sqrt{-5})/\mathbf{Q}$ (resp. $\mathbf{Q}(\sqrt{-1})/\mathbf{Q}$). We have

$$L_F(s,\chi) = L(s,\psi)L(s,\eta), \qquad L(0,\psi) = 2, \qquad L(0,\eta) = 1/2.$$

Hence we get

$$L_F(0,\chi) = 1, \qquad L_F'(0,\chi) = 2L'(0,\eta) + \frac{1}{2}L'(0,\psi).$$

By (2.11), we have

$$L'(0,\psi) = \sum_{a=1}^{19}\psi(a)\log\Gamma(\frac{a}{20}) - 2\log 20, \qquad L'(0,\eta) = \log\Gamma(\frac{1}{4}) - \log\Gamma(\frac{3}{4}) - \frac{1}{2}\log 4.$$

Put

$$G = \frac{\Gamma(\frac{1}{4})}{\Gamma(\frac{3}{4})}\prod_{a=1}^{19}\Gamma(\frac{a}{20})^{\psi(a)/4}. \tag{6.8}$$

Then we obtain

$$L_F'(0,\chi) = -\log 80 + 2\log G. \tag{6.9}$$

Since $L_F(s,\chi) = \zeta_F(s,c_1) + \zeta_F(s,c_2) - \zeta_F(s,c_3) - \zeta_F(s,c_4)$, we have

$$\zeta_F'(0,c_1) + \zeta_F'(0,c_2) = -\frac{1}{2}\log 80 + \log G, \tag{6.10}$$

$$\zeta_F'(0,c_3) = \zeta_F'(0,c_4) = \frac{1}{4}\log 80 - \frac{1}{2}\log G, \tag{6.11}$$

in view of (6.6) and (6.7). By (6.5), we obtain

$$\zeta_F'(0,c_1) = -\frac{1}{4}\log 80 + \frac{1}{8}\log\epsilon + \frac{1}{2}\log(1+\sqrt{\epsilon_0}) + \frac{1}{2}\log G, \tag{6.12}$$

$$\zeta_F'(0,c_2) = -\frac{1}{4}\log 80 - \frac{1}{8}\log\epsilon - \frac{1}{2}\log(1+\sqrt{\epsilon_0}) + \frac{1}{2}\log G. \tag{6.13}$$

Thus we have determined $\zeta_F'(0,c_i)$ explicitly.

We take $\mathfrak{a}_1 = (1)$ as a representative of the narrow ideal class of F. Then we have

$$R(\epsilon,c_1) = \left\{\frac{1}{4},\ 1+\frac{1}{4}\epsilon,\ \frac{3}{4}+\frac{3}{4}\epsilon\right\}, \qquad R(\epsilon,c_2) = \left\{\frac{3}{4},\ 1+\frac{3}{4}\epsilon,\ \frac{1}{4}+\frac{1}{4}\epsilon\right\},$$
$$R(\epsilon,c_3) = \left\{\frac{1}{4}+\frac{1}{2}\epsilon,\ \frac{1}{2}+\frac{3}{4}\epsilon,\ \frac{1}{4}+\frac{3}{4}\epsilon\right\}, \quad R(\epsilon,c_4) = \left\{\frac{3}{4}+\frac{1}{2}\epsilon,\ \frac{1}{2}+\frac{1}{4}\epsilon,\ \frac{3}{4}+\frac{1}{4}\epsilon\right\}.$$

We note that

$$c_1^\sigma = c_1, \qquad c_2^\sigma = c_2, \qquad c_3^\sigma = c_4, \qquad c_4^\sigma = c_3,$$
$$\zeta_F(0,c_1) = \zeta_F(0,c_2) = 1/4, \qquad \zeta_F(0,c_3) = \zeta_F(0,c_4) = -1/4,$$

the equalities on the second line being immediate consequences of (4.9) and (5.2). By (5.68), we obtain

$$X(c_1) = \frac{1}{2}\zeta_F'(0,c_1), \tag{6.14}$$

$$X(c_2) = \frac{1}{2}\zeta_F'(0,c_2) - \frac{1}{4}\log\epsilon, \tag{6.15}$$

$$X(c_3) + X(c_4) = \zeta_F'(0,c_3) + \frac{1}{4}\log\epsilon. \tag{6.16}$$

Regarding K as an abelian extension of F, we have

$$g_K(\mathrm{id},\mathrm{id}) = \pi^{-1/2}\exp\Big(\frac{1}{2}\sum_{i=1}^{4}\chi(c_i)X(c_i)\Big) = \pi^{-1/2}\exp\Big(\frac{1}{4}L_F'(0,\chi) - \frac{1}{4}\log\epsilon\Big).$$

By (6.9), we obtain

$$g_K(\mathrm{id},\mathrm{id}) = (80\epsilon)^{-1/4}\pi^{-1/2}G^{1/2}.$$

By the Chowla-Selberg formula (1.4), we have

$$\Gamma\Big(\frac{1}{4}\Big)\Gamma\Big(\frac{3}{4}\Big)^{-1} \sim \pi^{1/2}p_{\mathbf{Q}(\sqrt{-1})}(\mathrm{id},\mathrm{id}),$$
$$\prod_{a=1}^{19}\Gamma\Big(\frac{a}{20}\Big)^{\psi(a)/4} \sim \pi^{1/2}p_{\mathbf{Q}(\sqrt{-5})}(\mathrm{id},\mathrm{id}).$$

Let α (resp. β) denote the generator of $\mathrm{Gal}(K/\mathbf{Q}(\sqrt{-1}))$ (resp. $\mathrm{Gal}(K/\mathbf{Q}(\sqrt{-5}))$). We have $\alpha\beta = \rho$. By Theorem 1.2, we get

$$\begin{aligned} G &\sim \pi p_{\mathbf{Q}(\sqrt{-1})}(\mathrm{id}, \mathrm{id}) p_{\mathbf{Q}(\sqrt{-5})}(\mathrm{id}, \mathrm{id}) \sim \pi p_K(\mathrm{id}, \mathrm{id}) p_K(\mathrm{id}, \alpha) p_K(\mathrm{id}, \mathrm{id}) p_K(\mathrm{id}, \beta) \\ &\sim \pi p_K(\mathrm{id}, \mathrm{id})^2. \end{aligned}$$

Thus we obtain $p_K(\mathrm{id}, \mathrm{id}) \sim g_K(\mathrm{id}, \mathrm{id})$, which follows also from Theorem 2.6 and Corollary 3.17.

By (6.5) , (6.14) and (6.15), we have

$$2(X(c_1) - X(c_2)) = \frac{3}{2} \log \epsilon_0 + \log(1 + \sqrt{\epsilon_0}). \tag{6.17}$$

Since $V(c_1) = V(c_2)$, we get

$$\frac{\Gamma_2(\frac{1}{4}, (1, \epsilon))^4 \Gamma_2(1 + \frac{1}{4}\epsilon, (1, \epsilon))^4 \Gamma_2(\frac{3}{4} + \frac{3}{4}\epsilon, (1, \epsilon))^4}{\Gamma_2(\frac{3}{4}, (1, \epsilon))^4 \Gamma_2(1 + \frac{3}{4}\epsilon, (1, \epsilon))^4 \Gamma_2(\frac{1}{4} + \frac{1}{4}\epsilon, (1, \epsilon))^4} = \epsilon_0^3 (1 + \sqrt{\epsilon_0})^2.$$

Taking account of the difference equation of the double gamma function, this equation is equivalent to that given in [Shi2], p. 190, line 7.

We find easily that

$$R(\epsilon_0, c_1) = \left\{ \frac{1}{4},\ 1 + \frac{1}{4}\epsilon_0,\ \frac{1}{4} + \frac{1}{4}\epsilon_0,\ \frac{1}{4} + \frac{1}{2}\epsilon_0,\ \frac{1}{2} + \frac{3}{4}\epsilon_0,\ \frac{3}{4} + \frac{1}{4}\epsilon_0 \right\}.$$

By Theorem 5.8, we have

$$X(c_1) + X(c_4) = \frac{1}{8} \log \epsilon. \tag{6.18}$$

Thus we have determined $X(c_i)$, $1 \leqq i \leqq 4$ completely. Since $\zeta_F(0, c_1) = -\zeta_F(0, c_4)$ and $V(c_1) = -V(c_4)$, (6.18) can be written as

$$\prod_{z \in R(\epsilon, c_1)} \frac{\Gamma_2(z, (1, \epsilon))}{\rho_2((1, \epsilon))} \prod_{z \in R(\epsilon, c_4)} \frac{\Gamma_2(z, (1, \epsilon))}{\rho_2((1, \epsilon))} = \epsilon^{1/8}. \tag{6.19}$$

We note that Shintani ([Shi2], p. 191) showed that $\rho_2((1, \epsilon)) = \sqrt{2\pi}(\epsilon')^{\epsilon/12}$.

EXAMPLE 6.4.. Let $F = \mathbf{Q}(\sqrt{5})$, $\mathfrak{f} = (7 + 2\sqrt{5})$. We have $N(\mathfrak{f}) = 29$. The conductor of the ideal class group $C_\mathfrak{f}$ is $\mathfrak{f}\infty_2$. Let M be the maximal ray class field of conductor $\mathfrak{f}\infty_2$ over F. We have $\mathrm{Gal}(M/F) \cong C_\mathfrak{f}$. This is the case considered by Stark in [St3]. We have $C_\mathfrak{f} \cong \mathbf{Z}/4\mathbf{Z}$ which is generated by the class of (2). For $1 \leqq i \leqq 4$, let $c_i \in C_\mathfrak{f}$ be the class of (2^{i-1}). Let ψ be the character of $C_\mathfrak{f}$ determined by $\psi(c_2) = i$. Then $\widehat{C_\mathfrak{f}}$ is generated by ψ. We have

$$\mathfrak{f}(\psi) = \mathfrak{f}(\psi^3) = \mathfrak{f}\infty_2, \qquad \mathfrak{f}(\psi^2) = \mathfrak{f}, \qquad s_1 = 1, \qquad s_2 = c_3.$$

By Lemma 5.2, we obtain

$$\zeta_F(0, c_i) = 0, \qquad 1 \leqq i \leqq 4, \tag{6.20}$$

$$\zeta_F'(0, c_3) = -\zeta_F'(0, c_1), \qquad \zeta_F'(0, c_4) = -\zeta_F'(0, c_2). \tag{6.21}$$

In [St3], Stark found numerical evidences that

$$\zeta_F'(0, c_1) = \frac{1}{2} \log \epsilon_+, \qquad \zeta_F'(0, c_2) = \frac{1}{2} \log \epsilon_-$$

where

$$\epsilon_{\pm} = \frac{(3+2\sqrt{5}) \pm \sqrt{7+2\sqrt{5}} + \sqrt{(20+14\sqrt{5}) \pm (6+4\sqrt{5})\sqrt{7+2\sqrt{5}}}}{4}. \tag{6.22}$$

We take $\mathfrak{a}_1 = (1)$ as a representative of the narrow ideal class of F. By calculation, we have $|R(\epsilon_0, c_i)| = 14$, $1 \leqq i \leqq 4$. By Theorem 5.8, we get

$$X(c_1) + X(c_3) = \frac{21}{2 \cdot 29} \log \epsilon_0, \tag{6.23}$$

$$X(c_2) + X(c_4) = \frac{5}{2 \cdot 29} \log \epsilon_0. \tag{6.24}$$

For $c \in C_{\mathfrak{f}}$, put $X(c') = X(c^{\sigma}; \{C_j^{\sigma}\}, \mathfrak{a}_{\mu}^{\sigma})$. By Lemma 3.8, we have

$$X(c) + X(c') = \zeta_F'(0, c). \tag{6.25}$$

We have $c^{\sigma} \in C_{\mathfrak{f}^{\sigma}}$, $\mathfrak{f}^{\sigma} = (7 - 2\sqrt{5})$. In $C_{\mathfrak{f}^{\sigma}}$, $s_2 = 1$ holds. We can calculate $X(c_i')$ by Theorem 5.8 and (5.66). We have $|R(\epsilon_0, c_i^{\sigma})| = 7$, $1 \leqq i \leqq 4$ and

$$\begin{aligned} X(c_1') = X(c_3') &= -\frac{21}{4 \cdot 29} \log \epsilon_0, \\ X(c_2') = X(c_4') &= -\frac{5}{4 \cdot 29} \log \epsilon_0. \end{aligned} \tag{6.26}$$

Thus, admitting (6.22), we obtain

$$X(c_1) = \frac{21}{4 \cdot 29} \log \epsilon_0 + \frac{1}{2} \log \epsilon_+, \tag{6.27}$$

$$X(c_2) = \frac{5}{4 \cdot 29} \log \epsilon_0 + \frac{1}{2} \log \epsilon_-, \tag{6.28}$$

$$X(c_3) = \frac{21}{4 \cdot 29} \log \epsilon_0 - \frac{1}{2} \log \epsilon_+, \tag{6.29}$$

$$X(c_4) = \frac{5}{4 \cdot 29} \log \epsilon_0 - \frac{1}{2} \log \epsilon_-. \tag{6.30}$$

Corrections to [Y4]

p. 1205, line 6: ${}^t x = (x_1, x_2, \dots, x_n)$ should be ${}^t x = (x_1, x_2, \dots, x_r)$.
p. 1206, line 17: $\mathbf{Q}^{r(j)}$ should be $\mathbf{R}^{r(j)}$.
p. 1207: In (2.11) and (2.13), $x(z)_m$ should be $x_m(z)$. The same change should also be made in (4.2), (4.4), (4.5), (5.4), (5.8), (5.10) and (5.11).
p. 1212, line 24: The definition of $\eta_i^{-1}\Phi_i^0$ is inaccurate. Write $\eta_i = \sum_{\alpha \in J_L} a_\alpha \cdot \alpha$, $a_\alpha \in \mathbf{Z}$ and define $\eta_i^{-1} = \sum_{\alpha \in J_L} a_\alpha \cdot \alpha^{-1}$, $\eta_i^{-1}\Phi_i^0 = \sum_{\delta \in \Phi_i^0} \eta_i^{-1}\delta$.
p. 1213, line 15: $L(m/2, \lambda_i)^{\epsilon_i}$ should be $L(m/2, \lambda_i)$.
p. 1218 $\sim$ 1219: The treatment of the operation

$$\mathfrak{a}_\mu \longrightarrow (\alpha_\mu)\mathfrak{a}_\mu, \qquad \{C_j\}_{j \in J} \longrightarrow \{(\alpha_\mu)^{-1} C_j\}_{j \in J}$$

is oversimplified. It should be discussed more carefully as in §3 of this chapter.
p. 1221, line 2: $x(z)_m$ should be $x_m(z)$.

p. 1221, line 6: $x(z)_1$ should be $x_1(z)$.

p. 1222: In (6.7), $\log \dfrac{\Gamma_2(z',(1,\epsilon))}{\rho_2((1,\epsilon'))}$ should be $\log \dfrac{\Gamma_2(z',(1,\epsilon'))}{\rho_2((1,\epsilon'))}$.

p. 1223, line 9: insert "if a and b are even."

p. 1229: $R = \epsilon^{-2/9} \cdot \dfrac{\sqrt{637530967 - 18592639\sqrt{205}}}{3^2 \cdot 31 \cdot 61}$ should be

$R = \epsilon^{-2/9} \cdot \dfrac{\sqrt{\frac{637530967-18592639\sqrt{205}}{2}}}{3^2 \cdot 31 \cdot 61}$. We have $\sqrt{\frac{637530967-18592639\sqrt{205}}{2}} = \frac{34883-533\sqrt{205}}{2}$. All double roots which appear in Examples 8 ~ 10 can be simplified similarly.

p. 1231, line 22: $\lambda^{(i)}_{a,b,}$ should be $\lambda^{(i)}_{a,b,c}$.

Exercises

1. Let K be a CM-field of degree $2n$ and F be the maximal real subfield of K. Let χ be the Hecke character of $F_A^\times$ which corresponds to the quadratic extension K/F.
   (a) Using II, (1.2), show that
   $$L(0,\chi) = \frac{2R_K}{w_K R_F} \cdot \frac{h_K}{h_F}.$$
   (b) From the definition of the regulator, deduce that
   $$\frac{R_K}{R_F} = \frac{2^{n-1}}{[E_K : W_K E_F]}.$$
   (c) Let f be the homomorphism $E_K \ni \epsilon \longrightarrow \epsilon^{1-\rho} \in E_K$. Here ρ denotes the complex conjugation. Show that $W_K^2 \subset \operatorname{Im}(f) \subset W_K$.
   (d) Show that $[E_K : W_K E_F] = [\operatorname{Im}(f) : W_K] = 1$ or 2.
   (e) Show that $K = F(\sqrt{-\epsilon})$ with a totally positive unit ϵ of F if $[E_K : W_K E_F] = 2$.
2. For a CM-field of degree $2n$, show that
   $$\prod_{\Phi} p_K(\Phi,\Phi) \sim \Big(\prod_{\sigma \in J_K} p_K(\sigma,\sigma)\Big)^{2^{n-1}}$$
   where Φ extends over all 2^n CM-types of K.
3. Let K, L, σ, τ be the same as above the formula (4.22). Show that
   $$p_K(\mathrm{id},\mathrm{id}) \sim p_L(\mathrm{id},\mathrm{id}) p_L(\mathrm{id},\sigma\tau), \qquad p_K(\sigma,\sigma) \sim p_L(\mathrm{id},\mathrm{id}) p_L(\mathrm{id},\sigma\tau)^{-1}.$$
4. Let K be a CM-field abelian over $\mathbf{Q}$ and let F be a totally real algebraic number field normal over $\mathbf{Q}$. We assume $K \cap F = \mathbf{Q}$ and put $L = K \vee F$. Take $\sigma \in \operatorname{Gal}(K/\mathbf{Q})$ and put $\operatorname{Inf}_{L/K}(\sigma) = \sum_{i=1}^n \sigma_i$, $\sigma_i \in J_L$. Define $g_L(\mathrm{id},\sigma_i)$ regarding L as an abelian extension of the fixed field of $\langle \rho, \sigma_i \rangle$ and set $g_L(\mathrm{id}, \operatorname{Inf}_{L/K}(\sigma)) = \prod_{i=1}^n g_L(\mathrm{id},\sigma_i)$. Show that
   $$g_K(\mathrm{id},\sigma) \sim g_L(\mathrm{id}, \operatorname{Inf}_{L/K}(\sigma)).$$

5. Let $F = \mathbf{Q}(\sqrt{29})$ and p be an odd prime which decomposes completely in F. Let $p\mathfrak{O}_F = \mathfrak{p}\mathfrak{p}'$ be the prime ideal decomposition. Show that there exists a totally imaginary quadratic extension K of F of relative discriminant $\mathfrak{p}$ if and only if

$$(\frac{-1}{\mathfrak{p}}) = 1, \qquad (\frac{\epsilon_0}{\mathfrak{p}}) = -1.$$

Here $\epsilon_0 = (5+\sqrt{29})/2$ and $(\frac{}{\mathfrak{p}})$ denotes the quadratic residue symbol in $\mathfrak{O}_F/\mathfrak{p}$. Show that the set of all such p in the range $p < 500$ is given by

$$\{13, 53, 149, 173, 257, 277, 313, 349, 353, 457\}.$$

6. In the first case of Example 4.1, show that

$$X(c_1) + X(c_2) = -\frac{4}{13}\log\epsilon_0, \qquad \epsilon_0 = \frac{5+\sqrt{29}}{2}$$

using Theorem 5.8.

7. Let F be a real quadratic field and ϵ_0 be the fundamental unit of F. Let C denote the ideal class group modulo $(1)\infty_1\infty_2$.
   (a) We assume $N(\epsilon_0) = -1$. The C is identical to the ideal class group of F. Using Theorem 5.8, show that

$$X(c; \{C_j\}, \mathfrak{a}_\mu) = \log\epsilon_0\Big[-\frac{3}{4} + \sum_{z=x+y\epsilon_0\in R(\epsilon_0,c),0<x<1,0<y<1} B_1(x)B_1(y)\Big]$$

   for an ideal class c.
   (b) Using (3.11), (5.66) and (a), show that $\zeta_F'(0,c) = -\frac{1}{2}\log\epsilon_0$.
   (c) When $F = \mathbf{Q}(\sqrt{5})$, derive, from (a) and (4.7), that

$$\log\frac{\Gamma_2(1,(1,\epsilon))}{\rho_2((1,\epsilon))} = \frac{\sqrt{5}}{24}\log\epsilon - \frac{3}{8}\log\epsilon,$$

   where $\epsilon = \epsilon_0^2 = \frac{3+\sqrt{5}}{2}$.
   (d) We assume $N(\epsilon_0) = 1$. Using Theorem 5.12, show that

$$\zeta_F'(0,c) + \zeta_F'(0,cs_2) = -\frac{1}{2}\log\epsilon_0, \qquad c\in C.$$

8. Let F be a field and F_{sep} be a separable closure of F. For a subfield L of F_{sep}, let L_{ab} denote the maximal abelian extension of L contained in F_{sep}. Assume that $K_{ab} = K \vee F_{ab}$ for every finite extension K ($\subset F_{sep}$) of F. Show that $F_{sep} = F_{ab}$.

CHAPTER IV

EXPLICIT CONE DECOMPOSITIONS AND APPLICATIONS

Let F be a totally real algebraic number field of degree $n+1$ and let E_F^+ be the group of all totally positive units of F. Let E be a subgroup of E_F^+ of finite index. In this chapter, we will study explicit cone decompositions of $\mathbf{R}_+^{n+1}$ with respect to the action of E. We will also give applications to the invariants $V(c)$ and $W(c)$ introduced in chapter III.

In §1, we will give a concrete process to decompose an n-dimensional cube into open simplexes. We will introduce the notion of *a special decomposition* of an n-dimensional cube. In §2, we will prove a theorem of topological nature which is necessary in the next section. In §3, we will prove a sufficient condition which guarantees that we can transform a special decomposition of an n-dimensional cube into a cone decomposition of $\mathbf{R}_+^{n+1}$. For this condition to be satisfied, we must be able to choose generators of E with certain special properties. In §4, we will illustrate the result in §3 by examples. In §5, we will show that we can find at least one E for which this condition holds. In §6, we will give applications to the invariants $V(c)$ and $W(c)$.

1. A special decomposition of a higher dimensional cube

1.1. Let n be a positive integer. We set

$$I^n = \{(x_1, x_2, \ldots, x_n) \in \mathbf{R}^n \mid 0 \leqq x_i \leqq 1 \quad \text{for} \quad 1 \leqq i \leqq n\}.$$

For $1 \leqq i \leqq n$, we put

$$I_i^+(n) = \{(x_1, x_2, \ldots, x_n) \in I^n \mid x_i = 0\},$$
$$I_i^-(n) = \{(x_1, x_2, \ldots, x_n) \in I^n \mid x_i = 1\}.$$

Let P_1, P_2, ..., $P_r \in \mathbf{R}^n$ be points such that $(r-1)$ vectors $P_2 - P_1$, $P_3 - P_1$, ..., $P_r - P_1$ are linearly independent. We put

$$S(P_1, P_2, \ldots, P_r) = \{t_1P_1 + t_2P_2 + \cdots + t_rP_r \mid 0 < t_1, t_2, \ldots, t_r,\ t_1 + t_2 + \cdots + t_r = 1\}$$

and call it *an* $(r-1)$*-dimensional open simplex with vertices* P_1, P_2, ..., P_r.

LEMMA 1.1. *Let* $S = S(P_1, P_2, \ldots, P_r) \subset \mathbf{R}^n$ *be an* $(r-1)$*-dimensional open simplex. Then we have*

$$\bar{S} = \sqcup_{s=1}^{r} \sqcup_{\{i_1, i_2, \ldots, i_s\} \subset \{1, 2, \ldots, r\}} S(P_{i_1}, P_{i_2}, \ldots, P_{i_s}).$$

PROOF. We have

$$\bar{S} = \{t_1P_1 + t_2P_2 + \ldots + t_rP_r \mid 0 \leqq t_1, t_2, \cdots, t_r,\ t_1 + t_2 + \cdots + t_r = 1\}.$$

Take any $x = t_1P_1 + t_2P_2 + \cdots + t_rP_r \in \bar{S}$ and put $\{i_1, i_2, \ldots, i_s\} = \{i \mid t_i \neq 0\}$. Then $s \geqq 1$ and we have $x \in S(P_{i_1}, P_{i_2}, \ldots, P_{i_s})$.

Next let $x \in S(P_{i_1}, P_{i_2}, \ldots, P_{i_s}) \cap S(P_{j_1}, P_{j_2}, \ldots, P_{j_u})$ and put

$$x = t_1P_{i_1} + t_2P_{i_2} + \cdots + t_sP_{i_s} = v_1P_{j_1} + v_2P_{j_2} + \cdots + v_uP_{j_u}.$$

Taking the difference, we get the relation of the form

$$\lambda_1P_1 + \lambda_2P_2 + \cdots + \lambda_rP_r = 0, \qquad \lambda_1 + \lambda_2 + \cdots + \lambda_r = 0,$$

where $\lambda_i \neq 0$ for some i if $\{i_1, \ldots, i_s\} \neq \{j_1, \ldots, j_u\}$. From this relation, we get

$$\lambda_2(P_2 - P_1) + \cdots + \lambda_r(P_r - P_1) = 0.$$

Therefore we obtain $\lambda_1 = \lambda_2 = \ldots = \lambda_r = 0$, $\{i_i, i_2, \ldots, i_s\} = \{j_1, j_2, \ldots, j_u\}$. This completes the proof.

We call an open simplex $S(P_{i_1}, P_{i_2}, \ldots, P_{i_s})$ as in Lemma 1 *a face* of S.

LEMMA 1.2. *Let $S = S(P_1, P_2, \ldots, P_r) \subset \mathbf{R}^n$ be an $(r-1)$-dimensional open simplex. Then we have*

$$\bar{S} \times [0,1] = \cup_{i=1}^{r} \overline{S(P_i', P_{i+1}', \ldots, P_r', P_1'', P_2'', \ldots, P_i'')}.$$

Here we put $P_i' = (P_i, 0)$, $P_i'' = (P_i, 1)$.

PROOF. After making a suitable affine transformation of $\mathbf{R}^n$, we may assume that $n = r - 1$ and that

$$\begin{aligned} &P_1 = (0, 0, \ldots, 0), \quad P_2 = (1, 0, \ldots, 0), \ldots, \\ &P_i = (0, \ldots, 0, 1, 0, \ldots, 0), \ldots, P_r = (0, \ldots, 0, 1) \in \mathbf{R}^{r-1}. \end{aligned}$$

Then we have

$$\bar{S} = \{(t_1, t_2, \ldots, t_{r-1}) \mid 0 \leqq t_i \quad \text{for} \quad 1 \leqq i \leqq r-1,\ t_1 + t_2 + \cdots + t_{r-1} \leqq 1\},$$

$$\begin{aligned} \bar{S} \times [0,1] = \{(t_1, t_2, \ldots, t_{r-1}, t_r) \mid 0 \leqq t_i \ \text{ for }\ 1 \leqq i \leqq r, \\ t_1 + t_2 + \cdots + t_{r-1} \leqq 1,\ t_r \leqq 1\}. \end{aligned}$$

We see easily that

$$\overline{S(P_1', P_2', \ldots, P_r', P_1'')} = \{(t_1, t_2, \ldots, t_r) \in \bar{S} \times [0,1] \mid t_1 + t_2 + \cdots + t_r \leqq 1\},$$

$$\begin{aligned} \overline{S(P_i', P_{i+1}', \ldots, P_r', P_1'', P_2'', \cdots, P_i'')} = \{(t_1, t_2, \ldots, t_r) \in \bar{S} \times [0,1] \mid \\ t_{i-1} + t_i + \cdots + t_r \geqq 1,\ t_i + t_{i+1} + \cdots + t_r \leqq 1\} \end{aligned}$$

for $2 \leqq i \leqq r$. Hence the assertion follows.

LEMMA 1.3. *The notation being the same as in Lemma 1.2, put*

$$S_i = S(P'_i, P'_{i+1}, \dots, P'_r, P''_1, P''_2, \dots, P''_i)$$

for $1 \leqq i \leqq r$. *Then we have* $S_i \cap S_j = \emptyset$ *if* $i \neq j$. *Set*

$$\{Q_1, Q_2, \dots Q_{2r}\} = \{P'_1, P'_2, \dots, P'_r, P''_1, P''_2, \dots, P''_r\}$$

and let

$$S(Q_{i_1}, Q_{i_2}, \dots, Q_{i_s}) \subset \partial \bar{S}_i, \qquad S(Q_{j_1}, Q_{j_2}, \dots, Q_{j_u}) \subset \partial \bar{S}_j$$

be open simplexes those are faces of S_i *and* S_j. *If*

$$S(Q_{i_1}, Q_{i_2}, \dots, Q_{i_s}) \cap S(Q_{j_1}, Q_{j_2}, \dots, Q_{j_u}) \neq \emptyset,$$

then

$$S(Q_{i_1}, Q_{i_2}, \dots, Q_{i_s}) = S(Q_{j_1}, Q_{j_2}, \dots, Q_{j_u}).$$

PROOF. Using the notation in the proof of Lemma 1.2, we see easily that

$$\begin{aligned} S_1 &= \{(t_1, t_2, \dots, t_r) \in S \times (0,1) \mid t_1 + t_2 + \dots + t_r < 1\}, \\ S_i &= \{(t_1, t_2, \dots, t_r) \in S \times (0,1) \mid t_{i-1} + t_i + \dots + t_r > 1, \\ &\qquad\qquad\qquad\qquad\qquad\qquad t_i + t_{i+1} + \dots + t_r < 1\} \end{aligned}$$

for $2 \leqq i \leqq r$. Hence $S_i \cap S_j = \emptyset$ if $i \neq j$. Suppose that

$$x \in S(Q_{i_1}, Q_{i_2}, \dots, Q_{i_s}) \cap S(Q_{j_1}, Q_{j_2}, \dots, Q_{j_u}).$$

Put $x = (x_1, x_2, \dots, x_r)$. If $i = j$, the assertion follows from Lemma 1.1. We may assume that $i < j$. If $P'_k \in \{Q_{i_1}, Q_{i_2}, \dots, Q_{i_s}\}$ for some $k \leqq j-1$, then we have $x_k + x_{k+1} + \dots + x_r < 1$. On the other hand, we have $x_{j-1} + x_j + \dots + x_r \geqq 1$ since $x \in \bar{S}_j$. This is a contradiction. Therefore we must have $P'_k \notin \{Q_{i_1}, Q_{i_2}, \dots, Q_{i_s}\}$ for every $k \leqq j-1$. Hence we obtain

$$\{Q_{i_1}, Q_{i_2}, \dots, Q_{i_s}\} \subset \{P'_j, \dots, P'_r, P''_1, \dots, P''_j\}.$$

This shows that $S(Q_{i_1}, Q_{i_2}, \dots, Q_{i_s}) \subset \partial \bar{S}_j$. Now the assertion follows from Lemma 1.1.

1.2. We use the same notation as in Lemmas 1.2 and 1.3. By Lemma 1.2, we have

$$\bar{S} \times [0,1] = \cup_{i=1}^r \bar{S}_i.$$

Including faces of S_i and removing repetitions, we obtain, by Lemma 1.3, a decomposition of the form

$$\bar{S} \times [0,1] = \sqcup_{l \in L} T_l \tag{1.1}$$

where T_l is an open simplex which is a face of some S_i. Suppose that $T_l \cap (\partial \bar{S} \times [0,1]) \neq \emptyset$. Put $T_l = S(Q_{i_1}, Q_{i_2}, \dots, Q_{i_s})$. Since

$$\begin{aligned} \partial \bar{S} \times [0,1] = \{&(t_1, t_2, \dots, t_{r-1}, t_r) \in \bar{S} \times [0,1] \mid \\ &t_i = 0 \quad \text{for some} \quad 1 \leqq i \leqq r-1 \quad \text{or} \quad t_1 + t_2 + \dots + t_{r-1} = 1\}, \end{aligned}$$

we have $P'_i \notin \{Q_{i_1}, Q_{i_2}, \dots, Q_{i_s}\}$ and $P''_i \notin \{Q_{i_1}, Q_{i_2}, \dots, Q_{i_s}\}$ for some $1 \leqq i \leqq r$. Hence we have $T_l \subset \partial\bar{S} \times [0,1]$. Omitting those T_l such that $T_l \cap \partial\bar{S} \times [0,1] \neq \emptyset$ in (1.1), we obtain

$$S \times [0,1] = \sqcup_{l \in L'} T_l. \tag{1.2}$$

In (1.2), we see that, if $T_l \cap (S \times \{0\}) \neq \emptyset$ (resp. $T_l \cap (S \times \{1\}) \neq \emptyset$), then $T_l \subset (S \times \{0\})$ (resp. $T_l \subset (S \times \{1\})$), similarly to the above. Therefore we obtain a decomposition of the form

$$S \times [0,1] = S \times \{0\} \sqcup S \times \{1\} \sqcup_{k \in K} T_k. \tag{1.3}$$

Here T_k is an open simplex which is a face of some S_i. We note that the decomposition (1.3) depends on the numbering of vertices of S.

PROPOSITION 1.4. *We have a decomposition of I^n into open simplexes*

$$I^n = \sqcup_{j \in \tilde{J}} S_j$$

which has the following properties.

(1) *S_j is an $r(j)$-dimensional open simplex with vertices taken from vertices of I^n.*

(2) *For every $j \in \tilde{J}$, there exists a subset I_j of $\tilde{J}$ such that $\partial\bar{S}_j = \sqcup_{\alpha \in I_j} S_\alpha$.*

We note an obvious fact which follows from (1) that if $S_j \cap I_i^{\pm}(n) \neq \emptyset$, then $S_j \subset I_i^{\pm}(n)$. For $1 \leqq i \leqq n-1$, put $J_i^{\pm} = \{j \in \tilde{J} \mid S_j \subset I_i^{\pm}(n)\}$.

(3) *There exists a one-to-one correspondence $J_i^+ \ni j \longrightarrow k \in J_i^-$ such that $S_k = S_j^1$. Here $S_j^1 = S_j + (0, \dots, 0, 1, 0, \dots, 0)$, the translation of S_j by a vector whose i-th component is 1 and all the other components are 0.*

PROOF. The assertion is obvious if $n = 1$. We proceed by induction on n. Let

$$I^n = \sqcup_{j \in \tilde{J}(n)} S_j \tag{1.4}$$

be a decomposition which satisfies (1) $\sim$ (3). For $1 \leqq i \leqq n$, we have

$$I_i^{\pm}(n) = \sqcup_{j \in J_i^{\pm}(n)} S_j. \tag{1.5}$$

I^n has 2^n vertices $(a_1, a_2, \dots, a_n)$, $a_i = 0$ or 1. For a vertex $P = (a_1, a_2, \dots, a_n)$, we put $n(P) = 1 + \sum_{i=1}^n a_i \cdot 2^{i-1}$, and define an ordering on the set of all vertices of I^n. According to this ordering, we decompose

$$S_j \times [0,1] = \sqcup_{k \in \Lambda_j} S_k^{(j)} \tag{1.6}$$

using (1.3). Thus we have

$$I^{n+1} = \sqcup_{j \in \tilde{J}(n)} \sqcup_{k \in \Lambda_j} S_k^{(j)}. \tag{1.7}$$

It suffices to show that (1.7) satisfies (2) and (3). First let us verify the property (2). We consider $S_k^{(j)}$ appearing in the decomposition of $S_j \times [0,1]$ in (1.6) and check (2) taking $S_k^{(j)}$ as S_j. We proceed by induction on $\dim S_j$. If $\dim S_j = 0$, then $S_j = S(P) = P$ and

$$S_j \times [0,1] = P \times \{0\} \sqcup P \times \{1\} \sqcup S(P', P'').$$

The assertion is obvious in this case. Let $T \subset \partial\bar{S}_k^{(j)}$ be a face of $S_k^{(j)}$. We put

$$S_j = S(P_1, P_2, \ldots, P_m), \qquad P_1 < P_2 < \ldots < P_m,$$
$$T = S(P'_{i_1}, P'_{i_2}, \ldots, P'_{i_s}, P''_{j_1}, P''_{j_2}, \ldots, P''_{j_t}),$$

We have $T \subset \bar{S}_j \times [0,1]$ and there exists an i, $1 \leqq i \leqq m$ such that

$$P'_i \notin \{P'_{i_1}, \ldots, P'_{i_s}, P''_{j_1}, \ldots, P''_{j_t}\}, \qquad P''_i \notin \{P'_{i_1}, \ldots, P'_{i_s}, P''_{j_1}, \ldots, P''_{j_t}\}.$$

Put $S_l = S(P_1, \ldots, \check{P}_i, \ldots, P_m)$. By the induction hypothesis, we have $l \in \tilde{J}(n)$. We have $T \subset \partial\overline{S(P'_u, \ldots, P'_m, P''_1, \ldots, P''_u)}$ for some $1 \leqq u \leqq m$. Put $S_\alpha = S(P'_u, \ldots, \check{P}'_i, \ldots, P'_m, P''_1, \ldots, \check{P}''_i, \ldots, P''_u)$. Then S_α appears in the decomposition of $S_l \times [0,1]$ and $\partial\bar{S}_\alpha \supset T$. Since $\dim S_l < \dim S_j$, the assertion follows by the induction hypothesis.

Next let us verify the property (3). We have

$$I_i^{\pm}(n+1) = I_i^{\pm}(n) \times [0,1], \quad 1 \leqq i \leqq n,$$
$$I_{n+1}^{+}(n+1) = I^n \times \{0\}, \qquad I_{n+1}^{-}(n+1) = I^n \times \{1\}.$$

Since $S_j \times \{0\}$ and $S_j \times \{1\}$ appear in (1.7) for $j \in \tilde{J}(n)$, the property (3) holds for $i = n+1$. Suppose that $1 \leqq i \leqq n$ and that S_j, $j \in J_i^+(n)$ corresponds to $S_{j'}$, $j' \in J_i^-(n)$. Let $S_j = S(P_1, P_2, \ldots, P_m)$, $P_1 < P_2 < \ldots < P_m$. Then we have $S_{j'} = S(P_1^*, P_2^*, \ldots, P_m^*)$ with $P_k^* = P_k + (0, \ldots, 0, 1, 0, \ldots, 0)$. By the definition of the ordering, we have $P_1^* < P_2^* < \ldots < P_m^*$. Therefore, when we decompose $S_j \times [0,1]$ and $S_{j'} \times [0,1]$ according to this ordering, all open simplexes which appear in this decomposition correspond each other as in (3). This completes the proof.

We call the decomposition of I^n as in Proposition 1.4 *a special decomposition* of I^n. Put

(1.8) $$F = \{(x_1, x_2, \ldots, x_n) \in \mathbf{R}^n \mid 0 \leqq x_i < 1 \quad \text{for} \quad 1 \leqq i \leqq n\}.$$

Clearly there exists a subset J of $\tilde{J}$ such that

$$F = \sqcup_{j \in J} S_j.$$

2. Topological preparations

2.1. The following theorem will play a key role in constructing explicit cone decompositions.

THEOREM 2.1. *Let h be a continuous injective mapping of I^n into $\mathbf{R}^n$. We assume that h satisfies*

(2.1) $$h(x+u) = h(x) + u \quad \textit{whenever} \quad x,\ x+u \in I^n,\ u \in \mathbf{Z}^n.$$

Define a mapping $\tilde{h}$ of $\mathbf{R}^n$ into $\mathbf{R}^n$ by

(2.2) $$\tilde{h}(x+u) = h(x) + u, \quad x \in I^n,\ u \in \mathbf{Z}^n.$$

Then $\tilde{h}$ gives a homeomorphism of $\mathbf{R}^n$ onto $\mathbf{R}^n$.

Though the statement of this theorem is "intuitively clear", detailed considerations will be required for a rigorous proof. The proof will be accomplished through a series of lemmas. First we note a consequence of the theorem.

COROLLARY 2.2. *We define F by (1.8). Then $h(F)$ gives a fundamental domain of $\mathbf{R}^n/\mathbf{Z}^n$.*

PROOF. Take any $x \in \mathbf{R}^n$. Take $y \in \mathbf{R}^n$ so that $\tilde{h}(y) = x$ and take $u \in \mathbf{Z}^n$ so that $y + u \in F$. Then $x + u = \tilde{h}(y + u) \in h(F)$. Next assume that x, $x' \in h(F)$ satisfy $x' = x + u$ with $u \in \mathbf{Z}^n$. Take y, $y' \in F$ so that $\tilde{h}(y) = x$, $\tilde{h}(y') = x'$. Then we have $\tilde{h}(y + u) = \tilde{h}(y')$. Hence $y + u = y'$ and $u = 0$. This completes the proof.

Let f be a continuous mapping from a Hausdorff topological space X into a locally compact topological space Y. We call f *proper* if $f^{-1}(C)$ is compact for every compact subset C of Y. If f is proper, then f is a closed mapping. (cf. Bourbaki [Bo1], Chap. I, §10.)

LEMMA 2.3. *The mapping $\tilde{h}$ is continuous and proper.*

PROOF. The well-definedness of $\tilde{h}$ is clear. First let us show that $\tilde{h}$ is continuous. It suffices to show the continuity of $\tilde{h}$ at $x \in F$. Since this is clear if x is an interior point of I^n, we may assume that $x \in F \cap \partial I^n$. Put $x = (x_1, x_2, \dots, x_n)$. Then $0 \leqq x_j < 1$ for $1 \leqq j \leqq n$. Suppose that $\{z_m\}_{m=1}^{\infty}$, $z_m = (z_m^{(1)}, z_m^{(2)}, \dots, z_m^{(n)}) \in \mathbf{R}^n$ converges to x. Since $-1 < z_m^{(i)} < 1$ holds for every i when m is sufficiently large, we may assume that $-1 < z_m^{(i)} < 1$ for all m and i. For z_m, there exists a unique $u = (u_1, u_2, \dots u_n) \in \mathbf{Z}^n$ such that $z_m + u \in F$. Then $u_i = 0$ or 1. Fixing such a $u \in \mathbf{Z}^n$, we form a subsequence $\{w_l\}$ collecting those z_m such that $z_m + u \in F$. It suffices to show that $\lim_{l\to\infty} \tilde{h}(w_l) = \tilde{h}(x)$ assuming $\{w_l\}$ is an infinite sequence. Put $w_l = (w_l^{(1)}, w_l^{(2)}, \dots, w_l^{(n)})$. Since $w_l + u \in F$, we have $0 \leqq w_l^{(i)} < 1$ if $u_i = 0$ and $-1 \leqq w_l^{(i)} < 0$ if $u_i = 1$. If $u_i = 1$, we have $x_i = 0$. Thus $x + u \in I^n$. Since $\tilde{h}$ is continuous on I^n, we have $\lim_{l\to\infty} \tilde{h}(w_l) = \lim_{l\to\infty} \tilde{h}(w_l + u) - u = \tilde{h}(x + u) - u = \tilde{h}(x)$.

Next let us show that $\tilde{h}$ is proper. Let C be a compact subset of $\mathbf{R}^n$. We can choose $M > 0$ so that if $(y_1, y_2, \dots, y_n) \in C$, then $|y_i| \leqq M$ for every i. Since $h(I^n)$ is compact, we can choose $m > 0$ so that if $(y_1, y_2, \dots, y_n) \in h(I^n)$, then $|y_i| \leqq m$ for every i. Suppose that $x \in \tilde{h}^{-1}(C)$. Put $x = z + u$ with $z \in I^n$, $u \in \mathbf{Z}^n$, $x = (x_1, x_2, \dots, x_n)$, $u = (u_1, u_2, \dots, u_n)$. By (2.2), we have $|u_i| \leqq M + m$ for every i. Hence $|x_i| \leqq M + m + 1$ for every i. Therefore $\tilde{h}^{-1}(C)$ is compact. This completes the proof.

By this lemma, Theorem 2.1 follows if we can show that $\tilde{h}$ is bijective.

LEMMA 2.4. *Let $f : \mathbf{R}^n \longrightarrow \mathbf{R}^n$ be a continuous closed injective mapping. Then f is surjective.*

PROOF. We consider the one point compactification $S^n = \mathbf{R}^n \cup \{\infty\}$ of $\mathbf{R}^n$. Since f gives a homeomorphism of $\mathbf{R}^n$ onto $f(\mathbf{R}^n)$ and $\mathbf{R}^n$ is an open subset of S^n, we see that $f(\mathbf{R}^n)$ is open in S^n by the Brower theorem on the invariance of domain (cf. Spanier [Sp], p. 199, Theorem 16). Hence $f(\mathbf{R}^n)$ is open and closed in $\mathbf{R}^n$ and we obtain $f(\mathbf{R}^n) = \mathbf{R}^n$.

By this lemma, Theorem 2.1 is reduced to show that $\tilde{h}$ is injective. This point is the heart of the proof.

2.2. For $v \in \mathbf{R}^n$, let $\|v\|$ denote the length of the vector v. Let $f : \partial I^n \longrightarrow \mathbf{R}^n$ be a continuous mapping. For $a \in \mathbf{R}^n$, $a \notin f(\partial I^n)$, we put

$$(r(a,f))(x) = \frac{f(x)-a}{\|f(x)-a\|}, \qquad x \in \partial I^n. \tag{2.3}$$

Then $r(a,f)$ defines a continuous mapping of ∂I^n into S^{n-1}. Choosing generators of $H^{n-1}(\partial I^n, \mathbf{Z})$ and of $H^{n-1}(S^{n-1}, \mathbf{Z})$ so that they correspond to orientations coming from the standard orientation of $\mathbf{R}^n$, we define

$$d(a,f) = \deg(r(a,f)), \tag{2.4}$$

the degree of the mapping $r(a,f)$. Then $d(a,f)$ depends only on the homotopy class of f as a mapping $\partial I^n \longrightarrow \mathbf{R}^n - \{a\}$. When f is injective, we denote by $X_1(f)$ (resp. $X_2(f)$) the interior (resp. exterior) domain with respect to $f(\partial I^n)$. We have

$$d(a,f) = \begin{cases} \pm 1 & \text{if} \quad a \in X_1(f) \\ 0 & \text{if} \quad a \in X_2(f) \end{cases} \tag{2.5}$$

if f is injective. For $1 \leqq i \leqq n$, let $e_i = (0, \dots, 0, 1, 0, \dots, 0) \in \mathbf{R}^n$ be the unit vector with 1 in the i-th component.

LEMMA 2.5. *Let $f : \partial I^n \longrightarrow \mathbf{R}^n$ be a continuous mapping which satisfies*

$$f(x+e_i) = f(x) + e_i \qquad \text{whenever} \quad x \quad \text{and} \quad x + e_i \in \partial I^n, \quad 1 \leqq i \leqq n. \tag{2.6}$$

If $y \in \mathbf{R}^n$ satisfies that $y + u \notin f(\partial I^n)$ for every $u \in \mathbf{Z}^n$, then we have

$$\sum_{u \in \mathbf{Z}^n} d(y+u, f) = 1. \tag{2.7}$$

PROOF. For $1 \leqq i \leqq n$, we put

$$I_i^+ = \{(x_1, x_2, \dots, x_n) \in I^n \mid x_i = 0\}, \qquad I_i^- = \{(x_1, x_2, \dots, x_n) \in I^n \mid x_i = 1\}.$$

Since f can be approximated arbitrarily well by a function g such that $g|I_i^+$ and $g|I_i^-$ are smooth for $1 \leqq i \leqq n$ and that g satisfies (2.6), we may assume that $f|I_i^+$ and $f|I_i^-$ are smooth for $1 \leqq i \leqq n$. For $a \in \mathbf{R}^n$, we consider the Newton potential:

$$\varphi_a(x) = \begin{cases} c_2 \log \frac{1}{\|x-a\|}, & n = 2 \\ c_n \frac{1}{\|x-a\|^{n-2}}, & n \geqq 3 \end{cases}$$

where $c_2 = -1/2\pi$ and $c_n = -\{(n-2) \times \text{area of } S^{n-1}\}^{-1}$ if $n \geqq 3$. Assume that f is injective for a moment. Put $\partial D = f(\partial I^n)$ and let $\mathbf{n}$ be the vector field of outward-drown normal vectors to ∂D of length 1. Then as is well known, using Stokes' theorem, we have

$$\int_{\partial D} (\text{grad } \varphi_a, \mathbf{n})\, d\Omega = \begin{cases} 1 & \text{if} \quad a \in X_1(f), \\ 0 & \text{if} \quad a \in X_2(f). \end{cases} \tag{2.8}$$

Here $(\ ,\ )$ denotes the standard inner product on $\mathbf{R}^n$ and $d\Omega$ denotes the volume element of ∂D. Write $f(x) = (f_1(x), f_2(x), \dots, f_n(x)), x \in \partial I^n$. Write $t \in I_i^+$ (resp. I_i^-) as $t = (t_1, \dots, t_{i-1}, 0, t_{i+1}, \dots, t_n)$ (resp. $t = (t_1, \dots, t_{i-1}, 1, t_{i+1}, \dots, t_n)$). For $t \in I_i^+$ or I_i^-, we consider $(n-1) \times n$-matrix $A = (\dfrac{\partial f_l}{\partial t_k})$, $1 \leqq k \leqq n$, $k \neq i$,

$1 \leqq l \leqq n$; a row of A is of the form $(\frac{\partial f_1}{\partial t_k}, \dots, \frac{\partial f_n}{\partial t_k})$, which we denote by $\mathbf{a}_k$. Let A_j be the matrix obtained from A by deleting the j-th column and put $\alpha_j = (-1)^{j-1} \det A_j$, $\mathbf{a} = (\alpha_1, \alpha_2, \dots, \alpha_n)$. By adding $\mathbf{a}_k$, $k \neq i$ to A as the n-th row and considering the expansion of the determinant of the matrix so obtained according to the n-th row, we see that $\mathbf{a}$ is orthogonal to $\mathbf{a}_k$. Hence we have $\mathbf{n} = \pm \frac{\mathbf{a}}{\|\mathbf{a}\|}$. On $I_i^{\pm}$, put $dt = dt_1 \dots dt_{i-1} dt_{i+1} \dots dt_n$. Then we see that $d\Omega = \|\mathbf{a}\| dt$. We obtain

$$(2.9) \quad \int_{\partial D} (\operatorname{grad} \varphi_a, \mathbf{n})\, d\Omega = c_n' (\sum_{i=1}^{n} \epsilon_i \int_{I_i^+} \frac{(f(t)-a, \mathbf{a})}{\|f(t)-a\|^n}\, dt + \epsilon_i' \int_{I_i^-} \frac{(f(t)-a, \mathbf{a})}{\|f(t)-a\|^n}\, dt).$$

Here $c_2' = -c_2$, $c_n' = -(n-2)c_n$ if $n \geqq 3$, $\epsilon_i = \pm 1$, $\epsilon_i' = \pm 1$. Considering orientations of I_i^+ and I_i^-, we see that $\epsilon_i' = -\epsilon_i$; ϵ_i depends only on the signature of $d(a,f)$. We choose ϵ_i for the case $d(a,f) = 1$ and fix it. Now dropping the assumption that f is injective, we define

$$(2.10) \quad \begin{aligned} d_i^+(a,f) &= \epsilon_i \int_{I_i^+} \frac{(f(t)-a, \mathbf{a})}{\|f(t)-a\|^n}\, dt, \qquad a \notin f(I_i^+), \\ d_i^-(a,f) &= -\epsilon_i \int_{I_i^-} \frac{(f(t)-a, \mathbf{a})}{\|f(t)-a\|^n}\, dt, \qquad a \notin f(I_i^-). \end{aligned}$$

By the assumption (2.6) on f, we have

$$(2.11) \qquad d_i^-(a+e_i, f) = -d_i^+(a,f), \quad 1 \leqq i \leqq n \quad \text{if} \quad a,\ a+e_i \notin f(\partial I^n).$$

We see that $\sum_{i=1}^n d_i^+(a,f) + d_i^-(a,f)$ depends only on the homotopy class of f as a mapping $\partial I^n \longrightarrow \mathbf{R}^n - \{a\}$. The homotopy classes of mappings $(\partial I^n, x) \longrightarrow (\mathbf{R}^n - \{a\}, y)$, $x \in \partial I^n$, $y \in \mathbf{R}^n - \{a\}$ are the same as the homotopy classes of mappings $(S^{n-1}, x_0) \longrightarrow (S^{n-1}, y_0)$, x_0, $y_0 \in S^{n-1}$, i.e., the homotopy group $\pi_{n-1}(S^{n-1}) \cong \mathbf{Z}$. Therefore, in view of (2.5) and (2.8), we have

$$(2.12) \qquad d(a,f) = \sum_{i=1}^{n} d_i^+(a,f) + d_i^-(a,f) \qquad \text{if} \quad a \notin f(\partial I^n).$$

Now let $y = (y_1, y_2, \dots, y_n)$ be as in our lemma. For a positive integer M, we put

$$S_M(y,f) = \sum_{u_1=-M}^{M} \cdots \sum_{u_n=-M}^{M} d(y+u, f).$$

By (2.11) and (2.12), we have

$$(2.13) \quad \begin{aligned} &S_M(y,f) \\ &= \sum_{i=1}^{n} (\sum_{u_1=-M}^{M} \cdots \sum_{u_n=-M}^{M} d_i^+(y + u_1 e_1 + \cdots + M e_i + \cdots + u_n e_n, f) \\ &\quad - d_i^+(y + u_1 e_1 + \cdots + (-M-1) e_i + \cdots + u_n e_n, f)), \end{aligned}$$

where $\sum_{u_i=-M}^{M}$ is omitted in the inner sum.

Choose a positive constant C so that $|x_j| \leqq C$ for all $x = (x_1, x_2, \dots, x_n) \in f(\partial I^n)$ and all j. Let $g : \partial I^n \longrightarrow \mathbf{R}^n$ be the identity mapping and we consider the homotopy $f_t = (1-t)f + tg$, $0 \leqq t \leqq 1$, which deforms f to g. Assume $M > C + 1 + \max_{1\leqq i\leqq n} |y_i|$. Put

$$\begin{aligned} T_M(y, f_t) = \sum_{i=1}^{n} \Big(\sum_{u_1=-M}^{M} \cdots \sum_{u_n=-M}^{M} & d_i^+(y + u_1e_1 + \cdots + Me_i + \cdots + u_ne_n, f_t) \\ & - d_i^+(y + u_1e_1 + \cdots + (-M-1)e_i + \cdots + u_ne_n, f_t)\Big), \end{aligned}$$

where $\sum_{u_i=-M}^{M}$ is omitted in the inner sum. Clearly f_t satisfies (2.6). Hence if $y + u \notin f_t(\partial I^n)$ for every $u \in \mathbf{Z}^n$, then, using (2.11) and (2.12) for f_t, we have

$$T_M(y, f_t) = \sum_{u_1=-M}^{M} \cdots \sum_{u_n=-M}^{M} d(y+u, f_t). \tag{2.14}$$

There exists a subset W of $\mathbf{R}^n$ such that $\mathbf{R}^n - W$ is dense in $\mathbf{R}^n$ and that if $y \notin W$, $y + u \in f_t(\partial I^n)$ for some $u \in \mathbf{Z}^n$ happens only for finitely many t. We can check this claim easily by considering the Jacobian matrix of $f_t|I_i^{\pm}$ as the function of n variables. The left-hand side of (2.7) does not change when we replace y by a sufficiently close point. Therefore we may assume that $y \notin W$ and that $y + u \notin \partial I^n$ for all $u \in \mathbf{Z}^n$. By the condition on M and (2.10), we see that $T_M(y, f_t)$ is a continuous function of t. By (2.14) it takes values in $\mathbf{Z}$ except for finitely many t. Therefore $T_M(y, f_t)$ is a constant function. Thus we obtain $S_M(y, f) = T_M(y, g)$. By the assumption $y + u \notin \partial I^n$ for all $u \in \mathbf{Z}^n$, we have $T_M(y, g) = S_M(y, g)$. Since $S_M(y, g) = 1$ is obvious, we complete the proof.

LEMMA 2.6. *Let $f : \partial I^n \longrightarrow \mathbf{R}^n$ be a continuous injective mapping which satisfies (2.6). Put $\partial D(f) = f(\partial I^n)$. If $y \in X_1(f)$, then $y + u \in X_2(f)$ for all $0 \neq u \in \mathbf{Z}^n$. If $y \in \partial D(f)$, put $y = f(z)$ with $z \in \partial I^n$. Then if $y + u \in \overline{X_1(f)}$ with $u \in \mathbf{Z}^n$, we have $z + u \in \partial I^n$.*

PROOF. Assume $y \in X_1(f)$. If $y + u \notin \partial D(f)$ for all $u \in \mathbf{Z}^n$, then Lemma 2.5 shows that $y + u \in X_2(f)$ for all $0 \neq u \in \mathbf{Z}^n$. Suppose that $y + u \in \partial D(f)$ for some $0 \neq u \in \mathbf{Z}^n$. Then we can find a sufficiently close point y' to y such $y' \in X_1(f)$, $y' + u \in X_1(f)$, $y' + v \notin \partial D(f)$ for all $v \in \mathbf{Z}^n$. This is a contradiction.

Next let $y = f(z) \in \partial D(f)$, $z \in \partial I^n$. To show the second assertion, we may assume that $z = (z_1, z_2, \ldots, z_n)$, $0 \leqq z_i < 1$ for $1 \leqq i \leqq n$. Assume that $y + u \in \overline{X_1(f)}$ with $0 \neq u \in \mathbf{Z}^n$. If $y + u \in X_1(f)$, this contradicts the first assertion. Hence $y + u \in \partial D(f)$. Put $y + u = f(w)$, $w \in \partial I^n$ and let $w = w' + v$, $v \in \mathbf{Z}^n$, $w' = (w_1', w_2', \ldots, w_n') \in \partial I^n$, $0 \leqq w_i' < 1$ for $1 \leqq i \leqq n$. We have $f(z) + u' = f(w')$, $u' = u - v$. It suffices to show that $u' = 0$. Assume $u' \neq 0$. Then $z \neq w'$. We can make a small deformation of f to f' so that $y \in X_1(f')$, $y + u' \in \partial D(f')$, contradicting the first assertion. This completes the proof.

Now let us show that $\tilde{h}$ is injective. Assume that $\tilde{h}(y) = \tilde{h}(z)$ for $y, z \in \mathbf{R}^n$. Put $y = y_0 + u$, $z = z_0 + v$ with $y_0, z_0 \in F$, $u, v \in \mathbf{Z}^n$. Then we have $h(y_0) + u - v = h(z_0)$. It suffices to show $u = v$. Put $f = h|\partial I^n$. Note that $h(z_0) \in \overline{X_1(f)}$. If $h(y_0) \in X_1(f)$, then we obtain $u - v = 0$ by Lemma 2.6. If $h(y_0) \in \partial D(f)$, then we have $y_0 + u - v \in \partial I^n$ by Lemma 2.6. This implies $h(y_0) + u - v = h(y_0 + u - v)$. Therefore we have $y_0 + u - v = z_0$, $u - v = 0$. This proves that $\tilde{h}$ is injective.

3. A sufficient condition for a cone decomposition

3.1. Let F be a totally real algebraic number field of degree $n+1$. We put

$$H = \{(x_1, x_2, \dots, x_{n+1}) \in \mathbf{R}_+^{n+1} \mid x_1 x_2 \cdots x_{n+1} = 1\}$$
$$P = \{(y_1, y_2, \dots, y_{n+1}) \in \mathbf{R}^{n+1} \mid y_1 + y_2 + \cdots + y_{n+1} = 0\},$$
$$f((x_1, x_2, \dots, x_{n+1})) = (\log x_1, \log x_2, \dots, \log x_{n+1}), \qquad (x_1, x_2, \dots, x_{n+1}) \in H.$$

We consider H (resp. P) as a real Lie group by the componentwise multiplication (resp. addition). Then f gives an isomorphism $H \cong P$ as real Lie groups. Let E be a subgroup of E_F^+ of finite index. Then E is a discrete subgroup of H (by $E \ni \epsilon \longrightarrow (\epsilon^{(1)}, \dots, \epsilon^{(n+1)})$) and H/E is compact. Take generators of E:

$$E = \langle \epsilon_1, \epsilon_2, \dots, \epsilon_n \rangle \quad (\cong \mathbf{Z}^n). \tag{3.1}$$

For $a = (a_1, a_2, \dots, a_n) \in \mathbf{R}^n$, we put

$$g(a) = (y_1, y_2, \dots, y_{n+1}), \qquad y_j = \sum_{i=1}^{n} a_i \log \epsilon_i^{(j)}, \quad 1 \leqq j \leqq n+1.$$

Then g is an isomorphism $\mathbf{R}^n \cong P$. We put $\varphi = g^{-1} \circ f$. Then we have $\varphi : H \cong \mathbf{R}^n$ and

$$\varphi(\epsilon_i x) = \varphi(x) + e_i, \qquad x \in H, \quad 1 \leqq i \leqq n, \tag{3.2}$$

where $e_i \in \mathbf{R}^n$ denotes the vector whose i-th component is 1 and other components are 0. Thus φ induces an isomorphism $H/E \cong \mathbf{R}^n/\mathbf{Z}^n$.

Now let $I^n = \sqcup_{j \in \tilde{J}} S_j$ be a special decomposition of I^n. We have $F = \sqcup_{j \in J} S_j$ for a subset J of $\tilde{J}$. For S_j, $j \in \tilde{J}$, we let correspond an open simplicial cone C_j in the following way. Let $\dim S_j = r(j) - 1$ and let $P_1, P_2, \dots, P_{r(j)}$ be the vertices of S_j. Let $P_i = (a_1^{(i)}, a_2^{(i)}, \dots, a_n^{(i)})$, $1 \leqq i \leqq r(j)$ and put $\xi_i = \epsilon_1^{a_1^{(i)}} \epsilon_2^{a_2^{(i)}} \cdots \epsilon_n^{a_n^{(i)}}$. We let $C_j = C(\xi_1, \xi_2, \dots, \xi_{r(j)})$.

THEOREM 3.1. *Let the notation be the same as above. Put $J' = \{j \in J \mid \dim S_j = n\}$. We assume that*
(1) *For every $j \in J'$, $\xi_1, \xi_2, \dots, \xi_{n+1}$ as above are linearly independent over $\mathbf{Q}$.*
(2) *Let j, $k \in J'$, $j \neq k$ and put*

$$C_j = C(\eta_1, \eta_2, \dots, \eta_{n+1}), \qquad C_k = C(\psi_1, \psi_2, \dots, \psi_{n+1}),$$
$$\{\omega_1, \omega_2, \dots, \omega_m\} = \{\eta_1, \eta_2, \dots, \eta_{n+1}\} \cap \{\psi_1, \psi_2, \dots, \psi_{n+1}\}.$$

Then there exists a linear form $H_{jk} \in Hom_{\mathbf{R}}(\mathbf{R}^{n+1}, \mathbf{R})$ such that

$$\begin{aligned} H_{jk}(\omega_i) &= 0, && 1 \leqq i \leqq m, \\ H_{jk}(\eta_i) &> 0 && \text{if} \quad \eta_i \notin \{\omega_1, \omega_2, \dots, \omega_m\}, \\ H_{jk}(\psi_i) &< 0 && \text{if} \quad \psi_i \notin \{\omega_1, \omega_2, \dots, \omega_m\}. \end{aligned}$$

If these two conditions are satisfied, then we have

$$\mathbf{R}_+^{n+1} = \sqcup_{\epsilon \in E} \epsilon \, (\sqcup_{j \in J} C_j).$$

3.2. For the proof of this theorem, we need some preparations.

LEMMA 3.2. *Let* $C = C(v_1, v_2, \dots, v_r)$, $v_i \in \mathbf{R}_+^{n+1}$ *be an* r*-dimensional open simplicial cone in* $\mathbf{R}_+^{n+1}$. *Then we have*

$$\bar{C} = \sqcup_{i=1}^{r} \sqcup_{\{i_1, i_2, \dots, i_s\} \subset \{1,2,\dots,r\}} C(v_{i_1}, v_{i_2}, \dots, v_{i_s}) \sqcup \{0\}.$$

Since the proof is similar to that of Lemma 1.1 and is easy, we omit it.

LEMMA 3.3. *Let* $C = C(v_1, v_2, \dots, v_r)$, $v_i \in \mathbf{R}_+^{n+1}$ *be an* r*-dimensional open simplicial cone. Then* $C \cap H$ *is homeomorphic to an* $(r-1)$*-dimensional open simplex.*

PROOF. Put

$$S = \{t_1 v_1 + t_2 v_2 + \cdots + t_r v_r \mid 0 < t_1, t_2, \dots, t_r,\ t_1 + t_2 + \cdots + t_r = 1\}.$$

It suffices to show that $C \cap H$ is homeomorphic to S. For $x = t_1 v_1 + t_2 v_2 + \cdots + t_r v_r \in C \cap H$, put

$$\pi(x) = \frac{1}{t_1 + t_2 + \cdots + t_r} x. \tag{3.3}$$

Then π defines a continuous mapping of $C \cap H$ into S. Let

$$v_j = (v_{j1}, v_{j2}, \dots, v_{j(n+1)}), \qquad 1 \leqq j \leqq n.$$

For $y = t_1 v_1 + t_2 v_2 + \cdots + t_r v_r \in S$, put

$$\phi(y) = ay, \qquad a = (\prod_{i=1}^{n+1} t_1 v_{1i} + t_2 v_{2i} + \cdots + t_r v_{ri})^{-1/(n+1)}.$$

Then ϕ defines a continuous mapping of S into $C \cap H$. We can verify easily that $\pi \circ \phi = \mathrm{id}$, $\phi \circ \pi = \mathrm{id}$. This completes the proof.

PROOF OF THEOREM 3.1. First we will show that C_j, $j \in \tilde{J}$ are mutually disjoint. Let m, $l \in \tilde{J}$, $m \neq l$. Clearly $\cup_{j \in J'} S_j$ is dense in I^n. The property (2) of the special decomposition implies that there exist j, $k \in J'$ such that $S_m \subset \bar{S}_j$, $S_l \subset \bar{S}_k$. Then we have $C_m \subset \bar{C}_j$, $C_l \subset \bar{C}_k$. If $j = k$, the assertion is clear by Lemma 3.2 and from the construction of C_j from S_j. Putting C_j and C_k in the form (2), Theorem 3.1, we let

$$C_m = C(\eta_{i_1}, \eta_{i_2}, \dots, \eta_{i_s}), \qquad C_l = C(\psi_{j_1}, \psi_{j_2}, \dots, \psi_{j_t}).$$

If there exists an $\eta_{i_a} \notin \{\omega_1, \omega_2, \dots, \omega_m\}$, $1 \leqq a \leqq s$, we have $H_{jk}(x) > 0$ for all $x \in C_m$. If there exists a $\psi_{j_b} \notin \{\omega_1, \omega_2, \dots, \omega_m\}$, $1 \leqq b \leqq t$, we have $H_{jk}(x) < 0$ for all $x \in C_l$. Therefore we may assume that $\{\eta_{i_1}, \eta_{i_2}, \dots, \eta_{i_s}\} \subset \{\omega_1, \omega_2, \dots, \omega_m\}$ and $\{\psi_{j_1}, \psi_{j_2}, \dots, \psi_{j_t}\} \subset \{\omega_1, \omega_2, \dots, \omega_m\}$. Then C_m, $C_l \subset C_j$ and the assertion reduces to the case $j = k$.

Let $C_j = C(\xi_1, \xi_2, \dots, \xi_{r(j)})$ and let

$$S'_j = \{t_1 \xi_1 + t_2 \xi_2 + \cdots + t_{r(j)} \xi_{r(j)} \mid t_1, t_2, \dots, t_{r(j)} > 0,\ t_1 + t_2 + \cdots + t_{r(j)} = 1\}.$$

Let π_j be the homeomorphism of $C_j \cap H$ onto S'_j given in Lemma 3.3. Clearly π_j extends to the homeomorphism $\bar{\pi}_j$ of $\bar{C}_j \cap H$ onto $\bar{S}'_j$. Let q_j be the simplicial isomorphism of S'_j onto S_j sending ξ_i to P_i where $P_i = (a_1^{(i)}, a_2^{(i)}, \dots, a_n^{(i)})$ is a vertex of I^n and $\xi_i = \epsilon_1^{a_1^{(i)}} \epsilon_2^{a_2^{(i)}} \cdots \epsilon_n^{a_n^{(i)}}$; q_j extends to the homeomorphism $\bar{q}_j$ of $\bar{S}'_j$

onto $\bar{S}_j$. We put $\psi_j = q_j \circ \pi_j$, $\bar{\psi}_j = \bar{q}_j \circ \bar{\pi}_j$. Since $C_j \cap H$ are mutually disjoint, we can define a mapping $\psi = \cup_{j \in \tilde{J}} \psi_j$ of $\cup_{j\in\tilde{J}} C_j \cap H$ into $\cup_{j\in\tilde{J}} S_j = I^n$ by setting $\psi | C_j \cap H = \psi_j$. Clearly ψ is a bijection. Let us show that ψ is continuous. By definition, it is clear that $\psi_m = \bar{\psi}_j | C_m$ if $C_m \subset C_j$. Hence it suffices to show that

$$\bar{\psi}_j | \bar{C}_j \cap \bar{C}_k \cap H = \bar{\psi}_k | \bar{C}_j \cap \bar{C}_k \cap H \tag{3.4}$$

for $j, k \in J'$, $j \neq k$. By the property (2) of the special decomposition and Lemmas 1.1 and 3.2, we have

$$\bar{C}_j \cap \bar{C}_k = \cup_{\alpha \in I_j \cap I_k} C_\alpha \cup \{0\}.$$

Hence (3.4) follows. We have $\cup_{j\in\tilde{J}} C_j \cap H = \cup_{j \in J} \bar{C}_j \cap H$. Since ψ is a one-to-one continuous mapping and $\cup_{j\in\tilde{J}} \bar{C}_j \cap H$ is compact, ψ gives a homeomorphism of $\cup_{j\in\tilde{J}} C_j \cap H$ onto I^n.

Now put $h = \varphi \circ \psi^{-1}$. Then h defines an injective continuous mapping of I^n into $\mathbf{R}^n$. Let us show that

$$h(x + e_i) = h(x) + e_i \quad \text{whenever} \quad x, x + e_i \in I^n, \quad 1 \leqq i \leqq n. \tag{3.5}$$

Assume that x, $x + e_i \in I^n$ for some i. Then we have $x \in I_i^+(n)$, $x + e_i \in I_i^-(n)$. By the property (3) of the special decomposition, we have $x \in S_j$ and $x + e_i \in S_k = S_j + e_i$ for some j, $k \in \tilde{J}$. By the definition of C_j, we have $C_k = \epsilon_i C_j$. If $\psi_j(y) = x$, $y \in C_j$, we have $\psi_k(\epsilon_i y) = x + e_i$. Hence $\psi^{-1}(x + e_i) = \epsilon_i \psi^{-1}(x)$. Now (3.5) follows from (3.2). By (3.5), we see easily that h satisfies (2.1). Therefore $h(F)$ is a fundamental domain of $\mathbf{R}^n/\mathbf{Z}^n$ by Corollary 2.2.

Now it follows easily that $\sqcup_{j \in J} C_j$ is a fundamental domain of $\mathbf{R}_+^{n+1}/E$. In fact, it suffices to show that

$$H = \sqcup_{\epsilon \in E}\ \epsilon\,(\sqcup_{j\in J}(C_j \cap H)). \tag{3.6}$$

First we note that $\psi^{-1}(F) = \sqcup_{j \in J}(C_j \cap H)$. Take any $x \in H$. We can find $y \in h(F)$ and $u = (u_1, u_2, \ldots, u_n) \in \mathbf{Z}^n$ so that $\varphi(x) = y + u$. We have $y = \varphi(z)$ with $z \in \sqcup_{j\in J}(C_j \cap H)$. By (3.2), we have $\varphi(\epsilon_1^{-u_1}\epsilon_2^{-u_2}\cdots\epsilon_n^{-u_n} x) = y$. Hence $\epsilon_1^{-u_1}\epsilon_2^{-u_2}\cdots\epsilon_n^{-u_n} x \in \sqcup_{j\in J}(C_j \cap H)$. Next let x, $x' \in \sqcup_{j \in J}(C_j \cap H)$ and $x' = \epsilon x$ with $\epsilon = \epsilon_1^{u_1}\epsilon_2^{u_2}\cdots\epsilon_n^{u_n}$. Then we have $\varphi(x') = \varphi(x) + (u_1, u_2, \ldots, u_n)$. Since $\varphi(x)$, $\varphi(x') \in h(F)$, we have $(u_1, u_2, \ldots, u_n) = 0$, $\epsilon = 1$. This completes the proof of Theorem 3.1.

The following Lemma will be used in the next section.

LEMMA 3.4. *Let the notation be the same as in Theorem 3.1. Let L_{jk} be a linear form such that*

$$\begin{aligned} L_{jk}(\omega_i) &= 0, && 1 \leqq i \leqq m, \\ L_{jk}(\eta_i) &\geqq 0, && 1 \leqq i \leqq n+1 \\ L_{jk}(\psi_i) &< 0 && \textit{if} \quad \psi_i \notin \{\omega_1, \omega_2, \ldots, \omega_m\}. \end{aligned}$$

Then there exists a linear form H_{jk} which satisfies the condition (2) *of Theorem* 3.1.

PROOF. Since $\eta_1, \eta_2, \dots, \eta_{n+1}$ are linearly independent and $\{\omega_1, \omega_2, \dots, \omega_m\}$ is a subset of $\{\eta_1, \eta_2, \dots, \eta_{n+1}\}$, there exists a linear form L'_{jk} such that $L'_{jk}(\eta_i) = 1$ if $\eta_i \notin \{\omega_1, \omega_2, \dots, \omega_m\}$, $1 \leqq i \leqq n+1$ and $L'_{jk}(\omega_i) = 0$, $1 \leqq i \leqq m$. Then the linear form $L_{jk} + uL'_{jk}$ with a small positive number u suffices for our purpose.

4. Examples

4.1. By the procedure described in §1, an explicit special decomposition of I^n can be inductively constructed as follows. Let $I^n = \sqcup_{j \in J} S_j$ be a special decomposition. We call an open simplex S_j *a big cell* if $\dim S_j = n$. Put $J' = \{j \in J \mid \dim S_j = n\}$. We call C_j *a big cone* for $j \in J'$. We number 2^n vertices of I^n by setting

$$P_m = (a_1, a_2, \dots, a_m), \qquad m = 1 + \sum_{i=1}^{n} a_i \cdot 2^{i-1}.$$

If $S(P_{m_1}, P_{m_2}, \dots, P_{m_{n+1}})$ is a big cell appearing in a special decomposition of I^n, let $n+1$ big cells $S(P_{m_i}, P_{m_{i+1}}, \dots, P_{m_{n+1}}, P_{m_1+2^n}, P_{m_2+2^n}, \dots, P_{m_i+2^n})$, $1 \leqq i \leqq n+1$ appear in the special decomposition of I^{n+1}. Thus there are $n!$ big cells in the special decomposition of I^n and adjoining all faces of big cells, we obtain a special decomposition of I^n. For simplicity, put $\{m_1, m_2, \dots, m_j\} = S(P_{m_1}, P_{m_2}, \dots, P_{m_j})$.

$n = 1$: $\{1,2\}$ is a big cell and $F = \{1,2\} \sqcup \{1\}$.
$n = 2$: $\{1,2,3\}$ and $\{2,3,4\}$ are big cells and

$$F = \{1,2,3\} \sqcup \{2,3,4\} \sqcup \{1,2\} \sqcup \{1,3\} \sqcup \{2,3\} \sqcup \{1\}.$$

$n = 3$: $\{1,2,3,5\}$, $\{2,3,5,6\}$, $\{3,5,6,7\}$, $\{2,3,4,6\}$, $\{3,4,6,7\}$ and $\{4,6,7,8\}$ are big cells and

$$\begin{aligned} F =& \{1,2,3,5\} \sqcup \{2,3,5,6\} \sqcup \{3,5,6,7\} \sqcup \{2,3,4,6\} \sqcup \{3,4,6,7\} \sqcup \{4,6,7,8\} \\ &\sqcup \{1,2,3\} \sqcup \{1,2,5\} \sqcup \{1,3,5\} \sqcup \{2,3,5\} \sqcup \{2,3,6\} \sqcup \{2,5,6\} \sqcup \{3,5,6\} \\ &\sqcup \{3,5,7\} \sqcup \{3,6,7\} \sqcup \{2,3,4\} \sqcup \{3,4,6\} \sqcup \{4,6,7\} \sqcup \{1,2\} \sqcup \{1,3\} \sqcup \{1,5\} \\ &\sqcup \{2,3\} \sqcup \{2,5\} \sqcup \{3,5\} \sqcup \{3,6\} \sqcup \{1\}. \end{aligned}$$

4.2. First we consider the case $n = 2$. Let $\epsilon = \epsilon_1$, $\eta = \epsilon_2$; thus $E = \langle \epsilon, \eta \rangle$. We let P_i correspond to a unit by $P_1 \to 1$, $P_2 \to \epsilon$, $P_3 \to \eta$, $P_4 \to \epsilon\eta$. Big cones are $C_1 = C(1, \epsilon, \eta)$, $C_2 = C(\epsilon, \eta, \epsilon\eta)$. The condition (1) of Theorem 3.1 is that $\{1, \epsilon, \eta\}$ and $\{\epsilon, \eta, \epsilon\eta\}$ are linearly independent. The condition (2) of Theorem 3.1 is that there exists a linear form H such that

$$H(\epsilon) = H(\eta) = 0, \qquad H(1) > 0, \qquad H(\epsilon\eta) < 0.$$

Let $\epsilon\eta = x + y\epsilon + z\eta$ with x, y, $z \in \mathbf{Q}$. Then such an H exists if and only if $x < 0$. Thus we recover Theorem 1 of Thomas and Vasquez[TV]: If 1, ϵ , η are linearly independent and if $x < 0$, then

$$C(1, \epsilon, \eta) \sqcup C(\epsilon, \eta, \epsilon\eta) \sqcup C(1, \epsilon) \sqcup C(1, \eta) \sqcup C(\epsilon, \eta) \sqcup C(1)$$

gives a fundamental domain of $\mathbf{R}_+^3/E$. In [TV], the existence of such generators ϵ and η for $E = E_F^+$ is shown by a theorem of Berwick [Be] (see Exercise 5).

4.3. Next let us consider the case $n = 3$. We treat a numerical example. Let $f(X) = X^4 - X^3 - 3X^2 + X + 1$. Then $f(X)$ is irreducible and has four real roots:

$$\rho = 2.09529\dots, \quad \rho' = 0.73764\dots, \quad \rho'' = -0.47725\dots, \quad \rho''' = -1.35567\dots.$$

Let $F = \mathbf{Q}(\rho)$. Then F is a totally real non-normal quartic extension of $\mathbf{Q}$. By Billevič [Bi], we have $\mathfrak{O}_F = \mathbf{Z}[\rho]$, the discriminant of F is $5^2 \cdot 29$ and $E_F = \langle \epsilon_1, \eta_1, \psi_1, -1 \rangle$, $\epsilon_1 = \rho^2 - 1$, $\eta_1 = \rho^3 + \rho^2 - \rho - 1$, $\psi_1 = \rho^3 + 2\rho^2 - 1$. We have $E_F^+ = \langle \epsilon_1^2, \eta_1^2, \psi_1^2 \rangle$, $[E_F : E_F^+] = 2^4$. We put

$$\epsilon = \epsilon_1^2, \quad \eta = \epsilon_1^{-2}\eta_1^2, \quad \psi = \epsilon_1^2\eta_1^2\psi_1^{-2}.$$

We let P_i correspond to a unit by $P_1 \to 1$, $P_2 \to \epsilon$, $P_3 \to \eta$, $P_4 \to \epsilon\eta$, $P_5 \to \psi$, $P_6 \to \epsilon\psi$, $P_7 \to \eta\psi$, $P_8 \to \epsilon\eta\psi$. We identify P_i's with these units. We let

$$C_1 = C(P_1, P_2, P_3, P_5), \quad C_2 = C(P_2, P_3, P_5, P_6), \quad C_3 = C(P_3, P_5, P_6, P_7),$$
$$C_4 = C(P_2, P_3, P_4, P_6), \quad C_5 = C(P_3, P_4, P_6, P_7), \quad C_6 = C(P_4, P_6, P_7, P_8).$$

We can verify easily that $\dim C_j = 4$, $1 \leqq j \leqq 6$. Our task is to verify the separation condition of Theorem 3.1, (2).

When P_{i_1}, P_{i_2}, P_{i_3} and P_{i_4} are linearly independent, let $L_{i_1|i_2i_3i_4}$ be the linear form on $\mathbf{R}^4$ such that

$$L_{i_1|i_2i_3i_4}(P_{i_1}) = 1, \qquad L_{i_1|i_2i_3i_4}(P_{i_2}) = L_{i_1|i_2i_3i_4}(P_{i_3}) = L_{i_1|i_2i_3i_4}(P_{i_4}) = 0.$$

For x, y, z, $w \in \mathbf{R}^4$, we put $\Delta[x, y, z, w] = \det(x\ y\ z\ w)$.

(1) Separation of C_1 with C_j, $2 \leqq j \leqq 6$.

Let

$$\epsilon\eta = x + y\epsilon + z\eta + w\psi, \qquad x, y, z, w \in \mathbf{Q}.$$

Then, by Cramer's formula, we have

$$L_{1|235}(P_4) = x = \frac{\Delta[\epsilon\eta, \epsilon, \eta, \psi]}{\Delta[1, \epsilon, \eta, \psi]}.$$

We obtain $L_{1|235}(P_4) = -\dfrac{11}{2}$. Similarly we obtain

$$L_{1|235}(P_6) = -\frac{5}{2}, \quad L_{1|235}(P_7) = -2, \quad L_{1|235}(P_8) = -25.$$

By Lemma 3.4, the separation condition holds for (C_1, C_j), $2 \leqq j \leqq 6$.

(2) Separation of C_2 with C_j, $3 \leqq j \leqq 6$.

We have

$$L_{2|356}(P_4) = \frac{1}{5}, \quad L_{2|356}(P_7) = -\frac{1}{5}, \quad L_{2|356}(P_8) = 0.$$

By the second value, the separation condition holds for (C_2, C_3). We have

$$L_{5|236}(P_4) = -\frac{8}{5}, \quad L_{5|236}(P_7) = -\frac{2}{5}, \quad L_{5|236}(P_8) = -7.$$

By these values, we see that the separation condition holds for (C_2, C_j), $j \geqq 4$.

(3) Separation of C_3 with C_j, $4 \leqq j \leqq 6$.

Using $-L_{2|356}$ and Lemma 3.4, we see that the separation condition holds for (C_3, C_4). We have

$$L_{5|367}(P_4) = -2, \qquad L_{5|367}(P_8) = -7.$$

By these values, we see that the separation condition holds for (C_3, C_j), $j = 5, 6$.

(4) Separation of C_4 with C_j, $5 \leqq j \leqq 6$.

We have

$$L_{2|346}(P_7) = -\frac{1}{4}, \qquad L_{2|346}(P_8) = -\frac{7}{8}.$$

Hence the separation condition holds for (C_4, C_j), $j = 5, 6$.

(5) Separation of C_5 with C_6.

We have

$$L_{3|467}(P_8) = -\frac{5}{2}.$$

Hence the separation condition holds for (C_5, C_6).

Therefore

$$\begin{aligned} &C(1,\epsilon,\eta,\psi) \sqcup C(\epsilon,\eta,\psi,\epsilon\psi) \sqcup C(\eta,\psi,\epsilon\psi,\eta\psi) \sqcup C(\epsilon,\eta,\epsilon\eta,\epsilon\psi) \\ &\sqcup C(\eta,\epsilon\eta,\epsilon\psi,\eta\psi) \sqcup C(\epsilon\eta,\epsilon\psi,\eta\psi,\epsilon\eta\psi) \sqcup C(1,\epsilon,\eta) \sqcup C(1,\epsilon,\psi) \sqcup C(1,\eta,\psi) \\ &\sqcup C(\epsilon,\eta,\psi) \sqcup C(\epsilon,\eta,\epsilon\psi) \sqcup C(\epsilon,\psi,\epsilon\psi) \sqcup C(\eta,\psi,\epsilon\psi) \sqcup C(\eta,\psi,\eta\psi) \\ &\sqcup C(\eta,\epsilon\psi,\eta\psi) \sqcup C(\epsilon,\eta,\epsilon\eta) \sqcup C(\eta,\epsilon\eta,\epsilon\psi) \sqcup C(\epsilon\eta,\epsilon\psi,\eta\psi) \sqcup C(1,\epsilon) \\ &\sqcup C(1,\eta) \sqcup C(1,\psi) \sqcup C(\epsilon,\eta) \sqcup C(\epsilon,\psi) \sqcup C(\eta,\psi) \sqcup C(\eta,\epsilon\psi) \sqcup C(1) \end{aligned} \tag{4.1}$$

gives a fundamental domain of $\mathbf{R}_+^4/E_F^+$.

5. Explicit cone decompositions for index finite subgroups

5.1. Let F be a totally real algebraic number field of degree $n+1$. We assume $n \geqq 2$.

DEFINITION. *Suppose that for positive integers m_1, m_2, $\dots$, m_n, there are given $\epsilon_1(m_1)$, $\epsilon_2(m_2)$, $\dots$, $\epsilon_n(m_n) \in \mathbf{R}_+^{n+1}$. We write $\epsilon_i(m_i) = {}^t(\epsilon_i^{(1)}(m_i), \dots, \epsilon_i^{(n+1)}(m_i))$. We call $\epsilon_1(m_1)$, $\epsilon_2(m_2)$, $\dots$, $\epsilon_n(m_n)$ a fine sequence in $\mathbf{R}_+^{n+1}$ if it satisfies the following conditions.*

$$\prod_{j=1}^{n+1} \epsilon_i^{(j)}(m_i) = 1, \qquad 1 \leqq i \leqq n. \tag{C1}$$

$$\text{(C2)}\quad \begin{aligned}
&\textit{For } 1 \leqq i \leqq n,\\
&\epsilon_i^{(i+1)}(m_i) > 1, \qquad \epsilon_i^{(k)}(m_i) < 1, \quad k \neq i+1.\\
&\epsilon_1^{(2)}(m_1) > \epsilon_1^{(1)}(m_1) > \epsilon_1^{(3)}(m_1), \qquad \lim_{m_1 \to \infty} \epsilon_1^{(2)}(m_1) = \infty,\\
&\epsilon_2^{(3)}(m_2) > \epsilon_2^{(2)}(m_2) > \epsilon_2^{(1)}(m_2), \qquad \lim_{m_2 \to \infty} \epsilon_2^{(3)}(m_2) = \infty.\\
&\textit{For } 3 \leqq i \leqq n,\\
&\lim_{m_i \to \infty} \epsilon_i^{(i+1)}(m_i) = \infty, \qquad \lim_{m_i \to \infty} \epsilon_i^{(k)}(m_i)/\epsilon_i^{(i+1)}(m_i) = 0, \quad \textit{if} \quad k \neq i+1,\\
&\lim_{m_i \to \infty} \epsilon_i^{(k)}(m_i)/\epsilon_i^{(l)}(m_i) = 1, \qquad \textit{if} \quad k, l \neq i+1.
\end{aligned}$$

PROPOSITION 5.1. *For positive integers $m_1, m_2, \cdots, m_n$, there exist units $\epsilon_1(m_1), \epsilon_2(m_2), \ldots, \epsilon_n(m_n) \in E_F^+$ which satisfy the following conditions.*

1. *When we regard $\epsilon_i(m_i) \in \mathbf{R}_+^{n+1}$, the sequence is fine.*
2. *For every $m_1, m_2, \ldots, m_n$, $\epsilon_1(m_1), \epsilon_2(m_2), \ldots, \epsilon_n(m_n)$ generate a subgroup of E_F^+ of finite index.*

PROOF. We choose $\Gamma = (\gamma_{ij}) \in M(n, n, \mathbf{R})$ so that

$$\text{(5.1)}\quad \begin{aligned}
&\gamma_{12} > \gamma_{11} > \gamma_{13}, \qquad \gamma_{12} > 0, \qquad \gamma_{1j} < 0 \quad \text{if} \quad j \neq 2,\\
&\gamma_{23} > \gamma_{22} > \gamma_{21}, \qquad \gamma_{23} > 0, \qquad \gamma_{2j} < 0 \quad \text{if} \quad j \neq 3,\\
&\gamma_{i\,i+1} = n, \qquad \gamma_{ij} = -1 \quad \text{if} \quad j \neq i, \qquad 3 \leqq i \leqq n-1,\\
&\gamma_{nj} = -1, \qquad 1 \leqq j \leqq n.
\end{aligned}$$

Let $\eta_1, \eta_2, \ldots, \eta_n$ be generators of $E_F^+ \cong \mathbf{Z}^n$. Define $B = (b_{ij}) \in M(n, n, \mathbf{R})$ by

$$B\,(\log \eta_i^{(j)}) = \Gamma.$$

Here $(\log \eta_i^{(j)})$ denotes the $n \times n$-matrix whose (i,j)-component is $\log \eta_i^{(j)}$. Let us show that $(b_{i1}, b_{i2}, \ldots, b_{in}) \notin c\mathbf{Q}^n$ for every $c \in \mathbf{R}^\times$ for $3 \leqq i \leqq n$. Suppose that

$$(b_{i1}, b_{i2}, \ldots, b_{in}) = c(b'_{i1}, b'_{i2}, \ldots, b'_{in}) \qquad \text{with} \quad c \in \mathbf{R}^\times, \quad b'_{i1}, b'_{i2}, \ldots, b'_{in} \in \mathbf{Z}.$$

We have

$$\sum_{k=1}^{n} b_{ik} \log \eta_k^{(j)} = \begin{cases} n & \text{if} \quad j = i+1, \\ -1 & \text{if} \quad j \neq i+1, \end{cases}$$

for $3 \leqq i \leqq n$, $1 \leqq j \leqq n+1$. Hence we get

$$\prod_{k=1}^{n} (\eta_k^{(j)})^{b'_{ik}} = \prod_{k=1}^{n} (\eta_k^{(l)})^{b'_{ik}} \qquad \text{if} \quad j, l \neq i+1.$$

Put $x = \prod_{k=1}^{n} \eta_k^{b'_{ik}}$ and let $f(X)$ be the minimal polynomial of x over $\mathbf{Q}$. Since $x \in F$ and $x^{(j)} = x^{(l)}$ if $j, l \neq i+1$, we clearly have $\deg f \leqq 2$. The case $\deg f = 2$ cannot occur since we assumed $[F : \mathbf{Q}] \geqq 3$. Therefore $x \in \mathbf{Q}$, hence $x = 1$. Then we have $b'_{ik} = 0$, $1 \leqq k \leqq n$, which is a contradiction.

We are going to define

$$\text{(5.2)}\qquad \epsilon_i(m_i) = \prod_{j=1}^{n} \eta_j^{c_{ij}(m_i)},$$

by choosing $c_{ij}(m_i) \in \mathbf{Z}$, $1 \leqq i, j \leqq n$. For every i, $1 \leqq i \leqq n$, we choose an index p_i so that $b_{ip_i} \neq 0$. For $i = 1, 2$, we choose $c_{ij} \in \mathbf{Z}$ so that c_{ip_i} has the same signature as that of b_{ip_i}, and that

$$|c_{ij}/c_{ip_i} - b_{ij}/b_{ip_i}| < \delta, \qquad 1 \leqq j \leqq n$$

for a sufficiently small $\delta > 0$. Put $\epsilon_i = \prod_{j=1}^n \eta_j^{c_{ij}}$. Then we have

$$\frac{\log \epsilon_i^{(k)}}{c_{ip_i}} = \sum_{j=1}^n \frac{c_{ij}}{c_{ip_i}} \log \eta_j^{(k)},$$

which is close to

$$\sum_{j=1}^n \frac{b_{ij}}{b_{ip_i}} \log \eta_j^{(k)} = \frac{\gamma_{ik}}{b_{ip_i}}.$$

Hence choosing δ sufficiently small, ϵ_1 and ϵ_2 satisfy the required conditions in (C2). It is obvious that we can obtain sequences $\epsilon_1(m_1)$ and $\epsilon_2(m_2)$ which satisfy (C2), by taking powers of them for example.

Let $3 \leqq i \leqq n$. As we have shown, one of b_{ij}/b_{ip_i}, $1 \leqq j \leqq n$ is irrational. Hence by a well known theorem on Diophantine approximation, we can find $c_{ij} \in \mathbf{Z}$ so that

$$(5.3) \qquad |c_{ij}/c_{ip_i} - b_{ij}/b_{ip_i}| < \frac{1}{|c_{ip_i}|^{n/(n-1)}}, \qquad 1 \leqq j \leqq n.$$

Furthermore there exist infinitely many solutions of the Diophantine approximation (5.3) (cf. Hardy and Wright [HW], Theorem 200). Therefore we can find a sequence $c_{ij}(m)$, $1 \leqq m \in \mathbf{Z}$, which satisfies (5.3) with $c_{ij}(m)$ in place of c_{ij} and $\lim_{m\to\infty} |c_{ip_i}(m)| = \infty$. We may also assume that $c_{ip_i}(m)$ has the same signature as that of b_{ip_i}. Put $\psi_i(m) = \prod_{j=1}^n \eta_j^{c_{ij}(m)}$. Then we have

$$(5.4) \qquad \begin{aligned} \log \psi_i(m)^{(k)} &= c_{ip_i}(m) \sum_{j=1}^n \frac{c_{ij}(m)}{c_{ip_i}(m)} \log \eta_j^{(k)}, \\ |\log \psi_i(m)^{(k)} - c_{ip_i}(m) \sum_{j=1}^n \frac{b_{ij}}{b_{ip_i}} \log \eta_j^{(k)}| &\leqq nM \cdot \frac{1}{|c_{ip_i}(m)|^{1/(n-1)}}, \end{aligned}$$

where $M = \max_{1 \leqq j \leqq n, 1 \leqq k \leqq n+1} |\log \eta_j^{(k)}|$. Since

$$c_{ip_i}(m) \sum_{j=1}^n \frac{b_{ij}}{b_{ip_i}} \log \eta_j^{(k)} = \frac{c_{ip_i}(m)}{b_{ip_i}} \gamma_{ik},$$

we see easily from (5.4) that $\epsilon_i(m_i) = \psi_i(m_i + N)$ satisfies the condition (C2) taking $N \in \mathbf{Z}$ sufficiently large.

It remains to prove that $\epsilon_1(m_1)$, $\epsilon_2(m_2)$, $\ldots$, $\epsilon_n(m_n)$ are independent units. This assertion follows from the well known fact (see Exercise 2, (a)) since

$$\epsilon_i^{(i+1)}(m_i) > 1, \qquad \epsilon_i^{(k)}(m_i) < 1, \quad k \neq i+1.$$

This completes the proof.

5.2. Our purpose is to prove the following theorem.

THEOREM 5.2. *There exist units $\epsilon_1, \epsilon_2, \ldots, \epsilon_n$ of E_F^+ with the following properties.*

(1) *$\epsilon_1, \epsilon_2, \ldots, \epsilon_n$ generate a subgroup E of E_F^+ of finite index.*

(2) *Let $I^n = \sqcup_{j\in \tilde{J}} S_j$ be a special decomposition of I^n and let $F = \sqcup_{j\in J} S_j$. By the procedure of Theorem 3.1, we assign the open simplicial cone C_j to S_j. Then the conditions (1) and (2) of Theorem 3.1 hold and we have*

$$\mathbf{R}_+^{n+1} = \sqcup_{\epsilon\in E}\ \epsilon(\sqcup_{j\in J} C_j).$$

(3) *For every $j \in J$, let $C_j = C(\psi_1, \psi_2, \ldots, \psi_{r(j)})$. Then for every $1 \leqq p, q \leqq r(j)$, $p \neq q$ and $1 \leqq j, k \leqq n+1$, $j \neq k$, $\psi_p^{(j)}/\psi_p^{(k)} \neq \psi_q^{(j)}/\psi_q^{(k)}$ holds.*

To use an induction argument on n, we will prove a more general result.

THEOREM 5.3. *Let $\epsilon_1(m_1), \epsilon_2(m_2), \ldots, \epsilon_n(m_n)$ be a fine sequence in $\mathbf{R}^{n+1}$. Then there exist positive integers N_i, $1 \leqq i \leqq n$, depending on $m_1, m_2, \ldots, m_{i-1}$, with the following properties.*

(P) *Put $\epsilon_i = \epsilon_i(m_i)$, $1 \leqq i \leqq n$. Let $I^n = \sqcup_{j\in \tilde{J}} S_j$ be a special decomposition of I^n and let $F = \sqcup_{j\in J} S_j$. By the procedure of Theorem 3.1, we assign the open simplicial cone C_j to S_j. Then the conditions (1) and (2) of Theorem 3.1 hold if $m_i \geqq N_i$, $1 \leqq i \leqq n$.*

Let us show that Theorem 5.2 follows from Theorem 5.3. We take units $\epsilon_1(m_1)$, $\epsilon_2(m_2), \ldots, \epsilon_n(m_n)$ constructed by Proposition 5.1 as the fine sequence in Theorem 5.3. Then, when all of m_i are sufficiently large, $\epsilon_i = \epsilon_i(m_i)$, $1 \leqq i \leqq n$ satisfy the conditions (1) and (2) of Theorem 5.2. It suffices to show that the condition (3) holds when all of m_i are sufficiently large. Put $\epsilon_i = \epsilon_i(m_i)$, $1 \leqq i \leqq n$. Suppose that

$$\psi_p = \epsilon_1^{a_1}\epsilon_2^{a_2}\cdots\epsilon_n^{a_n}, \qquad \psi_q = \epsilon_1^{b_1}\epsilon_2^{b_2}\cdots\epsilon_n^{b_n}$$

with a_i, $b_j \in \mathbf{Z}$ satisfy $\psi_p^{(j)}/\psi_p^{(k)} = \psi_q^{(j)}/\psi_q^{(k)}$ for $j \neq k$. Put $c_i = a_i - b_i$, $1 \leqq i \leqq n$ and $\eta = \psi_p\psi_q^{-1}$. Then $\eta = \epsilon_1^{c_1}\epsilon_2^{c_2}\cdots\epsilon_n^{c_n}$ and $\eta^{(j)} = \eta^{(k)}$. It suffices to show $c_1 = c_2 = \ldots = c_n = 0$ when m_i, $1 \leqq i \leqq n$ are sufficiently large. If $\eta^{(j)} = \eta^{(k)}$ holds for sufficiently large m_n, $n \geqq 3$, we have $c_n = 0$ or $c_n \neq 0$, j, $k \neq n+1$, since $\lim_{m_n\to\infty} \epsilon_n^{(l)}(m_n)/\epsilon_n^{(n+1)}(m_n) = 0$, $l \neq n+1$. Suppose $c_n \neq 0$. Put $\eta_1 = \epsilon_1^{c_1}\epsilon_2^{c_2}\cdots\epsilon_{n-1}^{c_{n-1}}$. Since $\lim_{m_n\to\infty} \epsilon_n^{(j)}(m_n)/\epsilon_n^{(k)}(m_n) = 1$, we must have $\eta_1^{(j)} = \eta_1^{(k)}$, taking m_n sufficiently large. Then we have $\epsilon_n^{(j)} = \epsilon_n^{(k)}$. Let $F' = \{x \in F \mid x^{(j)} = x^{(k)}\}$. Then F' is a proper subfield of F. We can take $\sigma_i \in J_F$ so that $i \neq n+1$, $\sigma_i|F' = \sigma_{n+1}|F'$. Then we get $\epsilon_n^{(i)} = \epsilon_n^{(n+1)}$, which contradicts $\epsilon_n^{(i)}(m_n) < 1 < \epsilon_n^{(n+1)}(m_n)$. Therefore $c_n = 0$. Repeating this procedure, we get $c_3 = \ldots = c_n = 0$. Since we may assume that $\epsilon_i(m_i) = \eta_i^{m_i}$ for $i = 1, 2$, we must have $\eta_i^{(j)} = \eta_i^{(k)}$ if $c_i \neq 0$. Then we can obtain a contradiction in the same manner as above.

5.3. Before beginning the proof of Theorem 5.3, let us prove a few preparatory lemmas. For $x_1, x_2, \ldots, x_{n+1} \in \mathbf{R}^{n+1}$, we put

$$\Delta[x_1, x_2, \ldots, x_{n+1}] = \det(x_1\ x_2 \ldots x_{n+1}).$$

LEMMA 5.4. *Let* $P_1, P_2, \dots, P_m \in \mathbf{R}_+^{n+1}$ *be linearly independent (viewed as vectors* $\in \mathbf{R}^{n+1}$*) points and let* L *be a linear form on* $\mathbf{R}^{n+1}$ *satisfying* $L(P_i) = 0$, $1 \leqq i \leqq m$. *We assume* $1 \leqq m \leqq n-1$. *Then there exist* $T_1, T_2, \dots, T_{n-m} \in \mathbf{R}_+^{n+1}$ *such that*

$$L(P) = \Delta[P, T_1, \dots, T_{n-m}, P_1, P_2, \dots, P_m] \qquad \text{for every} \quad P \in \mathbf{R}^{n+1}.$$

PROOF. The assertion is obvious if L is trivial. We assume that L is non-trivial. Let $H = \{P \in \mathbf{R}^{n+1} \mid L(P) = 0\}$. Then H is an n-dimensional subspace of $\mathbf{R}^{n+1}$. We take points $U_1, \dots, U_{n-m} \in \mathbf{R}^{n+1}$ so that $\{U_1, \dots, U_{n-m}, P_1, \dots, P_m\}$ is a basis of H. Then $\Delta[P, U_1, \dots, U_{n-m}, P_1, \dots, P_m] = 0$ if and only if $L(P) = 0$. Replacing U_1 by cU_1, $c \in \mathbf{R}$ if necessary, we may assume that $L(P) = \Delta[P, U_1, \dots, U_{n-m}, P_1, \dots, P_m]$ for every $P \in \mathbf{R}^{n+1}$. Then we have $L(P) = \Delta[P, T_1, \dots, T_{n-m}, P_1, P_2, \dots, P_m]$ for every $P \in \mathbf{R}^{n+1}$ with $T_1, T_2, \dots, T_{n-m} \in \mathbf{R}_+^{n+1}$ by taking $T_i = U_i + tP_1$, $1 \leqq i \leqq n-m$ for a sufficiently large t. This completes the proof.

LEMMA 5.5. *Let* $\epsilon_1(m_1), \epsilon_2(m_2), \dots, \epsilon_n(m_n)$ *be a fine sequence in* $\mathbf{R}_+^{n+1}$. *Put* $\epsilon_i = \epsilon_i(m_i)$, $1 \leqq i \leqq n$. *For every* $P \in \mathbf{R}_+^{n+1}$, *there exist* $\lambda > 0$ *and* $a_1, a_2, \dots, a_n \in \mathbf{R}$ *such that* $\epsilon_1^{a_1}\epsilon_2^{a_2}\cdots\epsilon_n^{a_n} = \lambda P$.

PROOF. By the assumption (C1) and $\epsilon_i^{(i+1)}(m_i) > 1$, $\epsilon_i^{(k)}(m_i) < 1$ if $k \neq i+1$, we have $\det(\log \epsilon_j^{(i+1)}) \neq 0$, $1 \leqq i, j \leqq n$. Put $P = {}^t(b_1, b_2, \dots, b_{n+1})$ and take $\lambda > 0$ so that $\prod_{i=1}^{n+1}(\lambda b_i) = 1$. We can find $a_1, a_2, \dots, a_n \in \mathbf{R}$ so that

$$\sum_{j=1}^{n} a_j \log \epsilon_j^{(i)} = \log \lambda b_i, \qquad 2 \leqq i \leqq n+1.$$

Then this equation holds also for $i = 1$ and we obtain $\epsilon_1^{a_1}\epsilon_2^{a_2}\cdots\epsilon_n^{a_n} = \lambda P$.

5.4. In this subsection, we will give:

PROOF OF THEOREM 5.3. Put $\epsilon_i = \epsilon_i(m_i)$, $1 \leqq i \leqq n$. We put

$$E(P) = \Big(\prod_{i=1}^{n}(\epsilon_i^{(j)})^{a_i}\Big)_{1 \leqq j \leqq n+1} \in \mathbf{R}_+^{n+1}$$

for $P = {}^t(a_1, a_2, \dots, a_{n+1}) \in \mathbf{R}^{n+1}$; the j-th component of $E(P)$ is $\prod_{i=1}^{n}(\epsilon_i^{(j)})^{a_i}$. For $P_1, P_2, \dots, P_{n+1} \in \mathbf{R}^n$, we put

$$\tilde{\Delta}[P_1, P_2, \cdots, P_{n+1}] = \det\begin{pmatrix} P_1 & P_2 & \cdots & P_{n+1} \\ 1 & 1 & \cdots & 1 \end{pmatrix}.$$

First we examine the case $n = 2$. We have $|J| = 2$, $S_1 = S(P_1, P_2, P_3)$, $S_2 = S(P_2, P_3, P_4)$ where

$$P_1 = \begin{pmatrix} 0 \\ 0 \end{pmatrix}, \qquad P_2 = \begin{pmatrix} 1 \\ 0 \end{pmatrix}, \qquad P_3 = \begin{pmatrix} 0 \\ 1 \end{pmatrix}, \qquad P_4 = \begin{pmatrix} 1 \\ 1 \end{pmatrix}.$$

We have

$$\tilde{\Delta}[P_1, P_2, P_3] = 1, \qquad \tilde{\Delta}[P_4, P_2, P_3] = -1. \tag{5.5}$$

By definition, we have

$$\Delta[E(P_1),E(P_2),E(P_3)] = \begin{vmatrix} 1 & \epsilon_1^{(1)} & \epsilon_2^{(1)} \\ 1 & \epsilon_1^{(2)} & \epsilon_2^{(2)} \\ 1 & \epsilon_1^{(3)} & \epsilon_2^{(3)} \end{vmatrix} = \begin{vmatrix} 0 & \epsilon_1^{(1)} - \epsilon_1^{(2)} & \epsilon_2^{(1)} - \epsilon_2^{(2)} \\ 0 & \epsilon_1^{(2)} - \epsilon_1^{(3)} & \epsilon_2^{(2)} - \epsilon_2^{(3)} \\ 1 & \epsilon_1^{(3)} & \epsilon_2^{(3)} \end{vmatrix}$$
$$= (\epsilon_1^{(1)} - \epsilon_1^{(2)})(\epsilon_2^{(2)} - \epsilon_2^{(3)}) - (\epsilon_1^{(2)} - \epsilon_1^{(3)})(\epsilon_2^{(1)} - \epsilon_2^{(2)}),$$

$$\Delta[E(P_4),E(P_2),E(P_3)] = \epsilon_1^{(1)}\epsilon_2^{(1)}\epsilon_1^{(2)}\epsilon_2^{(2)}\epsilon_1^{(3)}\epsilon_2^{(3)}$$
$$\times \Big\{((\epsilon_2^{(1)})^{-1} - (\epsilon_2^{(2)})^{-1})((\epsilon_1^{(2)})^{-1} - (\epsilon_1^{(3)})^{-1})$$
$$- ((\epsilon_2^{(2)})^{-1} - (\epsilon_2^{(3)})^{-1})((\epsilon_1^{(1)})^{-1} - (\epsilon_1^{(2)})^{-1})\Big\}.$$

By the assumption, we have

$$\epsilon_1^{(2)} > \epsilon_1^{(1)} > \epsilon_1^{(3)}, \qquad \epsilon_2^{(3)} > \epsilon_2^{(2)} > \epsilon_2^{(1)}.$$

Hence

$$\Delta[E(P_1),E(P_2),E(P_3)] > 0, \qquad \Delta[E(P_4),E(P_2),E(P_3)] < 0. \tag{5.6}$$

Therefore the assertion holds when $n = 2$.

Now we assume that Theorem 5.3 holds for n and consider the case $n+1$. Let $\epsilon_1(m_1)$, $\epsilon_2(m_2)$, $\ldots$, $\epsilon_{n+1}(m_{n+1})$ be a fine sequence in $\mathbf{R}_+^{n+2}$. For $1 \leqq i \leqq n$, take $\lambda_i(m_i) > 0$ so that $\lambda_i(m_i)^{n+1}\prod_{j=1}^{n+1}\epsilon_i^{(j)}(m_i) = 1$. Put $\epsilon_i'(m_i)^{(j)} = \lambda_i(m_i)\epsilon_i(m_i)^{(j)}$, $1 \leqq j \leqq n+1$, $\epsilon_i'(m_i) = {}^t(\epsilon_i'(m_i)^{(1)}, \epsilon_i'(m_i)^{(2)}, \ldots, \epsilon_i'(m_i)^{(n+1)}) \in \mathbf{R}_+^{n+1}$. Then $\lambda_i(m_i) > 1$ and we see easily that $\epsilon_1'(m_1)$, $\epsilon_2'(m_2)$, $\ldots$, $\epsilon_n'(m_n)$ is a fine sequence in $\mathbf{R}_+^{n+1}$. For the sake of simplicity, we denote $\epsilon_i(m_i)$ and $\epsilon_i'(m_i)$ as ϵ_i and ϵ_i' respectively. For $P = {}^t(a_1, a_2, \ldots, a_n) \in \mathbf{R}^n$, we put

$$E'(P) = (\prod_{i=1}^n (\epsilon_i'^{(j)})^{a_i})_{1\leqq j\leqq n+1} \in \mathbf{R}_+^{n+1},$$
$$\tilde{P} = {}^t(a_1, a_2, \ldots, a_n, 0), \qquad \tilde{P}' = {}^t(a_1, a_2, \ldots, a_n, 1).$$

For $Q = {}^t(a_1, a_2, \ldots, a_{n+1}) \in \mathbf{R}^{n+1}$, we put

$$E(Q) = (\prod_{i=1}^{n+1} (\epsilon_i^{(j)})^{a_i})_{1\leqq j\leqq n+2} \in \mathbf{R}_+^{n+2}.$$

Let P_1, P_2, $\ldots$, P_{n+1}, $P_{n+2} \in \mathbf{R}^n$ and we are going to consider the quantity

$$\Delta[E(\tilde{P}_1), \ldots, E(\tilde{P}_i), E(\tilde{P}_{i+1}'), \ldots, E(\tilde{P}_{n+1}'), E(\tilde{P}_{n+2}')], \qquad 1 \leqq i \leqq n+1,$$

when m_{n+1} is sufficiently large. Let Δ denote this quantity, which is the determinant of the $(n+2)\times(n+2)$-matrix. Put $P_k = {}^t(a_{1k}, a_{2k}, \ldots, a_{nk})$, $1 \leqq k \leqq n+2$. We have

$$\Delta = \det((\prod_{l=1}^n \epsilon_l^{a_{lk}})^{(j)},\ (\prod_{l=1}^n \epsilon_l^{a_{lk}})^{(j)}\epsilon_{n+1}^{(j)},\ (\prod_{l=1}^n \epsilon_l^{a_{ln+2}})^{(j)}\epsilon_{n+1}^{(j)}). \tag{5.7}$$

Here the first (resp. the second, resp. the third) entry denotes the (j,k)-component for $1 \leqq k \leqq i$ (resp. $i+1 \leqq k \leqq n+1$, resp. $k = n+2$). Dividing the $(n+2)$-th row by $\epsilon_{n+1}^{(n+2)}$, we obtain

$$\begin{aligned}\Delta = &\epsilon_{n+1}^{(n+2)} \times \\ &\det \begin{pmatrix} (\prod_{l=1}^n \epsilon_l^{a_{lk}})^{(j)} & \prod_{l=1}^n \epsilon_l^{a_{lk}})^{(j)} \epsilon_{n+1}^{(j)} & (\prod_{l=1}^n \epsilon_l^{a_{ln+2}})^{(j)} \epsilon_{n+1}^{(j)}) \\ (\prod_{l=1}^n \epsilon_l^{a_{lk}})^{(n+2)} / \epsilon_{n+1}^{(n+2)} & (\prod_{l=1}^n \epsilon_l^{a_{lk}})^{(n+2)} & (\prod_{l=1}^n \epsilon_l^{a_{ln+2}})^{(n+2)} \end{pmatrix},\end{aligned} \tag{5.8}$$

where the second row denotes the $(n+2)$-th row divided by $\epsilon_{n+1}^{(n+2)}$ of the matrix in (5.7). For $i+1 \leqq k \leqq n+1$, we substruct the $(n+2)$-th column multiplied by c_k from the k-th column, where

$$c_k = (\prod_{l=1}^n \epsilon_l^{a_{lk}})^{(n+2)} / (\prod_{l=1}^n \epsilon_l^{a_{ln+2}})^{(n+2)}.$$

Then divide the k-th column by $\epsilon_{n+1}^{(1)}$ for $i+1 \leqq k \leqq n+1$. We obtain

$$\begin{aligned}\Delta = &\epsilon_{n+1}^{(n+2)} (\epsilon_{n+1}^{(1)})^{n+1-i} \times \\ &\det \begin{pmatrix} (\prod_{l=1}^n \epsilon_l^{a_{lk}})^{(j)} & E & (\prod_{l=1}^n \epsilon_l^{a_{ln+2}})^{(j)} \epsilon_{n+1}^{(j)}) \\ (\prod_{l=1}^n \epsilon_l^{a_{lk}})^{(n+2)} / \epsilon_{n+1}^{(n+2)} & 0 & (\prod_{l=1}^n \epsilon_l^{a_{ln+2}})^{(n+2)} \end{pmatrix},\end{aligned} \tag{5.9}$$

where

$$E = (\{(\prod_{l=1}^n \epsilon_l^{a_{lk}})^{(j)} - c_k (\prod_{l=1}^n \epsilon_l^{a_{ln+2}})^{(j)}\} \frac{\epsilon_{n+1}^{(j)}}{\epsilon_{n+1}^{(1)}}), \qquad 1 \leqq j \leqq n+1,\ i+1 \leqq k \leqq n+1$$

is an $(n+1) \times (n+1-i)$-matrix. We note that

$$\lim_{m_{n+1} \to \infty} \epsilon_{n+1}^{(n+2)}(m_{n+1}) (\epsilon_{n+1}^{(1)}(m_{n+1}))^{n+1-i} = \infty, \tag{5.10}$$

by the last condition of (C2), since $i \geqq 1$. Now we fix $m_1, m_2, \ldots, m_n$ and let $m_{n+1} \to \infty$. Then we see that

$$\begin{aligned}\Delta = &\epsilon_{n+1}^{(n+2)} (\epsilon_{n+1}^{(1)})^{n+1-i} (\prod_{l=1}^n \epsilon_l^{a_{ln+2}})^{(n+2)} \\ &\times \det \left(\prod_{l=1}^n \epsilon_l^{a_{lk}})^{(j)} \quad \{(\prod_{l=1}^n \epsilon_l^{a_{lk}})^{(j)} - c_k (\prod_{l=1}^n \epsilon_l^{a_{ln+2}})^{(j)}\} \right) \\ &+ o(\epsilon_{n+1}^{(n+2)} (\epsilon_{n+1}^{(1)})^{n+1-i}).\end{aligned} \tag{5.11}$$

Here the matrix inside of the determinant is of size $n+1$; the first (resp. second) entry denotes the (j,k)-component for $1 \leqq k \leqq i$ (resp. $i+1 \leqq k \leqq n+1$). In particular, we obtain:

LEMMA 5.6. *When $m_1, m_2, \ldots, m_n$ are fixed, we have*

$$\begin{aligned}&\Delta[E(\tilde{P}_1), \ldots, E(\tilde{P}_i), E(\tilde{P}'_{i+1}), \ldots, E(\tilde{P}'_{n+1}), E(\tilde{P}'_{n+2})] \\ &= O(\epsilon_{n+1}^{(n+2)}(m_{n+1}) (\epsilon_{n+1}^{(1)}(m_{n+1}))^{n+1-i}) \qquad \textit{for } m_{n+1} \to \infty.\end{aligned}$$

Next assume that $P_{n+2} = P_1$. For the matrix appearing in (5.11), we add the first column multiplied by c_k to the k-th column for $i+1 \leqq k \leqq n+1$. Then we get

$$(5.12)\qquad \begin{aligned} \Delta =& \epsilon_{n+1}^{(n+2)} (\epsilon_{n+1}^{(1)})^{n+1-i} (\prod_{l=1}^{n} \epsilon_l^{a_{l n+2}})^{(n+2)} \det((\prod_{l=1}^{n} \epsilon_l^{a_{lk}})^{(j)}) \\ &+ o(\epsilon_{n+1}^{(n+2)} (\epsilon_{n+1}^{(1)})^{n+1-i}). \end{aligned}$$

We note that

$$\det((\prod_{l=1}^{n} \epsilon_l^{a_{lk}})^{(j)}) = c_1 \Delta[E'(P_1), \dots, E'(P_{n+1})]$$

with $c_1 > 0$. Therefore we obtain

LEMMA 5.7. *We fix $m_1, m_2, \dots, m_n$. If $\Delta[E'(P_1), \dots, E'(P_{n+1})] \neq 0$, then we have*

$$\begin{aligned} &\Delta[E(\tilde{P}_1), \dots, E(\tilde{P}_i), E(\tilde{P}'_{i+1}), \dots, E(\tilde{P}'_{n+1}), E(\tilde{P}'_1)] \\ &\sim c\, \epsilon_{n+1}^{(n+2)}(m_{n+1}) (\epsilon_{n+1}^{(1)}(m_{n+1}))^{n+1-i} \qquad \textit{for } m_{n+1} \to \infty, \end{aligned}$$

where c is a non-zero constant which has the same signature as that of $\Delta[E'(P_1), \dots, E'(P_{n+1})]$. If $\Delta[E'(P_1), \dots, E'(P_{n+1})] = 0$, then

$$\begin{aligned} &\Delta[E(\tilde{P}_1), \dots, E(\tilde{P}_i), E(\tilde{P}'_{i+1}), \dots, E(\tilde{P}'_{n+1}), E(\tilde{P}'_1)] \\ &= o(\epsilon_{n+1}^{(n+2)}(m_{n+1}) (\epsilon_{n+1}^{(1)}(m_{n+1}))^{n+1-i}) \qquad \textit{for } m_{n+1} \to \infty. \end{aligned}$$

Using Lemma 5.7, we can show:

LEMMA 5.8. *Let $I^n = \sqcup_{j \in \bar{J}} S_j$, $F = \sqcup_{j \in J} S_j$ be as in Theorem 5.3. Let $S_j = S(P_1, P_2, \dots, P_{n+1})$, $j \in J$ be a big cell. Then there exist positive integers $N_1, \dots, N_n$, $N_i = N_i(m_1, \dots, m_{i-1})$ depending on $m_1, \dots, m_{i-1}$, such that $\Delta[E(P_1), E(P_2), \dots, E(P_{n+1})]$ is non-zero and has the same signature as that of $(-1)^n \tilde{\Delta}[P_1, P_2, \dots, P_{n+1}]$ if $m_i \geqq N_i$, $1 \leqq i \leqq n$.*

PROOF. By (5.5) and (5.6), we see that the assertion holds for $n = 2$ taking $N = 1$. We proceed by induction on n. Let $\tilde{S}_j$ be a big cell appearing in the special decomposition of I^{n+1}. Then, by construction, we have $\tilde{S}_j = S(\tilde{P}_i, \dots, \tilde{P}_{n+1}, \tilde{P}'_1, \dots, \tilde{P}'_i)$, $1 \leqq i \leqq n+1$ for a big cell $S_j = S(P_1, P_2, \dots, P_{n+1})$ appearing in the special decomposition of I^n. We have

$$\begin{aligned} &\tilde{\Delta}[\tilde{P}_i, \dots, \tilde{P}_{n+1}, \tilde{P}'_1, \dots, \tilde{P}'_i] \\ &= (-1)^n \Delta[\tilde{P}_{i+1} - \tilde{P}_i, \dots, \tilde{P}_{n+1} - \tilde{P}_i, \tilde{P}'_1 - \tilde{P}_i, \dots, \tilde{P}'_i - \tilde{P}_i] \\ &= (-1)^n \Delta[\tilde{P}_{i+1} - \tilde{P}_i, \dots, \tilde{P}_{n+1} - \tilde{P}_i, \tilde{P}'_1 - \tilde{P}'_i, \dots, \tilde{P}'_i - \tilde{P}_i] \\ &= (-1)^n \Delta[P_{i+1} - P_i, \dots, P_{n+1} - P_i, P_1 - P_i, \dots, P_{i-1} - P_i] \\ &= -\tilde{\Delta}[P_i, \dots, P_{n+1}, P_1, \dots, P_{i-1}]. \end{aligned}$$

By the induction hypothesis, there exist positive integers N_i such that $\Delta[E'(P_i), \dots, E'(P_{n+1}), E'(P_1), \dots, E'(P_{i-1})]$ is non-zero and has the same signature as that of $(-1)^n \tilde{\Delta}[P_i, \dots, P_{n+1}, P_1, \dots, P_{i-1}]$ if $m_i \geqq N_i$, $1 \leqq i \leqq n$. By Lemma 5.7, $\Delta[E(\tilde{P}_i), \dots, E(\tilde{P}_{n+1}), E(\tilde{P}'_1), \dots, E(\tilde{P}'_i)]$ is non-zero and has the same signature as that of $\Delta[E'(P_i),\dots,E'(P_{n+1}),E'(P_1),\dots,E'(P_{i-1})]$ if m_{n+1} is sufficiently large. This completes the proof.

By Lemma 5.8, we see that the condition (1) of Theorem 3.1 holds if $m_i \geqq N_i$, $1 \leqq i \leqq n+1$. To prove the separation condition (2), it suffices to show the following assertion (S).

(S). *Let $S_k = S(P_1, P_2, \dots, P_{n+1})$, $S_l = S(Q_1, Q_2, \dots, Q_{n+1})$, k, $l \in J$ be two distinct big cells appearing in the special decomposition of I^n. Put $V_1 = \{P_1, P_2, \dots, P_{n+1}\}$, $V_2 = \{Q_1, Q_2, \dots, Q_{n+1}\}$, $I = V_1 \cap V_2$. Then there exist positive integers N_1, N_2, $\dots$, N_n, N_i depending on m_1, $\dots$, m_{i-1}, such that if $m_i \geqq N_i$, $1 \leqq i \leqq n$, there exist T_1, T_2, $\dots$, $T_n \in \mathbf{R}^n$ which satisfy the following properties. Put $L(X) = \Delta[X, E(T_1), \dots, E(T_n)]$ for $X \in \mathbf{R}^{n+1}$. Then*

$$\{T_1, T_2, \dots, T_n\} \supset I, \qquad L(E(P_s)) > 0 \quad \text{if } P_s \notin I, \qquad L(E(Q_t)) < 0 \quad \text{if } Q_t \notin I.$$

If $n = 2$, the assertion holds taking $N_1 = N_2 = 1$, $T_1 = P_2$, $T_2 = P_3$ in view of (5.6). We proceed by induction on n. We assume that (S) holds for n and consider the case $n+1$.

We first consider the case where two big cells $\tilde{S}_k$ and $\tilde{S}_l$ of I^{n+1} are derived from *one* big cell $S_k = S(P_1, P_2, \dots, P_{n+1})$ of I^n. We may write

$$\tilde{S}_k = S(\tilde{P}_i, \dots, \tilde{P}_{n+1}, \tilde{P}'_1, \dots, \tilde{P}'_i), \qquad \tilde{S}_l = S(\tilde{P}_j, \dots, \tilde{P}_{n+1}, \tilde{P}'_1, \dots, \tilde{P}'_j)$$

with $1 \leqq i < j \leqq n+1$. Put

$$\begin{aligned}\tilde{I} &= \{\tilde{P}_i, \dots, \tilde{P}_{n+1}, \tilde{P}'_1, \dots, \tilde{P}'_i\} \cap \{\tilde{P}_j, \dots, \tilde{P}_{n+1}, \tilde{P}'_1, \dots, \tilde{P}'_j\} \\ &= \{\tilde{P}_j, \dots, \tilde{P}_{n+1}, \tilde{P}'_1, \dots, \tilde{P}'_i\}.\end{aligned}$$

For $X \in \mathbf{R}^{n+2}$, put

$$\begin{aligned}L_1(X) &= \Delta[X, E(\tilde{P}_{i+1}), \dots, E(\tilde{P}_{n+1}), E(\tilde{P}'_1), \dots, E(\tilde{P}'_i)], \\ L_2(X) &= \Delta[X, E(\tilde{P}_j), \dots, E(\tilde{P}_{n+1}), E(\tilde{P}'_1), \dots, E(\tilde{P}'_{j-1})].\end{aligned}$$

We choose $\mu_1 = \pm 1$, $\mu_2 = \pm 1$ so that

$$\begin{aligned}(-1)^n \mu_1 \tilde{\Delta}[P_i, \dots, P_{n+1}, P_1, \dots, P_{i-1}] > 0, \\ \mu_2 \tilde{\Delta}[P_j, \dots, P_{n+1}, P_1, \dots, P_{j-1}] > 0.\end{aligned}$$

By Lemma 5.8, there exist positive integers N_1, N_2, $\dots$, N_n such that

$$\begin{aligned}\mu_1 \Delta[E'(P_i), \dots, E'(P_{n+1}), \dots, E'(P_1), \dots, E'(P_{i-1})] > 0, \\ (-1)^n \mu_2 \Delta[E'(P_j), \dots, E'(P_{n+1}), \dots, E'(P_1), \dots, E'(P_{j-1})] > 0\end{aligned} \tag{5.13}$$

if $m_q \geqq N_q$, $1 \leqq q \leqq n$. By Lemma 5.7, we have

$$\begin{aligned}\mu_1 L_1(E(\tilde{P}_i)) > 0, \qquad \mu_1 L_1(E(\tilde{P}'_t)) < 0 \qquad \text{for } i < t \leqq j, \\ \mu_2 L_2(E(\tilde{P}'_j)) < 0, \qquad \mu_2 L_2(E(\tilde{P}_s)) > 0 \qquad \text{for } i \leqq s < j.\end{aligned}$$

if m_{n+1} is sufficiently large. Hence the linear form $L(X) = \mu_1 L_1(X) + \mu_2 L_2(X)$ satisfies

$$L(\tilde{P}_s) > 0, \quad i \leqq s < j, \qquad L(\tilde{P}'_t) < 0, \quad i < t \leqq j$$

and $L(X) = 0$ if $X = E(T)$ for $T \in \tilde{I}$. By Lemmas 5.4 and 5.5, if m_{n+1} is sufficiently large, there exist T_1, T_2, $\dots$, $T_{n+1} \in \mathbf{R}^{n+2}$ such that $L(X) =$

$\Delta[X, E(T_1), \ldots, E(T_{n+1})]$, $X \in \mathbf{R}^{n+2}$ and that $\{T_1, T_2, \ldots, T_{n+1}\} \supset \tilde{I}$. Therefore (S) holds for $\tilde{S}_k$ and $\tilde{S}_l$.

Next we consider the case where two big cells $\tilde{S}_k$ and $\tilde{S}_l$ of I^{n+1} are derived from *two distinct* big cells $S_k = S(P_1, P_2, \ldots, P_{n+1})$, $S_l = S(Q_1, Q_2, \ldots, Q_{n+1})$. By the induction hypothesis, we may assume that (S) holds for S_k and S_l with the symbol E replaced by E' if $m_i \geqq N_i$, $1 \leqq i \leqq n$. We fix $m_1, m_2, \ldots, m_n$. We may assume that there is an ordering of vertices of I^n and that

$$P_1 < P_2 < \ldots < P_{n+1}, \qquad Q_1 < Q_2 < \ldots < Q_{n+1},$$
$$\tilde{S}_k = S(\tilde{P}_i, \ldots, \tilde{P}_{n+1}, \tilde{P}'_1, \ldots, \tilde{P}'_i),$$
$$\tilde{S}_l = S(\tilde{Q}_j, \ldots, \tilde{Q}_{n+1}, \tilde{Q}'_1, \ldots, \tilde{Q}'_j) \qquad \text{with } 1 \leqq i, j \leqq n+1.$$

We put

$$\tilde{V}_1 = \{\tilde{P}_i, \ldots, \tilde{P}_{n+1}, \tilde{P}'_1, \ldots, \tilde{P}'_i\},$$
$$\tilde{V}_2 = \{\tilde{Q}_j, \ldots, \tilde{Q}_{n+1}, \tilde{Q}'_1, \ldots, \tilde{Q}'_j\}, \qquad \tilde{I} = \tilde{V}_1 \cap \tilde{V}_2.$$

We define $T_1^*, \ldots, T_n^* \in \mathbf{R}^{n+1}$ as follows. Let $1 \leqq \alpha \leqq n$. If $T_\alpha \notin I$, we put $T_\alpha^* = \tilde{T}_\alpha$. Suppose that $T_\alpha \in I$. We have $T_\alpha = Q_t$ for some t. If $\tilde{Q}_t \in \tilde{V}_2$ and $\tilde{Q}'_t \notin \tilde{V}_2$, we put $T_\alpha^* = \tilde{Q}_t$. If $\tilde{Q}_t \notin \tilde{V}_2$ and $\tilde{Q}'_t \in \tilde{V}_2$, we put $T_\alpha^* = \tilde{Q}'_t$. If $\tilde{Q}_t \in \tilde{V}_2$ and $\tilde{Q}'_t \in \tilde{V}_2$, we put $T_\alpha^* = \tilde{Q}_t$. In the last case, we have $t = j$. We define T_{n+1}^* as follows. If $Q_j \notin I$, we put $T_{n+1}^* = (T_1^*)'$. Here we put $(T_1^*)' = \tilde{Q}'$ if $T_1^* = \tilde{Q}$ with $Q \in \mathbf{R}^n$, and $(T_1^*)' = \tilde{Q}$ if $T_1^* = \tilde{Q}'$ with $Q \in \mathbf{R}^n$. If $Q_j \in I$, we put $T_{n+1}^* = \tilde{Q}'_j$. By construction, we have $\{T_1^*, T_2^*, \ldots, T_{n+1}^*\} \supset \tilde{I}$.

We first consider the linear form on $\mathbf{R}^{n+2}$ defined by

$$L_0(X) = \Delta[X, E(T_1^*), \ldots, E(T_{n+1}^*)], \qquad X \in \mathbf{R}^{n+2}. \tag{5.14}$$

If $Q_j \notin I$, we let a denote the number of T_α^*, $2 \leqq \alpha \leqq n$ for which the $(n+1)$-th coordinate is 1. If $Q_j \in I$, we have $Q_j = T_\beta$ for some β. We let a denote the number of T_α^*, $T_\alpha^* \in \{T_1^*, \ldots, T_n^*\} \setminus \{T_\beta^*\}$ for which the $(n+1)$-th coordinate is 1. We have $a \leqq n-1$ in both cases. We put $\mu_0 = 1$ (resp. -1) if the $(n+1)$-th coordinate of T_{n+1}^* is 1 (resp. 0). By Lemma 5.7 and the induction hypothesis, we have

$$\mu_0 L_0(E(\tilde{P}_s)) > c\,\epsilon_{n+1}^{(n+2)}(\epsilon_{n+1}^{(1)})^a \quad \text{for } m_{n+1} \to \infty \qquad \text{if } P_s \notin I, \tag{5.15}$$

$$\mu_0 L_0(E(\tilde{P}'_s)) > c\,\epsilon_{n+1}^{(n+2)}(\epsilon_{n+1}^{(1)})^{a+1} \quad \text{for } m_{n+1} \to \infty \qquad \text{if } P_s \notin I, \tag{5.16}$$

$$\mu_0 L_0(E(\tilde{Q}_t)) < -c\,\epsilon_{n+1}^{(n+2)}(\epsilon_{n+1}^{(1)})^a \quad \text{for } m_{n+1} \to \infty \qquad \text{if } Q_t \notin I, \tag{5.17}$$

$$\mu_0 L_0(E(\tilde{Q}'_t)) < -c\,\epsilon_{n+1}^{(n+2)}(\epsilon_{n+1}^{(1)})^{a+1} \quad \text{for } m_{n+1} \to \infty \qquad \text{if } Q_t \notin I. \tag{5.18}$$

Here c is a positive constant which does not depend on m_{n+1}. By the choice of T_α^*, we have

$$\mu_0 L_0(E(\tilde{Q}_t)) = 0 \qquad \text{if } \tilde{Q}_t \in \tilde{V}_2,\ Q_t \in I, \tag{5.19}$$

$$\mu_0 L_0(E(\tilde{Q}'_t)) = 0 \qquad \text{if } \tilde{Q}'_t \in \tilde{V}_2,\ Q_t \in I. \tag{5.20}$$

Thus $\mu_0 L_0$ vanishes for $E(T), T \in \tilde{I}$ and, roughly speaking, the separation condition for cones $\tilde{C}_k$ and $\tilde{C}_l$ holds except for $E(\tilde{P}_s)$ (resp. $E(\tilde{P}'_s)$) such that $P_s \in I$ and $\tilde{P}_s \in \tilde{V}_1$, $\tilde{P}_s \notin \tilde{I}$ (resp. $\tilde{P}'_s \in \tilde{V}_1$, $\tilde{P}'_s \notin \tilde{I}$). Obviously the case $\tilde{P}_s$, $\tilde{P}'_s \in \tilde{V}_1$ and $\tilde{P}_s \notin \tilde{I}$, $\tilde{P}'_s \notin \tilde{I}$ cannot happen. For such vertices, we have, by Lemma 5.7,

$$\mu_0 L_0(E(\tilde{P}_s)) = o(\epsilon_{n+1}^{(n+2)}(\epsilon_{n+1}^{(1)})^a) \quad \text{for } m_{n+1} \to \infty \qquad \text{if } P_s \in I, \tag{5.21}$$

$$\mu_0 L_0(E(\tilde{P}'_s)) = o(\epsilon_{n+1}^{(n+2)}(\epsilon_{n+1}^{(1)})^{a+1}) \quad \text{for } m_{n+1} \to \infty \qquad \text{if } P_s \in I. \tag{5.22}$$

Let $\mathfrak{S}_1$ (resp. $\mathfrak{S}_2$) denote the set of vertices P_s such that $P_s \in I$, $\tilde{P}_s \in \tilde{V}_1$, $\tilde{P}_s \notin \tilde{I}$ (resp. $P_s \in I$, $\tilde{P}'_s \in \tilde{V}_1$, $\tilde{P}'_s \notin \tilde{I}$). Let us show that at least one of $\mathfrak{S}_1$ and $\mathfrak{S}_2$ is empty. This is a crucial observation. Suppose that $P_{s_1} \in \mathfrak{S}_1$, $P_{s_2} \in \mathfrak{S}_2$. Since $P_{s_i} \in I$, we have $P_{s_i} = Q_{t_i}$ for some Q_{t_i}, $i = 1, 2$. Since $\tilde{P}_{s_1} \in \tilde{V}_1$, we have $i \leqq s_1 \leqq n+1$. Since $\tilde{P}_{s_1} \notin \tilde{I}$, we have $\tilde{Q}_{t_1} \notin \tilde{V}_2$, $1 \leqq t_1 < j$. Since $\tilde{P}'_{s_2} \in \tilde{V}_1$, we have $1 \leqq s_2 \leqq i$. Since $\tilde{P}'_{s_2} \notin \tilde{I}$, we have $\tilde{Q}'_{t_2} \notin \tilde{V}_2$, $j < t_2 \leqq n+1$. By $s_2 \leqq s_1$, we get $Q_{t_2} = P_{s_2} \leqq P_{s_1} = Q_{t_1}$. This contradicts $t_1 < t_2$, $Q_{t_1} < Q_{t_2}$. Therefore it cannot happen that both of $\mathfrak{S}_1$ and $\mathfrak{S}_2$ are non-empty.

We consider the case $\mathfrak{S}_1 \neq \emptyset$, $\mathfrak{S}_2 = \emptyset$. Take u, $1 \leqq u \leqq n+1$ so that $Q_u \notin I$. We consider the linear form on $\mathbf{R}^{n+2}$ defined by

$$L_1(X) = -\Delta[X, E(T_1^*), \dots, E(T_n^*), E(\tilde{Q}_u)], \qquad X \in \mathbf{R}^{n+2}. \tag{5.23}$$

Let b denote the number of T_α^*, $1 \leqq \alpha \leqq n$ for which the $(n+1)$-th coordinate is 1. By Lemma 5.6, we have

$$L_1(E(\tilde{P}_v)) = O(\epsilon_{n+1}^{(n+2)}(\epsilon_{n+1}^{(1)})^{b-1}), \qquad 1 \leqq v \leqq n+1, \tag{5.24}$$

$$L_1(E(\tilde{P}'_v)) = O(\epsilon_{n+1}^{(n+2)}(\epsilon_{n+1}^{(1)})^{b}), \qquad 1 \leqq v \leqq n+1, \tag{5.25}$$

$$L_1(E(\tilde{Q}_w)) = O(\epsilon_{n+1}^{(n+2)}(\epsilon_{n+1}^{(1)})^{b-1}), \qquad 1 \leqq w \leqq n+1, \tag{5.26}$$

$$L_1(E(\tilde{Q}'_w)) = O(\epsilon_{n+1}^{(n+2)}(\epsilon_{n+1}^{(1)})^{b}), \qquad 1 \leqq w \leqq n+1, \tag{5.27}$$

for $m_{n+1} \to \infty$. We also see that

$$L_1(E(\tilde{Q}_w)) = 0 \qquad \text{if } \tilde{Q}_w \in \tilde{V}_2, \ Q_w \in I, \tag{5.28}$$

$$L_1(E(\tilde{Q}'_w)) \leqq 0 \qquad \text{if } \tilde{Q}'_w \in \tilde{V}_2, \ Q_w \in I \tag{5.29}$$

if m_{n+1} is sufficiently large. In fact, (5.28) is obvious from the definition of T_α^*; (5.29) is easy to see except for the case $Q_j \in I$, $w = j$. Assume $Q_j \in I$ and put $Q_j = T_\alpha$. Then $T_\alpha^* = \tilde{Q}_j$ and we have

$$\begin{aligned}
&L_1(E(\tilde{Q}'_j)) \\
&= -\Delta[E(\tilde{Q}'_j), E(T_1^*), \dots, E(T_{\alpha-1}^*), E(\tilde{Q}_j), E(T_{\alpha+1}^*), \dots, E(T_n^*), E(\tilde{Q}_u)] \\
&= -\Delta[E(\tilde{Q}_j), E(T_1^*), \dots, E(T_{\alpha-1}^*), E(\tilde{Q}_u), E(T_{\alpha+1}^*), \dots, E(T_n^*), E(\tilde{Q}'_j)] \\
&\sim -\Delta[E'(Q_j), E'(T_1), \dots, E'(T_{\alpha-1}), E'(Q_u), E'(T_{\alpha+1}), \dots, E'(T_n)]\ \epsilon_{n+1}^{(n+2)}(\epsilon_{n+1}^{(1)})^b \\
&= \Delta[E'(Q_u), E'(T_1), \cdots, E'(T_n)]\ \epsilon_{n+1}^{(n+2)}(\epsilon_{n+1}^{(1)})^b
\end{aligned}$$

for $m_{n+1} \to \infty$, by Lemma 5.7 and the induction hypothesis. Hence (5.29) holds.

Now let $P_s \in \mathfrak{S}_1$. We have $P_s = Q_t = T_\alpha$ for some t and α. Since $\tilde{P}_s \in \tilde{V}_1$, $\tilde{P}_s \notin \tilde{I}$, we have $\tilde{Q}_t \notin \tilde{V}_2$, $T_\alpha^* = \tilde{Q}_t' = \tilde{P}_s'$. We get

$$\begin{aligned}
&L_1(E(\tilde{P}_s))\\
=&-\Delta[E(\tilde{P}_s), E(T_1^*), \dots, E(T_{\alpha-1}^*), E(\tilde{P}_s'), E(T_{\alpha+1}^*), \dots, E(T_n^*), E(\tilde{Q}_u)]\\
=&\Delta[E(\tilde{P}_s), E(T_1^*), \dots, E(T_{\alpha-1}^*), E(\tilde{Q}_u), E(T_{\alpha+1}^*), \dots, E(T_n^*), E(\tilde{P}_s')]\\
\sim&\Delta[E'(P_s), E'(T_1), \dots, E'(T_{\alpha-1}), E'(Q_u), E'(T_{\alpha+1}), \dots, E'(T_n)]\ \epsilon_{n+1}^{(n+2)}(\epsilon_{n+1}^{(1)})^{b-1}\\
=&-\Delta[E'(Q_u), E'(T_1), \cdots, E'(T_n)]\ \epsilon_{n+1}^{(n+2)}(\epsilon_{n+1}^{(1)})^{b-1}
\end{aligned}$$

for $m_{n+1} \to \infty$, by Lemma 5.7 and the induction hypothesis. We obtain

$$L_1(E(\tilde{P}_s)) > c_1 \epsilon_{n+1}^{(n+2)}(\epsilon_{n+1}^{(1)})^{b-1} \qquad \text{for} \quad P_s \in \mathfrak{S}_1, \tag{5.30}$$

with a positive constant c_1 which does not depend on m_{n+1}, by the induction hypothesis.

Next we consider the case $\mathfrak{S}_1 = \emptyset$, $\mathfrak{S}_2 \neq \emptyset$. In this case, we slightly modify the choice of T_1^*, T_2^*, ..., T_n^* as follows. If $Q_j \notin I$, they remain the same. Assume $Q_j \in I$, $Q_j = T_\alpha$. Then we put $T_\alpha^* = \tilde{Q}_j'$. Letting Q_u be the same as above, we consider the linear form on $\mathbf{R}^{n+2}$ defined by

$$L_2(X) = \Delta[X, E(T_1^*), \dots, E(T_n^*), E(\tilde{Q}_u)], \qquad X \in \mathbf{R}^{n+2}. \tag{5.31}$$

Let b' denote the number of T_α^*, $1 \leqq \alpha \leqq n$ for which the $(n+1)$-th coordinate is 1; $b' = b$ if $Q_j \notin I$ and $b' = b+1$ if $Q_j \in I$. Then (5.24) $\sim$ (5.27) hold with L_2 (resp. b') in place of L_1 (resp. b). For (5.28) and (5.29), we see that

$$L_2(E(\tilde{Q}_w)) \leqq 0 \qquad \text{if } \tilde{Q}_w \in \tilde{V}_2,\ Q_w \in I, \tag{5.28$'$}$$

$$L_2(E(\tilde{Q}_w')) = 0 \qquad \text{if } \tilde{Q}_w' \in \tilde{V}_2,\ Q_w \in I \tag{5.29$'$}$$

if m_{n+1} is sufficiently large, in a similar way to the above. Let $P_s \in \mathfrak{S}_2$. Similarly to the other case, we get

$$L_2(E(\tilde{P}_s')) > c_2 \epsilon_{n+1}^{(n+2)}(\epsilon_{n+1}^{(1)})^{b'} \qquad \text{for} \quad P_s \in \mathfrak{S}_2, \tag{5.32}$$

with a positive constant c_2 which does not depend on m_{n+1}.

We put $\mu_1 = \kappa(\epsilon_{n+1}^{(1)})^{a-b+1}$, $L = \mu_0 L_0 + \mu_1 L_1$ if $\mathfrak{S}_1 \neq \emptyset$, $\mathfrak{S}_2 = \emptyset$, $\mu_2 = \kappa(\epsilon_{n+1}^{(1)})^{a-b'+1}$, $L = \mu_0 L_0 + \mu_2 L_2$ if $\mathfrak{S}_1 = \emptyset$, $\mathfrak{S}_2 \neq \emptyset$, with a sufficiently small $\kappa > 0$. Then, when m_{n+1} is sufficiently large, the linear form L satisfies

$$L(E(\tilde{P}_s)) > 0 \qquad \text{if} \quad \tilde{P}_s \in \tilde{V}_1 \backslash \tilde{I}, \tag{5.33}$$

$$L(E(\tilde{P}_s')) > 0 \qquad \text{if} \quad \tilde{P}_s \in \tilde{V}_1 \backslash \tilde{I}, \tag{5.34}$$

$$L(E(\tilde{Q}_t)) \leqq 0 \qquad \text{if} \quad \tilde{Q}_t \in \tilde{V}_2, \tag{5.35}$$

$$L(E(\tilde{Q}_t')) \leqq 0 \qquad \text{if} \quad \tilde{Q}_t' \in \tilde{V}_2. \tag{5.36}$$

We change the roles of $\tilde{S}_k$ and $\tilde{S}_l$ and construct the linear form L' similarly to the above. Then the linear form $\tilde{L} = L + L'$ on $\mathbf{R}^{n+2}$ satisfies

$$\tilde{L}(E(X)) > 0 \qquad \text{if} \quad X \in \tilde{V}_1 \backslash \tilde{I}, \tag{5.37}$$

$$\tilde{L}(E(X)) < 0 \quad \text{if} \quad X \in \tilde{V}_2 \backslash \tilde{I}, \tag{5.38}$$

$$\tilde{L}(E(X)) = 0 \quad \text{if} \quad X \in \tilde{I} \tag{5.39}$$

when m_{n+1} is sufficiently large. By (5.37) and (5.38), it is clear that $\tilde{L}(P) = 0$ for some $P \in \mathbf{R}_+^{n+2}$. Then, by Lemmas 5.4 and 5.5, we see that (S) holds for $\tilde{S}_k$ and $\tilde{S}_l$. This completes the proof of Theorem 5.3.

6. Applications

In this section, we will present two applications of Theorem 5.2 to the invariants $V(c)$ and $W(c)$ introduced in III, §3. In this section, we let F be a totally real algebraic number field of degree n.

6.1. Let us first recall the definition of $V(c)$ in a slightly more general situation than in chapter III. Let $\mathfrak{a}_1, \mathfrak{a}_2, \dots, \mathfrak{a}_{h_0}$ be integral ideals which represent narrow ideal classes of F. Let $\mathfrak{f}$ be an integral ideal of F and $C_\mathfrak{f}$ be the ideal class group defined modulo $\mathfrak{f}\infty_1\infty_2\cdots\infty_n$ of F. Let C be an r-dimensional open simplicial cone in $\mathbf{R}_+^n$ with basis in $\mathfrak{O}_F$. We fix $c \in C_\mathfrak{f}$. We choose $\mathfrak{a}_\mu$ so that c and $\mathfrak{a}_\mu\mathfrak{f}$ belong to the same narrow ideal class. We define

$$\zeta^{(i)}(s, C, \mathfrak{a}_\mu) = \sum_{z \in (\mathfrak{a}_\mu\mathfrak{f})^{-1} \cap C, (z)\mathfrak{a}_\mu\mathfrak{f} = c \text{ in } C_\mathfrak{f}} (z^{(i)})^{-s}, \tag{6.1}$$

$$\zeta^{(i,k)}(s, C, \mathfrak{a}_\mu) = \sum_{z \in (\mathfrak{a}_\mu\mathfrak{f})^{-1} \cap C, (z)\mathfrak{a}_\mu\mathfrak{f} = c \text{ in } C_\mathfrak{f}} (z^{(i)}z^{(k)})^{-s}, \tag{6.2}$$

where $1 \leqq i, k \leqq n$, $i \neq k$ and $\Re(s)$ is sufficiently large. These functions can be continued to meromorphic functions of s which are holomorphic at $s = 0$.

We write

$$C = C(v_1, v_2, \dots, v_r), \qquad v_1, v_2, \dots, v_r \in \mathfrak{O}_F$$

and define an $n \times r$-matrix $A = A(C)$ by $A = (v_m^{(l)})$. For $z \in C$, we put

$$z = x_1(z)v_1 + x_2(z)v_2 + \cdots + x_r(z)v_r, \qquad {}^t x(z) = (x_1(z), x_2(z), \dots, x_r(z)) \in \mathbf{R}^r.$$

For a fractional ideal $\mathfrak{a}$ of F, we set

$$R(C, \mathfrak{a}) = \{z \in \mathfrak{a} \cap C \mid 0 < x_1(z), x_2(z), \dots, x_r(z) \leqq 1\}. \tag{6.3}$$

We set

$$R(C, c) = \{z \in R(C, (\mathfrak{a}_\mu\mathfrak{f})^{-1}) \mid (z)\mathfrak{a}_\mu\mathfrak{f} = c \text{ in } C_\mathfrak{f}\}. \tag{6.4}$$

By Theorems II, 2.1 and 2.2 in view of II, Lemma 3.2, we obtain

$$\zeta^{(i)}(0, C, \mathfrak{a}_\mu) = (-1)^r \sum_{z \in R(C,c)} \sum_l \prod_{m=1}^r \frac{B_{l_m}(x_m(z))}{l_m!} (v_m^{(i)})^{l_m - 1}, \tag{6.5}$$

$$\left.\frac{d}{ds}\zeta^{(i)}(s, C, \mathfrak{a}_\mu)\right|_{s=0} = \sum_{z \in R(C,c)} \log \frac{\Gamma_r(z^{(i)}, A^{(i)})}{\rho_r(A^{(i)})}, \tag{6.6}$$

$$\zeta^{(i,k)}(0,C,\mathfrak{a}_\mu) = \frac{(-1)^r}{2} \sum_{z\in R(C,c)} \sum_l \prod_{m=1}^r \frac{B_{l_m}(x_m(z))}{l_m!} \{(v_m^{(i)})^{l_m-1} + (v_m^{(k)})^{l_m-1}\}, \tag{6.7}$$

$$\begin{aligned} \frac{d}{ds}\zeta^{(i,k)}(s,C,\mathfrak{a}_\mu)\Big|_{s=0} &= \sum_{z\in R(C,c)} \Big[\log \frac{\Gamma_r(z^{(i)},A^{(i)})}{\rho_r(A^{(i)})} + \log \frac{\Gamma_r(z^{(k)},A^{(k)})}{\rho_r(A^{(k)})} \\ &+ \frac{(-1)^r}{2}\sum_l (C_{l,i,k}(A) + C_{l,k,i}(A)) \prod_{m=1}^r \frac{B_{l_m}(x_m(z))}{l_m!}\Big]. \end{aligned} \tag{6.8}$$

In (6.5) $\sim$ (6.8), $\sum_l$ is taken over all r-tuples of non-negative integers $l = (l_1, l_2, \ldots, l_r)$ such that $l_1 + l_2 + \cdots + l_r = r$. Let E be a subgroup of E_F^+ of finite index. Let

$$\mathbf{R}_+^n = \sqcup_{\epsilon\in E}\ \epsilon(\sqcup_{j\in J} C_j) \tag{6.9}$$

be a cone decomposition of $\mathbf{R}_+^n$, where C_j is an $r(j)$-dimensional open simplicial cone with basis in $\mathfrak{O}_F$. We define

$$V_E(c) = \sum_{j\in J}\Big(\frac{2}{n}\sum_{k=2}^n v_{1,k}^{(j)} - \frac{1}{n^2}\sum_{1\leqq i,k\leqq n, i\neq k} v_{i,k}^{(j)}\Big), \tag{6.10}$$

where

$$v_{i,k}^{(j)} = \frac{d}{ds}\zeta^{(i,k)}(s,C_j,\mathfrak{a}_\mu)\Big|_{s=0} - \frac{d}{ds}\zeta^{(i)}(s,C_j,\mathfrak{a}_\mu)\Big|_{s=0} - \frac{d}{ds}\zeta^{(k)}(s,C_j,\mathfrak{a}_\mu)\Big|_{s=0}.$$

By (6.5) $\sim$ (6.8), we have

$$\begin{aligned} V_E(c) = &\sum_{j\in J}(-1)^{r(j)}\sum_l \Big[\frac{1}{n}\sum_{k=2}^n (C_{l,1,k}(A_j) + C_{l,k,1}(A_j)) \\ &- \frac{1}{n^2}\sum_{1\leqq i,k\leqq n, i\neq k} C_{l,i,k}(A_j)\Big] \times \sum_{z\in R(C_j,c)} \prod_{m=1}^{r(j)} \frac{B_{l_m}(x(z)_m)}{l_m!}. \end{aligned} \tag{6.11}$$

Here $A_j = A(C_j)$ and $\sum_l$ is taken over all $r(j)$-tuples of non-negative integers $l = (l_1, l_2, \ldots, l_{r(j)})$ such that $l_1 + l_2 + \cdots + l_{r(j)} = r(j)$. In III, §3, we defined $V_E(c)$ for the case $E = E_F^+$ and denoted it by $V(c)$; it depends on the choice of (6.9) and also on the choice of $\mathfrak{a}_1, \mathfrak{a}_2, \ldots, \mathfrak{a}_{h_0}$.

6.2. In this subsection, we will give an application to the invariant $W(c)$ defined by III, (3.8) (cf. also III, (3.31)) using properties of $V_E(c)$.

LEMMA 6.1. *There exists an index finite subgroup E_0 of E_F^+ such that $V_{E_0}(c)$ is of the form*

$$V_{E_0}(c) = \sum_{i=1}^m a_i \log \epsilon_i \tag{6.12}$$

with $a_i \in \tilde{F}$ and a totally positive units ϵ_i of $\tilde{F}$, where $\tilde{F}$ is the normal closure of F over $\mathbf{Q}$; $V_{E_0}(c)$ is always of the form (6.12) independently of the choices of (6.9) and of $\mathfrak{a}_1, \mathfrak{a}_2, \ldots, \mathfrak{a}_{h_0}$.

PROOF. We fix any choice of $\mathfrak{a}_1, \mathfrak{a}_2, \ldots, \mathfrak{a}_{h_0}$. We take E_0 as the subgroup E given in Theorem 5.2 and use the cone decomposition constructed there. Then we have $C_j = C(\psi_1, \psi_2, \ldots, \psi_{r(j)})$ with $\psi_i \in E_F^+$, $1 \leqq i \leqq r(j)$. By I, Lemma 2.3, we obtain

$$\begin{aligned}
&C_{l,i,k}(A_j) + C_{l,k,i}(A_j) \\
&= \sum_{p \in T_2(l)} \frac{1}{\psi_p^{(i)}\psi_p^{(k)}} \frac{\prod_{t \in T_1(l)} (\psi_t^{(i)}\psi_p^{(k)} - \psi_p^{(i)}\psi_t^{(k)})^{l_t - 1}}{\prod_{q \in T_2(l), q \neq p} (\psi_q^{(i)}\psi_p^{(k)} - \psi_p^{(i)}\psi_q^{(k)})} \log \frac{\psi_p^{(i)}}{\psi_p^{(k)}},
\end{aligned} \tag{6.13}$$

the condition II, (2.15) being satisfied. By (6.11), we see that $V_{E_0}(c)$ is of the form (6.12) for this choice of the cone decomposition.

If we replace the cone decomposition by its refinement, $V_{E_0}(c)$ does not change. If we replace a member C_j of a cone decomposition by ϵC_j, $\epsilon \in E_0$, $V_{E_0}(c)$ changes to

$$V_{E_0}(c) + \log \epsilon^{(1)} \left[\zeta^{(1)}(0, C_j, \mathfrak{a}_\mu) - \frac{1}{n} \sum_{k=1}^{n} \zeta^{(k)}(0, C_j, \mathfrak{a}_\mu) \right]$$

as shown in III, §3.[1] Therefore, by III, Lemma 3.13, we see that $V_{E_0}(c)$ is always of the form of (6.12). This completes the proof.

THEOREM 6.2. *Let E be a subgroup of finite index of E_F^+. For any choice of the cone decomposition and for any choice of $\mathfrak{a}_1, \mathfrak{a}_2, \ldots, \mathfrak{a}_{h_0}$, we have*

$$\sum_{j \in J} \zeta^{(1)}(0, C_j, \mathfrak{a}_\mu) \in \mathbf{Q}.$$

PROOF. First take $E = E_0$, E_0 being as above. Let α be a totally positive element of F. Replace $\mathfrak{a}_\mu$ by $(\alpha)\mathfrak{a}_\mu$ and $\{C_j\}_{j \in J}$ by $\{\alpha^{-1} C_j\}_{j \in J}$. Then, by the same calculation as in III, §3, $V_E(c)$ changes to (cf. III, (3.44))

$$\begin{aligned}
&V_E(c; \{\alpha^{-1} C_j\}, (\alpha)\mathfrak{a}_\mu) = V_E(c; \{C_j\}, (\alpha)\mathfrak{a}_\mu) \\
&- \sum_{j \in J} \left[(\log \alpha^{(1)} - \frac{1}{n} \log N(\alpha))(\zeta^{(1)}(0, C_j, \mathfrak{a}_\mu) - \frac{1}{n} \sum_{k=1}^{n} \zeta^{(k)}(0, C_j, \mathfrak{a}_\mu) \right].
\end{aligned}$$

Denote this quantity by $V_E'(c)$. We choose $\alpha \gg 0$ so that (α) is a prime ideal of degree 1. Put

$$d = \sum_{j \in J} (\zeta^{(1)}(0, C_j, \mathfrak{a}_\mu) - \frac{1}{n} \sum_{k=1}^{n} \zeta^{(k)}(0, C_j, \mathfrak{a}_\mu)).$$

Then $d \in \overline{\mathbf{Q}}$ and

$$V_E'(c) = V_E(c) - (\log \alpha^{(1)} - \frac{1}{n} \log N(\alpha)) d.$$

Let us show $d = 0$. By Lemma 6.1, both of $V_E(c)$ and $V_E'(c)$ are of the form (6.12). Hence we have

$$(\log \alpha^{(1)} - \frac{1}{n} \log N(\alpha)) d = \sum_{i=1}^{l} a_i \log \epsilon_i$$

[1]Strictly speaking, the calculation was done for cone decompositions of $E_F^+ \backslash \mathbf{R}_+^n$, but the same calculation yields III, (3.34) for the present case.

with $a_i \in \tilde{F}$ and totally positive units $\epsilon_1, \epsilon_2, \ldots, \epsilon_l$ of $\tilde{F}$. We may assume that $l \geqq 1$ and $\log \epsilon_1, \ldots, \log \epsilon_l$ are linearly independent over $\mathbf{Q}$. Then $\log(\alpha^{(1)}/\sqrt[n]{N(\alpha)})$, $\log \epsilon_1, \ldots, \log \epsilon_l$ are linearly independent over $\mathbf{Q}$. By a theorem of Baker ([Bak], Theorem 1), $\log(\alpha^{(1)}/\sqrt[n]{N(\alpha)})$, $\log \epsilon_1, \ldots, \log \epsilon_l$ are linearly independent over $\overline{\mathbf{Q}}$. Therefore we must have $d = 0$. Since $\sum_{k=1}^{n} \zeta^{(k)}(0, C_j, \mathfrak{a}_\mu) \in \mathbf{Q}$, we have $\sum_{j \in J} \zeta^{(1)}(0, C_j, \mathfrak{a}_\mu) \in \mathbf{Q}$. This proves our assertion completely when $E = E_0$.

Next let E be any subgroup of E_F^+ of finite index and consider the cone decomposition (6.9). If we replace (6.9) by its refinement, $\sum_{j \in J} \zeta^{(1)}(0, C_j, \mathfrak{a}_\mu)$ does not change. If we replace a member C_j by ϵC_j, $\epsilon \in E$, $\sum_{j \in J} \zeta^{(1)}(0, C_j, \mathfrak{a}_\mu)$ does not change. By III, Lemma 3.13, $\sum_{j \in J} \zeta^{(1)}(0, C_j, \mathfrak{a}_\mu)$ does not depend on the cone decomposition. Let $E_F^+ = \sqcup_{l=1}^{q} \epsilon_l E$ be a coset decomposition and let

$$\mathbf{R}_+^n = \sqcup_{\epsilon \in E_F^+} \epsilon(\sqcup_{m \in M} D_m)$$

be a cone decomposition with respect to E_F^+. Then

$$\mathbf{R}_+^n = \sqcup_{\epsilon \in E} \epsilon(\sqcup_{l=1}^{q} \sqcup_{m \in M} \epsilon_l D_m) \tag{6.14}$$

is a cone decomposition with respect to E. Since

$$\zeta^{(i)}(s, \epsilon_l D_m, \mathfrak{a}_\mu) = (\epsilon_l^{(i)})^{-s} \zeta^{(i)}(s, D_m, \mathfrak{a}_\mu),$$

we obtain the relation

$$\sum_{j \in J} \zeta^{(1)}(0, C_j, \mathfrak{a}_\mu) = q \sum_{m \in M} \zeta^{(1)}(0, D_m, \mathfrak{a}_\mu). \tag{6.15}$$

Using (6.15) first for E_0 and E_F^+ and next to E and E_F^+, we see that the assertion holds in general. This completes the proof.

COROLLARY 6.3. *Let*

$$\mathbf{R}_+^n = \sqcup_{\epsilon \in E_F^+} \epsilon(\sqcup_{j \in J} C_j)$$

be a cone decomposition of $\mathbf{R}_+^n$ *with respect to* E_F^+. *Then:*

1. *For the invariant* $W(c)$ *defined by* III, (3.8), *we have*

$$W(c) = -\frac{1}{n} \log N(\mathfrak{a}_\mu \mathfrak{f}) \zeta_F(0, c).$$

2. *Put*

$$\zeta_F^{(1)}(s, c) = \sum_{z \in (\mathfrak{a}_\mu \mathfrak{f})^{-1} \cap (\sqcup_{j \in J} C_j), (z)\mathfrak{a}_\mu \mathfrak{f} = c \text{ in } C_\mathfrak{f}} (z^{(1)})^{-s}.$$

Then we have

$$\zeta_F^{(1)}(0, c) = \sum_{j \in J} \zeta^{(1)}(0, C_j, \mathfrak{a}_\mu) = \zeta_F(0, c).$$

PROOF. We have (II, (3.5))

$$\zeta_F(s,c) = N(\mathfrak{a}_\mu\mathfrak{f})^{-s}\sum_{j\in J}\sum_{z\in R(C_j,c)}\zeta(s,A_j,x(z)).$$

Applying I, Theorem 2.1, we see that

$$\zeta_F(0,c) = \frac{1}{n}\sum_{j\in J}\sum_{k=1}^{n}\zeta^{(k)}(0,C_j,\mathfrak{a}_\mu).$$

The same theorem shows that $\zeta^{(k)}(0,C_j,\mathfrak{a}_\mu) = \zeta^{(1)}(0,C_j,\mathfrak{a}_\mu)^{\sigma_k}$. By Theorem 6.2, we have $\zeta^{(k)}(0,C_j,\mathfrak{a}_\mu) = \zeta^{(1)}(0,C_j,\mathfrak{a}_\mu)$ for $1 \leqq k \leqq n$. Now the second assertion is clear. The first assertion follows from III, (3.31).

6.3. In this subsection, we will consider the invariant $V_E(c)$.

THEOREM 6.4. *Let E be a subgroup of E_F^+ of finite index. For any choice of the cone decomposition and for any choice of $\mathfrak{a}_1, \mathfrak{a}_2, \ldots, \mathfrak{a}_{h_0}$, $V_E(c)$ is of the form*

$$V_E(c) = \sum_{i=1}^{m} a_i \log\epsilon_i \tag{6.16}$$

where $a_i \in \tilde{F}$ and ϵ_i is a totally positive unit of $\tilde{F}$, $1 \leqq i \leqq m$.

PROOF. We first examine the dependence of $V_E(c)$ on the cone decomposition and on $\mathfrak{a}_\mu$. If we replace the cone decomposition (6.9) by its refinement, $V_E(c)$ does not change. If we replace a member C_j by ϵC_j, $\epsilon\in E$, $V_E(c)$ changes to

$$V_E(c) + \log\epsilon^{(1)}\left[\zeta^{(1)}(0,C_j,\mathfrak{a}_\mu) - \frac{1}{n}\sum_{k=1}^{n}\zeta^{(k)}(0,C_j,\mathfrak{a}_\mu)\right].$$

Hence, by III, Lemma 3.13, the fact that $V_E(c)$ is of the form (6.16) does not depend on the cone decomposition. Next replace $\mathfrak{a}_\mu$ by $(\alpha)\mathfrak{a}_\mu$ and $\{C_j\}_{j\in J}$ by $\{\alpha^{-1}C_j\}_{j\in J}$ with a totally positive element α of F. Then $V_E(c)$ changes to

$$\begin{aligned} &V_E(c;\{\alpha^{-1}C_j\},(\alpha)\mathfrak{a}_\mu) = V_E(c;\{C_j\},(\alpha)\mathfrak{a}_\mu)\\ &-\sum_{j\in J}\left[(\log\alpha^{(1)} - \frac{1}{n}\log N(\alpha))(\zeta^{(1)}(0,C_j,\mathfrak{a}_\mu) - \frac{1}{n}\sum_{k=1}^{n}\zeta^{(k)}(0,C_j,\mathfrak{a}_\mu)\right].\end{aligned}$$

By Theorem 6.2, we see that $V_E(c)$ does not change by this operation. Therefore the fact that $V_E(c)$ is of the form (6.16) does not depend on the choice of $\mathfrak{a}_\mu$.

Let $E_F^+ = \sqcup_{l=1}^{q}\epsilon_l E$ be a coset decomposition and the notation be the same as in the proof of Theorem 6.2. Let $V_E'(c)$ denote the invariant defined by using (6.14). Since

$$\zeta^{(i,k)}(s,\epsilon_l D_m,\mathfrak{a}_\mu) = (\epsilon_l^{(i)}\epsilon_l^{(k)})^{-s}\zeta^{(i,k)}(s,D_m,\mathfrak{a}_\mu),$$

we have

$$\begin{aligned}V'_E(c) = qV_{E_F^+}(c) + \sum_{m\in M}\sum_{l=1}^{q}\Big[\big(\frac{2}{n}\sum_{k=2}^{n}(-\log\epsilon_l^{(1)}\epsilon_l^{(k)})\zeta^{(1,k)}(0,D_m,\mathfrak{a}_\mu)\\ +\log\epsilon_l^{(1)}\zeta^{(1)}(0,D_m,\mathfrak{a}_\mu)+\log\epsilon_l^{(k)}\zeta^{(k)}(0,D_m,\mathfrak{a}_\mu)\big)\\ -\big(\frac{1}{n^2}\sum_{1\leqq i,k\leqq n, i\neq k}(-\log\epsilon_l^{(i)}\epsilon_l^{(k)})\zeta^{(i,k)}(0,D_m,\mathfrak{a}_\mu)\\ +\log\epsilon_l^{(i)}\zeta^{(i)}(0,D_m,\mathfrak{a}_\mu)+\log\epsilon_l^{(k)}\zeta^{(k)}(0,D_m,\mathfrak{a}_\mu)\big)\Big].\end{aligned}$$

By a direct computation using

$$\zeta^{(i,k)}(0,D_m,\mathfrak{a}_\mu) = \frac{1}{2}(\zeta^{(i)}(0,D_m,\mathfrak{a}_\mu)+\zeta^{(k)}(0,D_m,\mathfrak{a}_\mu)),$$

we obtain

$$V'_E(c) = qV_{E_F^+}(c) + \sum_{l=1}^{q}\log\epsilon_l^{(1)}\sum_{m\in M}(\zeta^{(1)}(0,D_m,\mathfrak{a}_\mu) - \frac{1}{n}\sum_{k=1}^{n}\zeta^{(k)}(0,D_m,\mathfrak{a}_\mu)).$$

By Theorem 6.2, we get

$$V'_E(c) = qV_{E_F^+}(c). \tag{6.17}$$

Hence we see that the assertion holds for E_F^+ by Lemma 6.1, applying (6.17) to E_0 and E_F^+. Then applying (6.17) to E and E_F^+, we see that the assertion holds for E. This completes the proof.

CONJECTURE 6.5. *Let $c\in C_{\mathfrak{f}}$ and define $V_E(c)$ by (6.10). Then $V_E(c)$ is of the form*

$$V_E(c) = \sum_{i=1}^{m} a_i\log\epsilon_i, \qquad a_i\in F, \quad \epsilon_i\in E_F^+. \tag{6.18}$$

The proof of Theorem 6.4 shows that if $V_E(c)$ is of the form (6.18) for one choice of E, (6.9) and $\mathfrak{a}_\mu$, then Conjecture 6.5 holds for all the other choices.

Exercises

1. Let F be a totally real algebraic number field of degree n and E be a subgroup of E_F^+ of finite index. Let

$$\mathbf{R}_+^n = \sqcup_{\epsilon\in E}\ \epsilon\ (\sqcup_{j\in J}C_j)$$

be a cone decomposition where J is a finite set of indices. Put

$$D=\sqcup_{j\in J}C_j, \qquad U=\{\epsilon\in E \mid \epsilon\bar{D}\cap\bar{D}\neq\emptyset\}, \qquad E^1=\langle U\rangle.$$

Put

$$H=\{x=(x_1,x_2,\dots,x_n)\mid x\in\mathbf{R}_+^n,\ x_1x_2\cdots x_n=1\},$$
$$\bar{D}^1=\bar{D}\cap H, \qquad V_1=\cup_{u\in E^1}u\bar{D}^1, \qquad V_2=\cup_{u\in E_F^+\setminus E^1}u\bar{D}^1.$$

(a) Show that U is a finite set.
(b) Suppose that $V_1 \neq H$ and take $x \in H \setminus V_1$. Let $X \subset H$ be an open neighbourhood of x such that $\bar{X}$ is compact. Put $T = \{\epsilon \in E^1 \mid \epsilon\bar{D}^1 \cap X \neq \emptyset\}$. Show that T is a finite set.
(c) Show that V_1 is a closed subset of H.
(d) Show that V_2 is a closed subset of H.
(e) Derive a contradiction from $V_2 \neq \emptyset$. Show that $E^1 = E$, i.e., U generates E. (See Siegel [Si1], §9 for a theorem which holds in a more general situation.)

2. Let $A = (a_{ij}) \in M(n, \mathbf{R})$ and suppose that

$$a_{ii} > 0, \quad 1 \leqq i \leqq n, \qquad a_{ij} < 0, \quad 1 \leqq i, j \leqq n,\ i \neq j,$$
$$\sum_{i=1}^{n} a_{ij} > 0, \quad 1 \leqq j \leqq n.$$

(a) Show that $\det A > 0$.
(b) Take $x_1, x_2 \ldots, x_n > 0$ and let c_i, $1 \leqq i \leqq n$ be the solution of the simultaneous equations

$$\sum_{j=1}^{n} c_j a_{ij} = x_i, \qquad 1 \leqq i \leqq n.$$

Show that $c_i > 0$, $1 \leqq i \leqq n$.

3. Let $A = (a_{ij}) \in M(n, \mathbf{R})$ and suppose that

$$a_{ii} > 0, \quad 1 \leqq i \leqq n, \qquad a_{ij} < 0, \quad 1 \leqq i, j \leqq n,\ i \neq j,$$
$$\sum_{i=1}^{n} a_{ij} = 0, \quad 1 \leqq j \leqq n.$$

Suppose that $\sum_{j=1}^{n} c_j a_{ij} = 0$, $1 \leqq i \leqq n$ holds for $c_1, c_2, \ldots, c_n \in \mathbf{R}$. Using Exercise 2, (b), show that either $(c_1, c_2, \ldots, c_n) = (0, 0, \ldots, 0)$ or all c_i are non-zero and have the same signature.

4. Let F be a totally real algebraic number field of degree n. Let $J_F = \{\sigma_1, \sigma_2, \ldots, \sigma_n\}$. For $x \in F$, put $x^{(i)} = x^{\sigma_i}$.
(a) Show that, for $1 \leqq i \leqq n$, there exists $\epsilon_i \in E_F^+$ such that

$$\epsilon_i^{(i)} > 1, \qquad \epsilon_i^{(j)} < 1, \quad \forall j \neq i$$

and that $\epsilon_i^{(i)}$ is the minimum among units in E_F^+ with this property.
(b) Show that $[E_F^+ : \langle \epsilon_1, \epsilon_2, \ldots, \epsilon_n \rangle] < \infty$.
(c) Using Exercise 3, show that $\epsilon_1, \epsilon_2, \ldots, \epsilon_n$ satisfy essentially one relation $\epsilon_1^{e_1} \epsilon_2^{e_2} \cdots \epsilon_n^{e_n} = 1$, $0 < e_1, \ldots e_n \in \mathbf{Z}$.
(d) Show that

$$\epsilon_i^{(i)} \epsilon_j^{(i)} > 1, \qquad 1 \leqq i, j \leqq n.$$

5. In the same notation as in Exercise 4, we assume $n = 3$. In this exercise, we will show that $\epsilon_1 \epsilon_2 \epsilon_3 = 1$, $E_F^+ = \langle \epsilon_1, \epsilon_2, \epsilon_3 \rangle$, which is a special case of a theorem of Berwick [Be].
(a) Renumber ϵ_i so that $0 < e_1 \leqq e_2 \leqq e_3$. Put $\eta = \epsilon_2^{-1} \epsilon_3^{-1}$. Using Exercise 3, (d), show that $\eta^{(1)} \geqq \epsilon_1^{(1)}$.
(b) Show that $\eta^{(1)} > \epsilon_1^{(1)}$ leads to a contradiction, hence $\epsilon_1 \epsilon_2 \epsilon_3 = 1$.

(c) Take any $\psi \in E_F^+$. Choose $0 < e \in \mathbf{Z}$ so that $\psi^e \in \langle \epsilon_1, \epsilon_2, \epsilon_3 \rangle$. Multiplying ψ by suitable powers of ϵ_i, write

$$\psi^e = \epsilon_1^{a_1} \epsilon_2^{a_2} \epsilon_3^{a_3}, \qquad 0 \leqq a_1, a_2, a_3 < e.$$

Assume $\psi \neq 1$, $\psi^{(1)} > \psi^{(2)} > \psi^{(3)}$. Show that $\epsilon_1^{(1)} > \psi^{(1)} > \psi^{(2)} > 1 > \psi^{(3)}$.

(d) Considering ψ^{-1}, derive a contradiction.

(e) Derive a similar result for E_F suitably modifying the definition of ϵ_i.

6. In the same notation as in Exercise 4, we assume $n = 4$. In this exercise, we will show that an obvious generalization of Berwick's theorem does not hold.

(a) Let $f(X) = X^4 - X^3 - 3X^2 + X + 1$. Then $f(X)$ is irreducible and has four real roots:

$$\rho = 2.09529\ldots, \quad \rho' = 0.73764\ldots, \quad \rho'' = -0.47725\ldots, \quad \rho''' = -1.35567\ldots.$$

Let $F = \mathbf{Q}(\rho)$. This is the field examined in §4.3. Put $\epsilon = (\rho^2 - 1)^2$, $\eta = (\rho^3 + \rho^2 - \rho - 1)^2$, $\psi = (\rho^3 + 2\rho^2 - 1)^2$. Then it is known that $E_F^+ = \langle \epsilon, \eta, \psi \rangle$. Define $\sigma_i \in J_F$ by $\rho^{(1)} = \rho$, $\rho^{(2)} = \rho'$, $\rho^{(3)} = \rho''$, $\rho^{(4)} = \rho'''$. Show that

$$\epsilon_1 = \epsilon, \qquad \epsilon_2 = \epsilon^{-2}\eta, \qquad \epsilon_3 = \epsilon^{-1}\eta^{-2}\psi^2, \qquad \epsilon_4 = \epsilon^2\psi^{-1}, \qquad \epsilon_1\epsilon_2^2\epsilon_3\epsilon_4^2 = 1,$$

$$E_F^+ = \langle \epsilon_1, \epsilon_2, \epsilon_3, \epsilon_4 \rangle, \qquad [E_F^+ : \langle \epsilon_1, \epsilon_2, \epsilon_3 \rangle] = 2.$$

(b) Let $f(X) = X^4 - 4X^2 + X + 1$. Then $f(X)$ is irreducible and has four real roots:

$$\rho = 1.76401\ldots, \quad \rho' = 0.69382\ldots, \quad \rho'' = -0.39633\ldots, \quad \rho''' = -2.06149\ldots.$$

Let $F = \mathbf{Q}(\rho)$. Put $\epsilon = (\rho^3 - \rho^2 - \rho)^2$, $\eta = (\rho^3 - 2\rho^2 + 1)^2$, $\psi = (2\rho^3 - 4\rho^2 + 1)^2$. Then it is known that $E_F^+ = \langle \epsilon, \eta, \psi \rangle$ (cf. Billevič [Bi]). Define $\sigma_i \in J_F$ by $\rho^{(1)} = \rho$, $\rho^{(2)} = \rho'$, $\rho^{(3)} = \rho''$, $\rho^{(4)} = \rho'''$. Show that

$$\epsilon_1 = \epsilon^{-1}\eta^{-2}\psi^2, \qquad \epsilon_2 = \epsilon^2\eta^2\psi^{-3}, \qquad \epsilon_3 = \epsilon^{-1}\eta^2\psi^{-1}, \qquad \epsilon_4 = \epsilon,$$

$$\epsilon_1^2\epsilon_2\epsilon_3\epsilon_4 = 1, \qquad [E_F^+ : \langle \epsilon_1, \epsilon_2, \epsilon_3, \epsilon_4 \rangle] = 2.$$

7. In the same notation as in Exercise 4, is it possible to give simple bounds for e_i, $1 \leqq i \leqq n$ and $[E_F^+ : \langle \epsilon_1, \epsilon_2, \ldots \epsilon_n \rangle]$? The author does not know the answer.

CHAPTER V

APPLICATIONS OF A LIMIT FORMULA OF KRONECKER'S TYPE

In this chapter, we will first derive a limit formula of Kronecker's type. In the limit formula, a certain special function naturally appears. We will study the arithmetical nature of the values of this special function at CM-points in the light of our conjectures in chapter III.

In §1, we will study the Fourier expansion of Eisenstein series and derive a limit formula. The formula was given by Asai [As] when the class number of the base field is 1. By the limit formula, we obtain a function h on $\mathfrak{H}^n$ which has a certain automorphic property. In §2, we will prove a generalization of the exact Chowla-Selberg formula. In §3, we will study L-functions of orders of an algebraic number field. In §4, we will study the arithmetical nature of $h(w)$ for a CM-point w on $\mathfrak{H}^n$ applying the results in §3. We will deepen our study of $h(w)$ by taking account of Stark's conjecture. Roughly speaking, it turns out that $h(w)$ can be written using CM-periods and generalized regulators, assuming a conjecture in chapter III and Stark's conjecture. We will see the reason why we can obtain simple results only in the one dimensional case. In §5, we will clarify the relation of the function h and a holomorphic function studied by Hecke. We will also present a direct construction of a cohomology class attached to a Hilbert modular form, using the period integral of Eichler-Shimura type. Main results of this chapter are Theorems 2.5, 4.2, 4.3, 5.2 and Corollary 2.6.

1. A limit formula of Kronecker's type

In this section, we will derive a generalization of Kronecker's limit formula using the results of Shimura [Sh13], section 18.

1.1. In this subsection, we will fix notations concerning partial zeta functions of an algebraic number field.

Let k be an algebraic number field of finite degree. Put

$$\zeta_k(s) = \frac{c_k}{s-1} + \gamma_k + O(s-1), \qquad s \to 1. \tag{1.1}$$

By the analytic class number formula (II, (1.1)), we have

$$c_k = \frac{2^{r_1+r_2}\pi^{r_2} h_k R_k}{w_k |D_k|^{1/2}}, \tag{1.2}$$

where h_k is the class number and R_k is the regulator of k; w_k denotes the number of roots of unity contained in k. If $k = \mathbf{Q}$, γ_k is Euler's constant γ but its property is quite mysterious in general.

Let χ be a Hecke character of $k_A^\times$ which is trivial on $k^\times \prod_{v\in\mathbf{h}} \mathfrak{O}_v^\times k_\infty^\times$, where $\mathfrak{O}_v = \mathfrak{O}_{k_v}$ is the maximal order of k_v. Let c be an ideal class of k and let

$$\zeta_k(s,c) = \sum_{\mathfrak{A}\in c} N(\mathfrak{A})^{-s}$$

be the partial zeta function of the class c. By abuse of notation, we put $\chi(c) = \chi(a)$ taking $a \in k_A^\times$ so that the class of $\operatorname{div}(a)$ is c. Since $\zeta_k(s,c) = \frac{1}{h_k}\sum_\chi \chi(c)^{-1}L(s,\chi)$, we see that $\zeta_k(s,c)$ can be continued meromorphically to the whole complex plane, holomorphic except for the simple pole at $s=1$ and that

$$\lim_{s\to 1}(s-1)\zeta_k(s,c) = \frac{c_k}{h_k}.$$

We put

$$\zeta_k(s,c) = \frac{c_k}{h_k}\frac{1}{s-1} + \gamma_k(c) + O(s-1), \qquad s\to 1 \tag{1.3}$$

with a constant $\gamma_k(c)$. Then we have $\sum_c \gamma_k(c) = \gamma_k$.

We put

$$L^*(s,\chi) = |D_k|^{s/2}(\pi^{-s/2}\Gamma(s/2))^{r_1}((2\pi)^{1-s}\Gamma(s))^{r_2}L(s,\chi).$$

Then the functional equation

$$L^*(s,\chi) = \chi(\mathfrak{d})L^*(1-s,\chi^{-1}) \tag{1.4}$$

holds. Here $\mathfrak{d}$ denotes the different of k.

Put

$$A = 2^{-r_2}|D_k|^{1/2}\pi^{-n/2}, \qquad F(s,c) = A^s\Gamma(s/2)^{r_1}\Gamma(s)^{r_2}\zeta_k(s,c)$$

and let c' be the ideal class of $\mathfrak{A}^{-1}\mathfrak{d}_k$ for $\mathfrak{A}\in c$. By (1.4), we see that the functional equation

$$F(s,c) = F(1-s,c').$$

holds.

It is well known that $\Gamma'(1) = -\gamma$ ([WW], p. 236). Hence $\Gamma'(2) = 1-\gamma$ and putting $s=1/2$ after differentiating the duplication formula

$$2^{2s}\Gamma(s)\Gamma(s+1/2) = 2\pi^{1/2}\Gamma(2s),$$

we get

$$\Gamma'(1/2) = -\sqrt{\pi}(\gamma + 2\log 2).$$

Hence we have

$$\frac{\Gamma((1-s)/2)}{\Gamma(s/2)} = \frac{\sqrt{\pi}}{2}s\{1+(\gamma+\log 2)s\} + O(s^3), \qquad s\to 0,$$

$$\frac{\Gamma(1-s)}{\Gamma(s)} = s(1+2\gamma s) + O(s^3), \qquad s\to 0.$$

We obtain

$$\zeta_k(s,c) = -\frac{R_k}{w_k}s^{r_1+r_2-1}(1+\delta_k(c)s) + O(s^{r_1+r_2+1}), \qquad s\to 0, \tag{1.5}$$

$$\delta_k(c) = n\gamma + n\log 2\pi - \log|D_k| - \frac{w_k|D_k|^{1/2}\gamma_k(c')}{2^{r_1+r_2}\pi^{r_2}R_k}. \tag{1.6}$$

1.2. Let F be a totally real algebraic number field of degree n. We set $J_F = \{\sigma_1, \sigma_2, \dots, \sigma_n\}$ and consider F as a subring of $\mathbf{R}^n$ by the embedding $F \ni x \longrightarrow (x^{\sigma_1}, x^{\sigma_2}, \dots, x^{\sigma_n}) \in \mathbf{R}^n$. Let $\mathbf{h}$ (resp. $\mathbf{a}$) denote the set of all finite (resp. infinite) places of F. We identify $\mathbf{a}$ with J_F. Then, for a set S, $(s_1, s_2, \dots, s_n) \in S^n$ can be identified with $(s_v)_{v \in \mathbf{a}} \in S^{\mathbf{a}}$ by $s_v = s_i$ if v corresponds to σ_i. Let $F_{\mathbf{h}}$ denote the finite part of F_A and let $\mathcal{S}(F_{\mathbf{h}})$ denote the space of all Schwartz-Bruhat functions (i.e., locally constant compactly supported functions) on $F_{\mathbf{h}}$. We take $k = (k_1, k_2, \dots, k_n) \in \mathbf{Z}^{\mathbf{a}}$ and $\kappa \in \mathcal{S}(F_{\mathbf{h}})$. For $x = (x_1, x_2, \dots, x_n) \in \mathbf{C}^n$, we put $x^k = \prod_{i=1}^n x_i^{k_i}$. For $x \in \mathbf{R}_+^n$, $a \in \mathbf{C}^{\mathbf{a}}$, we define x^a similarly

We can find a subgroup U of E_F of finite index such that

$$\kappa(ud) = \kappa(d), \qquad u^{-k}|u|^k = 1 \quad \text{for every } u \in U \text{ and every } d \in F.$$

Then we put

$$D_k(s, \kappa) = [E_F : U]^{-1} \sum_{d \in F^\times / U} \kappa(d) d^{-k} |d|^{k - s\mathbf{a}}, \qquad s \in \mathbf{C}. \tag{1.7}$$

Here $s\mathbf{a} = (s, s, \dots, s) \in \mathbf{C}^{\mathbf{a}}$. This series converges when $\Re(s) > 1$.

Let $G = GL(2)$. We consider G as an algebraic group defined over F. Set $GL(2, \mathbf{R})_+ = \{g \in GL(2, \mathbf{R}) \mid \det g > 0\}$. In the usual manner, $GL(2, \mathbf{R})_+^n$ acts on $\mathfrak{H}^n = \mathfrak{H}^{\mathbf{a}}$. For $g = (g_v)_{v \in \mathbf{a}} \in GL(2, \mathbf{R})_+^n$ and $z = (z_v)_{v \in \mathbf{a}} \in \mathfrak{H}^{\mathbf{a}}$, we define the automorphic factor by $j(g, z) = (c_v z_v + d_v)_{v \in \mathbf{a}} \in \mathbf{C}^{\mathbf{a}}$, where $g_v = \begin{pmatrix} a_v & b_v \\ c_v & d_v \end{pmatrix}$. We take $k \in \mathbf{Z}^{\mathbf{a}}$ and $\lambda \in \mathcal{S}(F_{\mathbf{h}} \oplus F_{\mathbf{h}})$. We can find a subgroup U of E_F of finite index such that

$$\lambda(uh) = \lambda(h), \qquad u^{-k}|u|^k = 1 \quad \text{for every } u \in U \text{ and every } h \in F \oplus F.$$

Put $H = F \oplus F \setminus \{(0,0)\}$. We define an Eisenstein series by

$$E_k(z, s; \lambda) = [E_F : U]^{-1} \sum_{(c,d) \in H/U} \lambda(c, d)(cz + d)^{-k} y^{s\mathbf{a} - k/2} |cz + d|^{k - 2s\mathbf{a}}. \tag{1.8}$$

Here $z \in \mathfrak{H}^{\mathbf{a}}$, $s \in \mathbf{C}$, $y = \Im(z)$, $cz + d = (c^{\sigma_1} z_1 + d^{\sigma_1}, \dots, c^{\sigma_n} z_n + d^{\sigma_n}) \in \mathbf{C}^n$. We note that the series (1.7) and (1.8) are defined independently of U. The sum in (1.8) converges when $\Re(s) > 1$. Put

$$\Gamma = \{\gamma \in G_F \mid \lambda(x\gamma) = \lambda(x) \quad \text{for all } x \in F \oplus F, \quad \det \gamma \gg 0\}.$$

Then Γ is a congruence subgroup of G_F and we can verify easily that

$$E_k(\gamma z, s; \lambda) = E_k(z, s; \lambda) j(\gamma, z)^k (\det \gamma)^{-k/2}, \qquad \gamma \in \Gamma. \tag{1.9}$$

We are going to derive the Fourier expansion of $E_k(z, s; \lambda)$.

1.3. Put $H' = \{z \in \mathbf{C} \mid \Re(z) > 0\}$. For $z \in H'$ and $\alpha \in \mathbf{C}$, put $z^\alpha = \exp(\alpha \log z)$ choosing the branch of $\log z$ so that it takes real values on $\mathbf{R}_+$. We put $i^\alpha = e^{\alpha \pi i/2}$. For $z \in \mathfrak{H}$, we put $z^\alpha = i^\alpha(-iz)^\alpha$, $\bar{z}^\alpha = i^{-\alpha}(i\bar{z})^\alpha$.

For $z \in H'$ and $(\alpha, \beta) \in \mathbf{C}^2$, we set

$$\zeta(z; \alpha, \beta) = \int_0^\infty e^{-zx}(x+1)^{\alpha-1} x^{\beta-1} dx. \tag{1.10}$$

The integral converges when $\Re(\beta) > 0$. This is a confluent hypergeometric function. Put $\omega(z;\alpha,\beta) = \Gamma(\beta)^{-1} z^{\beta} \zeta(z;\alpha,\beta)$. It can be shown that $\omega(z;\alpha,\beta)$ can be continued to a holomorphic function on $H' \times \mathbf{C}^2$ and satisfies the functional equation ([Sh2], p. 85, [Sh9]):

$$\omega(z; 1-\beta, 1-\alpha) = \omega(z;\alpha,\beta). \tag{1.11}$$

Next, for $0 < g \in \mathbf{R}$, $h \in \mathbf{R}$ and $(\alpha,\beta) \in \mathbf{C}^2$, set

$$\xi(g,h;\alpha,\beta) = \int_{-\infty}^{\infty} e^{-2\pi i h x}(x+gi)^{-\alpha}(x-gi)^{-\beta} dx. \tag{1.12}$$

The integral converges when $\Re(\alpha+\beta) > 1$. It can be shown that (cf. [Sh9], (1.29), (3.3))

$$\begin{aligned} &\xi(g,h;\alpha,\beta) \\ &= i^{\beta-\alpha}(2\pi)(2\pi h)^{\alpha+\beta-1}\Gamma(\alpha)^{-1}\Gamma(\beta)^{-1} e^{-2\pi g h}\zeta(4\pi g h;\alpha,\beta), \qquad h > 0. \end{aligned} \tag{1.13}$$

For $z = (z_v)_{v\in\mathbf{a}} \in \mathbf{C}^{\mathbf{a}}$, we put $\mathbf{e}_{\mathbf{a}}(z) = \exp(2\pi i \sum_{v\in\mathbf{a}} z_v)$.

LEMMA 1.1. *Let $\mathfrak{m}$ be a fractional ideal of F and $z = x+iy \in \mathfrak{H}^{\mathbf{a}}$, $\alpha \in \mathbf{C}^{\mathbf{a}}$, $\beta \in \mathbf{C}^{\mathbf{a}}$. Put $\Xi(y,b;\alpha,\beta) = \prod_{v\in\mathbf{a}} \xi(y_v, b_v; \alpha_v, \beta_v)$. Then, if $\Re(\alpha_v+\beta_v) > 1$ for every $v \in \mathbf{a}$, we have*

$$D_F^{1/2} N(\mathfrak{m}) \sum_{a\in\mathfrak{m}} (z+a)^{-\alpha}(\bar{z}+a)^{-\beta} = \sum_{b\in(\mathfrak{d}\mathfrak{m})^{-1}} \mathbf{e}_{\mathbf{a}}(bx)\Xi(y,b;\alpha,\beta).$$

This lemma follows applying the Poisson summation formula to $f(x) = (x+iy)^{-\alpha}(x-iy)^{-\beta}$ with a fixed $y \in \mathbf{R}_+^{\mathbf{a}}$. See [Sh13], Lemma 18.4 for a detailed proof.

1.4. Let $\mathfrak{a}$ and $\mathfrak{b}$ be fractional ideals of F and take a_0, $b_0 \in F$. We take λ as the characteristic function of the closure of $(a_0+\mathfrak{a}) \oplus (b_0+\mathfrak{b})$. Then

$$U = \{u \in E_F \mid u \gg 0,\ (u-1)a_0 \in \mathfrak{a},\ (u-1)b_0 \in \mathfrak{b}\}$$

satisfies our condition. Put

$$A = [(a_0+\mathfrak{a}) \cap F^{\times}]/U, \qquad B = [(b_0+\mathfrak{b}) \cap F^{\times}]/U.$$

Let κ_A (resp. κ_B) be the characteristic function of the closure of $a_0+\mathfrak{a}$ (resp. $b_0+\mathfrak{b}$). We have

$$\begin{aligned} &[E_F : U] y^{k/2 - s\mathbf{a}} E_k(z,s;\lambda) \\ &= \sum_{d\in B} \lambda(0,d) d^{-k}|d|^{k-2s\mathbf{a}} + \sum_{c\in A}\sum_{d\in b_0+\mathfrak{b}} (cz+d)^{-k}|cz+d|^{k-2s\mathbf{a}}. \end{aligned}$$

The first term equals $\epsilon(a_0,\mathfrak{a})[E_F : U] D_k(2s,\kappa_A)$, where $\epsilon(a_0,\mathfrak{a}) = 1$ (resp. 0) if $a_0 \in \mathfrak{a}$ (resp. $a_0 \notin \mathfrak{a}$). The second term equals, when $\Re(s) > 1$,

$$\begin{aligned} &\sum_{c\in A} c^{-k}|c|^{k-2s\mathbf{a}} \sum_{a\in c^{-1}\mathfrak{b}} (z+c^{-1}b_0+a)^{-k}|z+c^{-1}b_0+a|^{k-2s\mathbf{a}} \\ &= \sum_{c\in A} c^{-k}|c|^{k-2s\mathbf{a}} D_F^{-1/2}|N(c)|N(\mathfrak{b})^{-1} \\ &\times \sum_{b\in\mathfrak{d}^{-1}c\mathfrak{b}^{-1}} \mathbf{e}_{\mathbf{a}}(bx+bc^{-1}b_0)\Xi(y,b;s\mathbf{a}+\frac{k}{2}, s\mathbf{a}-\frac{k}{2}) \end{aligned}$$

by Lemma 1.1. Therefore we obtain the Fourier expansion:

$$
\begin{aligned}
E_k(z,s;\lambda) &= \epsilon(a_0,\mathfrak{a})y^{s\mathbf{a}-k/2}D_k(2s,\kappa_B)\\
&+D_F^{-1}N(\mathfrak{b})^{-1}y^{s\mathbf{a}-k/2}\Xi(y,0;s\mathbf{a}+\frac{k}{2},s\mathbf{a}-\frac{k}{2})D_k(2s-1,\kappa_A)\\
&+[E_F:U]^{-1}D_F^{-1/2}N(\mathfrak{b})^{-1}y^{s\mathbf{a}-k/2}\\
&\times\sum_{b\in F^\times}\mathbf{e}_{\mathbf{a}}(bx)\Xi(y,b;s\mathbf{a}+\frac{k}{2},s\mathbf{a}-\frac{k}{2})\varphi(b,s)
\end{aligned}
\tag{1.14}
$$

with $\varphi(b,s)=\sum_{a,c}\mathbf{e}_{\mathbf{a}}(ab_0)c^{-k}|c|^{k+(1-2s)\mathbf{a}}$. Here the sum for φ is taken over all $(a,c)\in\mathfrak{b}^{-1}\mathfrak{d}^{-1}\times A$ such that $ac=b$.

1.5. We specialize (1.14) to the following situation. Let $\mathfrak{a}$ and $\mathfrak{b}$ be fractional ideals of F and let λ be the characteristic function of the closure of $\mathfrak{ab}\oplus\mathfrak{b}$. We further assume $k=0$. Then we may take $U=E_F$. We denote $E_k(z,s;\lambda)$ as $E(z,s;\mathfrak{a},\mathfrak{b})$. By (1.14), we have

$$
\begin{aligned}
E(z,s;\mathfrak{a},\mathfrak{b}) &= y^{s\mathbf{a}}D_0(2s,\kappa_B)\\
&+D_F^{-1}N(\mathfrak{b})^{-1}y^{s\mathbf{a}}\Xi(y,0;s\mathbf{a},s\mathbf{a})D_0(2s-1,\kappa_A)\\
&+D_F^{-1/2}N(\mathfrak{b})^{-1}y^{s\mathbf{a}}\sum_{b\in F^\times}\mathbf{e}_{\mathbf{a}}(bx)\Xi(y,b;s\mathbf{a},s\mathbf{a})\varphi(b,s).
\end{aligned}
\tag{1.15}
$$

Put

$$
\begin{aligned}
\Gamma_{\mathfrak{a}} = \Big\{\gamma=\begin{pmatrix} a & b\\ c & d\end{pmatrix}\in GL(2,F)\\
\mid a,d\in\mathfrak{O}_F,\ b\in\mathfrak{a}^{-1},\ c\in\mathfrak{a},\ \det\gamma\in E_F^+\Big\}.
\end{aligned}
\tag{1.16}
$$

Then $\lambda(x\gamma)=\lambda(x)$ for $x\in F\oplus F$, $\gamma\in\Gamma_{\mathfrak{a}}$. By (1.9), we have

$$
E(\gamma z,s;\mathfrak{a},\mathfrak{b})=E(z,s;\mathfrak{a},\mathfrak{b}),\qquad \gamma\in\Gamma_{\mathfrak{a}}.
\tag{1.17}
$$

We find easily that

$$
D_0(s,\kappa_A)=N(\mathfrak{ab})^{-s}\zeta_F(s,c_{(\mathfrak{ab})^{-1}}),\qquad D_0(s,\kappa_B)=N(\mathfrak{b})^{-s}\zeta_F(s,c_{\mathfrak{b}^{-1}}),
$$

where $c_{(\mathfrak{ab})^{-1}}$ (resp. $c_{\mathfrak{b}^{-1}}$) denotes the class of $(\mathfrak{ab})^{-1}$ (resp. $\mathfrak{b}^{-1}$) in the ideal class group of F and $\zeta_F(s,c_{(\mathfrak{ab})^{-1}})$ (resp. $\zeta_F(s,c_{\mathfrak{b}^{-1}})$) denotes the partial zeta function of the class $c_{(\mathfrak{ab})^{-1}}$ (resp. $c_{\mathfrak{b}^{-1}}$).

We denote by $K_s(z)$ the usual K-Bessel function. We take the formula

$$
K_s(z)=\frac{1}{2}\int_0^\infty \exp(-\frac{z}{2}(t+\frac{1}{t}))t^{s-1}dt
\tag{1.18}
$$

as the definition of $K_s(z)$. (cf. Watson, [Wat], p. 182.) The integral converges locally uniformly on the domain $\{(z,s)\mid \Re(z)>0,\ s\in\mathbf{C}\}$, hence defines an analytic function there. Changing the variable t to t^{-1}, we see immediately that

$$
K_s(z)=K_{-s}(z).
\tag{1.19}
$$

LEMMA 1.2. *For $g > 0$, $\Re(s) > 1/2$, we have*

$$\xi(g,h;s,s) = 2\cdot\frac{\pi^s|h|^{s-1/2}}{\Gamma(s)g^{s-1/2}}K_{s-1/2}(2\pi|h|g) \qquad \text{if } \; 0 \neq h \in \mathbf{R},$$

$$\xi(g,0;s,s) = g^{1-2s}\sqrt{\pi}\frac{\Gamma(s-1/2)}{\Gamma(s)}.$$

PROOF. For $g > 0$, $\Re(s) > 1/2$, $h \in \mathbf{R}$, we consider the integral

$$\int_{-\infty}^{\infty}\int_0^{\infty} \exp(-\pi(g^2+x^2)\,t)\exp(-2\pi i h x)\,t^{s-1}dtdx. \tag{1.20}$$

We have

$$\int_0^{\infty}\exp(-\pi(g^2+x^2)\,t)\,t^{s-1}dt = \pi^{-s}(g^2+x^2)^{-s}\Gamma(s), \qquad \Re(s) > 0.$$

Hence (1.20) converges absolutely if $\Re(s) > 1/2$ and is equal to

$$\pi^{-s}\Gamma(s)\xi(g,h;s,s).$$

Next we see easily that

$$\int_{-\infty}^{\infty}\exp(-\pi(g^2+x^2)\,t)\exp(-2\pi i h x)dx = \frac{1}{\sqrt{t}}\exp(-\pi(g^2t+\frac{h^2}{t})).$$

Hence (1.20) is equal to

$$\int_0^{\infty}\exp(-\pi(g^2t+\frac{h^2}{t}))t^{s-3/2}dt. \tag{1.21}$$

If $h = 0$, (1.21) is equal to $\pi^{-s+1/2}g^{1-2s}\Gamma(s-1/2)$. Hence we obtain the second formula in our Lemma. If $h \neq 0$, then changing the variable t to $gt/|h|$, we find that (1.21) is equal to

$$2(\frac{|h|}{g})^{s-1/2}K_{s-1/2}(2\pi|h|g).$$

Hence the first formula follows. This completes the proof.

By Lemma 1.2, (1.8) can be written as

$$\begin{aligned} E(z,s;\mathfrak{a},\mathfrak{b}) &= N(\mathfrak{b})^{-2s}y^{s\mathbf{a}}\zeta_F(2s,c_{\mathfrak{b}^{-1}}) \\ &+ D_F^{-1}N(\mathfrak{b})^{-1}y^{(1-s)\mathbf{a}}(\frac{\sqrt{\pi}\Gamma(s-1/2)}{\Gamma(s)})^nN(\mathfrak{ab})^{-(2s-1)}\zeta_F(2s-1,c_{(\mathfrak{ab})^{-1}}) \\ &+ 2^nD_F^{-1/2}N(\mathfrak{b})^{-1}(\frac{\pi^s}{\Gamma(s)})^ny^{(1/2)\mathbf{a}}\sum_{b\in F^\times}|N(b)|^{s-1/2}\varphi_{\mathfrak{b}}(b,x)\mathbf{e}_{\mathbf{a}}(bx) \\ &\times\prod_{v\in\mathbf{a}}K_{s-1/2}(2\pi|b_v|y_v). \end{aligned} \tag{1.22}$$

Here we write $\varphi(b,s)$ as $\varphi_{\mathfrak{b}}(b,s)$ to stress the dependence on $\mathfrak{b}$.

1.6. Let X denote the character group of the ideal class group of F. For $\chi \in X$, we set

$$E_\chi(z, s; \mathfrak{a}) = \chi(\mathfrak{a})^{-1} N(\mathfrak{a})^s \sum_{\mathfrak{b}} \chi(\mathfrak{b})^{-1} N(\mathfrak{b})^{2s} E(z, s; \mathfrak{a}, \mathfrak{b}) \tag{1.23}$$

where $\mathfrak{b}$ extends over a complete set of representatives of the ideal classes of F. By (1.17), we have

$$E_\chi(\gamma z, s; \mathfrak{a}) = E_\chi(z, s; \mathfrak{a}) \qquad \text{for} \quad \gamma \in \Gamma_{\mathfrak{a}}. \tag{1.24}$$

By definition, it is immediate to see that

$$E(z, s; \mathfrak{a}, (\alpha)\mathfrak{b}) = N(\alpha)^{-2s} E(z, s; \mathfrak{a}, \mathfrak{b}).$$

Hence $N(\mathfrak{b})^{2s} E(z, s; \mathfrak{a}, \mathfrak{b})$ depends only on the ideal class of $\mathfrak{b}$. We see that

$$\begin{aligned} &\frac{1}{h_F} \sum_{\chi \in X} \chi(\mathfrak{a}) N(\mathfrak{a})^{-s} E_\chi(z, s; \mathfrak{a}) \\ &= \frac{1}{h_F} \sum_{\mathfrak{b}} \sum_{\chi \in X} \chi(\mathfrak{b}) N(\mathfrak{b})^{2s} E(z, s; \mathfrak{a}, \mathfrak{b}) = E(z, s; \mathfrak{a}, (1)). \end{aligned}$$

Therefore we have

$$\sum_{(c,d) \in H/E_F,\, c \in \mathfrak{a},\, d \in \mathfrak{O}_F} y^{s\mathbf{a}} |cz + d|^{-2s\mathbf{a}} = \frac{1}{h_F} \sum_{\chi \in X} \chi(\mathfrak{a}) N(\mathfrak{a})^{-s} E_\chi(z, s; \mathfrak{a}). \tag{1.25}$$

By (1.22), we easily obtain

$$\begin{aligned} E_\chi(z,s; \mathfrak{a}) &= \chi(\mathfrak{a})^{-1} N(\mathfrak{a})^s y^{s\mathbf{a}} L(2s, \chi) \\ &+ D_F^{-1} N(\mathfrak{a})^{1-s} y^{(1-s)\mathbf{a}} \Big(\frac{\sqrt{\pi}\Gamma(s - 1/2)}{\Gamma(s)}\Big)^n L(2s - 1, \chi) \\ &+ 2^n D_F^{-1/2} \Big(\frac{\pi^s}{\Gamma(s)}\Big)^n \chi(\mathfrak{a})^{-1} N(\mathfrak{a})^s y^{(1/2)\mathbf{a}} \\ &\times \sum_{b \in F^\times} \sum_{\mathfrak{b}} \chi(\mathfrak{b})^{-1} N(\mathfrak{b})^{2s-1} \varphi_{\mathfrak{b}}(b, s) |N(b)|^{s-1/2} \\ &\times \mathbf{e_a}(bx) \prod_{v \in \mathbf{a}} K_{s-1/2}(2\pi |b_v| y_v). \end{aligned} \tag{1.26}$$

This formula gives the analytic continuation of $E_\chi(z, s; \mathfrak{a})$. For $x \in F_A^\times$ and $s \in \mathbf{C}$, we put

$$\sigma_{s,\chi}(x) = \prod_{v \in \mathbf{h}} \begin{cases} 1 + \chi_v(\varpi_v) q_v^s + \cdots + (\chi_v(\varpi_v) q_v^s)^{\mathrm{ord}_v(x_v)} & \text{if} \quad x_v \in \mathfrak{O}_v, \\ 0 \quad \text{if} \quad x_v \notin \mathfrak{O}_v. \end{cases} \tag{1.27}$$

Here $\mathfrak{O}_v$ denotes the ring of integers of F_v, ϖ_v is a prime element of F_v and $q_v = |\mathfrak{O}_v / \varpi \mathfrak{O}_v|$. We see easily that

$$\sigma_{-s,\chi}(x) = \chi(x_f) |x_f|_A^s \sigma_{s,\chi^{-1}}(x) \qquad x \in F_A^\times. \tag{1.28}$$

LEMMA 1.3. *Take* $a,\ d \in F_A^\times$ *so that* $\mathrm{div}(a^{-1}) = \mathfrak{a}$, $\mathrm{div}(d) = \mathfrak{d}$. *For* $0 \neq b \in \mathfrak{d}^{-1}\mathfrak{a}$, *we have*

$$\sum_{\mathfrak{b}} \chi(\mathfrak{b})^{-1} N(\mathfrak{b})^{2s-1} \varphi_{\mathfrak{b}}(b,s) = \chi(\mathfrak{a}) N(\mathfrak{a})^{2s-1} \sigma_{1-2s,\chi}(bda).$$

PROOF. Recall that $\varphi_{\mathfrak{b}}(b,s) = \sum_{a,c} |N(c)|^{1-2s}$ where a and c extend over

$$\mathfrak{b}^{-1}\mathfrak{d}^{-1} \times (\mathfrak{a}\mathfrak{b} \setminus \{0\})/E_F$$

such that $ac = b$. We see that $N(\mathfrak{b})^{2s-1}\varphi_{\mathfrak{b}}(b,s)$ depends only on the ideal class of $\mathfrak{b}$. Let

$$(b)\mathfrak{d}\mathfrak{a}^{-1} = \prod_i \mathfrak{p}_i^{e_i}$$

be the prime ideal decomposition. Suppose that $a \in \mathfrak{b}^{-1}\mathfrak{d}^{-1}$ satisfies $ac = b$ with $c \in \mathfrak{a}\mathfrak{b}$. Then we have $(a) = \mathfrak{b}^{-1}\mathfrak{d}^{-1}\mathfrak{q}$ with an integral ideal $\mathfrak{q}$ which divides $\prod_i \mathfrak{p}_i^{e_i}$. Conversely if $\mathfrak{q}$ is such an ideal, then, for $\mathfrak{b}$ in the same ideal class as $\mathfrak{d}^{-1}\mathfrak{q}$, there exists $a \in \mathfrak{b}^{-1}\mathfrak{d}^{-1}$ such that $(a) = \mathfrak{b}^{-1}\mathfrak{d}^{-1}\mathfrak{q}$, $ac = b$ with $c \in \mathfrak{a}\mathfrak{b}$. Therefore we have

$$\begin{aligned}
&\sum_{\mathfrak{b}} \chi(\mathfrak{b})^{-1} N(\mathfrak{b})^{2s-1} \varphi_{\mathfrak{b}}(b,s) \\
&= \sum_{\mathfrak{q}|\mathfrak{c}} \chi(\mathfrak{d})\chi(\mathfrak{q})^{-1} |N(a)|^{1-2s} N(\mathfrak{d})^{1-2s} N(\mathfrak{q})^{2s-1} |N(b)|^{1-2s} |N(a)|^{2s-1} \\
&= \chi(\mathfrak{d}) N(\mathfrak{d})^{1-2s} \sum_{\mathfrak{q}|\mathfrak{c}} \chi(\mathfrak{q})^{-1} N(\mathfrak{q})^{2s-1} \\
&= \chi(\mathfrak{d}) N(\mathfrak{d})^{1-2s} |N(b)|^{1-2s} \sigma_{2s-1,\chi^{-1}}(bda),
\end{aligned}$$

where $\mathfrak{c} = \prod_i \mathfrak{p}_i^{e_i} = (b)\mathfrak{d}\mathfrak{a}^{-1}$. By (1.28), we have

$$\sigma_{2s-1,\chi^{-1}}(bda) = \chi^{-1}(\mathfrak{d})\chi(\mathfrak{a}) |N(b)|^{2s-1} N(\mathfrak{d})^{2s-1} N(\mathfrak{a})^{1-2s} \sigma_{1-2s,\chi}(bda).$$

Hence our assertion follows.

By Lemma 1.3, (1.26) can be written as

$$\begin{aligned}
E_\chi(z,s;\mathfrak{a}) &= \chi(\mathfrak{a})^{-1} N(\mathfrak{a})^s y^{s\mathbf{a}} L(2s,\chi) \\
&+ D_F^{-1} N(\mathfrak{a})^{1-s} y^{(1-s)\mathbf{a}} \left(\frac{\sqrt{\pi}\,\Gamma(s-1/2)}{\Gamma(s)}\right)^n L(2s-1,\chi) \\
&+ 2^n D_F^{-1/2} \left(\frac{\pi^s}{\Gamma(s)}\right)^n N(\mathfrak{a})^{1-s} y^{(1/2)\mathbf{a}} \\
&\times \sum_{0\neq b\in\mathfrak{d}^{-1}\mathfrak{a}} \sigma_{1-2s}(bda) |N(b)|^{s-1/2} \mathbf{e}_{\mathbf{a}}(bx) \prod_{v\in\mathbf{a}} K_{s-1/2}(2\pi|b_v|y_v).
\end{aligned} \tag{1.29}$$

1.7. For $\chi \in X$, we put

$$\begin{aligned}
h_\chi(z;\mathfrak{a}) = \frac{D_F N(\mathfrak{a})}{2^{n-2}\pi^n h_F R_F} \Big[&\chi(\mathfrak{d}) L(2,\chi^{-1}) y^{\mathbf{a}} \\
+ \pi^n D_F^{-3/2} &\sum_{0\neq b\in\mathfrak{d}_F^{-1}\mathfrak{a}} |N(b)|^{-1} \sigma_{1,\chi}(bda) \exp(2\pi i(\sum_{v\in\mathfrak{a}} b_v x_v + i|b_v y_v|)) \Big]
\end{aligned} \tag{1.30}$$

Let us consider the behavior of $E_\chi(z, s; \mathfrak{a})$ at $s = 1$. Since

$$K_{1/2}(z) = \sqrt{\frac{\pi}{2z}} e^{-z}, \qquad \sigma_{-1,\chi}(bda) = \chi(\mathfrak{d})\chi(\mathfrak{a})^{-1}|N(b)|^{-1}N(\mathfrak{d})^{-1}N(\mathfrak{a})\sigma_{1,\chi}(bda),$$

we see that the first term and the third term of (1.29) give

$$\frac{2^{n-2}\pi^n h_F R_F}{D_F}\chi(\mathfrak{a})^{-1}\chi(\mathfrak{d})h_{\chi^{-1}}(z;\mathfrak{a}) + O(s-1), \qquad s \to 1.$$

Since $\Gamma'(1/2) = -\sqrt{\pi}(\gamma + 2\log 2)$, we have

$$\frac{\Gamma(s-1/2)}{\Gamma(s)} = \sqrt{\pi}\,\{1 - (2\log 2)(s-1)\} + O((s-1)^2), \qquad s \to 1. \tag{1.31}$$

Therefore we obtain, when $\chi = 1$,

$$\begin{aligned} E_1(z, s; \mathfrak{a}) = \frac{2^{n-2}\pi^n h_F R_F}{D_F}\Bigg[\frac{1}{s-1} - 2n\log 2 + \frac{D_F^{1/2}\gamma_F}{2^{n-2}h_F R_F} \\ - \log N(\mathfrak{a}) - \log y^{\mathbf{a}} + h_1(z;\mathfrak{a})\Bigg] + O(s-1), \qquad s \to 1. \end{aligned} \tag{1.32}$$

If $\chi \neq 1$, $E_\chi(z, s; \mathfrak{a})$ is holomorphic at $s = 1$ and we obtain

$$E_\chi(z, 1; \mathfrak{a}) = D_F^{-1/2}\pi^n L(1,\chi) + \frac{2^{n-2}\pi^n h_F R_F}{D_F}\chi(\mathfrak{a})^{-1}\chi(\mathfrak{d})h_{\chi^{-1}}(z;\mathfrak{a}). \tag{1.33}$$

Traditionally the limit formula of Kronecker's type deals with the behavior of Eisenstein series at $s = 1$. For our later applications, it is more convenient to consider the behavior at $s = 0$. By the functional equation (1.4), we see easily that the second term and the third term of (1.29) gives

$$2^{n-2}\pi^n h_F R_F h_\chi(z;\mathfrak{a})s^n + O(s^{n+1}), \qquad s \to 0.$$

We have

$$\zeta_F(s) = -\frac{R_F}{2}s^{n-1}\left[1 + (2n\gamma + 2n\log 2\pi - n\log D_F - \frac{D_F^{1/2}\gamma_F}{2^{n-1}R_F})s\right] + O(s^{n+1}), \quad s \to 0$$

$$L(s,\chi) = \frac{D_F^{1/2}}{2^n}\chi(\mathfrak{d})s^n \sum_c \chi(c)^{-1}\gamma_F(c) + O(s^{n+1}), \quad s \to 0.$$

Therefore, for $s \to 0$, we obtain

$$\begin{aligned} E_1(z,s;\mathfrak{a}) = -2^{n-2}h_F R_F s^{n-1}\Bigg[1 + (2n\gamma + 2n\log 2\pi - 2\log D_F \\ - \frac{D_F^{1/2}\gamma_F}{2^{n-2}h_F R_F} + \log N(\mathfrak{a}) + \log y^{\mathbf{a}} - h_1(z;\mathfrak{a}))s\Bigg] + O(s^{n+1}), \end{aligned} \tag{1.34}$$

$$\begin{aligned} E_\chi(z, s; \mathfrak{a}) = -2^{n-2}h_F R_F s^n\Bigg[- \frac{D_F^{1/2}}{2^{n-2}h_F R_F}\chi(\mathfrak{a})^{-1}\chi(\mathfrak{d})\sum_c \chi(c^{-1})\gamma_F(c) \\ - h_\chi(z;\mathbf{a}))\Bigg] + O(s^{n+1}), \qquad \text{if } \chi \neq 1. \end{aligned} \tag{1.35}$$

PROPOSITION 1.4. *Define a congruence subgroup* $\Gamma_{\mathfrak{a}}$ *by* (1.16). *If* $\chi \neq 1$, *then we have* $h_\chi(\gamma z, \mathfrak{a}) = h_\chi(z; \mathfrak{a})$ *for* $\gamma \in \Gamma_{\mathfrak{a}}$. *If* $\chi = 1$, *then we have* $h_1(\gamma z, \mathfrak{a}) - \log(y(\gamma z))^{\mathbf{a}} = h_1(z; \mathfrak{a}) - \log y(z)^{\mathbf{a}}$ *for* $\gamma \in \Gamma_{\mathfrak{a}}$. *Here* $y(\gamma z)$ *and* $y(z)$ *denote the imaginary parts of* γz *and* z *respectively.*

The assertion is obvious in view of (1.17), (1.34) and (1.35).

1.8. As an example, take $F = \mathbf{Q}$, $\mathfrak{a} = (1)$. Put $h(z) = h_1(z; (1))$. For $n \in \mathbf{Z}$, $n \neq 0$, put

$$\sigma_1(n) = \sum_{0<d||n|} 1.$$

By (1.30), we have

$$\begin{aligned} h(z) &= \frac{2}{\pi}\left[\zeta(2)y + \pi \sum_{0\neq n \in \mathbf{Z}} \frac{1}{|n|}\sigma_1(n)\exp(2\pi i(nx + i|ny|))\right] \\ &= \frac{\pi}{3}y + 4\sum_{n=1}^{\infty} \frac{\sigma_1(n)}{n}\cos(2\pi nx)e^{-2\pi ny}, \end{aligned}$$

where $z = x + iy$. Let

$$\eta(z) = e^{2\pi iz/24}\prod_{n=1}^{\infty}(1 - e^{2\pi inz})$$

be the Dedekind η-function. Then we get

$$\log \eta(z) = \frac{\pi iz}{12} - \sum_{n=1}^{\infty}\sum_{m=1}^{\infty}\frac{1}{m}e^{2\pi imnz} = \frac{\pi iz}{12} - \sum_{n=1}^{\infty}\frac{\sigma_1(n)}{n}e^{2\pi inz},$$

taking the branch of $\log \eta(z)$ so that it takes real values on the imaginary axis. We see

$$\log|\eta(z)|^2 = \log\eta(z) + \overline{\log\eta(z)} = -\frac{\pi y}{6} - 2\sum_{n=1}^{\infty}\frac{\sigma_1(n)}{n}\cos(2\pi nx)e^{-2\pi ny}.$$

Therefore we have

$$h(z) = -\log|\eta(z)|^4. \tag{1.36}$$

Put $E(z, s) = E_1(z, s; (1))$. By (1.25), we have

$$E(z, s) = \frac{1}{2}\sum_{(c,d)\in\mathbf{Z}^2\setminus\{(0,0)\}}\frac{y^s}{|cz + d|^{2s}}. \tag{1.37}$$

By (1.32), we have

$$\lim_{s\to 1} E(z, s) - \frac{\pi}{2}\frac{1}{s-1} = \pi(\gamma - \log 2 - \log(\sqrt{y}\,|\eta(z)|^2)), \tag{1.38}$$

which is Kronecker's first limit formula.

2. A generalization of the exact Chowla-Selberg formula

2.1. Let R be a Dedekind domain, K be its quotient field and V be a finite dimensional vector space over K. By *an R-lattice in V*, we mean a finitely generated R-module in V which spans V over K. We begin with the following proposition.

PROPOSITION 2.1. *Let k be an algebraic number field and K be an algebraic extension of degree n of k. Let $\mathfrak{A}$ be a fractional ideal of K. Then there exist ω_1, ω_2, ..., $\omega_n \in K$ and a fractional ideal $\mathfrak{a}$ of k such that*

$$\mathfrak{A} = \mathfrak{a}\omega_1 \oplus \mathfrak{O}_k\omega_2 \oplus \cdots \oplus \mathfrak{O}_k\omega_n.$$

Moreover the ideal class of $\mathfrak{a}$ is equal to the class of $\mathfrak{c}N_{K/k}(\mathfrak{A})$, where $\mathfrak{c}$ is a fractional ideal of k, independent of $\mathfrak{A}$, such that the square of the class of $\mathfrak{c}$ is the class of $D(K/k)$.

For the proof, we use the next Lemma.

LEMMA 2.2. *Let V be a vector space of finite dimension n over k. Let L be an $\mathfrak{O}_k$-lattice. Then there exist v_1, v_2, ..., $v_n \in V$ and a fractional ideal $\mathfrak{b}$ of k such that*

$$L = \mathfrak{b}v_1 \oplus \mathfrak{O}_k v_2 \oplus \cdots \oplus \mathfrak{O}_k v_n.$$

Moreover the ideal class of $\mathfrak{b}$ is uniquely determined by L.

This lemma is well known as the structure theorem for a finitely generated torsion free module over a Dedekind domain. See, for example, Proposition 24 in Bourbaki [Bo2], Chapitre 7, §4. We omit the proof.

PROOF OF PROPOSITION 2.1. Take $V = K$ in Lemma 2.2. The first assertion is then obvious. We can write

$$\mathfrak{O}_K = \mathfrak{c}\omega_1 \oplus \mathfrak{O}_k\omega_2 \oplus \ldots \oplus \mathfrak{O}_k\omega_n, \qquad \mathfrak{A} = \mathfrak{a}v_1 \oplus \mathfrak{O}_k v_2 \oplus \ldots \oplus \mathfrak{O}_k v_n$$

applying Lemma 2.2. Here $\{\omega_1, \ldots, \omega_n\}$ and $\{v_1, \ldots, v_n\}$ are basis of K over k and $\mathfrak{c}$ and $\mathfrak{a}$ are fractional ideals of k. Take $\gamma \in GL(n, k)$ so that $\gamma\omega_i = v_i$, $1 \leqq i \leqq n$. Take $x \in k_A^\times$ so that $\mathrm{div}(x) = \mathfrak{c}^{-1}\mathfrak{a}$ and put $y = \mathrm{diag}[x, 1, \ldots, 1]$. Then we have $y\gamma\mathfrak{O}_K = \mathfrak{A}$. On the other hand, we have $\mathfrak{A} = a\mathfrak{O}_K$ with $a \in K_A^\times$ such that $\mathrm{div}(a) = \mathfrak{A}$. Therefore we have

$$(*) \qquad a^{-1}y\gamma\mathfrak{O}_K = \mathfrak{O}_K.$$

When the multiplication by a is regarded as the linear transformation on $K \otimes_k k_v$ for a place v of k, its determinant is the v-component of $N_{K/k}(a)$. Considering the determinant of $a^{-1}y\gamma$, we see that $(*)$ implies $N_{K/k}(\mathfrak{A}) \equiv \mathfrak{c}^{-1}\mathfrak{a}$, the equivalence $\equiv$ being understood in the ideal class group of k. Therefore the last assertion is reduced to the case $\mathfrak{A} = \mathfrak{O}_K$.

Let $\{\omega_1', \omega_2', \ldots, \omega_n'\}$ be the dual basis of K over k with respect to the bilinear form $\mathrm{Tr}_{K/k}$. It is well known that ([We1], p. 154)

$$\mathfrak{d}_{K/k}^{-1} = \{x \in K \mid \mathrm{Tr}_{K/k}(xy) \in \mathfrak{O}_k \quad \text{for every } y \in \mathfrak{O}_K\}.$$

Clearly we have

$$\mathfrak{d}_{K/k}^{-1} = \mathfrak{c}^{-1}\omega_1' \oplus \mathfrak{O}_k\omega_2' \oplus \cdots \oplus \mathfrak{O}_k\omega_n'.$$

By what we have shown, we must have $\mathfrak{c}^{-1} \equiv \mathfrak{c}N_{K/k}(\mathfrak{d}_{K/k})^{-1} = \mathfrak{c}D(K/k)^{-1}$. Hence we get $D(K/k) \equiv \mathfrak{c}^2$. This completes the proof.

We write $\mathfrak{c} = \mathfrak{c}(K/k)$. The ideal class of $\mathfrak{c}(K/k)$ depends only on K and k.

2.2. Now we turn to our theme of this section. Let F be a totally real algebraic number field of degree n. We fix integral ideals $\mathfrak{a}_1, \mathfrak{a}_2, \ldots, \mathfrak{a}_{h_F}$ of F which represent ideal classes of F. Let X denote the character group of the ideal class group of F. For $\chi \in X$, we put $E_{\chi,i}(z,s) = E_\chi(z,s;\mathfrak{a}_i)$, $h_{\chi,i}(z) = h_\chi(z;\mathfrak{a}_i)$. Let K be a totally imaginary quadratic extension of F. Let $\mathfrak{A}_1, \mathfrak{A}_2, \ldots, \mathfrak{A}_{h_K}$ be integral ideals of K which represent ideal classes of K. Let c_j be the ideal class of $\mathfrak{A}_j^{-1}$. Let

$$\zeta_K(s, c_j) = \sum_{\mathfrak{A} \in c_j,\, \mathfrak{A}:\text{integral}} N(\mathfrak{A})^{-s}$$

be the partial zeta function of the class c_j. We have

$$\zeta_K(s) = \sum_{j=1}^{h_K} \zeta_K(s, c_j). \tag{2.1}$$

We see easily that

$$\zeta_K(s,c_j) = \sum_{(\alpha) \subseteqq \mathfrak{A}_j} N(\mathfrak{A}_j^{-1}(\alpha))^{-s} = N(\mathfrak{A}_j)^s \sum_{(\alpha),\alpha\in\mathfrak{A}_j} N((\alpha))^{-s}.$$

By Proposition 2.1, taking $\mathfrak{a}_{i(j)}$ so that $\mathfrak{a}_{i(j)}$ and $\mathfrak{c}N_{K/F}(\mathfrak{A}_j)$, $\mathfrak{c} = \mathfrak{c}(K/F)$ belong to the same ideal class, we can write

$$\mathfrak{A}_j = \mathfrak{a}_{i(j)}\omega_1^{(j)} \oplus \mathfrak{O}_F \omega_2^{(j)}, \qquad 1 \leqq j \leqq h_K \tag{2.2}$$

with $\omega_1^{(j)}, \omega_2^{(j)} \in K$. For simplicity, we suppress the index j for a moment. We find

$$\sum_{(\alpha),\alpha\in\mathfrak{A}} N((\alpha))^{-s} = [E_K : E_F]^{-1} \sum_{(c,d)/\sim,\, c\in\mathfrak{a}_i,\, d\in\mathfrak{O}_F} N((c\omega_1 + d\omega_2))^{-s}.$$

Here $(0,0)$ is excluded in the sum and

$$(c,d) \sim (c_1,d_1) \Longleftrightarrow \exists\epsilon \in E_F, \quad c_1 = \epsilon c,\ d_1 = \epsilon d.$$

Put $w = \omega_1/\omega_2$. We choose a CM-type Φ of K so that $\Im(w^\sigma) > 0$ for every $\sigma \in \Phi$. We have

$$\begin{aligned} N((c\omega_1 + d\omega_2))^{-s} &= N((\omega_2))^{-s} N((cw+d))^{-s} \\ &= N((\omega_2))^{-s} \prod_{\sigma\in\Phi} ((c^\sigma w^\sigma + d^\sigma)(c^{\sigma\rho} w^{\sigma\rho} + d^{\sigma\rho}))^{-s} \\ &= N((\omega_2))^{-s} \prod_{\sigma\in\Phi} |c^\sigma w^\sigma + d^\sigma|^{-2s}. \end{aligned}$$

By (1.25), we obtain, writing Φ as Φ_j,

$$\begin{aligned} \zeta_K(s,c_j) = [E_K : E_F]^{-1} N(\mathfrak{A}_j)^s N((\omega_2^{(j)}))^{-s} \prod_{\sigma\in\Phi_j} \Im(w_j^\sigma)^{-s} \\ \times \frac{1}{h_F} \sum_{\chi\in X} \chi(a_{i(j)})^{-1} (N(\mathfrak{a}_{i(j)}))^{-s} E_{\chi,i(j)}(w_j, s). \end{aligned} \tag{2.3}$$

Here, by w_j, we denote the CM-point $(w_j^\sigma)_{\sigma\in\Phi_j} \in \mathfrak{H}^n$, $w_j^\sigma = (\omega_1^{(j)}/\omega_2^{(j)})^\sigma$.

Using the notation of (1.5) and (1.6), we have

$$\zeta_K(s,c_j) = -\frac{R_K}{w_K}s^{n-1}\{1 + (2n\gamma + 2n\log 2\pi - \log|D_K| - \frac{w_K|D_K|^{1/2}\gamma_K(c'_j)}{2^n\pi^n R_K})s\} + O(s^{n+1}).$$

By (2.3), (1.34) and (1.35), we have

$$\begin{aligned}\zeta_K(s,c_j) = & \frac{1}{h_F[E_K : E_F]} \\ & \times \{1 + (\log N(\mathfrak{A}_j) - \log N((\omega_2^{(j)})) - \log \prod_{\sigma\in\Phi_j} \Im(w_j^\sigma) - \log N(\mathfrak{a}_{i(j)}))s\} \\ & \times (-2^{n-2}h_F R_F s^{n-1}) \\ & \times \{1 + (2n\gamma + 2n\log 2\pi - 2\log|D_F| - \frac{D_F^{1/2}\gamma_F}{2^{n-2}h_F R_F} + \log N(\mathfrak{a}_{i(j)}) \\ & + \log \prod_{\sigma\in\Phi_j} \Im(w_j^\sigma) - h_{1,i(j)}(w_j))s \\ & - \sum_{1\neq\chi\in X} \chi(a_{i(j)})^{-1}(\frac{D_F^{1/2}}{2^{n-2}h_F R_F}\chi(d)\sum_c \chi(c^{-1})\gamma_F(c)\chi(a_{i(j)}) + h_{\chi,i(j)}(w_j))s\} \\ & + O(s^{n+1}).\end{aligned}$$

Since

$$\frac{R_K}{R_F} = \frac{2^{n-1}}{[E_K : W_K E_F]} = \frac{2^{n-2}w_K}{[E_K : E_F]},$$

we have

$$\sum_{\chi\in X}\chi(a_{i(j)})^{-1}\chi(d)\sum_c \chi(c)^{-1}\gamma_F(c)\chi(a_{i(j)}) = \sum_c \gamma_F(c)\sum_{\chi\in X}\chi(dc^{-1}) = h_F\gamma_F(d),$$

Hence we obtain

$$\begin{aligned}\zeta_K(s,c_j) = & -\frac{R_K}{w_K}s^{n-1}\{1 + (2n\gamma + 2n\log 2\pi - 2\log D_F + \log N(\mathfrak{A}_j) \\ & - \log N((\omega_2^{(j)})) - \sum_{\chi\in X}\chi(a_{i(j)}))^{-1}h_{\chi,i(j)}(w_j) - \frac{D_F^{1/2}}{2^{n-2}R_F}\gamma_F(d))s\} \\ & + O(s^{n+1}), \qquad s\to 0.\end{aligned} \tag{2.4}$$

To write this formula in an invariant form, let us prove:

LEMMA 2.3. *For $1 \leqq j \leqq h_K$, we have*

$$N(\mathfrak{A}_j)N((\omega_2^{(j)}))^{-1}(\prod_{\sigma\in\Phi_j}\Im(w_j^\sigma))^{-1}N(\mathfrak{a}_{i(j)})^{-1} = \frac{2^n D_F}{|D_K|^{1/2}}.$$

PROOF. We use the limit formulas (1.32) and (1.33). If χ is non-trivial, $E_{\chi,i(j)}(z,s)$ is holomorphic at $s=1$ and if $\chi=1$, we see

$$\mathrm{Res}_{s=1}E_{\chi,i(j)}(z,s)=\frac{2^{n-2}\pi^n h_F R_F}{D_F}.$$

Hence, by (2.3), we obtain

$$\begin{aligned}&\mathrm{Res}_{s=1}\zeta_K(s,c_j)\\ &=[E_K:E_F]^{-1}N(\mathfrak{A}_j)N((\omega_2^{(j)}))^{-1}\prod_{\sigma\in\Phi_j}\Im(w_j^\sigma)^{-1}N(\mathfrak{a}_{i(j)})^{-1}\times 2^{n-2}\pi^n\frac{R_F}{D_F}.\end{aligned}$$

Since

$$\mathrm{Res}_{s=1}\zeta_K(s,c_j)=\frac{2^n\pi^n R_K}{w_K|D_K|^{1/2}},$$

we get

$$\begin{aligned}&N(\mathfrak{A}_j)N((\omega_2^{(j)}))^{-1}\prod_{\sigma\in\Phi_j}\Im(w_j^\sigma)^{-1}N(\mathfrak{a}_{i(j)})^{-1}\\ &=\frac{D_F}{|D_K|^{1/2}}\cdot\frac{4[E_K:E_F]}{w_K}\cdot\frac{R_K}{R_F}=\frac{2^nD_F}{|D_K|^{1/2}}.\end{aligned}$$

This completes the proof.

Using Lemma 2.3, (2.4) can be written as

$$\begin{aligned}\zeta_K(s,c_j)=&-\frac{R_K}{w_K}s^{n-1}\{1+(2n\gamma+2n\log 2\pi+n\log 2-\log D_F\\ &-\frac{1}{2}\log|D_K|-\frac{D_F^{1/2}}{2^{n-2}R_F}\gamma_F(d)+\log N(\mathfrak{a}_{i(j)})+\log\prod_{\sigma\in\Phi_j}\Im(w_j^\sigma)\\ &-\sum_{\chi\in X}\chi(a_{i(j)}))^{-1}h_{\chi,i(j)}(w_j))s\}+O(s^{n+1}),\qquad s\to 0.\end{aligned}\tag{2.5}$$

In this form, we can check easily, using Proposition 1.4, that the right-hand side does not depend (mod $O(s^{n+1})$) on the choice of $\mathfrak{A}_j$, $\mathfrak{a}_i$, $\omega_1^{(j)}$, $\omega_2^{(j)}$, as it should be. Comparing (2.5) with the formula below (2.3), we obtain

$$\begin{aligned}\frac{w_K|D_K|^{1/2}\gamma_K(c_j')}{2^n\pi^nR_K}=&-n\log 2+\log D_F-\frac{1}{2}\log|D_K|+\frac{D_F^{1/2}}{2^{n-2}R_F}\gamma_F(d)\\ &-\log N(\mathfrak{a}_{i(j)})\log\prod_{\sigma\in\Phi_j}\Im(w_j^\sigma)+\sum_{\chi\in X}\chi(a_{i(j)})^{-1}h_{\chi,i(j)}(w_j).\end{aligned}\tag{2.6}$$

By (1.5), we obtain

$$\zeta_K(s)=-\frac{R_K}{w_K}s^{n-1}(h_K+\sum_{\tilde{c}}\delta_K(\tilde{c})s)+O(s^{n+1}),\qquad s\to 0,$$

$$\zeta_F(s)=-\frac{R_F}{2}s^{n-1}(h_F+\sum_{c}\delta_F(c)s)+O(s^{n+1}),\qquad s\to 0$$

where $\tilde{c}$ (resp. c) extends over ideal classes of K (resp. F), $\delta_K(\tilde{c})$ and $\delta_F(c)$ being given by (1.6). Let χ denote the Hecke character of $F_A^\times$ which corresponds to the quadratic extension K/F. We have[1]

$$\zeta_K(s) = \zeta_F(s)L_F(s,\chi).$$

Hence we have

$$L_F(0,\chi) = \frac{2R_K h_K}{w_K R_F h_F} = \frac{2^{n-1}}{[E_K : E_F]}\cdot\frac{h_K}{h_F},$$

$$\frac{L'_F(0,\chi)}{L_F(0,\chi)} = \frac{1}{h_K}\sum_{\tilde{c}}\delta_K(\tilde{c}) - \frac{1}{h_F}\sum_{c}\delta_F(c).$$

By (1.6), we obtain

$$\frac{L'_F(0,\chi)}{L_F(0,\chi)} = n\gamma + n\log 2\pi + \log D_F - \log|D_K| + \frac{D_F^{1/2}\gamma_F}{2^{n-1}h_F R_F} - \frac{w_K|D_K|^{1/2}\gamma_K}{2^n\pi^n h_K R_K}. \tag{2.7}$$

Then using (2.6), we obtain

$$\begin{aligned}\frac{L'_F(0,\chi)}{L_F(0,\chi)} &= n\gamma + n\log 4\pi - \frac{1}{2}\log|D_K| + \frac{D_F^{1/2}}{2^{n-1}h_F R_F}(\gamma_F - 2h_F\gamma_F(d))\\ &+ \frac{1}{h_K}\Big(\sum_{j=1}^{h_K}(\log N(\mathfrak{a}_{i(j)}) + \log\prod_{\sigma\in\Phi_j}\Im(w_j^\sigma)\Big)\\ &- \frac{1}{h_K}\sum_{j=1}^{h_K}\sum_{\eta\in X}\eta(a_{i(j)})^{-1}h_{\eta,i(j)}(w_j).\end{aligned} \tag{2.8}$$

LEMMA 2.4. *The norm map from the ideal class group of K to that of F is surjective.*

PROOF. Let C_K (resp. C_F) denote the ideal class group of K. Let D be the image of the norm map from C_K to C_F and let L be the abelian extension of F which corresponds to D. By the well known theorem of class field theory ([Wel], p. 276, Corollary 5), we have $L \vee K = K$. Hence $L \subset K$. Since all places of F are unramified in L and ∞_1 is ramified in K, we must have $L = F$. Therefore $D = C_F$ and we complete the proof.

By this Lemma, (2.8) can be formulated as the following theorem.

THEOREM 2.5. *Let $\mathfrak{a}_1, \mathfrak{a}_2, \ldots, \mathfrak{a}_{h_F}$ be integral ideals of F which represent ideal classes of F. For $\mathfrak{a}_i$, let $\mathfrak{A}_{i1}, \mathfrak{A}_{i2}, \ldots, \mathfrak{A}_{i\,h_K/h_F}$ be integral ideals of K, belonging to distinct ideal classes of K, such that the ideal class of $\mathfrak{c}N_{K/F}(\mathfrak{A}_{ij})$, $\mathfrak{c} = \mathfrak{c}(K/F)$ is that of $\mathfrak{a}_i$ for $1 \leqq j \leqq h_K/h_F$. Write*

$$\mathfrak{A}_{ij} = \mathfrak{a}_i\omega_1^{(ij)} \oplus \mathfrak{O}_F\omega_2^{(ij)}, \qquad w_{ij} = \omega_1^{(ij)}/\omega_2^{(ij)}$$

[1]Though we use the letter χ in another context, no confusion should be possible since the character χ corresponding to K/F will always appear in $L_F(s,\chi)$.

and choose the CM-type Φ_{ij} of K so that $\Im(w_{ij}^\sigma) > 0$ for every $\sigma \in \Phi_{ij}$. Then we have

$$\frac{L'_F(0,\chi)}{L_F(0,\chi)} = n\gamma + n\log 4\pi - \frac{1}{2}\log|D_K|$$

$$(2.9) \qquad + \frac{D_F^{1/2}}{2^{n-1}h_F R_F}(\gamma_F - 2h_F\gamma_F(d)) + \frac{1}{h_F}\sum_{j=1}^{h_F}\log N(\mathfrak{a}_i)$$

$$- \frac{1}{h_K}\left(\sum_{i=1}^{h_F}\sum_{j=1}^{h_K/h_F}\left(\sum_{\eta\in X}\eta(a_i)^{-1}h_{\eta,i}(w_{ij})\right) - \log\prod_{\sigma\in\Phi_{ij}}\Im(w_{ij}^\sigma)\right).$$

Here, for the argument of the function $h_{\eta,i}$, w_{ij} denotes the CM-point $(w_{ij}^\sigma)_{\sigma\in\Phi_{ij}} \in \mathfrak{H}^n$.

By III, Proposition 3.3, our conjecture on absolute CM-periods (III, Conjecture 3.2) implies that

$$(2.10) \qquad \exp\left(\frac{L'_F(0,\chi)}{L_F(0,\chi)}\right) \sim \pi^n \prod_{\sigma\in J_K} p_K(\sigma,\sigma).$$

COROLLARY 2.6. *The relation (2.10) is equivalent to*

$$\pi^n \prod_{\sigma\in J_K} p_K(\sigma,\sigma)$$

$$\sim \exp\left(n\gamma + \frac{D_F^{1/2}}{2^{n-1}h_F R_F}(\gamma_F - 2h_F\gamma_F(d)) - \frac{1}{h_K}\sum_{i=1}^{h_F}\sum_{j=1}^{h_K/h_F}\sum_{\eta\in X}\eta(a_i)^{-1}h_{\eta,i}(w_{ij})\right).$$

2.3. As an example, let us derive the exact Chowla-Selberg formula (cf. Weil [We2], p. 92). For a lattice L in $\mathbf{C}$, we choose ω_1, ω_2 so that $L = \mathbf{Z}\omega_1 \oplus \mathbf{Z}\omega_2$, $\Im(\omega_1/\omega_2) > 0$ and put $\eta^{24}(L) = \eta^{24}(\omega_1/\omega_2)\omega_2^{-12}$. We see easily that this is well defined and that $\eta(\alpha L) = \alpha^{-12}\eta^{24}(L)$ for $\alpha \in \mathbf{C}^\times$. Let K be an imaginary quadratic field and let $\mathfrak{A}_1, \mathfrak{A}_2, \ldots, \mathfrak{A}_{h_K}$ be a complete set of representatives of the ideal classes of K. Then the exact Chowla-Selberg formula is

$$(2.11) \qquad \exp\left(\frac{12h_K L'(0,\chi)}{L(0,\chi)}\right) = (2\pi)^{12h_K}\prod_{i=1}^{h_K} N(\mathfrak{A}_i)^{12}|\eta^{24}(\mathfrak{A}_i)|^2,$$

where χ is the Dirichlet character which corresponds to the quadratic extension $K/\mathbf{Q}$. To prove (2.11), we may assume that $\mathfrak{A}_i$, $1 \leqq i \leqq h_K$ are integral ideals. Choose $\mathfrak{a}_1 = (1)$. Then Theorem 2.5 specialized to the case $F = \mathbf{Q}$ gives

$$\frac{L'(0,\chi)}{L(0,\chi)} = \log 4\pi - \frac{1}{2}\log d - \frac{1}{h_K}\left(\sum_{i=1}^{h_K} h(w_i) - \log\Im(w_i)\right).$$

Here we put $d = -D_K$, $\mathfrak{A}_i = \mathbf{Z}\omega_1^{(i)} \oplus \mathbf{Z}\omega_2^{(i)}$, $w_i = \omega_1^{(i)}/\omega_2^{(i)}$, $\Im(w_i) > 0$. By (1.36), we obtain

$$\exp\left(\frac{12h_K L'(0,\chi)}{L(0,\chi)}\right) = (4\pi)^{12h_K} d^{-6h_K}\prod_{i=1}^{h_K}|\eta^{24}(w_i)|^2\Im(w_i)^{12}.$$

Therefore (2.11) follows if we can show

$$N(\mathfrak{A}_i)|\omega_2^{(i)}|^{-2}\Im(w_i)^{-1} = 2d^{-1/2}, \qquad 1 \leqq i \leqq h_K.$$

This equality follows from Lemma 2.3 noting $N((\omega_2^{(i)})) = |\omega_2^{(i)}|^2$. Thus (2.11) is a special case of Theorem 2.5.

2.4. As the next example, let us consider the case where F is a real quadratic field. Let ϵ_0 denote the fundamental unit of F. Let ψ be the Dirichlet character which corresponds to the quadratic extension $F/\mathbf{Q}$. By $\zeta_F(s) = \zeta(s)L(s,\psi)$, (1.5) and (1.6), we obtain

$$L''(0,\psi) = 2h_F \log \epsilon_0 (2\gamma + \log 2\pi - \log D_F) - \sqrt{D_F}\gamma_F.$$

Hence we get

$$\begin{aligned} &2\gamma + 2\log 4\pi - \frac{1}{2}\log|D_K| + \frac{\sqrt{D_F}}{2h_F \log \epsilon_0}(\gamma_F - 2h_F\gamma_F(d)) \\ &= \log D_F - \frac{1}{2}\log|D_K| + \log 8\pi + \frac{1}{2h_F \log \epsilon_0}L''(0,\psi) \\ &+ \frac{\sqrt{D_F}}{h_F \log \epsilon_0}(\sum_c(\gamma_F(c) - \gamma_F(1))), \end{aligned}$$

noting that $\gamma_F(d) = \gamma_F(1)$, where $\gamma_F(1)$ denotes the constant attached to the identity class 1. By (1.5) and (1.6), we have

$$\gamma_F(c) - \gamma_F(1) = \frac{2}{\sqrt{D_F}}(\zeta_F''(0,c^{-1}) - \zeta_F''(0,1)). \tag{2.12}$$

By Theorem 2.5, we obtain

$$\begin{aligned} \frac{L_F'(0,\chi)}{L_F(0,\chi)} &= \log D_F - \frac{1}{2}\log|D_K| + \log 8\pi + \frac{1}{2h_F\log\epsilon_0}L''(0,\psi) \\ &+ \frac{\sqrt{D_F}}{h_F\log\epsilon_0}(\sum_c(\gamma_F(c) - \gamma_F(1))) + \frac{1}{h_F}\sum_{i=1}^{h_F}\log N(\mathfrak{a}_i) \\ &- \frac{1}{h_K}(\sum_{i=1}^{h_F}\sum_{j=1}^{h_K/h_F}(\sum_{\eta\in X}\eta(a_i)^{-1}h_{\eta,i}(w_{ij})) - \log\prod_{\sigma\in\Phi_{ij}}\Im(w_{ij}^{\sigma})). \end{aligned} \tag{2.13}$$

EXAMPLE 2.7. We take $F = \mathbf{Q}(\sqrt{5})$, $K = \mathbf{Q}(\sqrt{\frac{13+\sqrt{5}}{2}}i)$. This example was considered in [Y3], §4 and [Y4], §6. We have $h_F = h_K = 1$, $\epsilon_0 = \frac{\sqrt{5}+1}{2}$. The prime ideal $(\frac{13+\sqrt{5}}{2})$ of F ramifies in K; put $\mathfrak{P}^2 = (\frac{13+\sqrt{5}}{2})$. Then we have $\mathfrak{d}_{K/F} = \mathfrak{P}$. Since $\mathfrak{d}_{F/\mathbf{Q}} = (\sqrt{5})$, we obtain $|D_K| = 5^2 \cdot 41$. We have

$$\mathfrak{O}_K = \mathfrak{O}_F(\sqrt{\frac{13+\sqrt{5}}{2}}i + \frac{1-\sqrt{5}}{2})/2 \oplus \mathfrak{O}_F.$$

Choose the CM-type Φ as $\Phi = \{\mathrm{id}, \sigma\}$, where $\sigma \in J_K$ is defined by $(\sqrt{\frac{13+\sqrt{5}}{2}}\, i)^\sigma = \sqrt{\frac{13-\sqrt{5}}{2}}\, i$. Take $\mathfrak{a}_1 = (1)$ and put $h_{11} = h$. Then (2.13) gives

$$\frac{L_F'(0,\chi)}{L_F(0,\chi)} = \log 2\pi + \frac{1}{2\log \epsilon_0} L''(0,\psi) - h(w), \tag{2.14}$$

where w denotes the CM-point

$$\left(\frac{1}{2}\left(\sqrt{\frac{13+\sqrt{5}}{2}}\, i + \frac{1-\sqrt{5}}{2}\right),\ \frac{1}{2}\left(\sqrt{\frac{13-\sqrt{5}}{2}}\, i + \frac{1+\sqrt{5}}{2}\right)\right) \in \mathfrak{H}^2.$$

The numerical value of the left-hand side can be computed by III, (4.9), (4.10). We can compute $L''(0,\psi)$ using the Euler-Maclaurin summation. It gives

$$L''(0,\psi) = -0.02119705158977416325\ldots .$$

Using (1.30), it is not difficult to compute $h(w)$. We obtain

$$h(w) = 1.94864260531062239563\ldots .$$

In this way, we can check (2.14) numerically.

EXAMPLE 2.8. We take $F = \mathbf{Q}(\sqrt{10})$, $K = \mathbf{Q}(\sqrt{9+2\sqrt{10}}\, i)$. We have $h_F = 2$, $\epsilon_0 = 3 + \sqrt{10}$. Put $\mathfrak{p}_5 = \mathbf{Z}\cdot\sqrt{10} + \mathbf{Z}\cdot 5$. Then $\mathfrak{p}_5^2 = (5)$ and we can take (1) and $\mathfrak{p}_5$ as representatives of the narrow ideal classes of F. Put $\mathfrak{p}_{41} = (9+2\sqrt{10})$, $\mathfrak{p}_{41}' = (9-2\sqrt{10})$. Then we have $\mathfrak{p}_{41}\mathfrak{p}_{41}' = (41)$. Put $\mathfrak{f} = \mathfrak{p}_{41}$ and let $C_\mathfrak{f}$ denote the ideal class group modulo $\mathfrak{f}\infty_1\infty_2$. We see easily that

$$C_\mathfrak{f} = \{c_1,\ c_2,\ c_3,\ c_4\} \cong \mathbf{Z}/2\mathbf{Z} \oplus \mathbf{Z}/2\mathbf{Z},$$

where

$$c_1 = \text{the class of } (1), \qquad c_2 = \text{the class of } (6),$$
$$c_3 = \text{the class of } \mathfrak{p}_5, \qquad c_4 = \text{the class of } (6)\mathfrak{p}_5.$$

Computing by III, (4.9), we get

$$\zeta_F(0,c_1) = 3, \qquad \zeta_F(0,c_2) = -3, \qquad \zeta_F(0,c_3) = -1, \qquad \zeta_F(0,c_4) = 1. \tag{2.15}$$

Let χ be the character of $C_\mathfrak{f}$ defined by

$$\chi(c_1) = 1, \qquad \chi(c_2) = -1, \qquad \chi(c_3) = 1, \qquad \chi(c_4) = -1.$$

Let us show that K is the class field over F which corresponds to χ. For this purpose, put

$$z = \frac{1}{2}(\sqrt{9+2\sqrt{10}}\, i + 1 - \sqrt{10}).$$

Then z satisfies the equation

$$z^2 - (1-\sqrt{10})z + 5 = 0.$$

This implies that only $\mathfrak{f}$ can ramify in K/F. But K cannot be unramified over F since $h_F = 2$, $N(\epsilon_0) = -1$. We conclude that the conductor of K as a class field over F is $\mathfrak{f}\infty_1\infty_2$ and also that

$$\mathfrak{O}_K = \frac{1}{2}(\sqrt{9+2\sqrt{10}}\,i + 1 - \sqrt{10}) \cdot \mathfrak{O}_F \oplus \mathfrak{O}_F. \tag{2.16}$$

Let χ' be the character of $C_{\mathfrak{f}}$ which corresponds to K/F. Since

$$\chi'((\alpha)) = (\frac{\alpha}{\mathfrak{f}})\,\mathrm{sgn}\,(\alpha)\,\mathrm{sgn}(\alpha'), \qquad (\alpha) \in C_{\mathfrak{f}},$$

we have $\chi'(c_2) = -1$. By $z^2 - z \equiv 0 \mod \mathfrak{p}_5$, we see that $\mathfrak{p}_5$ splits completely in K. Therefore we get $\chi'(c_3) = 1$, $\chi = \chi'$. By (2.15), we obtain $L(0,\chi) = 4$, $h_K = 4$. Since $(D_K) = N_{K/F}(D(K/F))(D_F)^2$, we obtain

$$|D_K| = 40^2 \cdot 41. \tag{2.17}$$

Let us show that $\mathbf{Q}(\sqrt{2},\sqrt{5})$ is the Hilbert class field of $F = \mathbf{Q}(\sqrt{10})$. Put $L = \mathbf{Q}(\sqrt{2},\sqrt{5})$. For a quadratic subfield M of L, we have

$$(D_L) = N_{M/\mathbf{Q}}(D(L/M))(D_M^2). \tag{2.18}$$

Take $M = \mathbf{Q}(\sqrt{5})$. Since only (2) can ramify in L/M, we have $D_L = 2^\alpha 5^2$, $\alpha \in \mathbf{Z}$. Taking $M = \mathbf{Q}(\sqrt{2})$, we see that $D_L = 2^6 5^\beta$, $\beta \in \mathbf{Z}$. Hence we get $D_L = 2^6 5^2$. Now take $M = \mathbf{Q}(\sqrt{10})$. We get

$$(2^6 5^2) = (N_{\mathbf{Q}(\sqrt{10})/\mathbf{Q}}(D(L/\mathbf{Q}(\sqrt{10})))40^2).$$

Hence $D(L/\mathbf{Q}(\sqrt{10})) = (1)$ and L is unramified over $\mathbf{Q}(\sqrt{10})$.

Let ψ, ψ_1 and ψ_2 be the Dirichlet characters which correspond to $\mathbf{Q}(\sqrt{10})$, $\mathbf{Q}(\sqrt{5})$ and $\mathbf{Q}(\sqrt{2})$ respectively. By II, (1.2) and $\zeta_{\mathbf{Q}}(0) = -1/2$, we have

$$L(0,\psi_1) = L(0,\psi_2) = 0, \qquad L'(0,\psi_1) = \log(\frac{1+\sqrt{5}}{2}), \qquad L'(0,\psi_2) = \log(1+\sqrt{2}).$$

Let c be the non-trivial class of the ideal class group of F. Then $\zeta(s,1) - \zeta_F(s,c)$ is the L-function of F which corresponds to L/F. Considering the induced representation, we get

$$\zeta(s,1) - \zeta_F(s,c) = L(s,\psi_1)L(s,\psi_2).$$

Therefore we get

$$\zeta_F''(0,1) - \zeta_F''(0,c) = 2\log(\frac{1+\sqrt{5}}{2})\log(1+\sqrt{2}).$$

By (2.12), we obtain

$$\gamma_F(c) - \gamma_F(1) = -\frac{2}{\sqrt{10}}\log(\frac{1+\sqrt{5}}{2})\log(1+\sqrt{2}).$$

Therefore (2.13) takes the form

$$\begin{aligned}\frac{L_F'(0,\chi)}{L_F(0,\chi)} =& \frac{1}{2}\log 5 - \frac{1}{2}\log 41 + \log 8\pi + \frac{1}{4\log(3+\sqrt{10})}L''(0,\psi) \\ &- \frac{2}{\log(3+\sqrt{10})}\log(\frac{1+\sqrt{5}}{2})\log(1+\sqrt{2}) \\ &- \frac{1}{4}(\sum_{i=1}^{2}\sum_{j=1}^{2}(\sum_{\eta\in X}\eta(a_i)^{-1}h_{\eta,i}(w_{ij}) - \log\prod_{\sigma\in\Phi_{ij}}\Im(w_{ij}^{\sigma}))),\end{aligned} \tag{2.19}$$

noting $\mathfrak{a}_1 = (1)$, $\mathfrak{a}_2 = \mathfrak{p}_5$.

We are going to give w_{ij} explicitly. Let $\mathfrak{p}_5 = \mathfrak{P}_5\mathfrak{P}_5^{\rho}$ be the decomposition of $\mathfrak{p}_5$ in K. Let us first show that $\mathfrak{p}_5\mathfrak{O}_K$ is not principal. Suppose that $\mathfrak{p}_5\mathfrak{O}_K = (\alpha)$, $\alpha \in \mathfrak{O}_K$. Since $(\alpha^{\rho}) = (\alpha)$, we have $\alpha^{\rho}/\alpha = \epsilon \in E_K$. For every $\sigma \in J_K$, we have $\epsilon^{\sigma} = \alpha^{\rho\sigma}/\alpha^{\sigma} = \alpha^{\sigma\rho}/\alpha^{\sigma}$, $|\epsilon^{\sigma}| = 1$. Therefore $\epsilon \in W_K = \{1,-1\}$. If $\epsilon = 1$, we get $\alpha \in \mathfrak{O}_F$. Then $\mathfrak{p}_5 = (\alpha)$ in F, which is a contradiction. If $\epsilon = -1$, we get $\alpha^{\rho} = -\alpha$, $\alpha = \sqrt{9+2\sqrt{10}}\,i\times\beta$ with $\beta \in \mathfrak{O}_F$. Since $\mathfrak{p}_5^2 = (5)$, this is a contradiction. Therefore $\mathfrak{p}_5\mathfrak{O}_K$ is not principal; hence both of $\mathfrak{P}_5$ and $\mathfrak{P}_5^{\rho}$ are not principal.

Since $zz^{\rho} = 5$, there are three possibilities:

$$(z) = \mathfrak{P}_5^2 \qquad \text{or} \qquad (\mathfrak{P}_5^{\rho})^2 \qquad \text{or} \qquad \mathfrak{P}_5\mathfrak{P}_5^{\rho}.$$

If $(z) = \mathfrak{P}_5\mathfrak{P}_5^{\rho}$, this contradicts the fact that $\mathfrak{p}_5\mathfrak{O}_K$ is not principal. Changing $\mathfrak{P}_5^{\rho}$ to $\mathfrak{P}_5$ if necessary, we may assume that $(z) = \mathfrak{P}_5^2$. The class of $\mathfrak{P}_5^{\rho}$ is different from the class of $\mathfrak{P}_5$. Since $h_K = 4$, we conclude that the ideal class group $C_K \cong \mathbf{Z}/2\mathbf{Z} \oplus \mathbf{Z}/2\mathbf{Z}$ and that $\{(1), \mathfrak{P}_5, \mathfrak{P}_5^{\rho}, \mathfrak{P}_5\mathfrak{P}_5^{\rho}\}$ gives a complete set of representatives of the ideal classes of K.

We have shown already that $\mathfrak{O}_K = \mathfrak{O}_F \cdot z \oplus \mathfrak{O}_F$. Hence we may take $\mathfrak{c} = (1)$. We may set

$$\mathfrak{A}_{11} = (1), \qquad \mathfrak{A}_{12} = \mathfrak{p}_5\mathfrak{O}_K = \mathfrak{P}_5\mathfrak{P}_5^{\rho}, \qquad \mathfrak{A}_{21} = \mathfrak{P}_5, \qquad \mathfrak{A}_{22} = \mathfrak{P}_5^{\rho}.$$

Put $\mathfrak{P} = \mathfrak{p}_5 \oplus \mathfrak{O}_F z$. We see easily that $\mathfrak{P}$ is an integral ideal of K. Since $z \equiv 0 \bmod \mathfrak{P}_5$ and $\mathfrak{p}_5 \subset \mathfrak{P}_5$, we have $\mathfrak{P} \subset \mathfrak{P}_5$. From $[\mathfrak{O}_K : \mathfrak{P}] = [\mathfrak{O}_K : \mathfrak{P}_5] = 5$, we get $\mathfrak{P} = \mathfrak{P}_5$. Thus we obtain

$$\mathfrak{P}_5 = \mathfrak{p}_5 \oplus \mathfrak{O}_F z, \qquad \mathfrak{P}_5^{\rho} = \mathfrak{p}_5 \oplus \mathfrak{O}_F z^{\rho}. \tag{2.20}$$

Next put $\mathfrak{A} = \mathfrak{O}_F(5z+\sqrt{10}) \oplus \mathfrak{O}_F(2\sqrt{10}\,z+5)$. We can verify easily that $\mathfrak{A}$ is an integral ideal of K. Since $5z$, $\sqrt{10}$, $2\sqrt{10}z$, $5 \in \mathfrak{p}_5\mathfrak{O}_K$, we have $\mathfrak{A} \subset \mathfrak{p}_5\mathfrak{O}_K$. We have

$$[\mathfrak{O}_K : \mathfrak{A}] = |N_{F/\mathbf{Q}}(\det\begin{pmatrix} 5 & \sqrt{10} \\ 2\sqrt{10} & 5\end{pmatrix}))| = 25.$$

Hence we get $\mathfrak{A} = \mathfrak{p}_5\mathfrak{O}_K$. Thus we obtain

$$\mathfrak{p}_5\mathfrak{O}_K = \mathfrak{P}_5\mathfrak{P}_5^{\rho} = \mathfrak{O}_F(5z+\sqrt{10}) \oplus \mathfrak{O}_F(2\sqrt{10}\,z+5). \tag{2.21}$$

Therefore we obtain

$$w_{11} = \left(\frac{1}{2}(\sqrt{9+2\sqrt{10}\,i}+1-\sqrt{10}),\ \frac{1}{2}(\sqrt{9-2\sqrt{10}\,i}+1+\sqrt{10})\right),$$

$$w_{12} = \left(\frac{5\cdot\frac{1}{2}(\sqrt{9+2\sqrt{10}\,i}+1-\sqrt{10})+\sqrt{10}}{2\sqrt{10}\cdot\frac{1}{2}(\sqrt{9+2\sqrt{10}\,i}+1-\sqrt{10})+5},\right.$$
$$\left.\frac{5\cdot\frac{1}{2}(\sqrt{9-2\sqrt{10}\,i}+1+\sqrt{10})-\sqrt{10}}{-2\sqrt{10}\cdot\frac{1}{2}(\sqrt{9-2\sqrt{10}\,i}+1+\sqrt{10})+5}\right),$$

$$w_{21} = \left(\frac{-1}{\frac{1}{2}(\sqrt{9+2\sqrt{10}\,i}+1-\sqrt{10})},\ \frac{-1}{\frac{1}{2}(\sqrt{9-2\sqrt{10}\,i}+1+\sqrt{10})}\right),$$

$$w_{22} = \left(\frac{1}{\frac{1}{2}(-\sqrt{9+2\sqrt{10}\,i}+1-\sqrt{10})},\ \frac{1}{\frac{1}{2}(-\sqrt{9-2\sqrt{10}\,i}+1+\sqrt{10})}\right).$$

We have

$$L(2,\psi) = (40)^{-3/2}\cdot 28\pi^2, \qquad L(2,\psi_1) = 5^{-5/2}\cdot 4\pi^2, \qquad L(2,\psi_2) = 8^{-3/2}\cdot 2\pi^2.$$

Hence, for $\eta \in X$, we have

$$\zeta_F(2) = L(2,\eta) = \frac{\pi^2}{6}\cdot(40)^{-3/2}\cdot 28\pi^2, \qquad \eta = 1,$$
$$L(2,\eta) = L(2,\psi_1)L(2,\psi_2) = 5^{-1}\cdot(40)^{-3/2}\cdot 8\pi^4, \qquad \eta \neq 1.$$

By (1.30), we can compute $h_{\eta,i}(w_{ij})$ numerically. The left-hand side of (2.19) can be computed by III, (4.9), (4.10). It gives

$$\frac{L_F'(0,\chi)}{L_F(0,\chi)} = -2.03945539998326080763\dots.$$

By the Euler-Maclaurin summation, we get

$$L''(0,\psi) = 3.16227766016837933199\dots.$$

In this way, we can check (2.19) numerically. (A technical remark: It is inconvenient to compute $h_{\eta,1}(w_{12})$ directly since $(\Im(w_{12}))^{\mathbf{a}}$ is small. It is better to compute $h_{\eta,1}(w_{12}')$ for $w_{12}' = \begin{pmatrix} 3+\sqrt{10} & 0 \\ 0 & (3+\sqrt{10})^{-1}\end{pmatrix}\begin{pmatrix} 3+\sqrt{10} & -1 \\ -(2+\sqrt{10}) & 1\end{pmatrix} w_{12}$ and recover $h_{\eta,1}(w_{12})$ using the automorphic property given by Proposition 1.4.)

3. L-functions of orders of an algebraic number field

3.1. By Corollary 2.6, it became clear that the sum of values of $h_{\chi,i}$ at certain CM-points are related to CM-periods, at least conjecturally. We wish to understand the arithmetical nature of $h_{\chi,i}$ at every CM-point. This motivates our investigation on the L-functions of orders.

Let K be an algebraic number field. A subring $\mathfrak{O}$ of K is called *an order* of K if $\mathfrak{O} \ni 1$ and $\mathfrak{O}$ is a $\mathbf{Z}$-lattice in K. A subset $\mathfrak{A}$ of K is called *an* $\mathfrak{O}$*-ideal* if $\mathfrak{A}$ is a $\mathbf{Z}$-lattice in K and $\mathfrak{O}\mathfrak{A} \subset \mathfrak{A}$; $\mathfrak{A}$ is called *proper* if

$$\mathfrak{O} = \{x \in K \mid x\mathfrak{A} \subset \mathfrak{A}\}.$$

As usual, two $\mathfrak{O}$-ideals $\mathfrak{A}_1$ and $\mathfrak{A}_2$ belong to the same class if and only if $\mathfrak{A}_1 = \alpha\mathfrak{A}_2$ with some $\alpha \in K^\times$.

Let F be an algebraic number field and K be a quadratic extension of F. In this section, we consider an order $\mathfrak{O}$ of K which contains $\mathfrak{O}_F$.

PROPOSITION 3.1. *Let $\mathfrak{O}$ be an order of K which contains $\mathfrak{O}_F$. Then there exists an integral ideal $\mathfrak{f}$ of F such that $\mathfrak{O} = \mathfrak{O}_F + \mathfrak{f}\mathfrak{O}_K$. A* **Z***-lattice $\mathfrak{A}$ is a proper $\mathfrak{O}$-ideal if and only if $\mathfrak{A} = a\mathfrak{O}$ with $a \in K_A^\times$.*

PROOF. This is well known (cf. Shimura [Sh1], Proposition 4.11, (5.4.2)). We sketch a proof for the convenience of the reader. Let v be a finite place of F and put $R = \mathfrak{O}_{F_v}$, $\mathfrak{O}_v = \mathfrak{O} \otimes_{\mathfrak{O}_F} R$, $K_v = K \otimes_F F_v$, $\mathfrak{O}_{K_v} = \mathfrak{O}_K \otimes_{\mathfrak{O}_F} R$. Let ϖ_v be a prime element of F_v. Since R is a principal ideal domain, we see that $\mathfrak{O}_v$ is a free R-module of rank 2. Hence we have $\mathfrak{O}_v = Rx_1 \oplus Rx_2$ with $x_1,\, x_2 \in \mathfrak{O}_v$. Since $1 \in \mathfrak{O}_v$, there exist $c,\, d \in R$ such that $cx_1 + dx_2 = 1$. If ϖ_v divides c and d, then $\mathfrak{O}_v$ contains ϖ_v^{-1}, hence $\mathfrak{O}_v \supset F_v$, which contradicts the fact that $\mathfrak{O}_v$ is finitely generated over R. Since c and d are coprime, we can find $\begin{pmatrix} a & b \\ c & d \end{pmatrix} \in SL(2, R)$. Then we have $\mathfrak{O}_v = Rx \oplus R$ with $x = ax_1 + bx_2$. Applying this argument to $\mathfrak{O}_{K_v}$, we have $\mathfrak{O}_{K_v} = Ry \oplus R$ with $y \in \mathfrak{O}_{K_v}$. Since $\mathfrak{O}_v \subset \mathfrak{O}_{K_v}$, we have $x = c_v y + d_v$ with $c_v,\, d_v \in R$. Then $\mathfrak{O}_v = Rc_v y \oplus R$. Therefore we obtain $\mathfrak{O}_v = R + c_v\mathfrak{O}_{K_v}$. Since $\mathfrak{O}_v = \mathfrak{O}_{K_v}$ for almost all v, we can take $c_v = 1$ for almost all v and define $c \in F_A^\times$ so that $c_\infty = 1$ and the v-component of c is c_v for every finite place v. Put $\mathfrak{f} = \mathrm{div}(c)$, which is an integral ideal of F. Then we see that $\mathfrak{O} = \mathfrak{O}_F + \mathfrak{f}\mathfrak{O}_K$.

Let $\mathfrak{A}$ be a proper $\mathfrak{O}$-ideal. Let v be a finite place of F. Put $\mathfrak{A}_v = \mathfrak{A} \otimes_{\mathfrak{O}} \mathfrak{O}_v$. It suffices to show that $\mathfrak{A}_v$ is a principal $\mathfrak{O}_v$-ideal. It is not difficult to see that

$$(*) \qquad \mathfrak{O}_v = \{x \in K_v \mid x\mathfrak{A}_v \subset \mathfrak{A}_v\}.$$

We have $\mathfrak{O}_{K_v}\mathfrak{A}_v = \alpha\mathfrak{O}_{K_v}$ with $\alpha \in K_v^\times$. Changing $\mathfrak{A}_v$ to $\alpha^{-1}\mathfrak{A}_v$, we may assume $\mathfrak{O}_{K_v}\mathfrak{A}_v = \mathfrak{O}_{K_v}$. Then $\mathfrak{A}_v \subset \mathfrak{O}_{K_v}$ and $\mathfrak{A}_v$ contains a unit $u \in \mathfrak{O}_{K_v}^\times$. Changing $\mathfrak{A}_v$ to $u^{-1}\mathfrak{A}_v$, we may assume that $\mathfrak{A}_v \subset \mathfrak{O}_{K_v}$, $1 \in \mathfrak{A}_v$. Then we have $\mathfrak{O}_v \subset \mathfrak{A}_v \subset \mathfrak{O}_{K_v}$. Since $1 \in \mathfrak{A}_v$, we can write $\mathfrak{A}_v = Rty \oplus R$, $\mathfrak{O}_{K_v} = Ry \oplus R$ with $t \in R$ by the same argument as above. Then we have $\mathfrak{A}_v = R + t\mathfrak{O}_{K_v}$. Therefore $\mathfrak{A}_v$ coincides with an order and we obviously have $\mathfrak{A}_v = \mathfrak{O}_v$ by $(*)$. This completes the proof.

We call the ideal $\mathfrak{f}$ *the conductor* of $\mathfrak{O}$. Let v be a finite place of F. Put $K_v = K \otimes_F F_v$. If v remains prime or ramifies in K, let w be the place of K lying above v. Then $K_w = K_v$, which is a quadratic extension of F_v and $\mathfrak{O}_v$ is an order of K_w. We have $\mathfrak{O}_{K_w} = \mathfrak{O}_{K_v}$. Assume that v splits in K and let w_1 and w_2 be two places of K lying above v. Then $K_{w_1} \cong K_{w_2} \cong F_v$, $K_v \cong F_v \oplus F_v$. By this isomorphism, $\mathfrak{O}_{K_v}$ is mapped onto $\mathfrak{O}_{F_v} \oplus \mathfrak{O}_{F_v}$. We can regard $\prod_v \mathfrak{O}_v^\times$ as a subgroup of $(K_A^\times)_f$.

PROPOSITION 3.2. *Let $h_{\mathfrak{O}}$ denote the number of proper $\mathfrak{O}$-ideal classes of $\mathfrak{O}$. We have*

$$h_{\mathfrak{O}} = \frac{h_K}{[E_K : E_K \cap \mathfrak{O}]} \prod_v [\mathfrak{O}_{K_v}^\times : \mathfrak{O}_v^\times],$$

where v extends over all finite places of F and $\mathfrak{O}_{K_v} = \mathfrak{O}_K \otimes_{\mathfrak{O}_F} \mathfrak{O}_{F_v}$, $\mathfrak{O}_v = \mathfrak{O} \otimes_{\mathfrak{O}_F} \mathfrak{O}_{F_v}$.

PROOF. Let $\mathbf{h}$ denote the set of all finite places of F. Consider the canonical homomorphism

$$K_A^\times/K^\times \prod_{v\in\mathbf{h}} \mathfrak{O}_v^\times K_\infty^\times \longrightarrow K_A^\times/K^\times \prod_{v\in\mathbf{h}} \mathfrak{O}_{K_v}^\times K_\infty^\times.$$

By Proposition 3.1, the cardinality of the left-hand side is $h_\mathfrak{O}$. Therefore we have

$$h_\mathfrak{O} = h_K \times |K^\times \prod_{v\in\mathbf{h}} \mathfrak{O}_{K_v}^\times K_\infty^\times / K^\times \prod_{v\in\mathbf{h}} \mathfrak{O}_v^\times K_\infty^\times|.$$

Define a mapping

$$\psi : \prod_{v\in\mathbf{h}} (\mathfrak{O}_{K_v}^\times/\mathfrak{O}_v^\times) \longrightarrow K^\times \prod_{v\in\mathbf{h}} \mathfrak{O}_{K_v}^\times K_\infty^\times / K^\times \prod_{v\in\mathbf{h}} \mathfrak{O}_v^\times K_\infty^\times$$

by $\psi((x_v \mod \mathfrak{O}_v^\times)_v) = (x_v)_v \mod K^\times \prod_{v\in\mathbf{h}} \mathfrak{O}_v^\times K_\infty^\times$, taking $x_w = 1$ for every infinite place w of K. Then ψ is a surjective homomorphism and we see that

$$\ker\psi \cong \frac{(\prod_{v\in\mathbf{h}} \mathfrak{O}_{K_v}^\times \times K_\infty^\times) \cap (K^\times \prod_{v\in\mathbf{h}} \mathfrak{O}_v^\times K_\infty^\times)}{(\prod_v \mathfrak{O}_{v\in\mathbf{h}}^\times \times K_\infty^\times) \cap (K^\times \prod_{v\in\mathbf{h}} \mathfrak{O}_v^\times K_\infty^\times)} \cong E_K/(E_K \cap \mathfrak{O}).$$

Hence the assertion follows.

3.2. For two proper $\mathfrak{O}$-ideals $\mathfrak{A}_1 = a_1\mathfrak{O}$, $\mathfrak{A}_2 = a_2\mathfrak{O}$, a_1, $a_2 \in K_A^\times$, we define their product by $\mathfrak{A}_1\mathfrak{A}_2 = a_1a_2\mathfrak{O}$. Then the classes of proper $\mathfrak{O}$-ideals make a group isomorphic to $K_A^\times/K^\times \prod_{v\in\mathbf{h}} \mathfrak{O}_v^\times K_\infty^\times$, where $\mathbf{h}$ denotes the set of all finite places of F. Let χ be a character of $K_A^\times$ which is trivial on $K^\times \prod_{v\in\mathbf{h}} \mathfrak{O}_v^\times K_\infty^\times$. We define the character χ_* of the group of proper $\mathfrak{O}$-ideal classes by

(3.1) $$\chi_*(a\mathfrak{O}) = \chi(a), \qquad a \in K_A^\times.$$

We define the L-function of χ_* by

(3.2) $$L(s,\chi_*) = \sum_{\mathfrak{A}} \chi_*(\mathfrak{A}) N(\mathfrak{A})^{-s}, \qquad \Re(s) > 1,$$

where $\mathfrak{A}$ extends over all proper $\mathfrak{O}$-ideals contained in $\mathfrak{O}$ and $N(\mathfrak{A}) = |\mathfrak{O}/\mathfrak{A}|$. We are going to compare $L(s,\chi_*)$ with $L(s,\chi)$. The convergence of (3.2) when $\Re(s) > 1$ follows from that of (3.3), which will become clear by our calculations below.

By Proposition 3.1, proper $\mathfrak{O}$-ideals contained in $\mathfrak{O}$ are in one-to-one correspondence with $(\prod_{v\in\mathbf{h}} \mathfrak{O}_v \cap K_A^\times)/\prod_{v\in\mathbf{h}} \mathfrak{O}_v^\times$. If $\mathfrak{A} = a\mathfrak{O}$ with $a = (a_v)_v \in \prod_{v\in\mathbf{h}} \mathfrak{O}_v \cap K_A^\times$, then we have $N(\mathfrak{A}) = \prod_{v\in\mathbf{h}} |\mathfrak{O}_v/a_v\mathfrak{O}_v| = \prod_{v\in\mathbf{h}} \|a_v\|_v^{-1}$. Here $\|a_v\|_v$ denotes the module of the automorphism $x \longrightarrow a_v x$ of K_v. Therefore we obtain

(3.3) $$L(s,\chi_*) = \prod_{v\in\mathbf{h}} \sum_{a\in\mathfrak{O}_v\cap K_v^\times/\mathfrak{O}_v^\times} \chi_v(a)\|a\|_v^s,$$

where χ_v denotes the restriction of χ to $K_v^\times \subset K_A^\times$. Put

(3.4) $$I_v = \sum_{a\in\mathfrak{O}_v\cap K_v^\times/\mathfrak{O}_v^\times} \chi_v(a)\|a\|_v^s.$$

If v does not divide $\mathfrak{f}$, then we have $\mathfrak{O}_v = \mathfrak{O}_{K_v}$ and I_v coincides with the v-factor of $L(s,\chi)$. Our task is to compute I_v when v divides $\mathfrak{f}$. We fix a finite place v of F which divides $\mathfrak{f}$. Let ϖ be a prime element of F_v. Put $R = \mathfrak{O}_{F_v}$ and let

$q = |R/\varpi R|$. Let $\mathfrak{f}R = \varpi^c R$ with a positive integer c. We must distinguish three cases.

Case I. The case where v is completely decomposed in K.

Let w_1 and w_2 be two distinct places of K lying above v. We may assume that $K \subset F_v$ and that the place w_1 is obtained by the restriction of v to K. We have

$$(*) \qquad K_v = K \otimes_F F_v \cong K_{w_1} \oplus K_{w_2} \cong F_v \oplus F_v,$$

where $a \otimes b$, $a \in K$, $b \in F_v$ is mapped to $(ab, a^\sigma b) \in F_v \oplus F_v$, σ being the non-trivial automorphism of K over F. By this isomorphism, $\mathfrak{O}_{K_v}$ is mapped onto $R \oplus R$. Let us show that $\mathfrak{O}_v$ is mapped onto

$$\{(x, y) \mid x, y \in R,\ x \equiv y \mod \varpi^c\}.$$

Call A this subring. For $(x, y) \in A$, we have

$$(x, y) = (x, x) + (0, y - x)$$

and $(x, x) = x(1, 1) = x \times$ the image of the unit of $\mathfrak{O}_v$, $(0, y - x) \in$ the image of $\varpi^c \mathfrak{O}_{K_v}$. Hence A is contained in the image of $\mathfrak{O}_v$. Considering the indices $[\mathfrak{O}_{K_v} : \mathfrak{O}_v]$ and $[R \oplus R : A]$, we see that A is the image of $\mathfrak{O}_v$ itself. By the isomorphism $(*)$, we identify K_v with $F_v \oplus F_v$ and $\mathfrak{O}_v$ with A.

For $(x, y) \in F_v^\times \times F_v^\times$, we have, by definition

$$\chi_v((x, y)) = \chi_{w_1}(x)\chi_{w_2}(y), \qquad \|(x, y)\|_v = |x|_v|y|_v.$$

Then we see that

$$I_v = \sum_{(x,y) \in R \oplus R,\, x \equiv y \mod \varpi^c,\, x \neq 0, y \neq 0, \mod \mathfrak{O}_v^\times} \chi_{w_1}(x)\chi_{w_2}(y)|x|_v^s|y|_v^s.$$

For a non-negative integer m, put

$$U_m = \{(x, y) \in R^\times \times R^\times \mid x \equiv y \mod \varpi^m\}.$$

Then we have $U_0 = \mathfrak{O}_{K_v}^\times$, $U_c = \mathfrak{O}_v^\times$ and $[U_0 : U_m] = (q - 1)q^{m-1}$ if $m \geqq 1$. Let m_0 be the least non-negative integer such that $\chi_v|U_{m_0}$ is trivial. We have $0 \leqq m_0 \leqq c$.

We can write $(x, y) \in \mathfrak{O}_v \cap K_v^\times$ as $(\varpi^\alpha x_0, \varpi^\beta y_0)$, x_0, $y_0 \in R^\times$, $0 \leqq \alpha, \beta \in \mathbf{Z}$. The condition

$$\varpi^\alpha x_0 \equiv \varpi^\beta y_0 \mod \varpi^c$$

must hold. If $\alpha < c$ or $\beta < c$, then we get $\alpha = \beta$. If $\alpha \geqq c$ or $\beta \geqq c$, then we have $\alpha \geqq c$ and $\beta \geqq c$ and the above congruence is satisfied automatically. Therefore we have

$$\begin{aligned} I_v = &\sum_{\alpha=0}^{c-1} \sum_{(x_0,y_0) \in U_{c-\alpha}/U_c} \chi_v((\varpi^\alpha x_0, \varpi^\alpha y_0))q^{-2\alpha s} \\ &+ \sum_{\alpha=c}^{\infty}\sum_{\beta=c}^{\infty} \sum_{(x_0,y_0) \in U_0/U_c} \chi_v((\varpi^\alpha x_0, \varpi^\beta y_0))q^{-(\alpha+\beta)s}. \end{aligned}$$

If $\alpha > c - m_0$, we have

$$\sum_{(x_0,y_0) \in U_{c-\alpha}/U_c} \chi_v((\varpi^\alpha x_0, \varpi^\alpha y_0)) = 0.$$

Hence the first term equals

$$\sum_{\alpha=0}^{c-m_0} \chi_v((\varpi^\alpha, \varpi^\alpha))q^{-2\alpha s}[U_{c-\alpha} : U_c]$$

if $m_0 \geqq 1$, and equals

$$\sum_{\alpha=0}^{c-1} \chi_v((\varpi^\alpha, \varpi^\alpha))q^{-2\alpha s}[U_{c-\alpha} : U_c]$$

if $m_0 = 0$. The second term vanishes if $m_0 \geqq 1$; if $m_0 = 0$, it equals

$$\sum_{\alpha=c}^{\infty}\sum_{\beta=c}^{\infty} \chi_v((\varpi^\alpha, \varpi^\beta))q^{-(\alpha+\beta)s}[U_0 : U_c].$$

Summing up, we obtain

$$(3.5)\quad I_v = \frac{1 - \chi_{w_1}(\varpi_{w_1})^{c-m_0+1}\chi_{w_2}(\varpi_{w_2})^{c-m_0+1}q^{(c-m_0+1)(1-2s)}}{1 - \chi_{w_1}(\varpi_{w_1})\chi_{w_2}(\varpi_{w_2})q^{1-2s}} \qquad \text{if} \quad m_0 \geqq 1,$$

$$(3.6)\quad \begin{aligned} I_v = &\frac{1 - \chi_{w_1}(\varpi_{w_1})^c\chi_{w_2}(\varpi_{w_2})^c q^{c(1-2s)}}{1 - \chi_{w_1}(\varpi_{w_1})\chi_{w_2}(\varpi_{w_2})q^{1-2s}} \\ &+ \frac{(q-1)q^{c-1}\chi_{w_1}(\varpi_{w_1})^c\chi_{w_2}(\varpi_{w_2})^c q^{-2cs}}{(1-\chi_{w_1}(\varpi_{w_1})q^{-s})(1-\chi_{w_2}(\varpi_{w_2})q^{-s})} \qquad \text{if} \quad m_0 = 0. \end{aligned}$$

Case II. The case where v is ramified in K.

Let w be the place of K lying above v and let π be a prime element of K_w. We have $\mathfrak{O}_{K_v} = \mathfrak{O}_{K_w}$, the maximal order of K_w. For a non-negative integer m, put

$$\mathfrak{O}_m = R + \varpi^m \mathfrak{O}_{K_v}.$$

Then $\mathfrak{O}_m$ is a subring of $\mathfrak{O}_{K_v}$ and we have $\mathfrak{O}_0 = \mathfrak{O}_{K_v}$, $\mathfrak{O}_c = \mathfrak{O}_v$. Put $U_m = \mathfrak{O}_m^\times$. Put, for $m \geqq 1$,

$$U_m^* = \{x \in \mathfrak{O}_{K_v}^\times \mid x \equiv 1 \mod \varpi^m\}.$$

Then we find $[U_0 : U_m^*] = (q-1)q^{2m-1}$, $[U_m : U_m^*] = (q-1)q^{m-1}$. Hence we have $[U_0 : U_m] = q^m$.

We can write $a \in \mathfrak{O}_v$, $a \neq 0$ as $a = \pi^\alpha x$, $0 \leqq \alpha \in \mathbf{Z}$, $x \in \mathfrak{O}_{K_v}^\times$. By (3.4), we have

$$I_v = \sum_{\alpha=0}^{\infty} J_\alpha, \qquad J_\alpha = \sum_{x \in \pi^{-\alpha}\mathfrak{O}_v \cap \mathfrak{O}_{K_v}^\times / U_c} \chi_w(\pi^\alpha x)q^{-\alpha s}.$$

We are going to calculate J_α. Let m_0 be the least non-negative integer such that $\chi_w | U_{m_0}$ is trivial. Let us first show that

$$(3.7)\qquad \mathfrak{O}_m \cap \mathfrak{O}_{K_v}^\times = \mathfrak{O}_m^\times, \qquad m \geqq 0.$$

Let σ be the generator of $\mathrm{Gal}(K_w/F_v)$. Let $x \in \mathfrak{O}_m \cap \mathfrak{O}_{K_v}^\times$ and put $y = xx^\sigma$. Then $y \in R^\times$. Hence $x^{-1} = x^\sigma y^{-1} \in \mathfrak{O}_m$ since $\mathfrak{O}_m$ is stable under σ. This proves (3.7). In particular, we see that $J_0 = 1$.

Assume $0 < \alpha < 2c$. Assume that $x \in \mathfrak{O}_{K_v}^\times$ and $\pi^\alpha x \in \mathfrak{O}_v$. Let us show that α must be even. Assume that α is odd and put $\alpha = 2\beta + 1$. Then $0 \leqq \beta \leqq c - 1$.

Put $\pi^\alpha x = r + \varpi^c y$ with $r \in R$, $y \in \mathfrak{O}_{K_v}$. Then we have $r \in \pi^\alpha \mathfrak{O}_{K_v} \cap R$. Hence we must have $r \equiv 0 \mod \varpi^{\beta+1}$. We get $x \in \pi \mathfrak{O}_{K_v}$, which contradicts $x \in \mathfrak{O}_{K_v}^\times$.

Now assume that α is even and put $\alpha = 2\beta$, $1 \leqq \beta < c$. Assume that $x \in \mathfrak{O}_{K_v}^\times$ and $\pi^\alpha x \in \mathfrak{O}_v$. Put $\pi^\alpha x = r + \varpi^c y$ with $r \in R$, $y \in \mathfrak{O}_{K_v}$. Put $\pi^2 = \varpi u$ with $u \in \mathfrak{O}_{K_v}^\times$. Then we find $r \equiv 0 \mod \varpi^\beta$, $u^\beta x \in \mathfrak{O}_{c-\beta}$. By (3.7), we see that $\pi^{-\alpha}\mathfrak{O}_v \cap \mathfrak{O}_{K_v}^\times = u^{-\beta} U_{c-\beta}$. Hence we get

$$J_\alpha = \sum_{x \in u^{-\beta} U_{c-\beta}/U_c} \chi_w(\varpi^\beta u^\beta x) q^{-2\beta s} = \chi_w(\varpi^\beta) q^{-2\beta s} \sum_{x \in U_{c-\beta}/U_c} \chi_w(x).$$

If $c - \beta < m_0$, then $J_\alpha = 0$. If $c - \beta \geqq m_0$, then $J_\alpha = \chi_w(\varpi)^\beta q^{-2\beta s}[U_{c-\beta} : U_c]$. Therefore $\sum_{\alpha=1}^{c-1} J_\alpha$ is equal to

$$\sum_{\beta=1}^{c-m_0} \chi_w(\varpi)^\beta q^{-2\beta s}[U_{c-\beta} : U_c] \qquad \text{if} \quad m_0 \geqq 1,$$

$$\sum_{\beta=1}^{c-1} \chi_w(\varpi)^\beta q^{-2\beta s}[U_{c-\beta} : U_c] \qquad \text{if} \quad m_0 = 0.$$

Next assume $\alpha \geqq 2c$. Then $\pi^\alpha x \in \mathfrak{O}_v$ for $x \in \mathfrak{O}_{K_v}$; hence $\pi^{-\alpha}\mathfrak{O}_v \cap \mathfrak{O}_{K_v}^\times = \mathfrak{O}_{K_v}^\times$. We have

$$J_\alpha = \chi_w(\pi)^\alpha q^{-\alpha s} \sum_{x \in \mathfrak{O}_{K_v}^\times / U_c} \chi_w(x).$$

If $m_0 \geqq 1$, then J_α vanishes. If $m_0 = 0$, then we have

$$J_\alpha = \chi_w(\pi)^\alpha q^{-\alpha s}[\mathfrak{O}_{K_v}^\times : U_c].$$

Summing up, we obtain

$$\text{(3.8)} \qquad I_v = \frac{1 - \chi_w(\varpi)^{c-m_0+1} q^{(c-m_0+1)(1-2s)}}{1 - \chi_w(\varpi) q^{1-2s}} \qquad \text{if} \quad m_0 \geqq 1,$$

$$\text{(3.9)} \qquad I_v = \frac{1 - \chi_w(\varpi)^c q^{c(1-2s)}}{1 - \chi_w(\varpi) q^{1-2s}} + \frac{\chi_w(\pi)^{2c} q^{c(1-2s)}}{1 - \chi_w(\pi) q^{-s}} \qquad \text{if} \quad m_0 = 0.$$

Case III. The case where v remains prime in K.

Let w be the place of K lying above v. We have $\mathfrak{O}_{K_v} = \mathfrak{O}_{K_w}$. For a non-negative integer m, put

$$\mathfrak{O}_m = R + \varpi^m \mathfrak{O}_{K_v}.$$

Then $\mathfrak{O}_m$ is a subring of $\mathfrak{O}_{K_v}$ and we have $\mathfrak{O}_0 = \mathfrak{O}_{K_v}$, $\mathfrak{O}_c = \mathfrak{O}_v$. Put $U_m = \mathfrak{O}_m^\times$. For $m \geqq 1$, put

$$U_m^* = \{x \in \mathfrak{O}_{K_v}^\times \mid x \equiv 1 \mod \varpi^m\}.$$

Then we have $[U_0 : U_m^*] = (q^2 - 1) q^{2(m-1)}$. Since the mapping

$$U_m \ni r + \varpi^m y \longrightarrow r \mod \varpi^m \in (R/\varpi^m)^\times, \qquad (r \in R, y \in \mathfrak{O}_{K_v})$$

is a surjective homomorphism with the kernel U_m^*, we have $[U_m : U_m^*] = (q-1)q^{m-1}$. Therefore we obtain

$$[U_0 : U_m] = (q+1) q^{m-1}, \qquad m \geqq 1.$$

Similarly to the Case II, we can show that

(3.10) $$\mathfrak{O}_m \cap \mathfrak{O}_{K_v}^\times = \mathfrak{O}_m^\times, \qquad m \geqq 0.$$

We can write $a \in \mathfrak{O}_v$, $a \neq 0$ as $a = \varpi^\alpha x$, $0 \leqq \alpha \in \mathbf{Z}$, $x \in \mathfrak{O}_{K_v}^\times$. By (3.4), we have

$$I_v = \sum_{\alpha=0}^{\infty} J_\alpha, \qquad J_\alpha = \sum_{x \in \varpi^{-\alpha}\mathfrak{O}_v \cap \mathfrak{O}_{K_v}^\times / U_c} \chi_w(\varpi^\alpha x) q^{-2\alpha s}.$$

By (3.10), we have $J_0 = 1$. Let m_0 be the least non-negative integer such that $\chi_w | U_{m_0}$ is trivial.

Assume $0 < \alpha < c$. If $\varpi^\alpha x \in \mathfrak{O}_v$ for $x \in \mathfrak{O}_{K_v}^\times$, we can write $\varpi^\alpha x = r + \varpi^c y$ with $r \in R$, $y \in \mathfrak{O}_{K_v}$. Since $r \equiv 0 \mod \varpi^\alpha$, we have $x \in \mathfrak{O}_{c-\alpha} \cap \mathfrak{O}_{K_v}^\times = U_{c-\alpha}$. Hence we see that $\varpi^{-\alpha}\mathfrak{O}_v \cap \mathfrak{O}_{K_v}^\times = U_{c-\alpha}$,

$$J_\alpha = \sum_{x \in U_{c-\alpha}/U_c} \chi_w(\varpi^\alpha x) q^{-2\alpha s}.$$

If $c - \alpha < m_0$, then $J_\alpha = 0$. If $c - \alpha \geqq m_0$, then $J_\alpha = \chi_w(\varpi)^\alpha q^{-2\alpha s}[U_{c-\alpha} : U_c]$. Therefore $\sum_{\alpha=1}^{c-1} J_\alpha$ is equal to

$$\sum_{\alpha=1}^{c-m_0} \chi_w(\varpi)^\alpha q^{\alpha(1-2s)} \qquad \text{if} \quad m_0 \geqq 1,$$

$$\sum_{\alpha=1}^{c-1} \chi_w(\varpi)^\alpha q^{\alpha(1-2s)} \qquad \text{if} \quad m_0 = 0.$$

Next assume $\alpha \geqq c$. Then $\varpi^{-\alpha}\mathfrak{O}_v \cap \mathfrak{O}_{K_v}^\times = \mathfrak{O}_{K_v}^\times$ and we have

$$J_\alpha = \sum_{x \in U_0/U_c} \chi_w(\varpi^\alpha x) q^{-2\alpha s}.$$

If $m_0 \geqq 1$, then J_α vanishes. If $m_0 = 0$, then we have

$$J_\alpha = \chi_w(\varpi)^\alpha q^{-2\alpha s}[U_0 : U_c] = \chi_w(\varpi)^\alpha q^{-2\alpha s}(q+1)q^{c-1}.$$

Summing up, we obtain

(3.11) $$I_v = \frac{1 - \chi_w(\varpi)^{c-m_0+1} q^{(c-m_0+1)(1-2s)}}{1 - \chi_w(\varpi) q^{1-2s}} \qquad \text{if} \quad m_0 \geqq 1,$$

(3.12) $$I_v = \frac{1 - \chi_w(\varpi)^c q^{c(1-2s)}}{1 - \chi_w(\varpi) q^{1-2s}} + (q+1)q^{c-1} \frac{\chi_w(\varpi)^c q^{-2cs}}{1 - \chi_w(\varpi) q^{-2s}} \qquad \text{if} \quad m_0 = 0.$$

In (3.5), I_v is a polynomial of q^{-s} and in (3.6), the first term is a polynomial of q^{-s} and the second term is a polynomial of q^{-s} times the v-factor of $L(s, \chi)$. Similar remark applies to (3.8), (3.9), (3.11), (3.12). We can put

(3.13) $$I_v = P_{\chi,v}(s) \times (\text{the } v\text{-factor of } L(s, \chi))$$

with a polynomial $P_{\chi,v}$ of q^{-s}. Put

$$P_\chi(s) = \prod_{v | \mathfrak{f}} P_{\chi,v}(s).$$

Then we obtain

$$L(s,\chi_*) = P_\chi(s)L(s,\chi). \tag{3.14}$$

We note that essentially the same results as in this subsection were independently obtained by Hida [Hi2], §4.1.

3.3. By the values of $[U_m : U_0]$ obtained in the course of the calculation in §3.2, we have

$$\prod_v [\mathfrak{O}_{K_v}^\times : \mathfrak{O}_v^\times] = N(\mathfrak{f})\prod_{\mathfrak{p}|\mathfrak{f}}(1 - \frac{\epsilon(\mathfrak{p})}{N(\mathfrak{p})}),$$

where $\mathfrak{p}$ extends over prime ideals of F dividing $\mathfrak{f}$ and $\epsilon(\mathfrak{p}) = 1, 0, -1$ according as $\mathfrak{p}$ decomposes completely, ramifies and remains prime in K. We also have $E_K \cap \mathfrak{O} = \mathfrak{O}^\times$. Therefore, by Proposition 3.2, we obtain (cf. [Sh1], Exercises 4.12 and 4.13)

$$h_{\mathfrak{O}} = h_K \cdot [E_K : \mathfrak{O}^\times]^{-1}N(\mathfrak{f})\prod_{\mathfrak{p}|\mathfrak{f}}(1 - \frac{\epsilon(\mathfrak{p})}{N(\mathfrak{p})}). \tag{3.15}$$

For the trivial character $\chi = 1$, we denote $L(s, 1_*)$ by $\zeta_{\mathfrak{O}}(s)$, which is the zeta function of the order $\mathfrak{O}$. Taking a basis of $\mathfrak{O}$ over $\mathbf{Z}$, we define the discriminant $D(\mathfrak{O})$ of $\mathfrak{O}$ in the ususal manner. Since $[\mathfrak{O}_K : \mathfrak{O}] = N(\mathfrak{f})$, we have

$$|D(\mathfrak{O})| = N(\mathfrak{f})^2|D_K|. \tag{3.16}$$

Let $w_{\mathfrak{O}}$ denote the number of roots of unity contained in $\mathfrak{O}$. We have $\mathfrak{O}^\times \cong \mathbf{Z}^{r_1+r_2-1} \times \mathbf{Z}/w_{\mathfrak{O}}\mathbf{Z}$ non-canonically. Taking $r_1 + r_2 - 1$ generators of a free part, we define the regulator $R(\mathfrak{O})$ in the usual manner. Then we see easily that

$$R(\mathfrak{O})/R_K = [E_K : W_K\mathfrak{O}^\times]. \tag{3.17}$$

PROPOSITION 3.3. *We have*

$$\mathrm{Res}_{s=1}\zeta_{\mathfrak{O}}(s) = \frac{2^{r_1}(2\pi)^{r_2}h_{\mathfrak{O}}R(\mathfrak{O})}{w_{\mathfrak{O}}|D(\mathfrak{O})|^{1/2}}.$$

PROOF. By II, (1.1), we have

$$\mathrm{Res}_{s=1}\zeta_K(s) = \frac{2^{r_1+r_2}\pi^{r_2}h_K R_K}{w_K|D_K|^{1/2}}. \tag{3.18}$$

By (3.14), we have

$$\mathrm{Res}_{s=1}\zeta_{\mathfrak{O}}(s) = P_1(1)\mathrm{Res}_{s=1}\zeta_K(s).$$

By (3.6), (3.9) and (3.12), we see that

$$P_1(1) = \prod_{\mathfrak{p}|\mathfrak{f}}(1 - \frac{\epsilon(\mathfrak{p})}{N(\mathfrak{p})}).$$

Using (3.15), (3.16) and (3.18), we obtain

$$\mathrm{Res}_{s=1}\zeta_{\mathfrak{O}}(s) = \frac{2^{r_1+r_2}\pi^{r_2}h_{\mathfrak{O}}[E_K : \mathfrak{O}^\times]R_K}{w_K|D(\mathfrak{O})|^{1/2}}.$$

Since $[E_K : \mathfrak{O}^\times] = [E_K : W_K\mathfrak{O}^\times]w_K/w_{\mathfrak{O}}$, the assertion follows in view of (3.17)

It is interesting to notice the close analogy to the usual analytic class number formula (3.18). Probably a similar formula holds for orders of more general type.

4. Toward the reciprocity law for the h-function

4.1. Let F be a totally real algebraic number field of degree n and let K be a CM-field such that $[K:F]=2$. Let $\mathfrak{O}$ be an order of K which contains $\mathfrak{O}_F$. Let c be a class of proper $\mathfrak{O}$-ideals. We define the partial zeta function of the class c by

$$\zeta_{\mathfrak{O}}(s,c)=\sum_{\mathfrak{A}\in c}N(\mathfrak{A})^{-s},\qquad \Re(s)>1, \tag{4.1}$$

where $\mathfrak{A}$ extends over all proper $\mathfrak{O}$-ideals whose class is c and $\mathfrak{A}\subset\mathfrak{O}$. Let ω be a character of $K_A^\times$ which is trivial on $K^\times\prod_v\mathfrak{O}_v^\times K_\infty^\times$. Define the character ω_* of the group of proper $\mathfrak{O}$-ideal classes by (3.1) taking $\chi=\omega$. Then we have

$$L(s,\omega_*)=\sum_c\omega_*(c)\zeta_{\mathfrak{O}}(s,c). \tag{4.2}$$

By (3.14), we see that $\zeta_{\mathfrak{O}}(s,c)$ can be continued meromorphically to the whole s-plane and holomorphic except for the simple pole at $s=1$.

We take a proper $\mathfrak{O}$-ideal $\mathfrak{A}$ so that the class of $\mathfrak{A}^{-1}$ is c. Then we have

$$\zeta_{\mathfrak{O}}(s,c)=N(\mathfrak{A})^s\sum_{\alpha\mathfrak{O},\,\alpha\mathfrak{O}\subset\mathfrak{A}}N(\alpha)^{-s}. \tag{4.3}$$

Here $N(\mathfrak{A})=|a|_A^{-1}$ if $\mathfrak{A}=a\mathfrak{O}$ with $a\in K_A^\times$, the infinite part of a being 1. Since $\mathfrak{A}$ is an $\mathfrak{O}_F$-lattice in K, by Lemma 2.2, we can write

$$\mathfrak{A}=\mathfrak{a}\omega_1\oplus\mathfrak{O}_F\omega_2 \tag{4.4}$$

with ω_1, $\omega_2\in K$ and a fractional ideal $\mathfrak{a}$ of F. We have

$$\sum_{\alpha\mathfrak{O},\,\alpha\mathfrak{O}\subset\mathfrak{A}}N(\alpha)^{-s}=[\mathfrak{O}^\times:E_F]^{-1}\sum_{(c,d)/\sim,\,c\in\mathfrak{a},\,d\in\mathfrak{O}_F}N((c\omega_1+d\omega_2))^{-s}.$$

Here $(0,0)$ is excluded in the sum and

$$(c,d)\sim(c_1,d_1)\Longleftrightarrow\exists\epsilon\in E_F,\quad c_1=\epsilon c,\ d_1=\epsilon d.$$

Put $w=\omega_1/\omega_2$ and choose the CM-type Φ of K so that $\Im(w^\sigma)>0$ for every $\sigma\in\Phi$. As in §2.2, we have

$$N((c\omega_1+d\omega_2))^{-s}=N((\omega_2))^{-s}\prod_{\sigma\in\Phi}|c^\sigma w^\sigma+d^\sigma|^{-2s}.$$

By (1.25), we obtain

$$\begin{aligned}\zeta_{\mathfrak{O}}(s,c)&=[\mathfrak{O}^\times:E_F]^{-1}N(\mathfrak{A})^sN((\omega_2))^{-s}\prod_{\sigma\in\Phi}\Im(w^\sigma)^{-s}\\&\quad\times\frac{1}{h_F}\sum_{\chi\in X}\chi(a)^{-1}N(\mathfrak{a})^{-s}E_\chi(w,s;\mathfrak{a}),\end{aligned} \tag{4.5}$$

where $w = (w^\sigma)_{\sigma\in\Phi} \in \mathfrak{H}^n$. Then by (1.34), (1.35) and (4.5), we obtain, similarly to §2.2,

$$\begin{aligned}\zeta_{\mathfrak{O}}(s,c) &= -\frac{R_K}{w_K}[E_K : \mathfrak{O}^\times]s^{n-1}\\ &\times\Big\{1 + (2n\gamma + 2n\log 2\pi - 2\log|D_F| + \log N(\mathfrak{A})\\ &- \log N((\omega_2)) + 2\log N(\mathfrak{a})\\ &- \sum_{\chi\in X}\chi(a)^{-1}h_\chi(w) - \frac{|D_F|^{1/2}}{2^{n-2}R_F}\gamma_F(d))s\Big\} + O(s^{n+1}), \qquad s\to 0.\end{aligned} \tag{4.6}$$

This formula is a generalization of (2.4). The next Lemma is a generalization of Lemma 2.3.

LEMMA 4.1. *We have*

$$N(\mathfrak{A})N((\omega_2))^{-1}(\prod_{\sigma\in\Phi}\Im(w^\sigma))^{-1}N(\mathfrak{a})^{-1} = \frac{2^n D_F}{|D(\mathfrak{O})|^{1/2}}.$$

PROOF. Let x denote the quantity on the left-hand side of our lemma. By (4.5) and (1.32), we obtain

$$\mathrm{Res}_{s=1}\zeta_{\mathfrak{O}}(s,c) = [\mathfrak{O}^\times : E_F]^{-1}x \times 2^{n-2}\pi^n\frac{R_F}{D_F}.$$

By (4.2), we have

$$\zeta_{\mathfrak{O}}(s,c) = \frac{1}{h_{\mathfrak{O}}}\sum_\omega \omega_*(c)^{-1}L(s,\omega_*),$$

where ω extends over all characters of $K_A^\times$ which is trivial on $K^\times\prod_v\mathfrak{O}_v^\times K_\infty^\times$. By (3.14), $L(s,\omega_*)$ is an entire function if $\omega\neq 1$. By Proposition 3.3, we obtain

$$\mathrm{Res}_{s=1}\zeta_{\mathfrak{O}}(s,c) = \frac{2^n\pi^n R(\mathfrak{O})}{w_{\mathfrak{O}}|D(\mathfrak{O})|^{1/2}}.$$

Therefore we get

$$x = \frac{D_F}{|D(\mathfrak{O})|^{1/2}}\cdot\frac{4}{w_{\mathfrak{O}}}\frac{R(\mathfrak{O})}{R_F}[\mathfrak{O}^\times : E_F].$$

Using (3.17), we get

$$\begin{aligned}\frac{R(\mathfrak{O})}{R_F} &= \frac{R_K}{R_F}\frac{R(\mathfrak{O})}{R_K} = \frac{2^{n-1}}{[E_K : W_K E_F]}[E_K : W_K\mathfrak{O}^\times]\\ &= \frac{2^{n-1}}{[W_K\mathfrak{O}^\times : W_K E_F]} = \frac{2^{n-1}}{[\mathfrak{O}^\times : E_F]}[\mathfrak{O}^\times\cap W_K E_F : E_F].\end{aligned}$$

Hence

$$x = \frac{2^n|D_F|}{|D(\mathfrak{O})|^{1/2}}\cdot\frac{2[\mathfrak{O}^\times\cap W_K E_F : E_F]}{w_{\mathfrak{O}}}.$$

We see easily that $\mathfrak{O}^\times\cap W_K E_F = (\mathfrak{O}^\times\cap W_K)E_F$, $[\mathfrak{O}^\times\cap W_K E_F : E_F] = w_{\mathfrak{O}}/2$. Hence our assertion follows.

Using Lemma 4.1, (4.6) can be written as

$$\begin{aligned}\zeta_{\mathfrak{O}}(s,c) = &-\frac{R(\mathfrak{O})}{w_{\mathfrak{O}}}s^{n-1}\\ &\times\Big\{1+(2n\gamma+2n\log 2\pi+n\log 2-\log D_F\\ &-\frac{1}{2}\log|D(\mathfrak{O})|-\frac{D_F^{1/2}}{2^{n-2}R_F}\gamma_F(d)+\log N(\mathfrak{a})\\ &+\log\prod_{\sigma\in\Phi}\Im(w^{\sigma})-\sum_{\chi\in X}\chi(a)^{-1}h_{\chi}(w;\mathfrak{a}))s\Big\}+O(s^{n+1}), \qquad s\to 0,\end{aligned}\tag{4.7}$$

noting that (cf. (3.17))

$$\frac{R_K}{w_K}[E_K:\mathfrak{O}^{\times}] = \frac{R(\mathfrak{O})}{w_{\mathfrak{O}}}.$$

(4.7) is a generalization of (2.5).

4.2. Let Y denote the set of all characters of $K_A^{\times}$ which are trivial on $K^{\times}\prod_v \mathfrak{O}_v^{\times}K_{\infty}^{\times}$. For $\omega\in Y$, we put

$$L(s,\omega)=c(\omega)s^{r(\omega)}+O(s^{r(\omega)+1}) \qquad \text{for} \quad s\to 0, \quad c(\omega)\neq 0 \tag{4.8}$$

as in II, §1. We have $r(\omega)=n$ if $\omega\neq 1$ and $r(1)=n-1$. By (3.14), we have $L(s,\omega_*)=P_{\omega}(s)L(s,\omega)$ with a holomorphic function $P_{\omega}(s)$. We have

$$L(s,\omega_*)=P_{\omega}(0)c(\omega)s^{r(\omega)}+O(s^{r(\omega)+1}) \qquad \text{for} \quad s\to 0.$$

Put

$$H(c)=\sum_{\chi\in X}\chi(\mathfrak{a})^{-1}h_{\chi}(w;\mathfrak{a})-\log\prod_{\sigma\in\Phi}\Im(w^{\sigma})-\log N(\mathfrak{a}). \tag{4.9}$$

Suppose $\omega\in Y$, $\omega\neq 1$. By (4.7), we obtain

$$P_{\omega}(0)c(\omega)=\frac{R(\mathfrak{O})}{w_{\mathfrak{O}}}\sum_c \omega_*(c)H_c. \tag{4.10}$$

If $\omega=1$, (4.7) yields

$$\begin{aligned}\zeta_{\mathfrak{O}}(s) = &-\frac{h_{\mathfrak{O}}R(\mathfrak{O})}{w_{\mathfrak{O}}}s^{n-1}\\ &\times\{1+(2n\gamma+2n\log 2\pi+n\log 2-\log D_F-\frac{1}{2}\log|D(\mathfrak{O})|\\ &-\frac{D_F^{1/2}}{2^{n-2}R_F}\gamma_F(d)-\frac{1}{h_{\mathfrak{O}}}\sum_c H(c))s\}+O(s^{n+1}), \qquad s\to 0.\end{aligned}\tag{4.11}$$

Since

$$\zeta_{\mathfrak{O}}(s)=P_1(s)\zeta_F(s)L_F(s,\chi),$$

$$\zeta_F(s)=-\frac{h_F R_F}{2}s^{n-1}\{1+(n\gamma+n\log 2\pi-\log D_F-\frac{D_F^{1/2}\gamma_F}{2^{n-1}h_F R_F})s\}+O(s^{n+1}),$$

we have

$$\begin{aligned}\zeta_{\mathfrak{O}}(s) &= -\frac{h_F R_F}{w_F} L_F(0,\chi) P_1(0) s^{n-1} \\ &\times \left\{1 + (n\gamma + n\log 2\pi - \log D_F - \frac{D_F^{1/2}\gamma_F}{2^{n-1}h_F R_F} + \frac{L_F'(0,\chi)}{L_F(0,\chi)} + \frac{P_1'(0)}{P_1(0)})s\right\} \\ &+ O(s^{n+1}) \qquad \text{for} \quad s \to 0.\end{aligned} \tag{4.12}$$

Here we note that

$$P_1(0) = N(\mathfrak{f}) \prod_{\mathfrak{p}|\mathfrak{f}} (1 - \frac{\epsilon(\mathfrak{p})}{N(\mathfrak{p})}) = N(\mathfrak{f}) P_1(1),$$

which follows from (3.6), (3.9) and (3.12).

THEOREM 4.2. *Let $\mathfrak{O}$ be an order of K which contains $\mathfrak{O}_F$. Let $\mathfrak{A}$ be a proper $\mathfrak{O}$-ideal and c be the class of $\mathfrak{A}^{-1}$. We write $\mathfrak{A}$ in the form* (4.4). *Define $H(c)$ by* (4.9). *Then we have*

$$\begin{aligned}\frac{1}{h_{\mathfrak{O}}} \sum_c H(c) &= n\gamma + n\log 4\pi - \frac{1}{2}\log|D(\mathfrak{O})| \\ &+ \frac{D_F^{1/2}}{2^{n-1}h_F R_F}(\gamma_F - 2h_F\gamma_F(d)) - \frac{P_1'(0)}{P_1(0)} - \frac{L_F'(0,\chi)}{L_F(0,\chi)},\end{aligned} \tag{4.13}$$

$$\begin{aligned}H(c) = &\frac{w_{\mathfrak{O}}}{h_{\mathfrak{O}} R(\mathfrak{O})} \sum_{\omega \in Y, \omega \neq 1} \omega_*(c)^{-1} P_\omega(0) c(\omega) \\ &+ n\gamma + n\log 4\pi - \frac{1}{2}\log|D(\mathfrak{O})| \\ &+ \frac{D_F^{1/2}}{2^{n-1}h_F R_F}(\gamma_F - 2h_F\gamma_F(d)) - \frac{P_1'(0)}{P_1(0)} - \frac{L_F'(0,\chi)}{L_F(0,\chi)}.\end{aligned} \tag{4.14}$$

PROOF. Comparing (4.11) with (4.12), we immediately obtain (4.13). Then (4.14) follows from (4.10) and (4.13).

This theorem gives a generalization of Theorem 2.5. By (4.14), we have, for two proper $\mathfrak{O}$-ideal classes c_1 and c_2, that

$$\frac{h_{\mathfrak{O}} R(\mathfrak{O})}{w_{\mathfrak{O}}}(H(c_1) - H(c_2)) = \sum_{\omega \in Y, \omega \neq 1} (\omega_*(c_1)^{-1} - \omega_*(c_2)^{-1}) P_\omega(0) c(\omega). \tag{4.15}$$

We are interested in the arithmetical nature of $\sum_{\chi \in X} \chi(\mathfrak{a})^{-1} h_\chi(z;\mathfrak{a})$ for a CM-point $z \in \mathfrak{H}^n$. Choose $w \in K$ and a CM-type of K so that $z = (w^\sigma)_{\sigma \in \Phi}$. Take ω_1, $\omega_2 \in K$ so that $w = \omega_1/\omega_2$ and put $\mathfrak{A} = \mathfrak{a}\omega_1 \oplus \mathfrak{O}_F\omega_2$. Then $\mathfrak{O} = \{\alpha \in K \mid \alpha\mathfrak{A} \subset \mathfrak{A}\}$ is an order of K which contains $\mathfrak{O}_F$ and $\mathfrak{A}$ is a proper $\mathfrak{O}$-ideal. Therefore $H(c)$ gives the desired information on $\sum_{\chi \in X} \chi(\mathfrak{a})^{-1} h_\chi(z;\mathfrak{a})$, where c is the class of $\mathfrak{A}^{-1}$.

In (4.14), the nature of $c(\omega)$ can be predicted by Stark's conjecture. Using this conjecture, we expect that we can derive a reciprocity law for $H(c)$.

4.3. Let us write $c(\omega)$ explicitly assuming Stark's conjecture. Let L be the abelian extension of K such that $K^\times N_{L/K}(L_A^\times) = K^\times \prod_v \mathfrak{O}_v^\times K_\infty^\times$. Put $G = \mathrm{Gal}(L/K)$. We identify $\widehat{G}$ with Y. By II, Lemma 1.2, there exist units ϵ_1, ϵ_2, $\dots$, $\epsilon_n \in E_L$ such that $\prod_{i=1}^n \prod_{\sigma\in G} \epsilon_i^\sigma = \pm 1$ and $\langle\, \epsilon_i^\sigma \mid 1 \leqq i \leqq n,\ \sigma \in G\,\rangle$ is of finite index in E_L. Then define $f : X \otimes_{\mathbf{Z}} \mathbf{Q} \cong E_L \otimes_{\mathbf{Z}} \mathbf{Q}$ as in II, §1,1, taking $K = L$, $k = K$ there. Put

$$r_{ij} = \sum_{\tau\in G} \omega(\tau) \log|\epsilon_i^{\tau v_j}|, \qquad 1 \leqq i, j \leqq n,$$

where $\mathbf{a}_K = \{v_1, v_2, \dots, v_n\}$. Then, for $\omega \neq 1$, $R(\omega, f) = 2^{2r_2} \det(r_{ij})$ by II, Proposition 1.3. Put

$$c(\omega) = A(\omega) \det(r_{ij}) \qquad \text{for} \quad \omega \neq 1. \tag{4.16}$$

Now Stark's conjecture (II, Conjecture 1.1) states that

$$A(\omega)^\alpha = A(\omega^\alpha) \quad \text{for every} \quad \alpha \in \mathrm{Aut}(\mathbf{C}). \tag{4.17}$$

Paying the cost of changing $\{\epsilon_1, \epsilon_2, \dots, \epsilon_n\}$ to another system $\{\eta_1, \eta_2, \dots, \eta_n\}$ of units of L, we can bring (4.16) to the form $A(\omega) \in \mathbf{Q}$, by the same device as in II, §2. In fact, we put $A(1) = 1$. For $\sigma \in G$, put

$$b(\sigma) = \frac{1}{|G|} \sum_{\omega\in\hat{G}} A(\omega)\omega(\sigma).$$

Then $b(\sigma) \in \mathbf{Q}$ and

$$\sum_{\sigma\in G} b(\sigma)\omega(\sigma)^{-1} = A(\omega) \qquad \text{for every} \quad \omega \in \widehat{G}.$$

Hence there exist a positive integer m and a $\mathbf{Z}$-valued function $a(\sigma)$ on G such that

$$\sum_{\sigma\in G} a(\sigma)\omega(\sigma)^{-1} = mA(\omega) \qquad \text{for every} \quad \omega \in \widehat{G}.$$

We have $\sum_{\sigma\in G} a(\sigma) = mA(1) = m$. We put

$$\eta_1 = \prod_{\sigma\in G} (\epsilon_1)^{a(\sigma)\sigma}, \qquad \eta_i = \epsilon_i^m, \quad 2 \leqq i \leqq n.$$

Then we have

$$\begin{aligned} \sum_{\tau\in G} \omega(\tau) \log|\eta_1^{\tau v_j}| &= \sum_{\tau\in G} \omega(\tau) \sum_{\sigma\in G} a(\sigma) \log|\epsilon_1^{\sigma\tau v_j}| \\ &= \sum_{\tau\in G}\sum_{\sigma\in G} a(\sigma)\omega(\sigma)^{-1}\omega(\sigma\tau) \log|\epsilon_1^{\sigma\tau v_j}| = mA(\omega) \sum_{\tau\in G} \omega(\tau)\log|\epsilon_1^{\tau v_j}|. \end{aligned}$$

Put

$$r_{ij}^* = \sum_{\tau\in G} \omega(\tau) \log|\eta_i^{\tau v_j}|, \qquad 1 \leqq i, j \leqq n.$$

Then, by (4.16), we obtain

$$c(\omega) = \frac{1}{m^n} \det(r_{ij}^*). \tag{4.18}$$

We see that

$$\prod_{\sigma\in G}\eta_1^\sigma = (\prod_{\sigma\in G}\epsilon_1^\sigma)^m.$$

Hence the relation $\prod_{i=1}^n\prod_{\sigma\in G}\eta_i^\sigma = \pm 1$ still holds. For $\sigma \in G$, we have

$$\eta_1^\sigma = \prod_{\tau\in G}(\epsilon_1^\tau)^{a(\tau\sigma^{-1})}.$$

By considering the group determinant, we see that

$$\det(a(\tau\sigma^{-1})_{\sigma,\tau\in G}) = \prod_{\omega\in\hat{G}}(\sum_{\sigma\in G}a(\sigma)\omega(\sigma)^{-1}) = \prod_{\omega\in\hat{G}} mA(\omega)$$

is non-zero; hence the index of the group $\langle\eta_1^\sigma \mid \sigma\in G\rangle$ in $\langle\epsilon_1^\sigma \mid \sigma \in G\rangle$ is finite. Therefore the index of the group $\langle\eta_i^\sigma \mid 1 \leqq i \leqq n,\ \sigma\in G\rangle$ in E_L is still finite. Let us write η_i as ϵ_i, m^n as m and r_{ij}^* as r_{ij}. Then (4.18) can be written as

$$c(\omega) = \frac{1}{m}\det(r_{ij}), \qquad r_{ij} = \sum_{\tau\in G}\omega(\tau)\log|\epsilon_i^{\tau v_j}|. \tag{4.19}$$

This proves our assertion that $A(\omega)\in\mathbf{Q}$ choosing another system of units.

For $\sigma \in \mathrm{Gal}(\overline{\mathbf{Q}}/\mathbf{Q})$, we put

$$c(\omega)^{\tilde{\sigma}} = \frac{1}{m}\det(\sum_{\tau\in G}\omega(\tau)\log|\epsilon_i^{\tau v_j\sigma}|). \tag{4.20}$$

Here the symbol $\tilde{\sigma}$ is employed because σ does not act on $c(\omega)$ as the field automorphism. Let us compute $c(\omega)^{\tilde{\sigma}}$. As in II, §1.2, let

$$\mathrm{Gal}(\overline{\mathbf{Q}}/\mathbf{Q}) = \sqcup_{j=1}^n(\mathrm{Gal}(\overline{\mathbf{Q}}/K)\beta_j \sqcup \mathrm{Gal}(\overline{\mathbf{Q}}/K)\beta_j\rho)$$

be a coset decomposition, where ρ denotes the complex conjugation. We may take $v_j = \beta_j|K$, $1\leqq j\leqq n$. Put

$$\beta_j\sigma = \sigma'\beta_{l(j)}\rho^{\mu(j)}, \qquad \sigma'\in\mathrm{Gal}(\overline{\mathbf{Q}}/K), \quad \mu(j) = 0 \text{ or } 1. \tag{4.21}$$

Since K is a CM-extension of a totally real field F, we have $\mathrm{Gal}(\overline{\mathbf{Q}}/K)\beta_j\rho = \mathrm{Gal}(\overline{\mathbf{Q}}/K)\rho\beta_j$ and

$$\mathrm{Gal}(\overline{\mathbf{Q}}/\mathbf{Q}) = \sqcup_{j=1}^n\mathrm{Gal}(\overline{\mathbf{Q}}/F)\beta_j.$$

Hence $l(j)$ is determined by the condition

$$\beta_j\sigma = \sigma''\beta_{l(j)}, \qquad \sigma''\in\mathrm{Gal}(\overline{\mathbf{Q}}/F). \tag{4.22}$$

Therefore $j \longrightarrow l(j)$ gives a permutation on n letters. We have

$$\begin{aligned}\sum_{\tau\in G}\omega(\tau)\log|\epsilon_i^{\tau v_j\sigma}| &= \sum_{\tau\in G}\omega(\tau)\log|\epsilon_i^{\tau\sigma'\beta_{l(j)}}|\\ &=\omega(\sigma')^{-1}r_{il(j)} = \omega(\beta_j\sigma\rho^{\mu(j)}\beta_{l(j)}^{-1})^{-1}r_{il(j)},\end{aligned}$$

regarding ω as a character of $\mathrm{Gal}(\overline{\mathbf{Q}}/K)$. Therefore we obtain

$$c(\omega)^{\tilde{\sigma}} = \mathrm{sgn}(l)\prod_{j=1}^n\omega(\beta_j\sigma\rho^{\mu(j)}\beta_{l(j)}^{-1})^{-1}c(\omega). \tag{4.23}$$

We note that

$$\sqrt{D_F}^{\,\sigma} = \mathrm{sgn}(l)\sqrt{D_F}, \tag{4.24}$$

which follows from (4.22) and the definition of the discriminant. We take $a \in K_A^\times$ so that $[a,K]|L = \prod_{j=1}^n(\beta_j\sigma\rho^{\mu(j)}\beta_{l(j)}^{-1})|L$ and let $c(\sigma)$ be the class of the proper $\mathfrak{O}$-ideal $a\mathfrak{O}$. Then we have

$$c(\omega)^{\tilde{\sigma}} = \mathrm{sgn}(l)\omega_*(c(\sigma))^{-1}c(\omega). \tag{4.25}$$

Now by (4.14), we obtain:

THEOREM 4.3. *Let the notation be the same as above. Assume Stark's conjecture* (II, *Conjecture* 1.1). *Then there exist units* $\epsilon_1, \epsilon_2, \ldots, \epsilon_n \in E_L$ *such that* $\prod_{i=1}^n\prod_{\sigma\in G}\epsilon_i^\sigma = \pm1$, $\langle\epsilon_i^\sigma \mid 1 \leqq i \leqq n,\ \sigma\in G\rangle$ *is of finite index in* E_L *and that* (4.19) *holds. Let* $\sigma \in \mathrm{Gal}(\overline{\mathbf{Q}}/\mathbf{Q})$. *Define* $c(\omega)^{\tilde{\sigma}}$ *by* (4.20) *and let* $c(\sigma)$ *be the class of the proper* $\mathfrak{O}$*-ideal* $a\mathfrak{O}$ *where* $[a,K]|L = \prod_{j=1}^n(\beta_j\sigma\rho^{\mu(j)}\beta_{l(j)}^{-1})|L$. *Then we have*

$$\begin{aligned}H(cc(\sigma)) &= \frac{w_{\mathfrak{O}}}{h_{\mathfrak{O}}R(\mathfrak{O})}\sum_{\omega\in Y,\omega\neq1} sgn(l)\omega_*(c)^{-1}P_\omega(0)c(\omega)^{\tilde{\sigma}}\\ &\quad + n\gamma + n\log 4\pi - \frac{1}{2}\log|D(\mathfrak{O})|\\ &\quad + \frac{D_F^{1/2}}{2^{n-1}h_FR_F}(\gamma_F - 2h_F\gamma_F(d)) - \frac{P_1'(0)}{P_1(0)} - \frac{L_F'(0,\chi)}{L_F(0,\chi)}.\end{aligned} \tag{4.26}$$

This formula can be regarded as expressing a sort of reciprocity law. (Compare (4.26) with (4.14).) It is proved under the assumption that $c(\omega)$ is of the form (4.19).

REMARK. As will be shown in the next subsection, we can unconditionally obtain a clear-cut result when $n = 1$. The drastic difference between the cases $n > 1$ and $n = 1$ can be conceptually explained by the nature of $c(\omega)$.

4.4. As an example, we take $F = \mathbf{Q}$. Let K be an imaginary quadratic filed of discriminant $-d$. We take $\Phi = \{\mathrm{id}\}$ and $w \in K$ so that $\Im(w) > 0$. We take $\mathfrak{a} = (1)$. Then, by (1.36), (4.14) can be written as

$$\begin{aligned}-\log|\eta(w)|^4 - \log\Im(w) &= \frac{w_{\mathfrak{O}}}{h_{\mathfrak{O}}}\sum_{\omega\neq1}\omega_*(c)^{-1}P_\omega(0)c(\omega)\\ &\quad + \log 4\pi - \frac{1}{2}\log|D(\mathfrak{O})| - \frac{P_1'(0)}{P_1(0)} - \frac{L'(0,\chi)}{L(0,\chi)},\end{aligned} \tag{4.27}$$

where χ is the Dirichlet character corresponding to $K/\mathbf{Q}$. Let us assume further that $\mathfrak{O} = \mathfrak{O}_K$. Then L is the Hilbert class field of K. Put $\epsilon = \epsilon_1$. We have

$$\begin{aligned}\sum_{\omega\neq1}\omega_*(c)^{-1}c(\omega) &= \frac{1}{m}\sum_{\omega\neq1}\omega_*(c)^{-1}\sum_{\tau\in G}\omega(\tau)\log|\epsilon^\tau|\\ &= \frac{1}{m}\sum_{\tau\in G}\log|\epsilon^\tau|(\sum_{\omega}\omega_*(c)^{-1}\omega(\tau) - 1) = \frac{h_K}{m}\log|\epsilon^{[\frac{L/K}{c}]}|,\end{aligned}$$

in view of (4.19) and $\prod_{\tau \in G} \epsilon^{\tau} = \pm 1$. Hence, by (4.27), we obtain

$$(4.28) \qquad |\epsilon^{[\frac{L/K}{c}]}|^{w_K/m} = \frac{\sqrt{|D_K|}}{4\pi \Im(w)} \exp(\frac{L'(0,\chi)}{L(0,\chi)}) |\eta(w)|^{-4}.$$

We note that c and w are related by (4.4) with $\mathfrak{a} = (1)$, $w = \omega_1/\omega_2$, the class of $\mathfrak{A}^{-1}$ being c.

We are going to show that (4.28) holds with a certain $\epsilon \in E_L$. This implies that Theorem 4.3 holds without assuming Stark's conjecture in this case. For a lattice L in $\mathbf{C}$, we define $\eta^{24}(L)$ as in §2.3. Let c be an ideal class of K. Let $\mathfrak{A}$ be a fractional ideal of K such that the ideal class of $\mathfrak{A}^{-1}$ is c. We have $\mathfrak{A}^{h_K} = (\alpha)$. We put

$$(4.29) \qquad \Delta(c) = \alpha^{12} \eta^{24 h_K}(\mathfrak{A}).$$

We see easily that $\Delta(c)$ does not depend on the choice of $\mathfrak{A}$. Since $|\alpha|^2 = N(\mathfrak{A})^{h_K}$, we have

$$(4.30) \qquad |\Delta(c)| = N(\mathfrak{A})^{6h_K} |\eta^{24h_K}(\mathfrak{A})|.$$

Put

$$(4.31) \qquad \eta_c = \Delta(1)/\Delta(c).$$

It is well known that η_c is a unit of L (cf. Siegel [Si1], p. 103). Put

$$X = \frac{\sqrt{|D_K|}}{4\pi \Im(w)} \exp(\frac{L'(0,\chi)}{L(0,\chi)}) |\eta(w)|^{-4},$$

the quantity on the right-hand side of (4.28). By definition, we have $\eta(w)^{24} = \eta^{24}(\mathfrak{A})\omega_2^{12}$. Using the exact Chowla-Selberg formula (2.11), we obtain

$$X^{12h_K} = (\frac{\sqrt{|D_K|}}{2\Im(w)|\omega_2|^2})^{12h_K} |\eta^{24}(\mathfrak{A})|^{-2h_K} \prod_{i=1}^{h_K} N(\mathfrak{A}_i)^{12} |\eta^{24}(\mathfrak{A}_i)|^2.$$

By Lemma 4.1, we obtain

$$\begin{aligned} X^{12h_K} &= N(\mathfrak{A})^{-12h_K} |\eta^{24}(\mathfrak{A})|^{-2h_K} \prod_{i=1}^{h_K} N(\mathfrak{A}_i)^{12} |\eta^{24}(\mathfrak{A}_i)|^2 \\ &= |\eta_c|^2 \prod_{c_1 \in C_K} (|\eta_{c_1}|^{2/h_K})^{-1}, \end{aligned}$$

where C_K denotes the ideal class group of K. Put

$$(4.32) \qquad \epsilon = \prod_{c_1 \in C_K} \eta_{c_1}^{-1}.$$

Then it is known that (cf. [Si1], p. 105)

$$(4.33) \qquad |\epsilon^{[\frac{L/K}{c}]}| = |\epsilon| |\eta_c|^{h_K}.$$

Hence we obtain

$$(4.34) \qquad X^{12h_K} = |\epsilon^{[\frac{L/K}{c}]}|^{2/h_K}$$

Therefore

$$|\epsilon^{[\frac{L/K}{c}]}|^{1/6h_K^2} = \frac{\sqrt{|D_K|}}{4\pi\Im(w)} \exp(\frac{L'(0,\chi)}{L(0,\chi)})|\eta(w)|^{-4} \tag{4.35}$$

holds and (4.28) is verified in this case with $m = 6w_K h_K^2$.

For the sake of completeness, let us prove the following proposition from which (4.33) follows. We can also deduce from it that $\prod_{\sigma\in G}\epsilon^\sigma = 1$ and that $\langle\epsilon^\sigma \mid \sigma \in G\rangle$ is of finite index in E_L.

PROPOSITION 4.4. *For ideal classes c_1, $c \in C_K$, we have*

$$(\eta_{c_1})^{[\frac{L/K}{c}]} = \eta_{c_1 c}(\eta_c)^{-1}.$$

PROOF. We prove the assertion using Shimura's reciprocity law ([Sh3], Main Theorem III) in the same way as in [Y3], p. 244 ~ 246. Let us recall some notations. Let $G = GL(2)$ considered as an algebraic group over $\mathbf{Q}$. We set

$$GL_+(2,\mathbf{R}) = \{g \in GL(2,\mathbf{R}) \mid \det(g) > 0\},$$
$$G_{\infty+} = \{g \in G_\infty \mid \det(g) > 0\} \cong GL_+(2,\mathbf{R}), \qquad G_{\mathbf{Q}+} = \{g \in G_{\mathbf{Q}} \mid \det(g) > 0\},$$
$$G_{A+} = \{g \in G_A \mid g_\infty \in G_{\infty+}\},$$
$$U = \prod_p GL(2,\mathbf{Z}_p) \times G_{\infty+},$$
$$\mathfrak{G} = \{(x,\sigma) \in G_{A+} \times \mathrm{Gal}(\overline{\mathbf{Q}}/\mathbf{Q}) \mid [\det(x)^{-1}, \mathbf{Q}] = \sigma|\mathbf{Q}_{ab}\}.$$

Then $\mathfrak{G}$ acts on the space of arithmetic automorphic forms. For a positive integer N, we set

$$U_N = \{u \in U \mid u_p \equiv \begin{pmatrix} 1 & 0 \\ 0 & 1 \end{pmatrix} \mod NM(2,\mathbf{Z}_p) \text{ for every } p \text{ which divides } N\}.$$

For $g = \begin{pmatrix} a & b \\ c & d \end{pmatrix} \in GL_+(2,\mathbf{R})$ and $z \in \mathfrak{H}$, we put $j(g,z) = cz + d$. For a function h on $\mathfrak{H}$, $g \in GL_+(2,\mathbf{R})$ and an integer t, we define a function $h \mid_t g$ on $\mathfrak{H}$ by $(h \mid_t g)(z) = h(g(z))j(g,z)^{-t}$. Take $\omega \in \mathfrak{O}_K$ so that

$$\mathfrak{O}_K = \mathbf{Z}\omega \oplus \mathbf{Z}, \qquad \Im(\omega) > 0.$$

We set

$$q(a)\begin{pmatrix} \omega \\ 1 \end{pmatrix} = \begin{pmatrix} a\omega \\ a \end{pmatrix}, \qquad a \in K,\ q(a) \in M(2,\mathbf{Q}).$$

Then q defines a homomorphism of $K^\times$ into $G_{\mathbf{Q}+}$. We can extend q to a homomorphism of $K_A^\times$ into G_{A+} and denote it by the same letter q.

Now take a fractional ideal $\mathfrak{a}$ so that the class of $\mathfrak{a}^{-1}$ is c_1 and let

$$\mathfrak{a} = \mathbf{Z}\omega_1 \oplus \mathbf{Z}\omega_2, \quad \Im(\omega_1/\omega_2) > 0.$$

Define $\alpha \in G_{\mathbf{Q}+}$ by $\begin{pmatrix} \omega_1 \\ \omega_2 \end{pmatrix} = \alpha \begin{pmatrix} \omega \\ 1 \end{pmatrix}$. Then, by definition, we have

$$\frac{\eta^{24}(\mathfrak{O}_K)}{\eta^{24}(\mathfrak{a})} = \frac{\eta^{24}(\omega)}{(\eta^{24}|_{12}\alpha)(\omega)}.$$

Let M be a positive integer such that $U_M \subset \alpha^{-1}U\alpha \cap U$. For $s \in K_A^\times$ such that $q(s)_p = 1$ whenever p divides M, choose $\sigma \in \mathrm{Gal}(\overline{\mathbf{Q}}/K)$ so that $\sigma|K_{ab} = [s, K]$. (Remark to [Y3]: In Proposition 6.1, σ should be confined to the elements of $\mathrm{Gal}(\overline{\mathbf{Q}}/K)$ such that $\sigma|K_{ab} = [s, K]$ with $s \in K_A^\times$ as above. We remark that such σ's form a dense subgroup of $\mathrm{Gal}(\overline{\mathbf{Q}}/K)$.) Put $\mathfrak{x} = (q(s)^{-1}, \sigma)$. Then $\mathfrak{x} \in \mathfrak{G}$ and Shimura's reciprocity law tells that

$$(\frac{\eta^{24}(\omega)}{(\eta^{24}|_{12}\alpha)(\omega)})^\sigma = \frac{(\eta^{24})^{\mathfrak{x}}(\omega)}{(\eta^{24}|_{12}\alpha)^{\mathfrak{x}}(\omega)}.$$

Put $\mathfrak{b} = s\mathfrak{O}_K$ and let

$$\mathfrak{b}^{-1} = \mathbf{Z}\omega_1' \oplus \mathbf{Z}\omega_2', \qquad \Im(\omega_1'/\omega_2') > 0.$$

Define $\xi \in G_{\mathbf{Q}+}$ by $\begin{pmatrix} \omega \\ 1 \end{pmatrix} = \xi \begin{pmatrix} \omega_1' \\ \omega_2' \end{pmatrix}$. Put $t = q(s)^{-1}\xi$. Then $t \in U$. For $u \in \mathbf{Q}_A^\times$, put $m(u) = \begin{pmatrix} 1 & 0 \\ 0 & u \end{pmatrix}$. Choose $\gamma \in SL(2, \mathbf{Z})$ so that $m(\det t)^{-1}t\gamma^{-1} \in U_M$. Then we have

$$(\eta^{24})^{\mathfrak{x}} = (\eta^{24})^{(m(\det t),\sigma)(m(\det t)^{-1}t\gamma^{-1},1)(\gamma\xi^{-1},1)} = \eta^{24}|_{12}\gamma\xi^{-1}.$$

Choose $\alpha' \in G_{\mathbf{Q}+}$ so that

$$\alpha\, m(\det t) = s'\alpha' \qquad \text{with} \quad s' \in U.$$

Then we have

$$\begin{aligned}(\eta^{24}|_{12}\alpha)^{\mathfrak{x}} &= (\eta^{24})^{(\alpha,1)(q(s)^{-1},\sigma)} \\ &= (\eta^{24})^{(s',\sigma)(\alpha',1)(m(\det t)^{-1}t\gamma^{-1},1)(\gamma\xi^{-1},1)} = \eta^{24}|_{12}\alpha'\gamma\xi^{-1}.\end{aligned}$$

As in [Y3], Lemma 6.2, we have

$$\mathbf{Z}^2\gamma\xi^{-1}\begin{pmatrix} \omega \\ 1 \end{pmatrix} = \mathfrak{b}^{-1}, \qquad \mathbf{Z}^2\alpha'\gamma\xi^{-1}\begin{pmatrix} \omega \\ 1 \end{pmatrix} = \mathfrak{a}\mathfrak{b}^{-1}.$$

Therefore we obtain

$$(\frac{\eta^{24}(\mathfrak{O}_K)}{\eta^{24}(\mathfrak{a})})^\sigma = \frac{\eta^{24}(\mathfrak{b}^{-1})}{\eta^{24}(\mathfrak{a}\mathfrak{b}^{-1})}.$$

We have $\sigma|L = (\frac{L/K}{\mathfrak{b}})$. This shows that $\eta^{24}(\mathfrak{O}_k)/\eta^{24}(\mathfrak{a}) \in L$ and we get

$$(\frac{\eta^{24}(\mathfrak{O}_K)}{\eta^{24}(\mathfrak{a})})^{[\frac{L/K}{\mathfrak{b}}]} = \frac{\eta^{24}(\mathfrak{b}^{-1})}{\eta^{24}(\mathfrak{a}\mathfrak{b}^{-1})}.$$

From this, we easily obtain

$$(4.36) \qquad (\frac{\Delta(1)}{\Delta(c_1)})^{[\frac{L/K}{c}]} = \frac{\Delta(c)}{\Delta(c_1 c)}$$

and the assertion follows.

5. A connection of automorphic forms with group cohomology

5.1. Let G be a group and M be a left G-module. For $0 \leqq n \in \mathbf{Z}$, let $C^n(G, M)$ denote the abelian group consisiting of all maps of G^n into M. (If $n = 0$, we put $C^0(G, M) = M$.) We define the cobounday operator

$$d : C^n(G, M) \longrightarrow C^{n+1}(G, M)$$

by the usual formula

$$\begin{aligned}(df)(g_1, \dots, g_{n+1}) &= g_1 f(g_2, \dots, g_{n+1}) \\ &+ \sum_{i=1}^{n} (-1)^n f(g_1, \dots, g_i g_{i+1}, \dots, g_{n+1}) + (-1)^{n+1} f(g_1, \dots, g_n).\end{aligned} \tag{5.1}$$

The n-th cohomology group $H^n(G, M)$, $n \geqq 0$ is defined as the n-th cohomology group of the complex $\{C^*(G, M), d\}$. An element of $C^n(G, M)$ is called *an n-chain* of G taking values in M and an n-chain which vanishes under the coboundary operator d is called *an n-cycle.*

Now let F be a totally real algebraic number field of degree n and put $J_F = \{\sigma_1, \sigma_2, \dots, \sigma_n\}$. For $\xi \in F$, we put $\xi^{(i)} = \xi^{\sigma_i}$. For $\gamma = \begin{pmatrix} a & b \\ c & d \end{pmatrix} \in GL(2, F)$, we put $\gamma^{(i)} = \begin{pmatrix} a^{(i)} & b^{(i)} \\ c^{(i)} & d^{(i)} \end{pmatrix}$. Let Γ be a congruence subgroup of $SL(2, \mathfrak{O}_F)$. We embed Γ into $SL(2, \mathbf{R})^n$ by $\gamma \longrightarrow (\gamma^{(1)}, \gamma^{(2)}, \dots, \gamma^{(n)})$. For $k = (k_1, k_2, \dots, k_n) \in \mathbf{Z}^n$, let $G_k(\Gamma)$ denote the space of all holomorphic Hilbert modular forms of weight k with respect to Γ (cf. III, §1.3).

For $0 \leqq l \in \mathbf{Z}$ and $\begin{bmatrix} u \\ v \end{bmatrix} \in \mathbf{C}^2$, put

$$\begin{bmatrix} u \\ v \end{bmatrix}^l = {}^t(u^l \; u^{l-1}v \dots uv^{l-1} \; v^l).$$

Define a representation $\rho_l : GL(2, \mathbf{C}) \longrightarrow GL(l+1, \mathbf{C})$ by

$$\rho_l(g) \begin{bmatrix} u \\ v \end{bmatrix}^l = (g \begin{bmatrix} u \\ v \end{bmatrix})^l.$$

Let $f \in G_k(\Gamma)$. We assume $k_i \geqq 2$, $1 \leqq i \leqq n$ and put $l_i = k_i - 2$. Put $\rho = \rho_{l_1} \otimes \rho_{l_2} \otimes \cdots \otimes \rho_{l_n}$ and let V be the representation space of ρ. Define a holomorphic V-valued n-form $\mathfrak{d}(f)$ on $\mathfrak{H}^n$ by

$$\mathfrak{d}(f) = f(z_1, z_2, \dots, z_n) \begin{bmatrix} z_1 \\ 1 \end{bmatrix}^{l_1} \otimes \begin{bmatrix} z_2 \\ 1 \end{bmatrix}^{l_2} \otimes \cdots \otimes \begin{bmatrix} z_n \\ 1 \end{bmatrix}^{l_n} dz_1 dz_2 \cdots dz_n. \tag{5.2}$$

Then we can verify easily that

$$\mathfrak{d}(f) \circ \gamma = \rho(\gamma) \mathfrak{d}(f), \qquad \gamma \in \Gamma. \tag{5.3}$$

Let $\mathcal{H}$ denote the vector space of all V-valued holomorphic functions on $\mathfrak{H}^n$. For $\varphi \in \mathcal{H}$ and $\gamma \in \Gamma$, we put

$$(\gamma\varphi)(z) = \rho(\gamma)\varphi(\gamma^{-1}z), \qquad z \in \mathfrak{H}^n. \tag{5.4}$$

Then $\mathcal{H}$ becomes a left Γ-module. Take "a base point" $w = (w_1, w_2, \dots, w_n) \in \mathfrak{H}^n$ and put

$$F(z) = \int_{w_1}^{z_1} \cdots \int_{w_n}^{z_n} \mathfrak{d}(f), \qquad z = (z_1, z_2, \dots, z_n) \in \mathfrak{H}^n. \tag{5.5}$$

This is a period integral of Eichler-Shimura type (cf. [Sh1], p. 233). We have $F(z) \in \mathcal{H}$ and

$$\frac{\partial}{\partial z_1}\frac{\partial}{\partial z_2} \cdots \frac{\partial}{\partial z_n} F(z) = f(z_1, z_2, \dots, z_n) \begin{bmatrix} z_1 \\ 1 \end{bmatrix}^{l_1} \otimes \begin{bmatrix} z_2 \\ 1 \end{bmatrix}^{l_2} \otimes \cdots \otimes \begin{bmatrix} z_n \\ 1 \end{bmatrix}^{l_n}. \tag{5.6}$$

By (5.3), we have, for $\gamma \in \Gamma$,

$$F(\gamma^{-1}(z)) = \rho(\gamma^{-1}) \int_{\gamma^{(1)}w_1}^{z_1} \cdots \int_{\gamma^{(n)}w_n}^{z_n} \mathfrak{d}(f). \tag{5.7}$$

Hence we get

$$\frac{\partial}{\partial z_1}\frac{\partial}{\partial z_2} \cdots \frac{\partial}{\partial z_n}(\gamma F - F) = 0. \tag{5.8}$$

LEMMA 5.1. *Let D be an open domain in $\mathbf{C}^n$ and let f be a holomorphic function on D. We assume that D is contractible.*

1. *If f satisfies*
$$\frac{\partial}{\partial z_1}\frac{\partial}{\partial z_2} \cdots \frac{\partial}{\partial z_n} f(z) = 0, \qquad z = (z_1, z_2, \dots, z_n) \in D$$
then we can express $f(z)$ as $f(z) = \sum_{i=1}^n f_i(z)$ with a holomorphic function $f_i(z)$ on D which does not depend on z_i.
2. *Assume $n \geqq 2$. If $f(z) = \sum_{i=2}^n f_i(z)$ with a holomorphic function $f_i(z)$ on D which does not depend on z_i and if $f(z)$ does not depend on z_1, then we have $f(z) = \sum_{i=2}^n g_i(z)$ with a holomorphic function $g_i(z)$ on D which does not depend on z_1 and z_i.*

PROOF. To show 1, we prove the following assertion by induction on k: If a holomorphic function $f(z)$ on D satisfies $\frac{\partial}{\partial z_1}\frac{\partial}{\partial z_2} \cdots \frac{\partial}{\partial z_k} f(z) = 0$, then we can write $f(z) = \sum_{i=1}^k f_i(k)$ with a holomorphic function $f_i(z)$ on D which does not depend on z_i. The assertion is obvious if $k = 1$. Assume $k \geqq 2$ and put $h(z) = \frac{\partial}{\partial z_1}\frac{\partial}{\partial z_2} \cdots \frac{\partial}{\partial z_{k-1}} f(z)$. Then $h(z)$ is a holomorphic function which does not depend on z_k. Let p_k be the projection map $\mathbf{C}^n \ni (z_1, z_2, \dots, z_n) \longrightarrow (z_1, \dots, z_{k-1}, z_{k+1}, \dots, z_n) \in \mathbf{C}^{n-1}$ and put $D_k = p_k(D)$. We can regard $h(z)$ as a holomorphic function on D_k. In other words, we have $h(z) = h_k(p_k(z))$, $z \in D$ with a holomorphic function h_k on D_k. The contractibility of D implies that of D_k. We see that D_k is a simply connected open domain in $\mathbf{C}^{n-1}$. Therefore there exists a holomorphic function $H_k(z)$ on D_k such that $\frac{\partial}{\partial z_1}\frac{\partial}{\partial z_2} \cdots \frac{\partial}{\partial z_{k-1}} H_k(z) = h_k(z)$. Put $f_k(z) = H_k(p_k(z))$, $z \in D$. Then we have $\frac{\partial}{\partial z_1}\frac{\partial}{\partial z_2} \cdots \frac{\partial}{\partial z_{k-1}} f_k(z) = h(z)$. Now we have

$$\frac{\partial}{\partial z_1}\frac{\partial}{\partial z_2} \cdots \frac{\partial}{\partial z_{k-1}}(f(z) - f_k(z)) = 0.$$

By the induction hypothesis, we obtain $f(z) - f_k(z) = \sum_{i=1}^{k-1} f_i(z)$ with a holomorphic function $f_i(z)$ on D which does not depend on z_i. This proves 1.

In the situation of 2, we have $\frac{\partial}{\partial z_2} \cdots \frac{\partial}{\partial z_n} f(z) = 0$. Applying 1 to the domain $p_1(D)$, where $p_1 : \mathbf{C}^n \ni (z_1, z_2, \dots, z_n) \longrightarrow (z_2, \dots, z_n) \in \mathbf{C}^{n-1}$ is the projection map, the assertion follows. This completes the proof.

By (5.8) and Lemma 5.1, 1, we see that $(\gamma F - F)(z) = \rho(\gamma) F(\gamma^{-1} z) - F(z)$ can be decomposed as[2]

$$(\gamma F - F)(z) = \sum_{i=1}^{n} f_1^{(i)}(\gamma; z), \qquad \gamma \in \Gamma, \tag{5.9}$$

where $f_1^{(i)}(\gamma; z) \in \mathcal{H}$ is a function of $z_1, \dots, \check{z}_i, \dots, z_n$. If $n = 1$, $f_1^{(1)}(\gamma; z)$ is a constant in V and is a 1-cocycle of Γ. Assume $n \geqq 2$. Regard $f_1^{(i)}$ as a 1-chain of Γ taking values in $\mathcal{H}$. Taking the coboundary $df_1^{(1)}$, we get, for $\gamma_1, \gamma_2 \in \Gamma$,

$$(df_1^{(1)})(\gamma_1, \gamma_2; z) = -\sum_{i=2}^{n} (df_1^{(i)})(\gamma_1, \gamma_2; z)$$

by (5.9). Since the right-hand side does not depend on z_1, by Lemma 5.1, 2, we can write

$$(df_1^{(1)})(\gamma_1, \gamma_2; z) = \sum_{i=2}^{n} f_2^{(i)}(\gamma_1, \gamma_2; z), \qquad \gamma_1, \gamma_2 \in \Gamma,$$

where $f_2^{(i)} \in \mathcal{H}$ is a function of $z_2, \dots, \check{z}_i, \dots, z_n$. By induction on m, let us show that we have a decomposition

$$\begin{aligned} &(df_m^{(m)})(\gamma_1, \gamma_2, \dots, \gamma_{m+1}; z_{m+1}, \dots, z_n) \\ &= \sum_{i=m+1}^{n} f_{m+1}^{(i)}(\gamma_1, \gamma_2, \dots, \gamma_{m+1}; z_{m+1}, \dots, \check{z}_i, \dots, z_n), \end{aligned} \tag{5.10}$$

$1 \leqq m \leqq n-1$. In (5.10), $f_{m+1}^{(i)}(\gamma_1, \gamma_2, \dots, \gamma_{m+1}; z_{m+1}, \dots, \check{z}_i, \dots, z_n)$ denotes a function in $\mathcal{H}$ which depends only on $z_{m+1}, \dots, z_{i-1}, z_{i+1}, \dots, z_n$. We have shown (5.10) when $m = 1$. Assume $n \geqq 3$ and that (5.10) holds for m, $1 \leqq m \leqq n-2$. By (5.10), we have

$$\begin{aligned} &(df_{m+1}^{(m+1)})(\gamma_1, \gamma_2, \dots, \gamma_{m+2}; z_{m+2}, \dots, z_n) \\ &= -\sum_{i=m+2}^{n} (df_{m+1}^{(i)})(\gamma_1, \gamma_2, \dots, \gamma_{m+2}; z_{m+1}, \dots, \check{z}_i, \dots, z_n). \end{aligned}$$

Since the right-hand side does not depend on z_{m+1}, we can write, by Lemma 5.1, 2,

$$\begin{aligned} &(df_{m+1}^{(m+1)})(\gamma_1, \gamma_2, \dots, \gamma_{m+2}; z_{m+2}, \dots, z_n) \\ &= \sum_{i=m+2}^{n} f_{m+2}^{(i)}(\gamma_1, \gamma_2, \dots, \gamma_{m+2}; z_{m+2}, \dots, \check{z}_i, \dots, z_n). \end{aligned}$$

---

[2]Here is a conflict of notation; but there is no fear to confuse the function F with the field F.

Thus we proved (5.10) for m replaced by $m+1$.

Now, when $m = n-2$, (5.10) gives

$$\begin{aligned}(df_{n-2}^{(n-2)})&(\gamma_1, \dots, \gamma_{n-1}; z_{n-1}, z_n)\\ &= f_{n-1}^{(n-1)}(\gamma_1, \dots, \gamma_{n-1}; z_n) + f_{n-1}^{(n)}(\gamma_1, \dots, \gamma_{n-1}; z_{n-1}).\end{aligned}$$

(If $n = 2$, this relation is understood to be (5.9).) Taking the coboundary, we get

$$(df_{n-1}^{(n-1)})(\gamma_1, \dots, \gamma_n; z_n) = -(df_{n-1}^{(n)})(\gamma_1, \dots, \gamma_n; z_{n-1}).$$

Therefore $(df_{n-1}^{(n-1)})(\gamma_1, \dots, \gamma_n; z_n) \in V$ is a constant and is an n-cocycle. Put $x(f)(\gamma_1, \dots, \gamma_n) = (df_{n-1}^{(n-1)})(\gamma_1, \dots, \gamma_n; z_n)$ and let $\xi(f)$ be the cohomology class of $x(f)$ in $H^n(\Gamma, V)$.

The decomposition (5.10) is not unique. The choice of $f_{m+1}^{(m+1)}(\gamma_1, \dots, \gamma_{m+1}; z_{m+2}, \dots, z_n)$ has the freedom of the addition of a function of the form

$$(*) \qquad \sum_{i=m+2}^{n} g_{m+1}^{(i)}(\gamma_1, \dots, \gamma_{m+1}; z_{m+2}, \dots, \check{z}_i, \dots, z_n), \qquad g_{m+1}^{(i)} \in \mathcal{H}.$$

This has the effect $f_{m+2}^{(m+2)} \longrightarrow f_{m+2}^{(m+2)} + dg_{m+1}^{(m+2)}$ and its contribution to $df_{n-1}^{(n-1)}$ vanishes if $m < n-2$. If $m = n-2$, $(*)$ is a constant $\in V$ and the cohomology class of $x(f)$ does not change.

If the base point is changed so that w_i is replaced by w_i' for one i, $1 \leqq i \leqq n$, then $F(z)$ changes to $F(z) + \psi(z_1, \dots, \check{z}_i, \dots, z_n)$ with $\psi \in \mathcal{H}$. By the above consideration, we see that the cohomology class of $x(f)$ does not change. Therefore we obtain the following theorem.

THEOREM 5.2. *Let the notation be the same as above. For $f \in G_k(\Gamma)$, define $F(z)$ by* (5.2) *and* (5.5). *For $0 \leqq m \leqq n-1$, $m+1 \leqq i \leqq n$, $\gamma_1, \dots, \gamma_{m+1} \in \Gamma$, we can find $f_{m+1}^{(i)}(\gamma_1, \dots, \gamma_{m+1}; z_{m+1}, \dots, \check{z}_i, \dots, z_n) \in \mathcal{H}$ so that* (5.9) *and* (5.10) *hold. Then $(df_{n-1}^{(n-1)})(\gamma_1, \dots, \gamma_n; z_n) = f_n^{(n)}(\gamma_1, \dots, \gamma_n)$ is a constant in V and is an n-cocycle of Γ. Its cohomology class in $H^n(\Gamma, V)$ does not depend on the choice of the base point $w \in \mathfrak{H}^n$ and the choice of the decompositions* (5.9) *and* (5.10).

REMARK 5.3. By a similar construction, we can directly attach cohomology classes to holomorphic automorphic forms of a wide class, defined on hermitian symmetric spaces. As far as the author knows, this construction did not appear in literatures explicitly in the higher dimensional case. Our construction is motivated by (5.16) in the next subsection.

EXAMPLE 5.4. Assume $n = 2$. By a direct calculation, we see easily that

$$(\gamma F - F)(z_1, z_2) = -\int_{\gamma^{(1)}w_1}^{z_1} \int_{w_2}^{\gamma^{(2)}w_2} \mathfrak{d}(f) - \int_{w_1}^{\gamma^{(1)}w_1} \int_{w_2}^{z_2} \mathfrak{d}(f).$$

Hence we may take

$$f_1^{(1)}(\gamma; z) = -\int_{\gamma^{(1)}w_1}^{z_1} \int_{w_2}^{\gamma^{(2)}w_2} \mathfrak{d}(f), \qquad f_2^{(2)}(\gamma; z) = -\int_{w_1}^{\gamma^{(1)}w_1} \int_{w_2}^{z_2} \mathfrak{d}(f).$$

A 2-cocycle $x(f)$ attached to f is given by

$$x(f)(\gamma_1,\gamma_2)=\rho(\gamma_1)f_1^{(1)}(\gamma_2;\gamma_1^{-1}z)-f_1^{(1)}(\gamma_1\gamma_2;z)+f_1^{(1)}(\gamma_1;z).$$

Hence we find

$$x(f)(\gamma_1,\gamma_2)=\int_{(\gamma_1\gamma_2)^{(1)}w_1}^{\gamma_1^{(1)}w_1}\int_{w_2}^{\gamma_1^{(2)}w_2}\mathfrak{d}(f),\qquad \gamma_1,\gamma_2\in\Gamma. \tag{5.11}$$

Assume that $f\in S_k(\Gamma)$, i.e., f is a cusp form. Let $p\in\Gamma$ be a parabolic element and $w^*=(w_1^*,w_2^*)$ be the fixed point of p. Using the condition that f is a cusp form, we may change the base point (w_1,w_2) to (w_1,w_2^*). Let $x^*(f)$ be the cocycle obtained from f using (w_1,w_2^*) as the base point. Then $x(f)$ and $x^*(f)$ define the same cohomology class. On the other hand, we have $x^*(f)(p,\gamma)=0$ for every $\gamma\in\Gamma$ by (5.11) . Therefore there exists a 1-chain $a(\gamma)\in C^1(\Gamma,V)$ such that

$$x(f)(p,\gamma)=\rho(p)a(\gamma)-a(p\gamma)+a(p),\qquad \gamma\in\Gamma. \tag{5.12}$$

A similar relation holds for $x(f)(\gamma,p)$. The condition (5.12) is an analogue of the parabolic condition on the cohomology of a Fuchsian group (cf. [Sh1], 8.1, 8.2).

5.2. Let us consider the relation among holomorphic Eisenstein series of weight 2, the h-function studied in the previous sections and a function studied by Hecke. For the sake of simplicity, we assume that the class number of F is 1. We write $h_1(z;(1))$ as $h(z)$. By (1.30), we have

$$\begin{aligned}h(z)=&\frac{D_F}{2^{n-2}\pi^nR_F}\Big[\zeta_F(2)y^{\mathbf{a}}\\&+\pi^nD_F^{-3/2}\sum_{0\neq b\in\mathfrak{d}_F^{-1}}|N(b)|^{-1}\sigma_1(bd)\exp(2\pi i(\sum_{v\in\mathfrak{a}}b_vx_v+i|b_vy_v|))\Big],\end{aligned} \tag{5.13}$$

where we put $\sigma_1=\sigma_{1,1}$.

Next we consider Eisenstein series of weight $(2,2,\dots,2)$. In (1.8), we take $k=(2,2,\dots,2)$ and λ as the characteristic function of the closure of $\mathfrak{O}_F\oplus\mathfrak{O}_F$, $U=E_F$. The Fourier expansion of the Eisenstein series $E_k(z,s;\lambda)$ is given by (1.14). We see that $E_k(z,s;\lambda)$ is holomorphic at $s=1$ and put $E_2(z)=E_k(z,1;\lambda)$. We assume $n\geqq 2$. In (1.14), we have

$$D_k(2s,\kappa_A)=D_k(2s,\kappa_B)=\zeta_F(2s).$$

By Lemma 2.4 of appendix I, we have

$$\lim_{s\to1}\Xi(y,0;(s+1)\mathbf{a},(s-1)\mathbf{a})\zeta_F(2s-1)=0$$

since $n\geqq 2$. For $b\in F^\times$, we have, by Lemma 2.6 of appendix I,

$$\Xi(y,b;2\mathbf{a},0\cdot\mathbf{a})=(-1)^n(2\pi)^{2n}\mathbf{e}_{\mathbf{a}}(iby)$$

if b is totally positive and 0 otherwise. Using Lemma 1.3 and (1.28), we obtain the Fourier expansion

$$E_2(z)=\zeta_F(2)+(2\pi i)^{2n}D_F^{-3/2}\sum_{0\ll b\in\mathfrak{d}_F^{-1}}\sigma_1(bd)\mathbf{e}_{\mathbf{a}}(bz). \tag{5.14}$$

By (1.9), we have $E_2(z)\in G_{(2,2,\dots,2)}(SL(2,\mathfrak{O}_F))$.

For $z = (z_1, z_2, \dots, z_n) \in \mathfrak{H}^n$, we put

$$
\begin{aligned}
H(z) = \frac{D_F}{2^{n-2}\pi^n R_F}\Bigg[&\frac{1}{(2\sqrt{-1})^n}\zeta_F(2)\prod_{i=1}^{n} z_i \\
&+ \pi^n D_F^{-3/2} \sum_{0\ll b\in\mathfrak{d}_F^{-1}} \frac{\sigma_1(bd)}{N(b)}\mathbf{e}_{\mathbf{a}}(bz)\Bigg].
\end{aligned}
\tag{5.15}
$$

Up to a constant multiple, this is the holomorphic function considered by Hecke for the case $n = 2$ (Werke [He], p. 398). We have

$$
\frac{\partial}{\partial z_1}\frac{\partial}{\partial z_2}\cdots\frac{\partial}{\partial z_n}H(z) = \frac{D_F}{2^{n-2}(2\pi i)^n R_F}E_2(z).
\tag{5.16}
$$

Let S denote the set of all maps from $\{1, 2, \dots, n\}$ to $\{1, -1\}$. For $\delta \in S$ and $b \in F$, we write $\operatorname{sgn} b = \delta$ if $b^{(i)}/\delta(i) > 0$, $1 \leqq i \leqq n$. For $\delta \in S$, $z = (z_1, z_2, \dots, z_n) \in \mathfrak{H}^n$, $z_i = x_i + \sqrt{-1}y_i$, we put

$$
C_\delta(z) = \prod_{i=1}^{n}(\delta(i)x_i + \sqrt{-1}y_i),
$$

$$
\begin{aligned}
H_\delta(z) = \frac{D_F}{2^{n-2}\pi^n R_F}\Bigg[&\frac{1}{(2\sqrt{-1})^n}\zeta_F(2)C_\delta(z) \\
&+ \pi^n D_F^{-3/2} \sum_{b\in\mathfrak{d}_F^{-1},\, \operatorname{sgn} b=\delta} \frac{\sigma_1(bd)}{|N(b)|}\exp(2\pi i(\sum_{v\in\mathbf{a}} b_v x_v + i|b_v y_v|))\Bigg].
\end{aligned}
\tag{5.17}
$$

Since $\sum_{\delta\in S} C_\delta(z) = (2\sqrt{-1})^n \prod_{i=1}^{n} y_i$, we have

$$
\sum_{\delta\in S} H_\delta(z) = h(z).
\tag{5.18}
$$

We have $H(z) = H_{\delta_0}(z)$ where $\delta_0 \in S$ is the map such that $\delta_0(i) = 1$, $1 \leqq i \leqq n$; (5.18) gives a relation of Hecke's $H(z)$ to $h(z)$. If $\delta(i) > 0$ (resp. $\delta(i) < 0$), then $H_\delta(z)$ is holomorphic (resp. anti-holomorphic) with respect to z_i. By (5.16), we have

$$
\frac{\partial}{\partial z_1}\frac{\partial}{\partial z_2}\cdots\frac{\partial}{\partial z_n}h(z) = \frac{D_F}{2^{n-2}(2\pi i)^n R_F}E_2(z),
\tag{5.19}
$$

where $\dfrac{\partial}{\partial z_i} = \dfrac{1}{2}(\dfrac{\partial}{\partial x_i} - i\dfrac{\partial}{\partial y_i})$ as usual. There are several relations among $H_\delta(z)$, $\delta \in S$. By the definition (5.17), we have

$$
\overline{H_\delta(z)} = H_\delta(-\bar{z}) = H_{-\delta}(z).
\tag{5.20}
$$

For $\epsilon \in E_F$, put $\underline{\epsilon}(i) = \operatorname{sgn} \epsilon^{(i)}$, $1 \leqq i \leqq n$ and regard $\underline{\epsilon} \in S$. For $z \in \mathfrak{H}^n$, we put

$$
C_\epsilon(z) = (\dots, x_i + \sqrt{-1}\underline{\epsilon}(i)y_i, \dots).
$$

Then we can verify easily that

$$
H_\delta(\epsilon C_\epsilon(z)) = H_{\delta\underline{\epsilon}}(z), \qquad z \in \mathfrak{H}^n,
\tag{5.21}
$$

where $\epsilon C_\epsilon(z) = (\dots, \epsilon^{(i)}(x_i + \sqrt{-1}\underline{\epsilon}(i)y_i), \dots)$ and $\delta\underline{\epsilon}(i) = \delta(i)\underline{\epsilon}(i)$, $1 \leqq i \leqq n$.

EXAMPLE 5.5. Let $n = 2$ and let ϵ_0 be the fundamental unit of F. Assume that $N(\epsilon_0) = -1$. By (5.18), (5.20) and (5.21), we have

$$h(z_1, z_2) = H(z_1, z_2) + \overline{H(z_1, z_2)} + H(\epsilon_0^{(1)} z_1, \epsilon_0^{(2)} \bar{z}_2) + \overline{H(\epsilon_0^{(1)} z_1, \epsilon_0^{(2)} \bar{z}_2)}. \tag{5.22}$$

Exercises

1. Let F be a totally real algebraic number field of degree n. We assume that the class number of F in the narrow sense is 1. Let $\Gamma = SL(2, \mathfrak{O}_F)$. For a prime ideal $\mathfrak{p}$ of F take $\pi \in F$ so that $\mathfrak{p} = (\pi)$, $\pi \gg 0$. Let

$$\Gamma \begin{pmatrix} 1 & 0 \\ 0 & \pi \end{pmatrix} \Gamma = \bigsqcup_{i=1}^{N(\mathfrak{p})+1} \Gamma \gamma_i$$

be a coset decomposition. Write $E_1(z, s; (1))$ of (1.23) as $E(z, s)$. Put

$$(E|T(\mathfrak{p}))(z, s) = \sum_{i=1}^{N(\mathfrak{p})+1} E(\gamma_i(z), s).$$

This is the action of the Hecke operator $T(\mathfrak{p})$. It does not depend on the choice of γ_i.

(a) Show that

$$(E|T(\mathfrak{p}))(z, s) = (N(\mathfrak{p})^{1-s} + N(\mathfrak{p})^s) E(z, s).$$

(b) Write $h_1(z; (1))$ of (1.30) as $h(z)$. From (a) and the limit formula (1.32), deduce that

$$\sum_{i=1}^{N(\mathfrak{p})+1} h(\gamma_i(z)) + 2 \log |J(\gamma_i, z)| = (N(\mathfrak{p}) + 1) h(z) + 2N(\mathfrak{p}) \log N(\mathfrak{p}).$$

Here for $g = (g_v)_{v \in \mathbf{a}} \in GL(2, \mathbf{R})_+^{\mathbf{a}}$ and $z = (z_v)_{v \in \mathbf{a}} \in \mathfrak{H}^{\mathbf{a}}$, we put $J(g, z) = \prod_{v \in \mathbf{a}} (c_v z_v + d_v)$.

(c) Specializing (b) to the case $F = \mathbf{Q}$, show that

$$\sum_{u=0}^{p-1} \log \left| \eta(\frac{z+u}{p}) \right| + \log |\eta(pz)| = (p+1) \log |\eta(z)|.$$

(d) Using (c) and the automorphic property of $\log |\eta(z)|$, show that

$$\frac{\eta(2i)}{\eta(i)} = 2^{-3/8}.$$

2. Using Exercise 1, (c), show that

$$\sum_{u=0}^{p-1} \log \eta(\frac{z+u}{p}) + \log \eta(pz) = (p+1) \log \eta(z) + i\theta$$

with a constant $\theta \in \mathbf{R}$. Using the definition of $\log \eta(z)$ (cf. §1.8), show that $\theta = \frac{\pi}{24}(p-1)$.

3. Let D be a connected open domain in $\mathbf{C}^2$. Let f be a holomorphic function on D such that $\dfrac{\partial}{\partial z_1}\dfrac{\partial}{\partial z_2} f(z_1, z_2) = 0$.
   (a) Show that f can be written as $f(z_1, z_2) = g(z_1) + h(z_2)$ with a holomorphic function $g(z_1)$ (resp. $h(z_2)$) on D which depends only on z_1 (resp. z_2) if $H^1(D, \mathbf{C}) = 0$. (Compare with Lemma 5.1.)
   (b) Find an example of D and f for which the expression as in (a) is impossible.

APPENDIX I

Eisenstein series on GL(2)

In this appendix, we will study Eisenstein series on $GL(2)$ over an arbitrary global field.

In §1, we introduce Eisenstein series on $GL(2)$. The Fourier expansion of Eisenstein series motivates our calculation of local integrals in §2. In §3, we will prove a functional equation of Eisenstein series on $GL(2)$ following the method of Jacquet [J]. This method is a natural extension of the proof of the functional equation of Hecke's L-function given in Weil [Wel], which can be regarded as the theory for $GL(1)$. In §4, we will study Eisenstein series of level 1 and prove a limit formula over an arbitrary algebraic number field, which is an adelic version of the formula given in chapter V, §1. Essentially the same formula was obtained by Asai [As] when the class number of the base field is 1.

1. Eisenstein series on $GL(2)$

1.1. Let k be a global field. We denote by $\mathbf{h}$ (resp. $\mathbf{a}$) the set of all finite (resp. infinite) places of k. By $\mathbf{a}_r$ (resp. $\mathbf{a}_c$), we denote the set of real (resp. imaginary) places of k. Thus $\mathbf{a} = \mathbf{a}_r \sqcup \mathbf{a}_c$. For $v \in \mathbf{h}$, $\mathfrak{O}_v$ (resp. ϖ_v) denotes the maximal compact subring (resp. a prime element) of k_v. For an algebraic group G defined over k and a field $K \supset k$, G_K denotes the group of K-rational points of G. For a place v of k, G_{k_v} will be abbreviated to G_v. We denote the adelization of G by G_A.

We put $G = GL(2)$. We consider G as an algebraic group defined over k. Let

$$P = \left\{ \begin{pmatrix} a & b \\ 0 & d \end{pmatrix} \in G \right\}, \qquad N = \left\{ \begin{pmatrix} 1 & b \\ 0 & 1 \end{pmatrix} \in G \right\}$$

be a Borel subgroup of G and its unipotent radical. Put $w = \begin{pmatrix} 0 & 1 \\ -1 & 0 \end{pmatrix} \in G$. Since

$$\begin{pmatrix} a & b \\ c & d \end{pmatrix} = \begin{pmatrix} b - ac^{-1}d & -a \\ 0 & -c \end{pmatrix} w \begin{pmatrix} 1 & c^{-1}d \\ 0 & 1 \end{pmatrix}, \qquad c \neq 0,$$

we have the Bruhat decomposition

$$G = P \sqcup PwP = P \sqcup PwN. \tag{1.1}$$

For a place v of k, we define a compact subgroup K_v of G_v as follows:

$$\begin{aligned} K_v &= GL(2, \mathfrak{O}_v), && v \in \mathbf{h}, \\ K_v &= SO(2, \mathbf{R}), && v \in \mathbf{a}_r, \\ K_v &= SU(2, \mathbf{C}), && v \in \mathbf{a}_c. \end{aligned}$$

We put $K = \prod_{v \in \mathbf{h} \cup \mathbf{a}} K_v$. Then we have the Iwasawa decomposition

$$G_A = P_A K. \tag{1.2}$$

Let F be a continuous function on G_A which satisfies the following conditions (A1) $\sim$ (A4).

(A1) F is left G_k-invariant.

(A2) For $v \in \mathbf{h}$, F is right invariant under an open compact subgroup of G_v.

(A3) For almost all (i.e., except for finitely many) $v \in \mathbf{h}$, F is right K_v-invariant.

(A4) For $v \in \mathbf{a}$ and $g \in G_A$, $F(gh)$ is a C^∞-function of $h \in G_v$.

Then we define a function H on $k_A \times k_A^\times$ by

$$H(x,y) = F\left(\begin{pmatrix} y & x \\ 0 & 1 \end{pmatrix}\right), \qquad x \in k_A, \quad y \in k_A^\times. \tag{1.3}$$

By (A1), we see that

$$H(x+\xi, y) = H(x,y) \qquad \text{for all} \quad \xi \in k, \tag{1.4}$$

$$H(\eta x, \eta y) = H(x,y) \qquad \text{for all} \quad \eta \in k^\times. \tag{1.5}$$

Let ψ be a non-trivial character of k_A which is trivial on k. To obtain the Fourier expansion of H, we need the following Lemma.

LEMMA 1.1. *Let φ be a complex valued continuous function on k_A. We assume that:*

(i) *φ is invariant under the translation by k.*
(ii) *For $v \in \mathbf{h}$ and $x \in k_A$, $\varphi(x+y)$, as a function of $y \in k_v$, is invariant under the translation by an open compact subgroup of k_v*
(iii) *For almost all $v \in \mathbf{h}$, φ is invariant under the translation by $\mathfrak{O}_v$.*
(iv) *For $v \in \mathbf{a}$ and $x \in k_A$, $\varphi(x+y)$ is a C^∞-function of $y \in k_v$.*

Then we have the Fourier expansion

$$\varphi(x) = \sum_{\xi \in k} c(\xi)\psi(\xi x), \qquad c(\xi) = \int_{k_A/k} \varphi(x)\psi(-\xi x)dx,$$

the Fourier series being absolutely uniformly convergent. Here dx denotes the Haar measure on k_A/k normalized so that $\mathrm{vol}(k_A/k) = 1$.

PROOF. Since k_A/k is compact and $x \longrightarrow \psi(\xi x)$ gives all characters of k_A/k when ξ extends over k, $\{\psi(\xi x)\}_{\xi \in k}$ is a complete orthonormal system of $L^2(k_A/k)$. Hence the Fourier series converges to φ in $L^2(k_A/k)$, when we regard φ as a function on k_A/k. Therefore it is sufficient to show that the Fourier series converges absolutely uniformly, which follows if we can show $\sum_{\xi \in k} |c(\xi)| < \infty$.

For $y \in k_A$, we have

$$\int_{k_A/k} \varphi(x+y)\psi(-\xi x)dx = \int_{k_A/k} \varphi(x+y)\psi(-\xi(x+y))\psi(\xi y)dx = \psi(\xi y)c(\xi). \tag{1.6}$$

By our assumption, for every $v \in \mathbf{h}$, there exists an integer $n_v \geqq 0$ such that φ is invariant under the translation by $\varpi_v^{n_v}\mathfrak{O}_v$. If $c(\xi) \neq 0$, then by (1.6), we see that $\psi_v(\xi y) = 1$ for all $y \in \varpi_v^{n_v}\mathfrak{O}_v$. Let δ_v be the order of ψ_v, i.e., $\varpi_v^{-\delta_v}\mathfrak{O}_v$

is the largest ideal on which ψ_v is trivial. Thus we see that $c(\xi) \neq 0$ implies $\xi \in k \cap \prod_{v \in \mathbf{h}} \varpi^{-n_v-\delta_v}\mathfrak{O}_v$. Put $\mathfrak{X} = k \cap \prod_{v \in \mathbf{h}} \varpi^{-n_v-\delta_v}\mathfrak{O}_v$. Since $n_v = \delta_v = 0$ for almost all v, $\mathfrak{X}$ is a fractional ideal of k if the characteristic of k is 0 and $\mathfrak{X}$ is a finite set if the characteristic of k is positive. To prove our assertion, we may assume that the characteristic of k is 0. By the approximation theorem, $k + k_\infty$ is dense in k_A. Hence we have $k_A = k + \prod_{v \in \mathbf{h}} \mathfrak{O}_v \times k_\infty$,

$$k_A/k \cong (\prod_{v \in \mathbf{h}} \mathfrak{O}_v \times k_\infty)/\mathfrak{O}_k. \tag{1.7}$$

Put $U = \prod_{v \in \mathbf{h}} \mathfrak{O}_v \times k_\infty$, $U' = \prod_{v \in \mathbf{h}} \varpi_v^{n_v}\mathfrak{O}_v \times k_\infty$, $L = U' \cap \mathfrak{O}_k$ and let

$$U = \sqcup_{i=1}^m (c_i + U'), \qquad c_i \in U \cap (k_A)_f$$

be a coset decomposition, where $(k_A)_f$ denotes the finite part of k_A. Let $\xi \in \mathfrak{X}$. Then $\varphi(x)$ and $\psi(-\xi x)$ are invariant under the translation by $\prod_{v \in \mathbf{h}} \varpi_v^{n_v}\mathfrak{O}_v$. Hence we obtain

$$\begin{aligned} c(\xi) &= \int_{U/\mathfrak{O}_k} \varphi(x)\psi(-\xi x)dx = \sum_{i=1}^m \int_{c_i+U'/L} \varphi(x)\psi(-\xi x)dx \\ &= \sum_{i=1}^m \psi(-c_i\xi) \int_{k_\infty/L} \varphi(c_i + x_\infty)\psi_\infty(-\xi x_\infty)dx_\infty. \end{aligned}$$

Here dx_∞ is a Haar measure on k_∞/L which does not depend on φ, ξ and i. Now k_∞ is isomorphic to the Euclidean space $\mathbf{R}^n$, $n = [k : \mathbf{Q}]$, L is a lattice in k_∞. By introducing suitable coordinates, we may assume that $k_\infty = \mathbf{R}^n$, $L = \mathbf{Z}^n$. Then $\sum_{\xi \in \mathfrak{X}} |c(\xi)| < \infty$ follows from the corresponding fact for the usual Fourier series for a C^∞-function on $\mathbf{R}^n/\mathbf{Z}^n$. This completes the proof.

Applying Lemma 1.1, we obtain the Fourier expansion

$$H(x,y) = \sum_{\xi \in k} c(\xi, y)\psi(\xi x),$$

$$c(\xi, y) = \int_{k_A/k} H(x,y)\psi(-\xi x)dx. \tag{1.8}$$

Put $c_0(y) = c(0,y)$. Since

$$H(\eta x, \eta y) = c_0(\eta y) + \sum_{\xi \in k^\times} c(\xi, \eta y)\psi(\xi\eta x) = c_0(\eta y) + \sum_{\xi \in k^\times} c(\eta^{-1}\xi, \eta y)\psi(\xi x)$$

for every $\eta \in k^\times$, we obtain

$$\begin{aligned} c_0(\eta y) &= c_0(y) \qquad \text{for every} \quad \eta \in k^\times,\ y \in k_A^\times, \\ c(\xi, y) &= c(\eta^{-1}\xi, \eta y) \qquad \text{for every} \quad \eta \in k^\times,\ y \in k_A^\times. \end{aligned}$$

Hence

$$c(\xi, y) = c(1, \xi y) \qquad \text{for every} \quad \xi \in k^\times,\ y \in k_A^\times.$$

Let $d \in k_A^\times$ be a differental idele attached to ψ and put

$$c(y) = c(1, d^{-1}y) \qquad \text{for} \quad y \in k_A^\times.$$

Then the Fourier expansion can be written as

$$H(x,y) = c_0(y) + \sum_{\xi \in k^\times} c(\xi dy)\psi(\xi x). \tag{1.9}$$

1.2. Let χ_1 and χ_2 be Hecke characters of $k_A^\times$. We have

$$|\chi_i(x)| = |x|_A^{\sigma_i}, \qquad x \in k_A^\times, \quad i = 1, 2 \tag{1.10}$$

with $\sigma_i \in \mathbf{R}$. Let f be a continuous function on G_A which satisfies

$$f(pg) = \chi_1(a)\chi_2(d)\left|\frac{a}{d}\right|_A^{1/2} f(g) \qquad \text{for every} \quad p = \begin{pmatrix} a & b \\ 0 & d \end{pmatrix} \in P_A,\ g \in G_A. \tag{1.11}$$

We further assume that f satisfies the following conditions (B1) $\sim$ (B3).

(B1) For $v \in \mathbf{h}$, f is right invariant under an open compact subgroup of G_v.

(B2) For almost all $v \in \mathbf{h}$, f is right K_v-invariant.

(B3) For $v \in \mathbf{a}$ and $g \in G_A$, $f(gh)$ is a C^∞-function of $h \in G_v$.

By (1.2) and (1.11), we see that f is completely determined by its restriction $f|K$ to K. We define Eisenstein series by

$$E(g,f) = \sum_{\gamma \in P_k \backslash G_k} f(\gamma g), \qquad g \in G_A. \tag{1.12}$$

LEMMA 1.2. *When g belongs to a compact subset of G_A and $\sigma_1 - \sigma_2 > 1 + \epsilon$ with $\epsilon > 0$, the series (1.12) converges absolutely uniformly.*

PROOF. For $v \in \mathbf{h} \cup \mathbf{a}$, we define a function φ_v on G_v by

$$\varphi_v(\begin{pmatrix} a & b \\ 0 & d \end{pmatrix} k) = |a|_v^{\sigma_1}|d|_v^{\sigma_2}\left|\frac{a}{d}\right|_v^{1/2}, \qquad \begin{pmatrix} a & b \\ 0 & d \end{pmatrix} \in P_v,\ k \in K_v.$$

We see easily that φ_v is well defined. Put $\varphi = \prod_{v \in \mathbf{h}\cup\mathbf{a}} \varphi_v$. Then φ satisfies (B1) $\sim$ (B3) with φ in place of f and

$$\varphi(pg) = |a|_A^{\sigma_1}|d|_A^{\sigma_2}\left|\frac{a}{d}\right|_A^{1/2} \varphi(g) \qquad \text{for every} \quad p = \begin{pmatrix} a & b \\ 0 & d \end{pmatrix} \in P_A,\ g \in G_A. \tag{1.11$'$}$$

Put $M = \max_{k \in K} |f(k)|$. Then we see that $|f(g)| \leqq M\varphi(g)$, $g \in G_A$. Hence it suffices to prove our assertion for φ. Let C be a compact subset of G_A. For $g_1 \in C$, let φ_1 be the right translation of φ by g_1, i.e., $\varphi_1(g) = \varphi(gg_1)$, $g \in G_A$. Since φ_1 satisfies (1.11$'$) with φ_1 in place of φ, we have

$$\varphi_1(g) \leqq (\max_{k \in K} \varphi_1(k)) \cdot \varphi(g), \qquad g \in G_A.$$

Put $M_1 = \max_{g \in KC} \varphi(g)$. Then we have

$$\sum_{\gamma \in P_k \backslash G_k} \varphi(\gamma g) \leqq M_1 \sum_{\gamma \in P_k \backslash G_k} \varphi(\gamma)$$

when $g \in C$. Therefore the assertion follows if we can show that $\sum_{\gamma \in P_k \backslash G_k} \varphi(\gamma)$ converges uniformly when $\sigma_1 - \sigma_2 > 1 + \epsilon$, $\epsilon > 0$.

By (1.1), a complete set of representatives of $P_k \backslash G_k$ is given by

$$\begin{pmatrix} 1 & 0 \\ 0 & 1 \end{pmatrix} \sqcup \left\{ w \begin{pmatrix} 1 & u \\ 0 & 1 \end{pmatrix} \mid u \in k \right\}.$$

Put $\sigma = \sigma_1 - \sigma_2$. As shown above, it suffices to show that $\sum_{u \in k} \varphi(w \begin{pmatrix} 1 & u \\ 0 & 1 \end{pmatrix})$ converges locally uniformly when $\sigma > 1$. We have

$$(1.13) \qquad w \begin{pmatrix} 1 & t \\ 0 & 1 \end{pmatrix} = \begin{pmatrix} -t^{-1} & 1 \\ 0 & -t \end{pmatrix} \begin{pmatrix} 1 & 0 \\ t^{-1} & 1 \end{pmatrix} \qquad \text{if} \quad t \neq 0.$$

Let $v \in \mathbf{h} \cup \mathbf{a}$. Since

$$\int_{t \in k_v, |t|_v \geqq 1} |t|_v^{-\sigma-1} dt < \infty$$

if $\sigma > 0$, we see that

$$\int_{k_v} \varphi_v(w \begin{pmatrix} 1 & t \\ 0 & 1 \end{pmatrix}) dt$$

converges if $\sigma > 0$. Furthermore Lemma 2.1 of §2 shows that

$$\int_{k_v} \varphi_v(w \begin{pmatrix} 1 & t \\ 0 & 1 \end{pmatrix}) dt = \frac{1 - q_v^{-\sigma-1}}{1 - q_v^{-\sigma}}$$

if $v \in \mathbf{h}$ when we normalize the Haar measure on k_v so that $\mathrm{vol}(\mathfrak{O}_v) = 1$. Hence, if $\sigma > 1$, the infinite product

$$\prod_{v \in \mathbf{h} \cup \mathbf{a}} \int_{k_v} \varphi_v(w \begin{pmatrix} 1 & t \\ 0 & 1 \end{pmatrix}) dt$$

converges locally uniformly and is equal to

$$\int_{k_A} \varphi(w \begin{pmatrix} 1 & x \\ 0 & 1 \end{pmatrix}) dx.$$

Assume $\sigma > 1$. Since $\varphi(w \begin{pmatrix} 1 & x \\ 0 & 1 \end{pmatrix}) \in L^1(k_A)$, $\sum_{u \in k} \varphi(w \begin{pmatrix} 1 & x+u \\ 0 & 1 \end{pmatrix})$ converges for almost all x and we have

$$\int_{k_A/k} \sum_{u \in k} \varphi(w \begin{pmatrix} 1 & x+u \\ 0 & 1 \end{pmatrix}) d\dot{x} = \int_{k_A} \varphi(w \begin{pmatrix} 1 & x \\ 0 & 1 \end{pmatrix}) dx,$$

where $d\dot{x}$ denotes the invariant measure on k_A/k determined by the Haar measure dx. For $x \in k_A$, we put

$$\varphi_x(g) = \varphi(g \begin{pmatrix} 1 & x \\ 0 & 1 \end{pmatrix}), \qquad g \in G_A, \qquad m(x) = \min_{k \in K} \varphi_x(k), \quad M(x) = \max_{k \in K} \varphi_x(k).$$

Since φ_x satisfies (1.11′) with φ_x in place of φ, we have

$$(1.14) \qquad m(x)\varphi(g) \leqq \varphi_x(g) \leqq M(x)\varphi(g), \qquad g \in G_A.$$

Hence $\sum_{u\in k}\varphi(w\begin{pmatrix}1 & x+u\\ 0 & 1\end{pmatrix})$ and $\sum_{u\in k}\varphi(w\begin{pmatrix}1 & u\\ 0 & 1\end{pmatrix})$ converges or diverges simultaneously. Thus we have shown that $\sum_{u\in k}\varphi(w\begin{pmatrix}1 & u\\ 0 & 1\end{pmatrix})$ converges if $\sigma > 1$. Taking a suitable fundamental domain F of $k_\infty/\mathfrak{O}_k$, we have

$$\begin{aligned}\int_{k_A/k}\sum_{u\in k}\varphi(w\begin{pmatrix}1 & x+u\\ 0 & 1\end{pmatrix})d\dot{x} &= \int_F\sum_{u\in k}\varphi(w\begin{pmatrix}1 & x+u\\ 0 & 1\end{pmatrix})dx\\ &\geqq \int_F m(x)\sum_{u\in k}\varphi(w\begin{pmatrix}1 & u\\ 0 & 1\end{pmatrix})dx.\end{aligned}$$

Hence

$$\sum_{u\in k}\varphi(w\begin{pmatrix}1 & u\\ 0 & 1\end{pmatrix}) \leqq \int_{k_A}\varphi(w\begin{pmatrix}1 & x\\ 0 & 1\end{pmatrix})dx \Big/ \int_F m(x)dx.$$

For $1+\epsilon_1 \leqq \sigma \leqq 1+\epsilon_2$ with $0<\epsilon_1<\epsilon_2$, $\int_{k_A}\varphi(w\begin{pmatrix}1 & x\\ 0 & 1\end{pmatrix})dx$ is bounded and $\int_F m(x)dx \geqq c_1$ with $c_1>0$ since $m(x)$, being a positive continuous function of σ and $x\in k_A$, takes the minimum positive value on the compact set $\{\tau \mid 1+\epsilon_1 \leqq \sigma \leqq 1+\epsilon_2\}\times\overline{F}$. Hence the convergence of $\sum_{u\in k}\varphi(w\begin{pmatrix}1 & u\\ 0 & 1\end{pmatrix})$ is locally uniform. This completes the proof.

1.3. We are going to compute the Fourier expansion of Eisenstein series. Assume $\sigma_1-\sigma_2>1$. Let dn denotes the Haar measure on $N_k\backslash N_A$ normalized so that $\mathrm{vol}(N_k\backslash N_A)=1$. For $g\in G_A$, we put

$$c_0(g) = \int_{N_k\backslash N_A} E(ng,f)dn. \tag{1.15}$$

In our previous notation, $c_0(y) = c_0(\begin{pmatrix}y & 0\\ 0 & 1\end{pmatrix})$. We have

$$\begin{aligned}c_0(g) &= \int_{N_k\backslash N_A}\sum_{\gamma\in P_k\backslash G_k} f(\gamma ng)dn\\ &= \int_{N_k\backslash N_A}\sum_{\gamma\in P_k\backslash G_k/N_k}\ \sum_{\delta\in N_k\cap\gamma^{-1}P_k\gamma\backslash N_k} f(\gamma\delta ng)dn.\end{aligned}$$

By Lemma 1.2, $\sum_{\gamma\in P_k\backslash G_k}|f(\gamma ng)|$ is a continuous function of $n\in N_A$. Since $N_k\backslash N_A$ is compact, we obtain

$$c_0(g) = \sum_{\gamma\in P_k\backslash G_k/N_k}\int_{N_k\backslash N_A}\sum_{\delta\in N_k\cap\gamma^{-1}P_k\gamma\backslash N_k} f(\gamma\delta ng)dn$$

by the dominated convergence theorem. Therefore we obtain

$$c_0(g) = \sum_{\gamma\in P_k\backslash G_k/N_k}\int_{N_k\cap\gamma^{-1}P_k\gamma\backslash N_A} f(\gamma ng)dn.$$

Here the choice of the measure on $N_k \cap \gamma^{-1}P_k\gamma\backslash N_A$ will be explained below. A complete set of representatives of $P_k\backslash G_k/N_k$ is given by $\{1, w\}$. If $\gamma = 1$, we have

$$\int_{N_k\backslash N_A} f(ng)dn = f(g).$$

If $\gamma = w$, we have $N_k \cap w^{-1}P_k w = \{1\}$. We determine the Haar measure dn on N_A by

$$\int_{N_k\backslash N_A} \sum_{\gamma\in N_k} \varphi(\gamma n)dn = \int_{N_A} \varphi(n)dn, \qquad \varphi \in L^1(N_A).$$

Then we obtain

$$c_0(g) = f(g) + \int_{N_A} f(wng)dn. \tag{1.16}$$

For $g = \begin{pmatrix} y & 0 \\ 0 & 1 \end{pmatrix}$, $\xi \in k^\times$, we have

$$\begin{aligned} c(\xi dy) &= \int_{N_k\backslash N_A} E(ng, f)\psi(-\xi u)dn = \int_{N_k\backslash N_A} \sum_{\gamma\in P_k\backslash G_k} f(\gamma ng)\psi(-\xi u)dn \\ &= \sum_{\gamma\in P_k\backslash G_k/N_k} \int_{N_k\cap\gamma^{-1}P_k\gamma\backslash N_A} f(\gamma ng)\psi(-\xi u)dn \end{aligned}$$

similarly to the above, where $n = \begin{pmatrix} 1 & u \\ 0 & 1 \end{pmatrix}$. For $\gamma = 1$, we obviously have

$$\int_{N_k\backslash N_A} f(ng)\psi(-\xi u)dn = 0.$$

Hence we obtain

$$c(\xi dy) = \int_{N_A} f(wng)\psi(-\xi u)dn, \qquad g = \begin{pmatrix} y & 0 \\ 0 & 1 \end{pmatrix}, \quad n = \begin{pmatrix} 1 & u \\ 0 & 1 \end{pmatrix}. \tag{1.17}$$

Let f_v be a function on G_v which satisfies

$$\begin{aligned} &f_v(pg) = (\chi_1)_v(a)(\chi_2)_v(d)\left|\frac{a}{d}\right|_v^{1/2} f_v(g) \\ &\text{for every} \quad p = \begin{pmatrix} a & b \\ 0 & d \end{pmatrix} \in P_v, \ g \in G_v. \end{aligned} \tag{1.18}$$

For $v \in \mathbf{h}$, we assume that f_v is invariant under the right translation by an open compact subgroup of G_v; furthermore we assume that for almost all $v \in \mathbf{h}$, $f_v(1) = 1$ and f_v is right K_v-invariant. For $v \in \mathbf{a}$, we assume that f_v is a C^∞-function. Then $f = \prod_{v\in\mathbf{h}\cup\mathbf{a}} f_v$ can be defined and satisfies (1.11), (B1) $\sim$ (B3). We assume that f is of this form. By (1.16) and (1.17), we obtain

$$c_0(g) = f(g) + \prod_v \int_{N_v} f_v(wng_v)dn, \tag{1.19}$$

$$c(\xi dy) = \prod_v \int_{N_v} f_v(wng_v)\psi_v(-\xi u)dn, \qquad n = \begin{pmatrix} 1 & u \\ 0 & 1 \end{pmatrix} \tag{1.20}$$

assuming the products of local integrals are absolutely convergent. Here the product of Haar mesures on N_v must coincide with the Haar measure on N_A.

2. Calculations of local integrals

In this section, we denote by k a local field. We simply denote G_k, P_k, N_k by G, P, N. If k is non-archimedean, $\mathfrak{O}$, ϖ, q denote the maximal compact subring of k, a prime element of k and the cardinality of the residue field $\mathfrak{O}/\varpi\mathfrak{O}$ respectively. We set $K = GL(2,\mathfrak{O})$ if k is non-archimedean, $K = SO(2,\mathbf{R})$ if $k = \mathbf{R}$, $K = SU(2,\mathbf{C})$ if $k = \mathbf{C}$. The absolute value of $x \in k$ is denoted by $|x|$. Thus $|x| = x\bar{x}$ if $k = \mathbf{C}$.

2.1. Let χ_1 and χ_2 be quasicharacters of $k^\times$. Let f be a complex valued function on G which satisfies

$$(2.1)\qquad f(pg) = \chi_1(a)\chi_2(d)\left|\frac{a}{d}\right|^{1/2} f(g) \qquad \text{for every} \quad p = \begin{pmatrix} a & b \\ 0 & d \end{pmatrix} \in P,\ g \in G.$$

When k is non-archimedean, we assume that f is invariant under the right translation by an open compact subgroup of G. When k is archimedean, we assume that f is right K-finite; then f is a C^∞-function. We denote by $\mathcal{B}(\chi_1,\chi_2)$ the space consisting of all such functions f. There exists $\sigma \in \mathbf{R}$ such that

$$|\chi_1\chi_2^{-1}(x)| = |x|^\sigma, \qquad x \in k^\times.$$

For $f \in \mathcal{B}(\chi_1,\chi_2)$, we put

$$(2.2)\qquad (T_w f)(g) = \int_N f(wng)dn, \qquad g \in G$$

where dn denotes a Haar measure on N. This integral is absolutely convergent if $\sigma > 0$. In fact, by (1.13) and (2.1), we get

$$f(w\begin{pmatrix} 1 & u \\ 0 & 1 \end{pmatrix} g) = |u|^{-1}\chi_1(-u^{-1})\chi_2(-u) f(\begin{pmatrix} 1 & 0 \\ u^{-1} & 1 \end{pmatrix} g).$$

Since

$$\int_{u\in k, |u|\geqq 1} |u|^{-1-\sigma} du < \infty, \qquad \sigma > 0,$$

the absolute convergence follows. Then by a direct computation, we see that $T_w(f) \in \mathcal{B}(\chi_2,\chi_1)$. We can regard $\mathcal{B}(\chi_1,\chi_2)$ as a representation space of G by the right translation. We see that T_w gives an intertwining operator from $\mathcal{B}(\chi_1,\chi_2)$ to $\mathcal{B}(\chi_2,\chi_1)$. When $f|K$ is fixed, $T_w(f)(g)$ can be continued meromorphically as a function of χ_1, χ_2. When k is archimedean, this fact will follow from our calculations given below. For the sake of completeness, let us sketch a proof when k is non-archimedean. (For the complex structure on the group of all quasicharacters of $k^\times$, see [Wel], p. 117. Here we simply note that a function $h(\chi_1,\chi_2)$ is holomorphic at (χ_1,χ_2) if $H(s_1,s_2) = h(\chi_1\omega^{s_1},\chi_2\omega^{s_2})$ is holomorphic at $(s_1,s_2) = (0,0)$, where $\omega(x) = |x|$, $x \in k^\times$.)

For a positive integer N, put $K_N = \{k \in K \mid k \equiv 1_2 \mod \varpi^N\}$, which is a congruence subgroup of K. Take N so that $f|K$ is right K_N-invariant. By $G = PK$, we may assume that $g \in K$, hence $g = 1$. The integral $\int_{\mathfrak{O}} f(w\begin{pmatrix} 1 & u \\ 0 & 1 \end{pmatrix})du$ is a constant. We have

$$\int_{k-\mathfrak{O}} f(w\begin{pmatrix}1 & u\\ 0 & 1\end{pmatrix})du = (\chi_1\chi_2)(-1)\int_{k-\mathfrak{O}} |u|^{-1}\chi_1(u^{-1})\chi_2(u)f(\begin{pmatrix}1 & 0\\ u^{-1} & 1\end{pmatrix})du$$

$$= (\chi_1\chi_2)(-1)\sum_{n=1}^{\infty}(\chi_1\chi_2^{-1})(\varpi)^n I_n,$$

where

$$I_n = \int_{\mathfrak{O}^\times}(\chi_1\chi_2^{-1})(u)f(\begin{pmatrix}1 & 0\\ \varpi^n u & 1\end{pmatrix})du.$$

Clearly we have

$$\sum_{n=N}^{\infty}(\chi_1\chi_2^{-1})(\varpi)^n I_n = \delta\,\mathrm{vol}(\mathfrak{O}^\times)\sum_{n=N}^{\infty}(\chi_1\chi_2^{-1})(\varpi)^n,$$

where $\delta = 1$ (resp. 0) if $\chi_1\chi_2^{-1}|\mathfrak{O}^\times = 1$ (resp. $\chi_1\chi_2^{-1}|\mathfrak{O}^\times \neq 1$). For $n < N$, we see that

$$I_n = \sum_{u\in\mathfrak{O}^\times/1+\varpi^{N-n}\mathfrak{O}}(\chi_1\chi_2^{-1})(u)f(\begin{pmatrix}1 & 0\\ \varpi^n u & 1\end{pmatrix})\int_{1+\varpi^{N-n}\mathfrak{O}}(\chi_1\chi_2^{-1})(t)dt.$$

If $\chi_1\chi_2^{-1}|(1+\varpi^{N-n}\mathfrak{O}) \neq 1$, we have $I_n = 0$. If $\chi_1\chi_2^{-1}|(1+\varpi^{N-n}\mathfrak{O}) = 1$, we have

$$I_n = \mathrm{vol}(1+\varpi^{N-n}\mathfrak{O})\sum_{u\in\mathfrak{O}^\times/1+\varpi^{N-n}\mathfrak{O}}(\chi_1\chi_2^{-1})(u)f(\begin{pmatrix}1 & 0\\ \varpi^n u & 1\end{pmatrix}).$$

This calculation shows that $T_w(f)(g)$ is a meromorphic function of χ_1, χ_2 which has a pole only when $\chi_1\chi_2^{-1}$ is unramified and $(\chi_1\chi_2^{-1})(\varpi) = 1$.

In the following subsections, we are going to calculate $T_w(f)$ and the integral $\int_N f(wng)\psi(-\xi u)du$ (cf. (1.20)) for a non-trivial additive character ψ of k.

2.2. In this subsection, we consider the case where k is non-archimedean.

LEMMA 2.1. *We assume that k is non-archimedean and that χ_1 and χ_2 are unramified. Define $f \in \mathcal{B}(\chi_1,\chi_2)$ by*

$$f(pk) = \chi_1(a)\chi_2(d)\left|\frac{a}{d}\right|^{1/2}, \qquad p = \begin{pmatrix}a & b\\ 0 & d\end{pmatrix} \in P,\ k \in K.$$

We normalize the Haar mesure on $N \cong k$ by taking $\mathrm{vol}(\mathfrak{O}) = 1$. *If $\sigma > 0$, then we have*

$$\int_N f(wng)dn = \frac{1-\chi_1(\varpi)\chi_2(\varpi)^{-1}q^{-1}}{1-\chi_1(\varpi)\chi_2(\varpi)^{-1}} \times \chi_1(d)\chi_2(a)\left|\frac{a}{d}\right|^{1/2},$$

for $g = pk$, $p = \begin{pmatrix}a & b\\ 0 & d\end{pmatrix} \in P$, $k \in K$.

PROOF. For $n = \begin{pmatrix}1 & u\\ 0 & 1\end{pmatrix}$, we have

$$wng = \begin{pmatrix}d & 0\\ 0 & a\end{pmatrix} w \begin{pmatrix}1 & a^{-1}du\\ 0 & 1\end{pmatrix}\begin{pmatrix}1 & a^{-1}b\\ 0 & 1\end{pmatrix} k.$$

Hence we get

$$\int_N f(wng)dn = \chi_1(d)\chi_2(a)\left|\frac{a}{d}\right|^{1/2}\int_N f(wn)dn.$$

If $u \in \mathfrak{O}$, we have $f(wn) = 1$. If $u \notin \mathfrak{O}$, we see that

$$f(wn) = \chi_1(u)^{-1}\chi_2(u)|u|^{-1}$$

by (1.13). Therefore, since $\mathrm{vol}(\varpi^{-n}\mathfrak{O}^\times) = q^n(1-q^{-1})$, we obtain

$$\int_N f(wn)dn = 1 + \sum_{n=1}^{\infty}(\chi_1(\varpi)/\chi_2(\varpi))^n(1-q^{-1}) = \frac{1-\chi_1(\varpi)\chi_2(\varpi)^{-1}q^{-1}}{1-\chi_1(\varpi)\chi_2(\varpi)^{-1}}$$

and complete the proof.

LEMMA 2.2. *Let the notation and assumptions be the same as in Lemma* 2.1. *For $z \in k^\times$, we put $I(z) = 0$ if $z \notin \mathfrak{O}$ and*

$$I(z) = \frac{(1-\chi_1(\varpi)\chi_2(\varpi)^{-1}q^{-1})(1-(\chi_1(\varpi)\chi_2(\varpi)^{-1})^{N+1})}{1-\chi_1(\varpi)\chi_2(\varpi)^{-1}}$$

if $z \in \mathfrak{O}$, where $N = \mathrm{ord}(z)$, ord being the normalized additive valuation of k. Let ψ be a non-trivial character of k and take $d \in k^\times$ so that $d^{-1}\mathfrak{O}$ is the largest $\mathfrak{O}$-submodule of k on which ψ is trivial. If $\sigma > 0$, then for $g = \begin{pmatrix} y & 0 \\ 0 & 1\end{pmatrix} \in G$ and $\xi \in k^\times$, we have

$$\int_N f(wng)\psi(-\xi u)dn = \chi_2(y)|y|^{1/2}I(d\xi y), \qquad n = \begin{pmatrix} 1 & u \\ 0 & 1\end{pmatrix}.$$

PROOF. Since

$$wng = \begin{pmatrix} 1 & 0 \\ 0 & y\end{pmatrix} w \begin{pmatrix} 1 & y^{-1}u \\ 0 & 1\end{pmatrix},$$

we have

$$\int_N f(wng)\psi(-\xi u)dn = \chi_2(y)|y|^{1/2}\int_N f(wn)\psi(-\xi yu)dn, \qquad n = \begin{pmatrix} 1 & u \\ 0 & 1\end{pmatrix}.$$

In a similar way to the proof of Lemma 2.1, we get

$$\int_N f(wn)\psi(-\xi yu)dn = \int_{\mathfrak{O}}\psi(-\xi yu)du + \int_{k-\mathfrak{O}}(\chi_1^{-1}\chi_2)(u)|u|^{-1}\psi(-\xi yu)du.$$

Put

$$\psi_0(x) = \psi(d^{-1}x),$$
$$J(z) = \int_{\mathfrak{O}}\psi_0(-zu)du + \int_{k-\mathfrak{O}}(\chi_1^{-1}\chi_2)(u)|u|^{-1}\psi_0(-zu)du$$

for $z \in k^\times$. Then

$$\int_N f(wn)\psi(-\xi yu)dn = J(d\xi y),$$

$$J(z) = \int_{\mathfrak{O}}\psi_0(-zu)du + \sum_{n=1}^{\infty}(\chi_1(\varpi)/\chi_2(\varpi))^n q^{-n}\int_{\varpi^{-n}\mathfrak{O}-\varpi^{-n+1}\mathfrak{O}}\psi_0(-zu)du.$$

If $z \notin \mathfrak{O}$, then $J(z) = 0$ since ψ_0 is trivial on $\mathfrak{O}$ and is non-trivial on $\varpi^{-1}\mathfrak{O}$. Assume $z \in \mathfrak{O}$ and put $z = \varpi^N z_1$, $z_1 \in \mathfrak{O}^\times$. Since $\psi_0(-zu)$ defines the trivial character of $\varpi^{-n}\mathfrak{O}$ if and only if $n \leqq N$, we have

$$\begin{aligned} J(z) &= 1 + \sum_{n=1}^{N} (\chi_1(\varpi)/\chi_2(\varpi))^n q^{-n} q^n (1-q^{-1}) - (\chi_1(\varpi)/\chi_2(\varpi))^{N+1} q^{-(N+1)} q^N \\ &= I(z). \end{aligned}$$

This completes the proof.

2.3. In this subsection, we consider the case where $k = \mathbf{R}$.

LEMMA 2.3. *We assume $k = \mathbf{R}$. Define $\tau \in \mathbf{C}$ by $(\chi_1\chi_2^{-1})(x) = |x|^\tau$ for $x \in \mathbf{R}_+$. Thus $\sigma = \Re(\tau)$. For $\theta \in \mathbf{R}$, put* $r(\theta) = \begin{pmatrix} \cos\theta & \sin\theta \\ -\sin\theta & \cos\theta \end{pmatrix} \in K$. *Let l be a non-negative integer such that $\chi_1(-1)\chi_2(-1) = (-1)^l$. Define $f \in \mathcal{B}(\chi_1, \chi_2)$ by*

$$f(pk) = \chi_1(a)\chi_2(d)\left|\frac{a}{d}\right|^{1/2} \exp(il\theta), \qquad p = \begin{pmatrix} a & b \\ 0 & d \end{pmatrix}, \ k = r(\theta) \in K.$$

We take the Haar measure on $N \cong \mathbf{R}$ as the measure corresponding to the usual Lebesgue measure on $\mathbf{R}$. If $\sigma > 0$, then, for $g = pk$ with p and k as above, we have

$$\begin{aligned} &\int_N f(wng)dn = \chi_1(d)\chi_2(a)\left|\frac{a}{d}\right|^{1/2} \exp(il\theta) \\ &\times \begin{cases} i^l \cdot \frac{\Gamma(\tau/2)\Gamma(1/2)}{\Gamma((\tau+l+1)/2)} \cdot (\tau/2 - 1/2)(\tau/2 - 3/2)\cdots(\tau/2 - (l-1)/2) & \text{if } l \text{ is even,} \\ i^l \cdot \frac{\Gamma((\tau+1)/2)\Gamma(1/2)}{\Gamma((\tau+l+1)/2)} \cdot (\tau/2 - 1)(\tau/2 - 2)\cdots(\tau/2 - (l-1)/2) & \text{if } l \text{ is odd.} \end{cases} \end{aligned}$$

Here the polynomial part of τ should be dropped when $l = 0$ or 1.

PROOF. We note that f is well defined since $\chi_1(-1)\chi_2(-1) = (-1)^l$. Similarly to the proof of Lemma 2.1, we obtain

$$\int_N f(wng)dn = \chi_1(d)\chi_2(a)\left|\frac{a}{d}\right|^{1/2} \exp(il\theta) \int_N f(wn)dn.$$

For $n = \begin{pmatrix} 1 & u \\ 0 & 1 \end{pmatrix}$,

$$wn = \begin{pmatrix} \frac{1}{\sqrt{u^2+1}} & \frac{-u}{\sqrt{u^2+1}} \\ 0 & \sqrt{u^2+1} \end{pmatrix} \begin{pmatrix} \frac{-u}{\sqrt{u^2+1}} & \frac{1}{\sqrt{u^2+1}} \\ \frac{-1}{\sqrt{u^2+1}} & \frac{-u}{\sqrt{u^2+1}} \end{pmatrix} \tag{2.3}$$

is the Iwasawa decomposition. Hence

$$f(wn) = (\chi_1^{-1}\chi_2)(\sqrt{u^2+1})(u^2+1)^{-1/2}\left(\frac{-u+i}{|-u+i|}\right)^l,$$

$$\begin{aligned} \int_N f(wn)dn &= \int_{\mathbf{R}} (u^2+1)^{-(\tau+l+1)/2}(-u+i)^l du \\ &= (-1)^l \sum_{j=0}^{l} \binom{l}{j} (-i)^{l-j} \int_{-\infty}^{\infty} (u^2+1)^{-(\tau+l+1)/2} u^j du, \end{aligned}$$

the integrals being absolutely convergent if $\Re(\tau) > 0$. Changing the variable by $u = \tan\theta$, we have

$$\int_{-\infty}^{\infty} (u^2+1)^{-\tau/2-l/2-1/2} u^j \, du = \int_{-\pi/2}^{\pi/2} (\cos\theta)^{\tau+l-j-1} (\sin\theta)^j \, d\theta.$$

When j is odd, this integral vanishes and when j is even, this integral is equal to

$$2\int_0^{\pi/2} (\cos\theta)^{\tau+l-j-1} (\sin\theta)^j \, d\theta.$$

This integral can be expressed by the beta function ([WW], p. 256) and is equal to

$$\frac{\Gamma((\tau+l-j)/2)\Gamma((j+1)/2)}{\Gamma((\tau+l+1)/2)}.$$

Thus we obtain

$$\int_N f(wn)dn = (-1)^l \sum_{j=0,\, j:\text{even}}^{l} \binom{l}{j} (-i)^{l-j} \cdot \frac{\Gamma((\tau+l-j)/2)\Gamma((j+1)/2)}{\Gamma((\tau+l+1)/2)}.$$

Now our assertion follows from the next Lemma.

LEMMA 2.4. *As meromorphic functions of* s, *we have*

$$(-1)^l \sum_{j=0,\, j:\text{even}}^{l} \binom{l}{j} (-i)^{l-j} \cdot \frac{\Gamma(s+l-j)/2)\Gamma((j+1)/2)}{\Gamma((s+l+1)/2)}$$

$$= \begin{cases} i^l \cdot \frac{\Gamma(s/2)\Gamma(1/2)}{\Gamma((s+l+1)/2)} \cdot (s/2-1/2)(s/2-3/2)\cdots(s/2-(l-1)/2) & \text{if } l \text{ is even,} \\ i^l \cdot \frac{\Gamma((s+1)/2)\Gamma(1/2)}{\Gamma((s+l+1)/2)} \cdot (s/2-1)(s/2-2)\cdots(s/2-(l-1)/2) & \text{if } l \text{ is odd.} \end{cases}$$

PROOF. We may assume $l \geqq 2$. The left-hand side can be written as

$$\frac{i^l}{\Gamma((s+l+1)/2)} \sum_{j=0}^{[l/2]} (-1)^j \binom{l}{2j} \Gamma((s+l-2j)/2)\Gamma(j+\frac{1}{2}). \tag{2.4}$$

When l is even, put $l = 2l_0$. Then (2.4) is equal to

$$\frac{i^l\Gamma(s/2)\Gamma(1/2)}{\Gamma((s+l+1)/2)} \sum_{j=0}^{l_0} (-1)^j \binom{l}{2j} (s/2)(s/2+1)\cdots(s/2+l/2-j-1) \cdot \frac{1}{2} \cdot \frac{3}{2} \cdots \frac{2j-1}{2}.$$

Hence we see that (2.4) can be written as

$$\frac{i^l\Gamma(s/2)\Gamma(1/2)}{\Gamma((s+l+1)/2)} 2^{-l_0} P_l(s)$$

with a monic polynomial $P_l(s)$ of degree l_0. When l is odd, put $l = 2l_0+1$. Then (2.4) is equal to

$$\frac{i^l\Gamma((s+1)/2)\Gamma(1/2)}{\Gamma((s+l+1)/2)}$$

$$\times \sum_{j=0}^{l_0} (-1)^j \binom{l}{2j} (s/2+1/2)(s/2+3/2)\cdots(s/2+l/2-j-1) \cdot \frac{1}{2} \cdot \frac{3}{2} \cdots \frac{2j-1}{2}.$$

Hence we see that (2.4) can be written as

$$\frac{i^l\Gamma((s+1)/2)\Gamma(1/2)}{\Gamma((s+l+1)/2)}2^{-l_0}P_l(s)$$

with a monic polynomial $P_l(s)$ of degree l_0. Therefore it suffices to show that

$$\int_{-\infty}^{\infty}(u^2+1)^{-(s+l+1)/2}(u-i)^l du = 0$$

for $s = 1, 3, \ldots, 2l_0 - 1$ when $l = 2l_0$ and for $s = 2, 4, \ldots, 2l_0$ when $l = 2l_0 + 1$. Suppose that $l = 2l_0$. Then for $s = 2s_0 - 1$, $s_0 \in \mathbf{Z}$, $1 \leqq s_0 \leqq l_0$, this integral is equal to

$$\int_{-\infty}^{\infty}(u+i)^{-s_0-l_0}(u-i)^{-s_0+l_0}du$$

and the integrand is holomorphic on the whole complex plane except for a pole at $u = -i$. For $R > 0$, we can deform the path of the integration so that it goes from $-\infty$ to $-R$ on the real axis, from $-R$ to R on the hemicircle lying in the upper half plane, from R to $+\infty$ on the real axis. By letting $R \to +\infty$, we see that the value of this integral is 0. The case $l = 2l_0 + 1$ can be handled similarly. This completes the proof.

To obtain analogues of Lemma 2.2 in the cases where $k = \mathbf{R}$ and $\mathbf{C}$, we need some properties of the Bessel function $K_s(z)$. Recall the formula V, (1.18):

$$K_s(z) = \frac{1}{2}\int_0^{\infty}\exp(-\frac{z}{2}(t+\frac{1}{t}))t^{s-1}dt, \qquad \Re(z) > 0, \quad s \in \mathbf{C}.$$

It is legitimate to differentiate under the integral sign. We obtain

$$\frac{d}{dz}K_s(z) = -\frac{1}{2}(K_{s+1}(z) + K_{s-1}(z)). \tag{2.5}$$

Next using integration by parts, we get

$$2sK_s(z) = z(K_{s+1}(z) - K_{s-1}(z)). \tag{2.6}$$

By (2.5) and (2.6), we obtain easily

$$z^2\frac{d^2}{dz^2}K_s(z) - (z^2+s^2)K_s(z) = 0,$$

which is a differential equation satisfied by $K_s(z)$. By (2.5) and (2.6), we find

$$\frac{1}{z}\frac{d}{dz}(z^sK_s(z)) = -z^{s-1}K_{s-1}(z).$$

Applying this operator m times, we obtain

$$(\frac{1}{z}\frac{d}{dz})^m(z^sK_s(z)) = (-1)^m z^{s-m}K_{s-m}(z), \qquad 1 \leqq m \in \mathbf{Z}. \tag{2.7}$$

Furthermore we have an symptotic formula ([Wat], p. 202)

$$K_s(z) \sim \sqrt{\frac{\pi}{2z}}e^{-z}, \qquad z \to +\infty.$$

For a non-negative integer l and $m \in \mathbf{Z}$, $0 \leqq m \leqq l$, we define a polynomial $f_m^{(l)}(a)$ recursively by $f_0^{(0)}(a) = 1$,

$$f_l^{(l)}(a) = a^l, \qquad 0 \leqq l,$$
$$f_m^{(l+1)}(a) = af_{m-1}^{(l)}(a) + \frac{d}{da}f_m^{(l)}(a) - f_m^{(l)}(a), \qquad 0 \leqq m \leqq l.$$

LEMMA 2.5. *Let the notation and assumptions be the same as in Lemma 2.3. Define a character ψ of $k = \mathbf{R}$ by $\psi(x) = \exp(2\pi ix)$. If $\sigma > 0$, then for $g = \begin{pmatrix} y & 0 \\ 0 & 1 \end{pmatrix} \in P$ and $\xi \in k^\times$, we have*

$$\int_N f(wng)\psi(-\xi u)dn = 2\chi_2(y)|y|^{1/2}(-i)^l \frac{\pi^{(\tau+l+1)/2}}{\Gamma((\tau+l+1)/2)}$$
$$\times \sum_{m=0}^{l} (-1)^m (2\pi)^{-m} f_m^{(l)}(2\pi\xi y)|\xi y|^{(\tau+l)/2-m} K_{(\tau+l)/2-m}(|2\pi\xi y|),$$

where $n = \begin{pmatrix} 1 & u \\ 0 & 1 \end{pmatrix}$. If $l \geqq 1$ and $\tau = l-1$, the above expression is equal to

$$\begin{cases} \chi_2(y)|y|^{1/2}(2\pi i)^l \cdot \frac{1}{(l-1)!}(\xi y)^{l-1}\exp(-2\pi\xi y) & \textit{if} \quad \xi y > 0, \\ 0 \qquad \textit{if} \quad \xi y < 0. \end{cases}$$

PROOF. Similarly to the proof of Lemma 2.2, we have

$$\int_N f(wng)\psi(-\xi yu)dn = \chi_2(y)|y|^{1/2} \int_N f(wn)\psi(-\xi yu)dn.$$

By (2.3), we have

$$\int_N f(wn)\psi(-\xi yu)dn = \int_{-\infty}^{\infty} (u^2+1)^{-(\tau+l+1)/2}(-u+i)^l \exp(-2\pi i\xi yu)du.$$

This integral converges absolutely if $\sigma > 0$. For $s \in \mathbf{C}$, $0 \leqq l \in \mathbf{Z}$ and $a \in \mathbf{R}^\times$, we put

$$(2.8) \qquad I_l(s,a) = \int_{-\infty}^{\infty} (u^2+1)^{-s-1/2}(u-i)^l \exp(-iau)du.$$

If $\Re(s) > l/2$, the integral converges absolutely. It is not difficult to see that the integral still converges if $\Re(s) > l/2 - 1$. We have

$$(2.9) \qquad \int_N f(wn)\psi(-\xi yu)dn = (-1)^l I_l(\frac{\tau+l}{2}, 2\pi\xi y).$$

Hence our assertion follows from the next Lemma.

LEMMA 2.6. *We have, for* $\Re(s) > l/2$,

$$I_l(s,a) = i^l \cdot 2^{1-s} \frac{\sqrt{\pi}}{\Gamma(s+1/2)} \sum_{m=0}^{l} (-1)^m f_m^{(l)}(a)|a|^{s-m} K_{s-m}(|a|).$$

For $l \geqq 1$, *we have*

$$I_l(l-1/2,a) = \begin{cases} (-i)^l \cdot \frac{2\pi}{(l-1)!} a^{l-1} \exp(-a), & a > 0, \\ 0, & a < 0. \end{cases}$$

PROOF. First let us show the latter assertion. We have

$$I_l(l-1/2,a) = \int_{-\infty}^{\infty} (u+i)^{-l} \exp(-iau) du.$$

Assume $l \geqq 2$. Then the integral converges absolutely. When $a < 0$, we deform the path of integration to the path which goes from $-\infty$ to $-R$ on the real axis, $-R$ to R on the hemicircle lying in the upper half plane, R to $+\infty$ on the real axis for $R > 0$. Letting $R \to +\infty$, we see that the integral vanishes. When $a > 0$, we deform the path of integration similarly to the above but now from $-R$ to R on the hemicircle lying in the lower half plane. Letting $R \to +\infty$, we see that

$$I_l(l-1/2,a) = -2\pi i \mathrm{Res}_{u=-i}(u+i)^{-l} \exp(-iau) = \frac{1}{(l-1)!}(-2\pi i)(-ia)^{l-1} \exp(-a).$$

Hence the latter assertion holds when $l \geqq 2$.

If $l = 1$, the integral converges conditionally. The same argument also applies to this case. The only delicate point is that the integral on the hemicircle from $-R$ to R tends to 0 when $R \to +\infty$. But this can be seen by dividing the hemicircle into two parts so that $|\arg(z)|$ is small for one part.

Next let us prove the first assertion. We have

$$I_0(s,a) = \xi(1, \frac{a}{2\pi}; s+1/2, s+1/2)$$

in the notation of V, §1.2. By V, Lemma 1.2, we get

$$I_0(s,a) = 2^{1-s} \frac{\sqrt{\pi}}{\Gamma(s+\frac{1}{2})} |a|^s K_s(|a|). \tag{2.10}$$

If $\Re(s) > (l+1)/2$, then it is easy to justify the differentiation of (2.8) with respect to a under the integral sign. We obtain

$$I_{l+1}(s,a) = i(\frac{\partial}{\partial a} I_l(s,a) - I_l(s,a)). \tag{2.11}$$

Define differential operators D and D_1 by

$$D = \frac{\partial}{\partial a}, \qquad D_1 = \frac{1}{a}\frac{\partial}{\partial a}.$$

By (2.11), we get

$$I_l(s,a) = i^l (D-1)^l I_0(s,a), \qquad \Re(s) > l/2. \tag{2.12}$$

We can write (2.7) as

$$D_1^m (a^s K_s(a)) = (-1)^m a^{s-m} K_{s-m}(a). \tag{2.13}$$

Let us verify that

$$(D-1)^l = \sum_{m=0}^{l} f_m^{(l)}(a) D_1^m \tag{2.14}$$

as differential operators. The case $l=0$ is trivial. For a polynomial $f(a)$, we have

$$D_1 f(a) = f(a) D_1 + a^{-1} f'(a)$$

where $f'(a)$ is the derivative of $f(a)$. Now assuming (2.14) for $(D-1)^l$, we have

$$(D-1)^{l+1} = (aD_1 - 1)\left(\sum_{m=0}^{l} f_m^{(l)}(a) D_1^m\right) = \sum_{m=0}^{l} (a f_m^{(l)}(a) D_1 + f_m^{(l)\prime}(a) - f_m^{(l)}(a)) D_1^m.$$

Hence we see that (2.14) for $(D-1)^{l+1}$ holds using the recursive relation satisfied by $f_m^{(l)}$.

Let us assume $a > 0$. By (2.10), (2.12), (2.13) and (2.14), we obtain

$$I_l(s,a) = i^l \cdot 2^{1-s} \frac{\sqrt{\pi}}{\Gamma(s+1/2)} \sum_{m=0}^{l} (-1)^m f_m^{(l)}(a) a^{s-m} K_{s-m}(a)$$

if $\Re(s) > l/2$. Hence the first assertion holds when $a > 0$. Next let us consider the case $a < 0$. Put $a' = -a$,

$$D' = \frac{\partial}{\partial a'}, \qquad D_1' = \frac{1}{a'} \frac{\partial}{\partial a'}.$$

Then $D' = -D$, $D_1' = D_1$,

$$(D-1)^l = \sum_{m=0}^{l} f_m^{(l)}(-a') D_1'^m.$$

By (2.10), we have

$$I_0(s,a) = 2^{1-s} \frac{\sqrt{\pi}}{\Gamma(s+1/2)} a'^s K_s(a').$$

By (2.12) and (2.13), we obtain

$$\begin{aligned} I_l(s,a) &= i^l \cdot 2^{1-s} \frac{\sqrt{\pi}}{\Gamma(s+1/2)} \sum_{m=0}^{l} f_m^{(l)}(-a') D_1'^m (a'^s K_s(a')) \\ &= i^l \cdot 2^{1-s} \frac{\sqrt{\pi}}{\Gamma(s+1/2)} \sum_{m=0}^{l} (-1)^m f_m^{(l)}(-a') a'^{s-m} K_{s-m}(a'). \end{aligned}$$

Therefore the first assertion holds for the case $a < 0$. This completes the proof.

REMARK. We can justify the differentiation of (2.8) with respect to a under the integral sign if $\Re(s) > (l-1)/2$. Thus Lemma 2.6 holds under the condition $\Re(s) > l/2 - 1$. When $l = 1$, $s = l - 1/2$, we have

$$I_1(1/2, a) = i \cdot 2^{1-s} \sqrt{\pi} \left[-|a|^{1/2} K_{1/2}(|a|) - a|a|^{-1/2} K_{-1/2}(|a|) \right].$$

Now the latter asserion for the case $l = 1$ also follows from the well known formulas

$$K_s(z) = K_{-s}(z), \qquad K_{1/2}(z) = \sqrt{\frac{\pi}{2z}}e^{-z}.$$

REMARK. We can express $I_l(s,a)$ using the hypergeometric function. In fact, in the notation of V, §1.2, we have

$$I_l(s, 2\pi a) = \xi(1, a; s+1/2, s+1/2-l).$$

Therefore, by V, (1.13), we have, for $a > 0$,

$$I_l(s, 2\pi a) = i^{-l} \cdot 2\pi\cdot(2\pi a)^{2s-l}\Gamma(s+1/2)^{-1}\Gamma(s+1/2-l)^{-1}e^{-2\pi a} \times \zeta(4\pi a; s+1/2, s+1/2-l).$$

2.4. In this subsection, we consider the case where $k = \mathbf{C}$. For $l \geqq 0$, let σ_l denote the symmetric tensor representation of degree l of $K = SU(2,\mathbf{C})$. Then $\dim \sigma_l = l+1$ and σ_l is irreducible. Any irreducible representation of K is equivalent to one of σ_l. In the regular representation of K realized on $L^2(K)$ by the right translation, σ_l occurs with multiplicity $l+1$. We have

$$SU(2,\mathbf{C}) = \left\{ \begin{pmatrix} \alpha & \beta \\ -\bar\beta & \bar\alpha \end{pmatrix} \,\middle|\, \alpha, \beta \in \mathbf{C},\ |\alpha|^2 + |\beta|^2 = 1 \right\}.$$

By considering matrix coefficients of σ_l, we see easily that the functions α^l, $\alpha^{l-1}\bar\beta$, $\ldots$, $\alpha\bar\beta^{l-1}$, $\bar\beta^l$ on K are mutually orthogonal and give the highest weight vectors which generate irreducible representations equivalent to σ_l. For $0 \leqq m \leqq l$, $m \in \mathbf{Z}$, we consider the function $f \in \mathcal{B}(\chi_1, \chi_2)$ which satisfies

$$f(\begin{pmatrix} a & b \\ 0 & d \end{pmatrix}\begin{pmatrix} \alpha & \beta \\ -\bar\beta & \bar\alpha \end{pmatrix}) = \chi_1(a)\chi_2(d)\left|\frac{d}{a}\right|^{1/2} \alpha^{l-m}\bar\beta^m, \qquad \begin{pmatrix} a & b \\ 0 & d \end{pmatrix} \in P, \quad \begin{pmatrix} \alpha & \beta \\ -\bar\beta & \bar\alpha \end{pmatrix} \in K. \tag{2.15}$$

We can write

$$(\chi_1\chi_2^{-1})(x) = (\frac{x}{(x\bar x)^{1/2}})^A |x|^\tau, \qquad x \in \mathbf{C}^\times \tag{2.16}$$

with $\tau \in \mathbf{C}$, $A \in \mathbf{Z}$. Thus $\sigma = \Re(\tau)$. Then we see easily that f is well defined if and only if

$$A = l - 2m. \tag{2.17}$$

LEMMA 2.7. *Let $k = \mathbf{C}$ and let the notation be the same as above. We take the Haar measure on $N \cong \mathbf{C}$ as the measure corresponding to the usual Lebesgue measure on $\mathbf{C}$. We assume* (2.17). *If $\sigma > 0$ and $g = pk$ with $p = \begin{pmatrix} a & b \\ 0 & d \end{pmatrix} \in P$, $k = \begin{pmatrix} \alpha & \beta \\ -\bar\beta & \bar\alpha \end{pmatrix} \in K$, then for f defined by* (2.15), *we have*

$$\int_N f(wng)dn = (-1)^{l-m}\chi_1(d)\chi_2(a)\left|\frac{a}{d}\right|^{1/2} \pi\alpha^m\bar\beta^{l-m} \cdot \frac{1}{\Gamma(\tau + l/2 + 1)}\varphi(\tau)$$

where

$$\varphi(\tau) = \Gamma(\tau + l/2 - m)(\tau - l/2)(\tau - l/2 + 1) \cdots (\tau - l/2 + m - 1)$$

if $l - 2m \geqq 0$,

$$\varphi(\tau) = \Gamma(\tau - l/2 + m)(\tau - l/2)(\tau - l/2 + 1) \cdots (\tau + l/2 - m - 1)$$

if $l-2m < 0$. *Here in the first case, the polynomial part should be dropped if* $m = 0$.

PROOF. Similarly to the proof of Lemma 2.1, we have

$$\int_N f(wng)dn = \chi_1(d)\chi_2(a)\left|\frac{d}{a}\right|^{1/2} \int_N f(wnk)dn.$$

For $n = \begin{pmatrix} 1 & u \\ 0 & 1 \end{pmatrix}$, write $u = re^{i\theta}$ in polar coordinates. Then

$$(2.18) \qquad wn = \begin{pmatrix} \frac{e^{-i\theta}}{\sqrt{r^2+1}} & -\frac{r}{\sqrt{r^2+1}} \\ 0 & \sqrt{r^2+1}e^{i\theta} \end{pmatrix} \begin{pmatrix} -\frac{r}{\sqrt{r^2+1}} & \frac{e^{i\theta}}{\sqrt{r^2+1}} \\ -\frac{e^{-i\theta}}{\sqrt{r^2+1}} & -\frac{r}{\sqrt{r^2+1}} \end{pmatrix}$$

is the Iwasawa decomposition. Hence

$$\begin{aligned} f(wnk) =& (\chi_1^{-1}\chi_2)(\sqrt{r^2+1}e^{i\theta})(r^2+1)^{-1} \\ & \times (-\frac{r}{\sqrt{r^2+1}}\alpha - \frac{e^{i\theta}}{\sqrt{r^2+1}}\bar{\beta})^{l-m} \overline{(-\frac{r}{\sqrt{r^2+1}}\beta + \frac{e^{i\theta}}{\sqrt{r^2+1}}\bar{\alpha})^{m}} \\ =& (-1)^{l-m}(r^2+1)^{-\tau-l/2-1}e^{-iA\theta}(r\alpha + e^{i\theta}\bar{\beta})^{l-m}(e^{-i\theta}\alpha - r\bar{\beta})^m. \end{aligned}$$

Thus we have

$$\begin{aligned} \int_N f(wnk)dn =& (-1)^{l-m} \sum_{a=0}^{l-m}\sum_{b=0}^{m} \binom{l-m}{a}\binom{m}{b} \alpha^{a+b}\bar{\beta}^{l-a-b} \\ & \times \int_0^\infty \int_0^{2\pi} (r^2+1)^{-\tau-l/2-1} r^a(-r)^{m-b} r e^{i\theta(-A+l-m-a-b)} d\theta dr. \end{aligned}$$

Integrating with respect to θ, we see that only the terms which satisfy $a+b=m$ contribute to the integral, in view of (2.17). We obtain

$$\begin{aligned} & \int_N f(wnk)dn = (-1)^{l-m} 2\pi \alpha^m \bar{\beta}^{l-m} \\ & \quad \times \sum_{b=\max(0,-A)}^{m} (-1)^{m-b} \binom{l-m}{m-b}\binom{m}{b} \int_0^\infty r^{2m-2b+1}(r^2+1)^{-\tau-l/2-1} dr, \end{aligned}$$

the integrals being convergent for $\sigma > 0$. Changing variables by $r = \tan\theta$ and reducing the integral to the beta function ([WW], p. 256), we get

$$\begin{aligned} & \int_N f(wnk)dn = (-1)^{l-m} \pi \alpha^m \bar{\beta}^{l-m} \\ & \times \sum_{b=\max(0,-A)}^{m} (-1)^{m-b} \binom{l-m}{m-b}\binom{m}{b} \frac{\Gamma(\tau + l/2 - m + b)\Gamma(m-b+1)}{\Gamma(\tau + l/2 + 1)}. \end{aligned}$$

Now our assertion follows from the next Lemma.

LEMMA 2.8. *As meromorphic functions of s, we have*

$$\sum_{b=\max(0,-A)}^{m}(-1)^{m-b}\binom{l-m}{m-b}\binom{m}{b}\Gamma(s+l/2-m+b)\Gamma(m-b+1)$$
$$=\begin{cases}\Gamma(s+l/2-m)(s-\frac{l}{2})(s-\frac{l}{2}+1)\cdots(s-\frac{l}{2}+m-1) & \text{if} \quad l-2m\geqq 0,\\ \Gamma(s-l/2+m)(s-\frac{l}{2})(s-\frac{l}{2}+1)\cdots(s+\frac{l}{2}-m-1) & \text{if} \quad l-2m<0.\end{cases}$$

Here, in the first case, the polynomial part should be dropped when $m=0$.

PROOF. Let us show the first case. We see easily that the left-hand side can be written as

$$\Gamma(s+l/2-m)P_m(s)$$

where $P_m(s)$ is a monic polynomial of degree m. Hence it suffices to show that $P_m(s)$ vanishes at $s=l/2,\ l/2-1,\ \ldots,\ l/2-m+1$. Since $\Gamma(s+l/2-m)^{-1}$ is non-vanishing for these values of s, it is enough to show

$$\sum_{b=0}^{m}(-1)^b\binom{m}{b}\frac{(l-m+b-\delta)!}{(l-2m+b)!}=0, \qquad \delta=1,2,\cdots,m.$$

Since $(l-m+b-\delta)!/(l-2m+b)!$ is a polynomial of b of degree $m-\delta$, this formula follows from

$$\sum_{b=0}^{m}(-1)^b\binom{m}{b}b^i=0, \qquad 0\leqq i\leqq m-1.$$

Changing b to $b+2m-l$, the left-hand side of the second formual can be written as

$$\sum_{b=0}^{l-m}(-1)^{l-m-b}\binom{l-m}{l-m-b}\binom{m}{b+2m-l}\Gamma(s-l/2+m+b)\Gamma(l-m-b+1).$$

This coincides with the first formula when we change m to $l-m$ there. Hence we obtain the second formula. This completes the proof.

LEMMA 2.9. *Let the notation and assumptions be the same as Lemma 2.8. Define a character ψ of $k=\mathbf{C}$ by $\psi(x)=\exp(2\pi i(x+\bar{x}))$. If $\sigma>0$, then for $g=\begin{pmatrix} y & 0\\ 0 & 1\end{pmatrix}\in P$ and $\xi\in k^\times$, we have*

$$\int_N f(wng)\psi(-\xi u)dn=\chi_2(y)|y|^{1/2}i^{l-m}\frac{(2\pi)^{\tau+l/2+1}}{\Gamma(\tau+l/2+1)}$$
$$\times(\xi y)^{l-m}|\xi y|^{(\tau-l/2+m)/2}K_{(\tau-l/2+m)/2}(4\pi|\xi y|^{1/2}),$$

where $n=\begin{pmatrix} 1 & u\\ 0 & 1\end{pmatrix}$.

PROOF. Similarly to the proof of Lemma 2.2, we have

$$\int_N f(wng)\psi(-\xi u)dn=\chi_2(y)|y|^{1/2}\int_N f(wn)\psi(-\xi yu)dn.$$

Write $\xi y=Re^{i\Theta}$, $u=re^{i\theta}$ in polar coordinates. Since

$$f(wn)=(-1)^{l-m}r^{l-m}(r^2+1)^{-(\tau+l/2+1)}e^{-i(A+m)\theta}$$

as we have shown in the proof of Lemma 2.7, we have

$$\int_N f(wn)\psi(-\xi yu)dn = (-1)^{l-m}\int_0^\infty\int_0^{2\pi} r^{l-m+1}(r^2+1)^{-(\tau+l/2+1)} \times \exp(-i(A+m)\theta - 4\pi i Rr\cos(\theta+\Theta))d\theta dr, \tag{2.19}$$

the integral being absolutely convergent if $\sigma > -m/2$. By a change of variables, we get

$$\int_0^{2\pi} \exp(-i(A+m)\theta - 4\pi i Rr\cos(\theta+\Theta))d\theta$$
$$= e^{i(A+m)\Theta}\int_0^{2\pi} \exp(-4\pi i Rr\cos\theta)\cos((A+m)\theta)d\theta.$$

We use the following integration formula ([Wat], p. 21, [GR], p. 482)

$$\int_0^\pi \exp(i\beta\cos\theta)\cos n\theta d\theta = i^n\pi J_n(\beta), \tag{2.20}$$

where $J_n(z)$ is the Bessel coefficient of order n. Since $A+m = l-m$ and $J_n(\beta) = (-1)^n J_n(-\beta)$, the integral of (2.19) is equal to

$$2i^{l-m}\pi e^{i(l-m)\Theta}\int_0^\infty r^{l-m+1}(r^2+1)^{-(\tau+l/2+1)}J_{l-m}(4\pi Rr)dr.$$

We further use the integration formula ([Wat], p. 434, [GR], p. 686, 6.565.4)

$$\int_0^\infty x^{\nu+1}(x^2+a^2)^{-(\mu+1)}J_\nu(bx)dx = a^{\nu-\mu}(\frac{b}{2})^\mu\frac{1}{\Gamma(\mu+1)}K_{\nu-\mu}(ab), \quad a>0,\quad b>0,\quad -1<\Re(\nu)<\Re(2\mu+\frac{3}{2}). \tag{2.21}$$

Then, noting $|z| = z\bar{z}$ for $z \in \mathbf{C}$, we obtain

$$\int_N f(wn)\psi(-\xi yu)dn$$
$$= 2i^{l-m}\pi e^{i(l-m)\Theta}\frac{(2\pi R)^{\tau+l/2}}{\Gamma(\tau+l/2+1)}K_{l-m-\tau-l/2}(4\pi R)$$
$$= i^{l-m}\frac{(2\pi)^{\tau+l/2+1}}{\Gamma(\tau+l/2+1)}(\xi y)^{l-m}|\xi y|^{(\tau-l/2+m)/2}K_{(\tau-l/2+m)/2}(4\pi|\xi y|^{1/2}).$$

This completes the proof.

3. The functional equation of Eisenstein series

In this section, we will treat the functional equation of Eisenstein series following the method of Jacquet [J].

3.1. Let k be a global field. Let E be a finite dimensional vector space over k. We define the space of Schwartz-Bruhat functions as follows. For $v \in \mathbf{h}$, let $\mathcal{S}(E_v)$ be the space of all locally constant and compactly supported complex valued functions on E_v. For $v \in \mathbf{a}$, E_v is isomorphic to an Euclidean space. We let $\mathcal{S}(E_v)$ be the space of all Schwartz functions on E_v. Taking a basis, we have $E \cong k^n$. We define $\mathcal{S}(E_A)$ to be the space of finite linear combinations of functions Φ on E_A

of the form $\Phi = \prod_v \Phi_v$, $\Phi_v \in \mathcal{S}(E_v)$, Φ_v is the characteristic function of $\mathfrak{O}_v^n$ for almost all $v \in \mathbf{h}$. Then $\mathcal{S}(E_A)$ does not depend on the choice of the basis.

Put $k_A^1 = \{x \in k_A^\times \mid |x|_A = 1\}$. Then $k_A^1/k^\times$ is compact and $k_A^\times/k_A^1 \cong \mathbf{R}_+$ if k is a number field, $k_A^\times/k_A^1 \cong \{q^n \mid n \in \mathbf{Z}\} \cong \mathbf{Z}$ if k is a function field, where q is the cardinality of the field of constants of k. Let Ω be the group of all Hecke characters of $k_A^\times$ and Ω_1 be the subgroup of Ω consisting of Hecke characters which are trivial on k_A^1. Then $\omega \in \Omega_1$ if and only if ω can be written as $\omega(x) = |x|_A^s$ with $s \in \mathbf{C}$. We see that $\Omega_1 \cong \mathbf{C}$ if k is a number field, $\Omega_1 \cong \mathbf{C}/(2\pi i/\log q)\mathbf{Z}$ if k is a function field. We define the complex structure on Ω_1 by this isomorphism. The topology of Ω is defined by assigning Ω_1 as an open subgroup and the complex structure of Ω is defined by translating the complex structure on Ω_1 to every coset of Ω_1 (cf. [Wel], p. 117). Define $\alpha \in \Omega_1$ by $\alpha(x) = |x|_A$. A function f on Ω is holomorphic at $\omega \in \Omega$ if and only if $f(\omega\alpha^s)$ is holomorphic at $s = 0$.

Let us first review basic facts on the zeta integral briefly. For details, see Weil [Wel]. For a place v of k, let $d^\times x_v$ be a Haar measure on $k_v^\times$. For $v \in \mathbf{h}$, we normalize $d^\times x_v$ so that $\mathrm{vol}(\mathfrak{O}_v^\times) = 1$. Then $d^\times x = \prod_v d^\times x_v$ defines a Haar measure on $k_A^\times$. Let $v \in \mathbf{h}$. For an unramified quasicharacter ω_v of $k_v^\times$ and the characteristic function Φ_v of $\mathfrak{O}_v$, we have the basic identiy

$$\int_{k_v^\times} \Phi_v(x)\omega_v(x)d^\times x = (1 - \omega_v(\varpi_v))^{-1} \tag{3.1}$$

if $|\omega_v(\varpi_v)| < 1$. By (3.1), we see that the local zeta integral

$$\int_{k_v^\times} \Phi_v(x)\omega_v(x)d^\times x$$

converges absolutely for any $\Phi_v \in \mathcal{S}(k_v)$ and for any quasicharacter ω_v of $k_v^\times$ if $\Re(\sigma_v) > 0$, where we define $\sigma_v \in \mathbf{R}$ by $|\omega_v(x)| = |x|_v^{\sigma_v}$, $x \in k_v^\times$.

Let ω be a Hecke character of $k_A^\times$. We have $\omega(x) = \prod_v \omega_v(x_v)$ for $x = (x_v) \in k_A^\times$ with quasicharacters ω_v of $k_v^\times$. Define $\sigma = \sigma(\omega) \in \mathbf{R}$ by

$$|\omega(x)| = |x|_A^\sigma, \qquad x \in k_A^\times.$$

We consider a Schwartz-Bruhat function on k_A of the form $\Phi = \prod_v \Phi_v$ where $\Phi_v \in \mathcal{S}(k_v)$ for all v and Φ_v is the characteristic function of $\mathfrak{O}_v$ for almost all $v \in \mathbf{h}$. Define the global zeta integral by

$$Z(\Phi, \omega) = \int_{k_A^\times} \Phi(x)\omega(x)d^\times x. \tag{3.2}$$

By (3.1), we see that the integral converges absolutely and is equal to

$$\prod_v \Phi_v(x)\omega_v(x)d^\times x$$

if $\sigma > 1$.

Take t_0 so that $0 < t_0 < 1$ and put $t_1 = t_0^{-1}$. Let F_0 be a continuous function on $\mathbf{R}_+$ such that

$$F_0(t) = 0 \quad \text{for} \quad 0 < t < t_0, \qquad 0 \leqq F_0(t) \leqq 1 \quad \text{for} \quad t_0 \leqq t < 1,$$
$$F_0(1) = 1/2, \qquad F_0(t) = 1 - F_0(t^{-1}) \quad \text{for} \quad 1 < t \leqq t_1, \qquad F_0(t) = 1 \quad \text{for} \quad t_1 \leqq t.$$

Put $F_1(t) = F_0(t^{-1})$, $t \in \mathbf{R}_+$. Then we have

$$F_0(t) + F_1(t) = 1, \quad t \in \mathbf{R}_+, \qquad F_1(t) = 0 \quad \text{for} \quad t_1 \leqq t.$$

Using F_0, F_1, we define the truncated zeta integrals by

$$(3.3) \qquad Z_i(\Phi, \omega) = \int_{k_A^\times} \Phi(x)\omega(x)F_i(|x|_A)d^\times x, \quad i = 0, 1.$$

By the trivial majoration, we see that this integral converges absolutely if $\sigma > 1$ and defines an analytic function of ω in the domain $\{\omega \mid \sigma(\omega) > 1\}$. We have

$$(3.4) \qquad Z(\Phi, \omega) = Z_0(\Phi, \omega) + Z_1(\Phi, \omega) \qquad \text{if} \quad \sigma > 1.$$

Since $F_0(t) \leqq (t/t_0)^\alpha$, $t \in \mathbf{R}_+$ for any $\alpha \geqq 0$, we see that $Z_0(\Phi, \omega)$ can be majorized by

$$t_0^{-\alpha} \int_{k_A^\times} |\Phi(x)||x|_A^{\sigma+\alpha} d^\times x.$$

Therefore $Z_0(\Phi, \omega)$ defines an entire function of ω.

Let ψ be a non-trivial character of k_A which is trivial on k. The Fourier transformation of $\Phi \in \mathcal{S}(k_A)$ is defined by

$$\hat{\Phi}(y) = \int_{k_A} \Phi(x)\psi(xy)dx.$$

We can normalize the Haar measure dx on k_A so that $\hat{\hat{\Phi}}(x) = \Phi(-x)$ holds. This measure is called *self-dual*. We have $\mathrm{vol}(k_A/k) = 1$ for the measure on k_A/k derived from the self-dual measure. Hereafter in this section, we use the self-dual measure exclusively. Then the Poisson summation formula holds:

$$(3.5) \qquad \sum_{\xi \in k} \Phi(\xi) = \sum_{\xi \in k} \hat{\Phi}(\xi).$$

Applying (3.5) to the function $y \to \Phi(xy)$, we obtain

$$(3.6) \qquad \sum_{\xi \in k} \Phi(\xi x) = |x|_A^{-1} \sum_{\xi \in k} \hat{\Phi}(\xi x^{-1}), \qquad x \in k_A^\times.$$

We can define the Haar measure $d^\times \dot{x}$ on $k_A^\times / k^\times$ so that

$$\int_{k_A^\times} \Psi(x)d^\times x = \int_{k_A^\times/k^\times} \sum_{\xi \in k^\times} \Psi(\xi x)d^\times \dot{x}$$

holds for any integrable function Ψ on $k_A^\times$. For simplicity, we denote $d^\times \dot{x}$ by $d^\times x$. For $\sigma > 1$, we have

$$\begin{aligned} Z_1(\Phi,\omega) &= \int_{k_A^\times/k^\times} \sum_{\xi\in k^\times} \Phi(\xi x)\omega(x)F_1(|x|_A)d^\times x \\ &= \int_{k_A^\times/k^\times} (\sum_{\xi\in k^\times} |x|_A^{-1}\hat{\Phi}(\xi x^{-1}) - \Phi(0) + |x|_A^{-1}\hat{\Phi}(0))\omega(x)F_1(|x|_A)d^\times x \\ &= \int_{k_A^\times} \hat{\Phi}(x^{-1})|x|_A^{-1}\omega(x)F_1(|x|_A)d^\times x - \Phi(0)\int_{k_A^\times/k^\times} \omega(x)F_1(|x|_A)d^\times x \\ &+ \hat{\Phi}(0)\int_{k_A^\times/k^\times} \omega(x)|x|_A^{-1}F_1(|x|_A)d^\times x. \end{aligned}$$

Changing x to x^{-1} in the first integral, we get

$$\begin{aligned} Z(\Phi,\omega) &= Z_0(\Phi,\omega) + Z_0(\hat{\Phi},\alpha\omega^{-1}) \\ &- \Phi(0)\int_{k_A^\times/k^\times} \omega(x)F_1(|x|_A)d^\times x + \hat{\Phi}(0)\int_{k_A^\times/k^\times} \omega(x)|x|_A^{-1}F_1(|x|_A)d^\times x. \end{aligned} \tag{3.7}$$

We define a Haar measure $d^\times \dot{x}$ on $k_A^\times/k_A^1$ as follows. If k is a number field, we let $d^\times \dot{x}$ be the measure which corresponds to dx/x on $\mathbf{R}_+$. If k is a function field, we let $d^\times \dot{x}$ be the measure which corresponds to the measure on $\mathbf{Z}$ assigning every point the mass 1. We take the invariant measure $d^\times y$ on $k_A^1/k^\times$ so that

$$\int_{k_A/k^\times} \varphi(x)d^\times x = \int_{k_A^\times/k_A^1} (\int_{k_A^1/k^\times} \varphi(xy)d^\times y)d^\times \dot{x}$$

holds for every integrable function φ on $k_A^\times/k^\times$. Then we have

$$\int_{k_A^\times/k^\times} \omega(x)|x|_A^s F_1(|x|_A)d^\times x = \int_{k_A^\times/k_A^1} (\int_{k_A^1/k^\times} \omega(xy)d^\times y)|x|_A^s F_1(|x|_A)d^\times \dot{x}$$

for $s \in \mathbf{C}$, $\Re(s)+\sigma > 0$. If $\omega|k_A^1$ is non-trivial, this integral vanishes. Assume $\omega|k_A^1$ is trivial. Then, for some $s(\omega) \in \mathbf{C}$, we have $\omega(x) = |x|_A^{s(\omega)}$, $x \in k_A^\times$. Put

$$I(s) = \int_{k_A^\times/k_A^1} |t|_A^s F_1(|t|_A)d^\times t, \tag{3.8}$$

$d^\times t$ being the measure on $k_A^\times/k_A^1$ as above. This integral converges absolutely if $\Re(s) > 0$.

Assume that k is a number field. If $\Re(s) > 0$, we have

$$I(s) = \int_0^\infty t^s F_1(t)d^\times t = \frac{1}{s}t_0^s + \int_{t_0}^{t_1} t^s F_1(t)d^\times t.$$

The last integral defines an entire function of s; hence $I(s)$ can be continued as a meromorphic function to the whole complex plane with the simple pole at $s = 0$.

We have

$$\int_{t_0}^{t_1} t^s F_1(t) d^\times t = \int_{t_0}^{t_1} t^{-s} F_0(t) d^\times t$$
$$= \int_{t_0}^{t_1} t^{-s}(1 - F_1(t)) d^\times t = -\frac{1}{s} t_1^{-s} - \int_0^\infty t^{-s} F_1(t) d^\times t,$$

when $\Re(s) < 0$. Hence we see that

$$I(s) = -I(-s). \tag{3.9}$$

If k is a function field, we have

$$I(s) = \sum_{n \in \mathbf{Z}} q^{ns} F_1(q^n) = \sum_{q^n \leqq t_0} q^{ns} + \sum_{t_0 < q^n < t_1} q^{ns} F_1(q^n).$$

Proceeding similarly to the above, we see that $I(s)$ can be continued as a meromorphic function to the whole complex plane with the simple poles at $s = 2\pi i n / \log q$, $n \in \mathbf{Z}$, holomorphic elsewhere and it satisfies (3.9).

By (3.7) and (3.9), we obtain

$$\begin{aligned} Z(\Phi, \omega) &= Z_0(\Phi, \omega) + Z_0(\hat{\Phi}, \alpha\omega^{-1}) \\ &\quad - \Phi(0)\mathrm{vol}(k_A^1/k^\times) I(s(\omega)) - \hat{\Phi}(0)\mathrm{vol}(k_A^1/k^\times) I(1 - s(\omega)). \end{aligned} \tag{3.10}$$

Here the third and the fourth terms on the right hand side should be ignored if $\omega | k_A^1$ is non-trivial. Now (3.10) shows that $Z(\Phi, \omega)$ can be continued as a meromorphic function of ω which is invariant under $\omega \to \alpha\omega^{-1}$, $\Phi \to \hat{\Phi}$. (Note that $\hat{\hat{\Phi}}(x) = \Phi(-x)$.) Thus we obtain

THEOREM 3.1. *$Z(\Phi, \omega)$ is a meromorphic function of ω which is holomorphic except at $\omega = 1$, α where it has at most a simple pole. Furthermore it satisfies the functional equation*

$$Z(\Phi, \omega) = Z(\hat{\Phi}, \alpha\omega^{-1}).$$

The functional equation of the Hecke L-function $L(s, \omega)$ can be obtained by applying Theorem 3.1 for a suitably chosen Φ.

3.2. Now let us consider Eisenstein series. Let $G = GL(2)$ considered as an algebraic group defined over k. For a Hecke character ω of $k_A^\times$ and $\Phi \in \mathcal{S}(k_A^2)$, we put

$$\mathcal{Z}(\omega, \Phi) = \int_{k_A^\times} \Phi((0, t)) \omega(t) d^\times t. \tag{3.11}$$

As shown above, this integral converges when $\sigma > 1$ and defines a meromorphic function of ω. We let G_A act on $\mathcal{S}(k_A^2)$ by

$$(g\Phi)(x) = \Phi(xg), \qquad g \in G_A, \quad \Phi \in \mathcal{S}(k_A^2).$$

Let χ_1 and χ_2 be Hecke characters of $k_A^\times$. We put

$$f(g) = \mathcal{Z}(\alpha\chi_1\chi_2^{-1}, g\Phi)\chi_1(\det g)|\det g|_A^{1/2}, \qquad g \in G_A. \tag{3.12}$$

We see easily that f satisfies (1.11):

$$f(\begin{pmatrix} a & b \\ 0 & d \end{pmatrix} g) = \chi_1(a)\chi_2(d)\left|\frac{a}{d}\right|_A^{1/2} f(g), \qquad \begin{pmatrix} a & b \\ 0 & d \end{pmatrix} \in P_A.$$

As in §1, we put

$$E(g, f) = \sum_{\gamma \in P_k \backslash G_k} f(\gamma g), \qquad E(\Phi, \chi_1, \chi_2) = E(1, f).$$

Now calculating formally, we have

$$\begin{aligned} E(\Phi, \chi_1, \chi_2) &= \sum_{\xi \in k} f(w \begin{pmatrix} 1 & \xi \\ 0 & 1 \end{pmatrix}) + f(1) \\ &= \sum_{\xi \in k} \int_{k_A^\times} \Phi((-t, -t\xi))(\alpha\chi_1\chi_2^{-1})(t) d^\times t + \int_{k_A^\times} \Phi((0, t))(\alpha\chi_1\chi_2^{-1})(t) d^\times t \\ &= \int_{k_A^\times / k^\times} \sum_{\xi \in k} \sum_{\eta \in k^\times} \Phi((t\eta, t\xi\eta))(\alpha\chi_1\chi_2^{-1})(t) d^\times t \\ &+ \int_{k_A^\times / k^\times} \sum_{\eta \in k^\times} \Phi((0, t\eta))(\alpha\chi_1\chi_2^{-1})(t) d^\times t. \end{aligned}$$

Hence

$$E(\Phi, \chi_1, \chi_2) = \int_{k_A^\times / k^\times} \sum_{(\xi, \eta) \in k^2 \backslash (0,0)} \Phi((\xi, \eta)t)(\alpha\chi_1\chi_2^{-1})(t) d^\times t \tag{3.13}$$

at least formally. For a Hecke character ω of $k_A^\times$, we put

$$\theta^i(\omega, \Phi) = \int_{k_A^\times / k^\times} \sum_{(\xi, \eta) \in k^2 \backslash (0,0)} \Phi((\xi, \eta)t)\omega(t) F_i(|t|_A) d^\times t, \quad i = 1, 2.$$

To justify the formal calculation above, we need the next lemma.

LEMMA 3.2. *Let $\varphi \in \mathcal{S}(k_A)$. For every positive integer n, there exists a positive constant c such that*

$$|\sum_{\xi \in k^\times} \varphi(\xi x)| \leqq c|x|_A^{-n}, \qquad x \in k_A^\times.$$

PROOF. By the Poisson summation formula, we have

$$\sum_{\xi \in k} \varphi(\xi x) = |x|_A^{-1} \sum_{\xi \in k} \hat{\varphi}(\xi x^{-1}).$$

Hence it suffices to show

$$|x|_A^n \left| \sum_{\xi \in k^\times} \varphi(\xi x) \right|$$

is bounded for $x \in k_A^\times$, $|x|_A \geqq 1$. In fact, we will show that $|x|_A^n \sum_{\xi \in k^\times} |\varphi(\xi x)|$ is bounded for $x \in k_A^\times$, $|x|_A \geqq 1$. To this end, we may assume that $\varphi = \prod_v \varphi_v$, $\varphi_v \in \mathcal{S}(k_v)$ where φ_v is the characteristic function of $\mathfrak{O}_v$ for almost all $v \in \mathbf{h}$. Replacing $\varphi(x)$ by $\varphi(xa)$ with a suitable $a \in (k_A^\times)_f$, we may assume that φ_v is the characteristic function of $\mathfrak{O}_v$ for all $v \in \mathbf{h}$. Put $h(x) = \sum_{\xi \in k^\times} |\varphi(\xi x)|$.

First assume that k is a function field. Then

$$h(x) = \sum_{\xi \in k^\times \cap x^{-1} \prod_{v \in \mathbf{h}} \mathfrak{O}_v} 1.$$

If $\xi \in k^\times \cap x^{-1} \prod_{v \in \mathbf{h}} \mathfrak{O}_v$, then $|\xi|_A = 1 \leqq |x|_A^{-1}$. Hence $h(x) = 0$ if $|x|_A > 1$. If $|x|_A = 1$, then $h(x)$ is bounded since $k_A^1/k^\times$ is compact.

Next we assume that k is a number field. Take any $v_0 \in \mathbf{a}$ and let Z be the subgroup of $k_A^\times$ which corresponds to $\mathbf{R}_+ \subset k_{v_0}^\times$. Then we have $k_A = k_A^1 \times Z$. Suppose that there exists a sequence $x_i \in k_A^\times$ such that $|x_i|_A^n |h(x_i)| \to \infty$ and we will derive a contradiction. Write $x_i = y_i z_i$ with $y_i \in k_A^1$ and $z_i \in Z$. Since $k_A^1/k^\times$ is compact, we can find a subsequence $\{y_{i_k}\}$ of $\{y_i\}$ such that $\xi_{i_k} y_{i_k}$ converges to $y \in k_A^1$ for some $\xi_{i_k} \in k^\times$. Replacing $\{x_i\}$ by $\{\xi_{i_k} x_{i_k}\}$, we may assume that

$$x_i = y_i z_i, \quad y_i \in k_A^1, \quad z_i \in Z, \quad y_i \text{ converges to } y \in k_A^1$$

and that $|z_i|_{v_0}^n |h(y_i z_i)| \to \infty$. We have

$$h(y_i z_i) = \sum_{\xi \in k^\times \cap y_i^{-1} \prod_{v \in \mathbf{h}} \mathfrak{O}_v} \prod_{v \in \mathbf{a}} |\varphi_v(\xi y_i z_i)|.$$

Since y_i converges to y, we have $k^\times \cap y_i^{-1} \prod_{v \in \mathbf{h}} \mathfrak{O}_v = k^\times \cap y^{-1} \prod_{v \in \mathbf{h}} \mathfrak{O}_v$ when i is sufficiently large. For every positive integer N and for every $v \in \mathbf{a}$, there exists $c_v > 0$ such that $|\varphi_v(w)| \leqq c_v |w|_v^{-N}$, $w \in k_v$. Hence we have

$$\begin{aligned} |h(y_i z_i)| &\leqq \sum_{\xi \in k^\times \cap y^{-1} \prod_{v \in \mathbf{h}} \mathfrak{O}_v} \prod_{v \in \mathbf{a}} c_v |\xi|_v^{-N} |y_i|_v^{-N} |z_i|_v^{-N} \\ &\leqq c' |z_i|_{v_0}^{-N} \sum_{\xi \in k^\times \cap y^{-1} \prod_{v \in \mathbf{h}} \mathfrak{O}_v} \prod_{v \in \mathbf{a}} |\xi|_v^{-N} \end{aligned}$$

when i is sufficiently large, where $c' = \prod_{v \in \mathbf{a}} c_v$. Since $k \cap y^{-1} \prod_{v \in \mathbf{h}} \mathfrak{O}_v$ is a lattice in $\prod_{v \in \mathbf{a}} k_v$, $\sum_{\xi \in k^\times \cap y^{-1} \prod_{v \in \mathbf{h}} \mathfrak{O}_v} \prod_{v \in \mathbf{a}} |\xi|_v^{-N}$ converges when $N \geqq 2$. This is a contradiction and completes the proof.

Returning to $\theta^i(\omega, \Phi)$, we assume that Φ is of the form $\Phi(x, y) = \Phi_1(x)\Phi_2(y)$ with $\Phi_i \in \mathcal{S}(k_A)$. Then, by Lemma 3.2 for the case $n = 1$, we have

$$\begin{aligned} |\theta^i(\omega, \Phi)| &\leqq c \int_{k_A^\times/k^\times} \left| \sum_{\xi \in k^\times} \Phi_1(\xi t) \right| |t|_A^{-1} \omega(t) F_i(|t|_A) d^\times t \\ &+ |\Phi_1(0)| \left| \int_{k_A^\times/k^\times} \sum_{\eta \in k^\times} \Phi_2(\eta t) \omega(t) F_i(|t|_A) d^\times t \right| \\ &+ |\Phi_2(0)| \left| \int_{k_A^\times/k^\times} \sum_{\xi \in k^\times} \Phi_1(\xi t) \omega(t) F_i(|t|_A) d^\times t \right|. \end{aligned}$$

Hence the integral defining $\theta^0(\omega, \Phi)$ always converges absolutely and that defining $\theta^1(\omega, \Phi)$ converges absolutely if $\sigma > 2$ where $|\omega(x)| = |x|_A^\sigma$. Since Φ can be majorized by a function of this form, the same conclusion holds generally. Checking the local uniformity of the convergence, we see that $\theta^0(\omega, \Phi)$ is an entire function of ω and $\theta^1(\omega, \Phi)$ is holomorphic when $\sigma > 2$.

Put $|\chi_1\chi_2^{-1}(x)| = |x|_A^\sigma$. When $\sigma > 1$, we can justify the formal calculation which led to (3.13) by a similar majoration. We obtain

$$E(\Phi, \chi_1, \chi_2) = \theta^0(\alpha\chi_1\chi_2^{-1}, \Phi) + \theta^1(\alpha\chi_1\chi_2^{-1}, \Phi). \tag{3.14}$$

An element $(x, y) \in k_A^2$ defines a character of k_A^2 by

$$(u\ v) \longrightarrow \psi(xv - yu) = \psi((x\ y)w\begin{pmatrix} u \\ v \end{pmatrix}).$$

In this way, we can identify k_A^2 with its character group. The Fourier transform of Φ with respect to the self-dual measure is defined by

$$\hat{\Phi}(x, y) = \int_{k_A}\int_{k_A} \Phi(u, v)\psi(yu - xv)dudv,$$

where du and dv are the self-dual measure on k_A. Since the paring $\langle(x, y), (u, v)\rangle = \psi(xv - yu)$ is skew-symmetric, we have $\hat{\hat{\Phi}} = \Phi$. By the Poisson summation formula, we have

$$\sum_{(\xi,\eta)\in k^2} \Phi((\xi, \eta)t) = |t|_A^{-2} \sum_{(\xi,\eta)\in k^2} \hat{\Phi}((\xi, \eta)t^{-1}) \qquad \text{for} \quad t \in k_A^\times.$$

First assume $\sigma > 1$. Then inserting this formula in the definition of $\theta^1(\alpha\chi_1\chi_2^{-1}, \Phi)$, we get

$$\begin{aligned}\theta^1(\alpha\chi_1\chi_2^{-1}, \Phi) = &\int_{k_A^\times/k^\times} \sum_{(\xi,\eta)\in k^2\setminus(0,0)} \hat{\Phi}((\xi, \eta)t^{-1})(\alpha^{-1}\chi_1\chi_2^{-1})(t)F_1(|t|_A)d^\times t \\ &- \Phi(0)\int_{k_A^\times/k^\times} (\alpha\chi_1\chi_2^{-1})(t)F_1(|t|_A)d^\times t \\ &+ \hat{\Phi}(0)\int_{k_A^\times/k^\times} (\alpha^{-1}\chi_1\chi_2^{-1})(t)F_1(|t|_A)d^\times t.\end{aligned}$$

Changing t to t^{-1} in the first integral, we obtain

$$\begin{aligned}\theta^1(\alpha\chi_1\chi_2^{-1}, \Phi) = \theta^0(\alpha\chi_1^{-1}\chi_2, \hat{\Phi}) &- \Phi(0)\int_{k_A^\times/k^\times} (\alpha\chi_1\chi_2^{-1})(t)F_1(|t|_A)d^\times t \\ &+ \hat{\Phi}(0)\int_{k_A^\times/k^\times} (\alpha^{-1}\chi_1\chi_2^{-1})(t)F_1(|t|_A)d^\times t.\end{aligned} \tag{3.15}$$

If $\chi_1\chi_2^{-1}|k_A^1$ is non-trivial, the last two integrals of (3.15) vanish. If $\chi_1\chi_2^{-1}|k_A^1$ is trivial, we have $\chi_1\chi_2^{-1}(x) = |x|_A^s$ with $s \in \mathbf{C}$. By (3.9) and (3.14), we obtain

$$\begin{aligned}E(\Phi, \chi_1, \chi_2) = &\theta^0(\alpha\chi_1\chi_2^{-1}, \Phi) + \theta^0(\alpha\chi_1^{-1}\chi_2, \hat{\Phi}) \\ &- \Phi(0)\mathrm{vol}(k_A^1/k^\times)I(1 + s) - \hat{\Phi}(0)\mathrm{vol}(k_A^1/k^\times)I(1 - s).\end{aligned} \tag{3.16}$$

This equation shows that $E(\Phi, \chi_1, \chi_2)$ is a meromorphic function of χ_1, χ_2 and remains invariant when we make the change $(\chi_1, \chi_2) \to (\chi_2, \chi_1)$, $\Phi \to \hat{\Phi}$. Thus we obtain

THEOREM 3.3. *$E(\Phi, \chi_1, \chi_2)$ is a meromorphic function of χ_1, χ_2 and satisfies the functional equation*

$$E(\Phi, \chi_1, \chi_2) = E(\hat{\Phi}, \chi_2, \chi_1).$$

Furthermore it is holomorphic at (χ_1, χ_2) if $\chi_1\chi_2^{-1} \neq \alpha, \alpha^{-1}$.

3.3. For a function f on G_A satisfying (1.11), we put

$$T_w(f)(g) = \int_{N_A} f(wng)dn, \qquad g \in G_A$$

whenever the integral is convergent. Here dn denotes the Haar measure on $N_A \cong k_A$ which corresponds to the self-dual measure on k_A. We are going to rewrite the functional equation of Theorem 3.3 using the global intertwining operator T_w. Let $\Phi \in \mathcal{S}(k_A^2)$. Let $f_\Phi(\chi_1, \chi_2)$ denote the function f defined by (3.12). We put

$$p_\Phi(t) = \int_{k_A} \Phi((t, u))du, \qquad t \in k_A.$$

Then $p_\Phi \in \mathcal{S}(k_A)$. Put $|\chi_1\chi_2^{-1}(t)| = |t|_A^\sigma$. When $\sigma > 1$, we have

$$\begin{aligned}(T_w(f_\Phi(\chi_1, \chi_2)))(1) &= \int_{k_A}\int_{k_A^\times} \Phi((-t, -tu))(\chi_1\chi_2^{-1})(t)|t|_A d^\times t du \\ &= \int_{k_A^\times} p_\Phi(t)(\chi_1\chi_2^{-1})(t)d^\times t = Z(p_\Phi, \chi_1\chi_2^{-1}).\end{aligned}$$

On the other hand, when $\sigma < 0$, we have

$$\begin{aligned}(f_{\hat{\Phi}}(\chi_2, \chi_1))(1) &= \int_{k_A^\times}\int_{k_A}\int_{k_A} \Phi((u, v))\psi(tu)dudv\,(\chi_2\chi_1^{-1})(t)|t|_A d^\times t \\ &= \int_{k_A^\times} \widehat{p_\Phi}(t)\chi_2\chi_1^{-1}(t)|t|_A d^\times t = Z(\widehat{p_\Phi}, \alpha\chi_2\chi_1^{-1}).\end{aligned}$$

By Theorem 3.1, we have $Z(p_\Phi, \chi_1\chi_2^{-1}) = Z(\widehat{p_\Phi}, \alpha\chi_2\chi_1^{-1})$ as meromorphic functions of χ_1, χ_2. Hence we get

$$(3.17) \qquad (T_w(f_\Phi(\chi_1, \chi_2)))(1) = f_{\hat{\Phi}}(\chi_2, \chi_1)(1).$$

Let $g \in G_A$. By definition and (3.17), we obtain

$$\begin{aligned}(T_w(f_\Phi(\chi_1, \chi_2)))(g) &= \chi_1(\det g)|\det g|_A^{1/2}(T_w(f_{g\Phi}(\chi_1, \chi_2)))(1) \\ &= \chi_1(\det g)|\det g|_A^{1/2}(f_{\widehat{g\Phi}}(\chi_2, \chi_1))(1).\end{aligned}$$

A simple calculation shows that

$$(\hat{g\Phi})((x\ y)) = |\det g|_A^{-1}\hat{\Phi}((x\ y)w\,{}^tg^{-1}w^{-1}) = |\det g|_A^{-1}\hat{\Phi}((x\ y)(\det g)^{-1}g).$$

Hence, by (3.12), we get

$$\begin{aligned}(f_{\hat{g\Phi}}(\chi_2, \chi_1))(1) &= |\det g|_A^{-1}\int_{k_A^\times} \hat{\Phi}((0, t)(\det g)^{-1}g)(\chi_2\chi_1^{-1})(t)|t|_A d^\times t \\ &= \chi_1(\det g)^{-1}|\det g|_A^{-1/2}(f_{\hat{\Phi}}(\chi_2, \chi_1))(g).\end{aligned}$$

Thus we obtain

$$(3.18)\qquad (T_w(f_\Phi(\chi_1,\chi_2)))(g) = (f_{\hat{\Phi}}(\chi_2,\chi_1))(g).$$

For a fixed $\Phi \in \mathcal{S}(k_A)$, (3.18) is always meaningful as an equality between meromorphic functions of χ_1 and χ_2. Now the functional equation of Theorem 3.3 can be written as

$$(3.19)\qquad E(g,f) = E(g,T_w(f)) \qquad \text{for} \quad f = f_\Phi(\chi_1,\chi_2).$$

We also note that

$$(3.20)\qquad T_w \circ T_w = 1 \qquad \text{for} \quad f = f_\Phi(\chi_1,\chi_2),$$

which follows from (3.18).

4. Eisenstein series of class 1

4.1. In this section, we will derive a generalization of Kronecker's limit formula over an arbitrary algebraic number field k. The result of V, §1 is a special case when k is totally real.

Let k be an algebraic number field. Let χ be a Hecke character of $k_A^\times$ which is trivial on $k^\times \prod_{v\in\mathbf{h}} \mathfrak{O}_v^\times k_\infty^\times$ and s be a complex number. For a place v of k, we define a function f_v on G_v by

$$f_v(pk) = \left|\frac{a}{d}\right|^s \chi_v(a), \qquad p = \begin{pmatrix} a & b \\ 0 & d \end{pmatrix} \in P_v,\ k \in K_v.$$

Then $f_v \in \mathcal{B}(\chi_v|\ |_v^{s-1/2}, |\ |_v^{1/2-s})$. Put $f = \prod_v f_v$. Since f depends on s and χ, we denote it by $f_{s,\chi}$ when it is necessary. As in §1, we define an Eisenstein series by

$$E(g,f_{s,\chi}) = \sum_{\gamma\in P_k\backslash G_k} f(\gamma g).$$

Let Z be the center of G and let $P' = \{\begin{pmatrix} x & y \\ 0 & 1 \end{pmatrix}\} \subset P$. We have $P = ZP'$. Let ω_χ be the character of Z_A defined by

$$\omega_\chi(\begin{pmatrix} a & 0 \\ 0 & a \end{pmatrix}) = \chi(a), \qquad a \in k_A^\times.$$

We have

$$E(zg,f_{s,\chi}) = \omega_\chi(z)E(g,f_{s,\chi}), \qquad z \in Z_A.$$

We put $E(g,s,\chi) = E(g,f_{s,\chi})$.

Let ψ_0 be the non-trivial character of $\mathbf{Q}_A$ which is trivial on $\mathbf{Q}$ such that $(\psi_0)_\infty(x) = \exp(2\pi ix)$, $x \in \mathbf{Q}_\infty = \mathbf{R}$. Then we have $(\psi_0)_p(x) = \exp(-2\pi i\mathrm{Fr}(x))$ for $x \in \mathbf{Q}_p$ where Fr denotes the fractional part of x. Let Tr denote the trace map from k_A to $\mathbf{Q}_A$ and we put $\psi(x) = \psi_0(\mathrm{Tr}(x))$, $x \in k_A$. Hereafter in this section, we take ψ in this way. Let d be a differental idele attached to ψ. As shown in §1, $E(g,s,\chi)$ has the Fourier expansion

$$(4.1)\qquad E(\begin{pmatrix} y & x \\ 0 & 1 \end{pmatrix},s,\chi) = c_0(y) + \sum_{\xi\in k^\times} c(\xi dy)\psi(\xi x), \qquad y \in k_A^\times,\quad x \in k_A$$

with $c_0(y)$ and $c(\xi dy)$ which are given by (1.19) and (1.20) respectively. By Lemmas 2.1, 2.3 and 2.7, we obtain immediately that

$$\begin{aligned} c_0(y) &= \chi(y)|y|_A^s \\ &+ |D_k|^{-1/2}|y|_A^{1-s}\frac{L(2s-1,\chi)}{L(2s,\chi)}\Big(\frac{\sqrt{\pi}\Gamma(s-1/2)}{\Gamma(s)}\Big)^{r_1}\Big(\frac{2\pi\Gamma(2s-1)}{\Gamma(2s)}\Big)^{r_2}. \end{aligned} \tag{4.2}$$

(The factor $|D_k|^{-1/2}$ and 2 in the numerator of the last term appear since we must calculate here using the self-dual measure on k_A.) By Lemmas 2.2, 2.5 and 2.9, we obtain

$$\begin{aligned} c(\xi dy) &= |D_k|^{-1/2}\prod_{v\in\mathbf{h}} I_v(d_v\xi_v y_v)|y_v|_v^{1-s}\prod_{v\in\mathbf{a}_r}\frac{2\pi^s}{\Gamma(s)}K_{s-1/2}(|2\pi\xi_v y_v|)|\xi_v|_v^{s-1/2}|y_v|_v^{1/2} \\ &\times\prod_{v\in\mathbf{a}_c}\frac{2(2\pi)^{2s}}{\Gamma(2s)}K_{2s-1}(|4\pi\xi_v y_v|)|\xi_v|_v^{s-1/2}|y_v|_v^{1/2}. \end{aligned}$$

Here $|\ |$ without the subscript v denotes the usual absolute value and ξ_v denotes the v-component of $\xi\in k_A^\times$; I_v, $v\in\mathbf{h}$ denotes the function on $k_v^\times$ given in Lemma 2.2. For $x\in k_A^\times$, put

$$\sigma_{s,\chi}(x) = \prod_{v\in\mathbf{h}}\begin{cases} 1+\chi_v(\varpi_v)q_v^s+\cdots+(\chi_v(\varpi_v)q_v^s)^{\mathrm{ord}_v(x_v)} & \text{if } x_v\in\mathfrak{O}_v, \\ 0 & \text{if } x_v\notin\mathfrak{O}_v. \end{cases} \tag{4.3}$$

By Lemma 2.2, we have

$$\prod_{v\in\mathbf{h}} I_v(d_v\xi_v y_v) = \frac{\sigma_{1-2s,\chi}(d\xi y)}{L(2s,\chi)}.$$

Hence we get

$$\begin{aligned} c(\xi dy) &= 2^{r_1+r_2}|D_k|^{-1/2}\Big(\frac{\pi^s}{\Gamma(s)}\Big)^{r_1}\Big(\frac{(2\pi)^{2s}}{\Gamma(2s)}\Big)^{r_2}\frac{|y_f|_A^{1-s}}{L(2s,\chi)}\sigma_{1-2s,\chi}(\xi dy) \\ &\times\prod_{v\in\mathbf{a}_r}K_{s-1/2}(|2\pi\xi_v y_v|)|\xi_v|_v^{s-1/2}|y_v|_v^{1/2} \\ &\times\prod_{v\in\mathbf{a}_c}K_{2s-1}(|4\pi\xi_v y_v|)|\xi_v|_v^{s-1/2}|y_v|_v^{1/2}. \end{aligned} \tag{4.4}$$

By (4.2) and (4.4), we obtain

$$\begin{aligned} &E\Big(\begin{pmatrix} y & x \\ 0 & 1\end{pmatrix}, s,\chi\Big) = \chi(y)|y|_A^s \\ &+|D_k|^{-1/2}|y|_A^{1-s}\frac{L(2s-1,\chi)}{L(2s,\chi)}\Big(\frac{\sqrt{\pi}\Gamma(s-1/2)}{\Gamma(s)}\Big)^{r_1}\Big(\frac{2\pi\Gamma(2s-1)}{\Gamma(2s)}\Big)^{r_2} \\ &+2^{r_1+r_2}|D_k|^{-1/2}\Big(\frac{\pi^s}{\Gamma(s)}\Big)^{r_1}\Big(\frac{(2\pi)^{2s}}{\Gamma(2s)}\Big)^{r_2}\frac{|y_f|_A^{1-s}}{L(2s,\chi)} \\ &\times\sum_{\xi\in k^\times}\sigma_{1-2s,\chi}(\xi dy)\prod_{v\in\mathbf{a}_r}K_{s-1/2}(|2\pi\xi_v y_v|)|\xi_v|_v^{s-1/2}|y_v|_v^{1/2} \\ &\times\prod_{v\in\mathbf{a}_c}K_{2s-1}(|4\pi\xi_v y_v|)|\xi_v|_v^{s-1/2}|y_v|_v^{1/2}\psi(\xi x). \end{aligned} \tag{4.5}$$

We put

$$L^*(s,\chi) = |D_k|^{s/2}(\pi^{-s/2}\Gamma(s/2))^{r_1}((2\pi)^{1-s}\Gamma(s))^{r_2}L(s,\chi).$$

Then the functional equation

$$L^*(s,\chi) = \chi(d)L^*(1-s,\chi^{-1}) \tag{4.6}$$

holds. Multiplying both sides of (4.5) by $L^*(2s,\chi)$ and using (4.6), we obtain

$$\begin{aligned} L^*(2s,\chi)E(\begin{pmatrix} y & x \\ 0 & 1 \end{pmatrix}, s,\chi) &= L^*(2s,\chi)\chi(y)|y|_A^s + \chi(d)L^*(2(1-s),\chi^{-1})|y|_A^{1-s} \\ &+2^{r_1+r_2}|D_k|^{s-1/2}(2\pi)^{r_2}|y_f|_A^{1-s}\sum_{\xi\in k^\times}\sigma_{1-2s,\chi}(\xi dy) \\ &\times\prod_{v\in\mathbf{a}_r}K_{s-1/2}(|2\pi\xi_v y_v|)|\xi_v|_v^{s-1/2}|y_v|_v^{1/2} \\ &\times\prod_{v\in\mathbf{a}_c}K_{2s-1}(|4\pi\xi_v y_v|)|\xi_v|_v^{s-1/2}|y_v|_v^{1/2}\psi(\xi x). \end{aligned}$$

On the right-hand side, change s to $1-s$ and χ to χ^{-1}. We see immediately that the sum of two constant terms is multiplied by $\chi(dy)^{-1}$. Concerning the Fourier terms, noting

$$\sigma_{-s,\chi}(x) = \chi(x_f)|x_f|_A^s\sigma_{s,\chi^{-1}}(x), \qquad |d_f|_A = |D_k|^{-1}, \tag{4.7}$$

and $K_s(z) = K_{-s}(z)$, we also see that every term is multiplied by $\chi(dy)^{-1}$. Therefore the functional equation for Eisenstein series is

$$L^*(2s,\chi)E(g,s,\chi) = \chi(\det g)\chi(d)L^*(2(1-s),\chi^{-1})E(g,1-s,\chi^{-1}). \tag{4.8}$$

(In fact, we proved (4.8) when $g \in P_A$. The general case follows from this case since $E(g,s,\chi)$ is right K-invariant.) By Lemmas 2.1, 2.3 and 2.7 and (4.6), we see that

$$(T_w(f_{s,\chi}))(g) = \chi(\det g)f_{1-s,\chi^{-1}}(g), \qquad g \in G_A.$$

Hence (4.8) is consistent with (3.19).

Now put

$$E^*(g,s,\chi) = L(2s,\chi)E(g,s,\chi). \tag{4.9}$$

By (4.5), we have

$$\begin{aligned} E^*(\begin{pmatrix} y & x \\ 0 & 1 \end{pmatrix}, s,\chi) &= L(2s,\chi)\chi(y)|y|_A^s \\ &+|D_k|^{-1/2}|y|_A^{1-s}L(2s-1,\chi)(\frac{\sqrt{\pi}\Gamma(s-1/2)}{\Gamma(s)})^{r_1}(\frac{2\pi\Gamma(2s-1)}{\Gamma(2s)})^{r_2} \\ &+2^{r_1+r_2}|D_k|^{-1/2}(\frac{\pi^s}{\Gamma(s)})^{r_1}(\frac{(2\pi)^{2s}}{\Gamma(2s)})^{r_2}|y_f|_A^{1-s} \\ &\times\sum_{\xi\in k^\times}\sigma_{1-2s,\chi}(\xi dy)\prod_{v\in\mathbf{a}_r}K_{s-1/2}(|2\pi\xi_v y_v|)|\xi_v|_v^{s-1/2}|y_v|_v^{1/2} \\ &\times\prod_{v\in\mathbf{a}_c}K_{2s-1}(|4\pi\xi_v y_v|)|\xi_v|_v^{s-1/2}|y_v|_v^{1/2}\psi(\xi x). \end{aligned} \tag{4.10}$$

4.2. The limit formula concerns the behavior of $E^*(g,s,\chi)$ at $s=1$. We consider the case $\chi=1$. We have

$$\frac{\Gamma(s-1/2)}{\Gamma(s)} = \sqrt{\pi}\,\{1-(2\log 2)(s-1)\} + O((s-1)^2), \qquad s\to 1, \quad (\mathrm{V},(1.31))$$
$$\frac{\Gamma(2s-1)}{\Gamma(2s)} = \frac{1}{2s-1} = 1-2(s-1)+O((s-1)^2), \qquad s\to 1,$$

$$\begin{aligned}&|y|_A^{1-s}\zeta_k(2s-1)\left(\frac{\sqrt{\pi}\Gamma(s-1/2)}{\Gamma(s)}\right)^{r_1}\left(\frac{2\pi\Gamma(2s-1)}{\Gamma(2s)}\right)^{r_2}\\ &=\pi^{r_1}(2\pi)^{r_2}\cdot\frac{c_k}{2(s-1)}\left\{1+\left(\frac{2\gamma_k}{c_k}-2r_1\log 2-2r_2-\log|y|_A\right)(s-1)\right\}\\ &\qquad + O((s-1)^2), \qquad s\to 1.\end{aligned}$$

Therefore, from (4.10), we obtain

$$(4.11)\qquad \begin{aligned}E^*(g,s,1) &= \frac{2^{n-1}\pi^n h_k R_k}{w_k|D_k|}\left[\frac{1}{s-1}-2r_1\log 2-2r_2\right.\\ &\left.+\frac{w_k|D_k|^{1/2}\gamma_k}{2^{r_1+r_2-1}\pi^{r_2}h_kR_k}-\log|\det g|_A+h(g)\right]+O(s-1), \qquad s\to 1,\end{aligned}$$

where $n=r_1+2r_2$ is the degree of k and $h(g)$ is the function given on P'_A by

$$(4.12)\qquad \begin{aligned}h\left(\begin{pmatrix} y & x\\ 0 & 1\end{pmatrix}\right) &= \frac{w_k|D_k|}{2^{n-1}\pi^n h_kR_k}\zeta_k(2)|y|_A\\ &+\frac{2^{r_2+1}w_k|D_k|^{1/2}}{h_kR_k}\sum_{\xi\in k^\times}\sigma_{-1}(\xi dy)\psi(\xi x)\\ &\times\prod_{v\in\mathbf{a}_r}K_{1/2}(|2\pi\xi_v y_v|)|\xi_v|_v^{1/2}|y_v|_v^{1/2}\prod_{v\in\mathbf{a}_c}K_1(|4\pi\xi_v y_v|)|\xi_v|_v^{1/2}|y_v|_v^{1/2}.\end{aligned}$$

Here we abbreviated $\sigma_{s,1}(\xi dy)$ to $\sigma_s(\xi dy)$.

4.3. We can analyze the behavior of $E^*(g,s,\chi)$ at $s=0$ using (4.10). We use the notation of V, §1. Concerning the first term, we have

$$\begin{aligned}L(2s,\chi)\chi(y)|y|_A^s &= \chi(y)(1+s\log|y|_A)\cdot -\frac{R_k}{w_k}\sum_c\chi(c)(1+2\delta_k(c)s)(2s)^{r_1+r_2-1}\\ &\quad + O(s^{r_1+r_2+1})\end{aligned}$$

If $\chi=1$, this can be written as

$$\begin{aligned}&-\frac{2^{r_1+r_2-1}h_kR_k}{w_k}s^{r_1+r_2-1}\\ &\times\{1+(\log|y|_A+2n\gamma+2n\log 2\pi-2\log|D_k|-\frac{w_k|D_k|^{1/2}\gamma_k}{2^{r_1+r_2-1}\pi^{r_2}h_kR_k})s\}\\ &\qquad + O(s^{r_1+r_2+1}),\end{aligned}$$

and if $\chi\neq 1$,

$$\frac{2^{r_1+r_2-1}R_k}{w_k}s^{r_1+r_2}\chi(y)\sum_c\frac{w_k|D_k|^{1/2}}{2^{r_1+r_2-1}\pi^{r_2}R_k}\chi(c)\gamma_k(c')+O(s^{r_1+r_2+1}).$$

The second term gives

$$\frac{L^*(2s-1,\chi)}{L^*(2s,\chi)}L(2s,\chi)|y|_A^{1-s} = \chi(d)L^*(2-2s,\chi^{-1})\cdot\frac{L(2s,\chi)}{L^*(2s,\chi)}|y|_A^{1-s}$$
$$= 2^{-r_2}\pi^{-n}|D_k|\chi(d)L(2,\chi^{-1})|y|_A s^{r_1+r_2} + O(s^{r_1+r_2+1}).$$

In view of (4.7), we see that the third term gives

$$2^n|D_k|^{1/2}\chi(d)\chi(y)\sum_{\xi\in k^\times}\sigma_{-1,\chi^{-1}}(\xi dy)\prod_{v\in\mathbf{a}_c}K_1(|4\pi\xi_v y_v|)|\xi_v|_v^{1/2}|y_v|_v^{1/2}$$
$$\times\prod_{v\in\mathbf{a}_r}K_{1/2}(|2\pi\xi_v y_v|)|\xi_v|_v^{1/2}|y_v|_v^{1/2}\psi(\xi x)s^{r_1+r_2} + O(s^{r_1+r_2+1}).$$

We define the function h_χ on G_A so that, if $\chi = 1$,

$$(4.13)\quad \begin{aligned} E^*(g,s,1) = &-\frac{2^{r_1+r_2-1}h_k R_k}{w_k}s^{r_1+r_2-1}\{1+(2n\gamma+2n\log 2\pi-2\log|D_k| \\ &-\frac{w_k|D_k|^{1/2}\gamma_k}{2^{r_1+r_2-1}\pi^{r_2}h_k R_k}+\log|\det g|_A - h_1(g))s\} \\ &+O(s^{r_1+r_2+1}), \qquad s\to 0 \end{aligned}$$

holds and that, if $\chi\neq 1$,

$$(4.14)\quad \begin{aligned} E^*(g,s,\chi) = &-\frac{2^{r_1+r_2-1}h_k R_k}{w_k}s^{r_1+r_2} \\ &\times(-\frac{w_k|D_k|^{1/2}}{2^{r_1+r_2-1}\pi^{r_2}h_k R_k}\chi(d)\sum_c\chi(c^{-1})\gamma_k(c)\chi(\det g) - h_\chi(g)) \\ &+O(s^{r_1+r_2+1}), \qquad s\to 0 \end{aligned}$$

holds. Then the above calculation shows that h_χ is given on P'_A by

$$(4.15)\quad \begin{aligned} h_\chi(\begin{pmatrix} y & x \\ 0 & 1\end{pmatrix}) = &\frac{w_k|D_k|}{2^{n-1}\pi^n h_k R_k}\chi(d)L(2,\chi^{-1})|y|_A \\ &+\frac{2^{r_2+1}w_k|D_k|^{1/2}}{h_k R_k}\chi(d)\chi(y)\sum_{\xi\in k^\times}\sigma_{-1,\chi^{-1}}(\xi dy)\psi(\xi x) \\ &\times\prod_{v\in\mathbf{a}_r}K_{1/2}(|2\pi\xi_v y_v|)|\xi_v|_v^{1/2}|y_v|_v^{1/2}\prod_{v\in\mathbf{a}_c}K_1(|4\pi\xi_v y_v|)|\xi_v|_v^{1/2}|y_v|_v^{1/2}. \end{aligned}$$

Here note that

$$\sum_c\chi(c)\gamma_k(c') = \chi(d)\sum_c\chi(c^{-1})\gamma_k(c).$$

We have $h = h_1$. Since $E(g,s,\chi)$ satisfies

$$E(\gamma zgk,s,\chi) = \omega_\chi(z)E(g,s,\chi), \qquad \gamma\in G_k,\ z\in Z_A,\ g\in G_A,\ k\in K,$$

we obtain the following automorphic property of $h_\chi(g)$ from (4.13) and (4.14).

PROPOSITION 4.1. *If $\chi = 1$, then $h(g) - \log|\det g|_A$ is left G_k-invariant and right K-invariant function on G_A, which transforms according to ω_χ under Z_A. If $\chi\neq 1$, then $h_\chi(g)$ is left G_k-invariant and right K-invariant function on G_A, which transforms according to ω_χ under Z_A*

REMARK. The formulas (4.5) and (4.10) hold even for the function field case dropping archimedean contributions. Hence we can obtain a limit formula also in the function field case; it may be interesting to investigate the function h_χ in detail.

APPENDIX II

ON HIGHER DERIVATIVES OF L-FUNCTIONS

In chapters III, IV, V, we were concerned with invariants related with the derivatives of L-functions at $s = 0$. In this appendix, we will explore second derivatives of L-functions at $s = 0$ using our results in chapter I, §3. This is the territory which was little cultivated before. The author feels to be excused for the meagreness of the results. He will be abundantly rewarded if the direction set forth in this chapter will bear some fruits in the future.

1. A search for new invariants

1.1. Let F be a totally real algebraic number field of degree n. Let $J_F = \{\sigma_1, \sigma_2, \dots, \sigma_n\}$. For an integral ideal $\mathfrak{f}$ of F, let $C_{\mathfrak{f}}$ denote the ideal class group of F defined modulo $\mathfrak{f}\infty_1\infty_2\cdots\infty_n$. Let K be an abelian extension of F. Let $\tilde{\mathfrak{f}}$ be the finite part of the conductor of K as a class field over F. We put $G = \mathrm{Gal}(K/F)$ and let $\widehat{G}$ denote the character group of G. By the Artin map, $\omega \in \widehat{G}$ can be regarded as the character of $C_{\tilde{\mathfrak{f}}}$. For $\omega \in \widehat{G}$, let $\mathfrak{f}(\omega)$ denote the finite part of the conductor of ω and $p(\omega)$ be the positive integer such that

$$L(s,\omega) = cs^{p(\omega)-1} + O(s^{p(\omega)}) \qquad \text{for} \quad s \to 0 \tag{1.1}$$

with $c \neq 0$. If $\omega = 1$, we have $p(\omega) = n$. Let $r(\omega)$ be the number of infinite places of F at which ω ramifies. Then we have

$$p(\omega) = n - r(\omega) + 1 \qquad \text{if} \quad \omega \neq 1. \tag{1.2}$$

For a positive integer p and an integral ideal $\mathfrak{f}$ of F, we put

$$\widehat{G}^{(p)} = \{\omega \in \widehat{G} \mid p(\omega) = p\},$$
$$\widehat{G}_{\mathfrak{f}} = \{\omega \in \widehat{G} \mid \mathfrak{f}(\omega) = \mathfrak{f}\}, \qquad \widehat{G}_{\mathfrak{f}}^{(p)} = \widehat{G}^{(p)} \cap \widehat{G}_{\mathfrak{f}}.$$

For $\tau \in G$, we put

$$Z_{K/F}(\tau) = \prod_{1\neq\omega\in\widehat{G}} \exp\Big(\frac{\omega(\tau)}{|G|}\frac{1}{p(\omega)}\frac{L^{(p(\omega))}(0,\omega)}{L^{(p(\omega)-1)}(0,\omega)}\Big), \tag{1.3}$$

$$Z_{K/F}^{(p)}(\tau) = \prod_{1\neq\omega\in\widehat{G}^{(p)}} \exp\Big(\frac{\omega(\tau)}{|G|}\frac{L^{(p)}(0,\omega)}{L^{(p-1)}(0,\omega)}\Big). \tag{1.4}$$

Suppose that $p = 1$, $1 \neq \omega \in \widehat{G}^{(p)}$. Then ω ramifies at every infinite place of F.

Hence K has the maximum CM-subfield K_0 and ω factors through $\mathrm{Gal}(K/F) \longrightarrow \mathrm{Gal}(K_0/F)$ (cf. III, Lemma 5.4). Put $G_0 = \mathrm{Gal}(K_0/F)$, $\bar{\tau} = \tau|K_0$. Then we have

$$Z^{(1)}_{K/F}(\tau) = \prod_{\omega \in (\hat{G}_0)_-} \exp(\frac{\omega(\bar{\tau})}{|G|}\frac{L'(0,\omega)}{L(0,\omega)}).$$

Assuming III, Conjecture 3.2, we have (cf. III, Corollary 3.6)

$$(1.5) \qquad Z^{(1)}_{K/F}(\tau)^{[K:K_0]} \sim \pi^{n\mu(\bar{\tau})/2} \prod_{i=1}^{n} p_{K_0^{\sigma_i}}(\mathrm{id}, \sigma_i^{-1}\bar{\tau}\sigma_i).$$

The symbols $Z_{K/F}(\tau)$ and $Z^{(p)}_{K/F}(\tau)$ are natural objects which have good functorial properties shown below.

LEMMA 1.1. *Let L be an abelian extension of F of finite degree which contains K. Let $\tau \in \mathrm{Gal}(K/F)$, $\mathrm{Inf}_{L/K}(\tau) = \sum_{i=1}^{l} \tau_i$, $\tau_i \in \mathrm{Gal}(L/F)$. We put*

$$Z_{L/F}(\mathrm{Inf}_{L/K}(\tau)) = \prod_{i=1}^{l} Z_{L/F}(\tau_i), \qquad Z^{(p)}_{L/F}(\mathrm{Inf}_{L/K}(\tau)) = \prod_{i=1}^{l} Z^{(p)}_{L/F}(\tau_i).$$

Then we have

$$Z_{K/F}(\tau) = Z_{L/F}(\mathrm{Inf}_{L/K}(\tau)), \qquad Z^{(p)}_{K/F}(\tau) = Z^{(p)}_{L/F}(\mathrm{Inf}_{L/K}(\tau)).$$

We leave the proof to the reader since the assertions follow directly from the definitions (1.3) and (1.4).

LEMMA 1.2. *Let F_0 and F be totally real algebraic number fields such that $F_0 \subset F$. Let K be an abelian extension of F_0 which contains F. Define $Z'_{K/F}(\tau)$ by (1.3) including $\omega = 1 \in \widehat{G}$. Then we have*

$$Z'_{K/F}(\tau) = Z'_{K/F_0}(\tau)^{[F:F_0]} \qquad \text{for} \quad \tau \in \mathrm{Gal}(K/F).$$

PROOF. Put $\widetilde{G} = \mathrm{Gal}(K/F_0)$, $G = \mathrm{Gal}(K/F)$. Take $\omega \in \widehat{G}$ and put $\mathrm{Ind}_G^{\tilde{G}}\omega = \oplus_{i=1}^{m}\omega_i$, $\omega_i \in \widehat{\widetilde{G}}$. We see that ω_i's are mutually different and that $L(s,\omega) = \prod_{i=1}^{m} L(s,\omega_i)$. Hence we obtain

$$\frac{1}{p(\omega)}\frac{L^{(p(\omega))}(0,\omega)}{L^{(p(\omega)-1)}(0,\omega)} = \sum_{i=1}^{m} \frac{1}{p(\omega_i)}\frac{L^{(p(\omega_i))}(0,\omega_i)}{L^{(p(\omega_i)-1)}(0,\omega_i)}.$$

Now the assertion follows from the definition.

1.2. Modeled after the case $p = 1$, we are going to define certain "invariants" factoring $Z^{(p)}_{K/F}(\tau)$. We assume $\sigma_1 = \mathrm{id}$. For $c \in C_{\mathfrak{f}}$, let $\zeta_F(s,c)$ be the partial zeta function of the class c. Since

$$(1.6) \qquad L(s,\omega) = \sum_{c \in C_{\mathfrak{f}}} \omega(c)\zeta_F(0,c)$$

for $\omega \in \widehat{G}_{\mathfrak{f}}$, (1.4) can be written as

$$Z^{(p)}_{K/F}(\tau) = \exp\Big(\frac{1}{|G|}\sum_{\mathfrak{f}|\tilde{\mathfrak{f}}}\sum_{1\neq\omega\in\hat{G}^{(p)}_{\mathfrak{f}}}\frac{\omega(\tau)}{L^{(p-1)}(0,\omega)}\sum_{c\in C_{\mathfrak{f}}}\omega(c)\zeta^{(p)}_F(0,c)\Big). \tag{1.7}$$

For $c \in C_{\mathfrak{f}}$, suppose that we have a "canonical" decomposition:

$$\zeta^{(p)}_F(0,c) = \sum_{i=1}^{n} X^{(p)}(c^{\sigma_i}). \tag{1.8}$$

Then we put

$$g^{(p)}_K(\mathrm{id},\tau) = \exp\Big(\frac{1}{|G|}\sum_{\mathfrak{f}|\tilde{\mathfrak{f}}}\sum_{1\neq\omega\in\hat{G}^{(p)}_{\mathfrak{f}}}\frac{\omega(\tau)}{L^{(p-1)}(0,\omega)}\sum_{c\in C_{\mathfrak{f}}}\omega(c)X^{(p)}(c)\Big). \tag{1.9}$$

As in chapter III, Lemma 3.8, we have

$$\begin{aligned} Z^{(p)}_{K/F}(\tau) &= \prod_{1\neq\omega\in\hat{G}^{(p)}}\exp\Big(\frac{\omega(\tau)}{|G|}\frac{L^{(p)}(0,\omega)}{L^{(p-1)}(0,\omega)}\Big) \\ &= \prod_{i=1}^{n} g^{(p)}_{K^{\sigma_i}}(\mathrm{id},\sigma_i^{-1}\tau\sigma_i), \qquad \tau\in G. \end{aligned} \tag{1.10}$$

In fact, as in III, Lemma 3.8, we have

$$g^{(p)}_{K^{\sigma_i}}(\mathrm{id},\sigma_i^{-1}\tau\sigma_i) = \exp\Big(\frac{1}{|G|}\sum_{\mathfrak{f}|\tilde{\mathfrak{f}}}\sum_{1\neq\omega\in\hat{G}^{(p)}_{\mathfrak{f}}}\frac{\omega(\tau)}{L^{(p-1)}(0,\omega)}\sum_{c\in C_{\mathfrak{f}}}\omega(c)X^{(p)}(c^{\sigma_i})\Big).$$

Then (1.10) follows from (1.7) and (1.8). We note that the following functoriality of Shimura type is also a formal consequence of the definition (1.9) (cf. III, Proposition 3.20).

LEMMA 1.3. *Let L be an abelian extension of F of finite degree which contains K. Let $\tau\in\mathrm{Gal}(K/F)$, $\mathrm{Inf}_{L/K}(\tau)=\sum_{i=1}^{l}\tau_i$, $\tau_i\in\mathrm{Gal}(L/F)$. We put*

$$g^{(p)}_L(\mathrm{id},\mathrm{Inf}_{L/K}(\tau)) = \prod_{i=1}^{l} g^{(p)}_L(\mathrm{id},\tau_i).$$

Then we have

$$g^{(p)}_K(\mathrm{id},\tau) = g^{(p)}_L(\mathrm{id},\mathrm{Inf}_{L/K}(\tau)). \tag{1.11}$$

PROOF. Put

$$\widetilde{G}=\mathrm{Gal}(L/F),\qquad G=\mathrm{Gal}(K/F),\qquad H=\mathrm{Gal}(L/K).$$

Take $\tilde{\tau}\in\widetilde{G}$ so that $\tau=\tilde{\tau}|K$. Let $\tilde{\mathfrak{h}}$ be the finite part of the conductor of L as the class field over F. Since $\mathrm{Inf}_{L/K}(\tau)=\{h\tilde{\tau}\mid h\in H\}$ by definition, we have

$$g^{(p)}_L(\mathrm{id},\mathrm{Inf}_{L/K}(\tau)) = \exp\Big(\frac{1}{|\widetilde{G}|}\sum_{\mathfrak{h}|\tilde{\mathfrak{h}}}\sum_{1\neq\omega\in\hat{\tilde{G}}^{(p)}_{\mathfrak{h}}}\frac{\sum_{h\in H}\omega(h\tilde{\tau})}{L^{(p-1)}(0,\omega)}\sum_{c\in C_{\mathfrak{h}}}\omega(c)X^{(p)}(c)\Big).$$

Let $\omega \in \widehat{\widetilde{G}}_{\mathfrak{h}}^{(p)}$, $\omega \neq 1$. If $\omega|H$ is non-trivial, then $\sum_{h\in H}\omega(h\tilde{\tau})=0$. If $\omega|H$ is trivial, then $\sum_{h\in H}\omega(h\tilde{\tau})$ is equal to $|H|\,\omega(\tilde{\tau})$. Therefore we obtain

$$\begin{aligned} g_L^{(p)}(\mathrm{id}, \mathrm{Inf}_{L/K}(\tau)) &= \exp\Big(\frac{|H|}{|\widetilde{G}|}\sum_{\mathfrak{f}|\tilde{\mathfrak{f}}}\sum_{1\neq\omega\in\hat{G}_{\mathfrak{f}}^{(p)}}\frac{\omega(\tau)}{L^{(p-1)}(0,\omega)}\sum_{c\in C_{\mathfrak{f}}}\omega(c)X^{(p)}(c)\Big) \\ &= g_K^{(p)}(\mathrm{id},\tau) \end{aligned}$$

and complete the proof.

1.3. Let $\mathfrak{a}_1, \mathfrak{a}_2, \dots, \mathfrak{a}_{h_0}$ be integral ideals which represent narrow ideal classes of F. Let $c \in C_{\mathfrak{f}}$ and choose $\mathfrak{a}_\mu$ so that c and $\mathfrak{a}_\mu\mathfrak{f}$ belong to the same narrow ideal class. For an open simplicial cone C in $\mathbf{R}_+^n$ with basis in $\mathfrak{O}_F$, we define

$$\zeta(s, C, \mathfrak{a}_\mu) = \sum_{z\in(\mathfrak{a}_\mu\mathfrak{f})^{-1}\cap C,\ (z)\mathfrak{a}_\mu\mathfrak{f}\equiv c \text{ in } C_{\mathfrak{f}}} N(z)^{-s}, \tag{1.12}$$

$$\zeta^{(i_1,i_2,\dots,i_k)}(s, C, \mathfrak{a}_\mu) = \sum_{z\in(\mathfrak{a}_\mu\mathfrak{f})^{-1}\cap C,\ (z)\mathfrak{a}_\mu\mathfrak{f}\equiv c \text{ in } C_{\mathfrak{f}}} (z^{(i_1)}z^{(i_2)}\cdots z^{(i_k)})^{-s}, \tag{1.13}$$

where $1 \leqq i_1, i_2, \dots, i_k \leqq n$ are integers. Thus

$$\zeta(s, C, \mathfrak{a}_\mu) = \zeta^{(1,2,\dots,n)}(s, C, \mathfrak{a}_\mu).$$

Let

$$\mathbf{R}_+^n = \sqcup_{\epsilon\in E_F^+}\epsilon\,(\sqcup_{j\in J} C_j) \tag{1.14}$$

be a cone decomposition of $\mathbf{R}_+^n$, where C_j is an open simplicial cone with basis in $\mathfrak{O}_F$. Then II, (3.4) can be written as

$$\zeta_F(s, c) = N(\mathfrak{a}_\mu\mathfrak{f})^{-s}\sum_{j\in J}\zeta(s, C_j, \mathfrak{a}_\mu). \tag{1.15}$$

Hence we have

$$\zeta_F^{(p)}(0, c) = \sum_{j\in J}\sum_{q=0}^{p}\binom{p}{q}(-\log N(\mathfrak{a}_\mu\mathfrak{f}))^{p-q}\cdot\Big(\frac{d}{ds}\Big)^q\zeta(s, C_j, \mathfrak{a}_\mu)\Big|_{s=0}. \tag{1.16}$$

Put $C_j = C(v_{j1}, v_{j2}, \dots, v_{jr(j)})$, $v_j \in \mathfrak{O}_F$ and define an $n\times r(j)$-matrix A_j by $A_j = (v_{jl}^{(i)})$ as in II, §3.2. By II, Lemma 3.2, we have

$$\zeta^{(i_1,i_2,\dots,i_k)}(s, C_j, \mathfrak{a}_\mu) = \sum_{z\in R(C_j,c)}\zeta\left(s, \begin{pmatrix} A_j^{(i_1)} \\ A_j^{(i_2)} \\ \vdots \\ A_j^{(i_k)} \end{pmatrix}, x(z)\right). \tag{1.17}$$

2. The case of the second derivatives of L-functions

2.1. Hereafter we will exclusively consider the case $p = 2$. Then (1.16) can be written as

$$\zeta_F^{(2)}(0,c) = \sum_{j\in J}\Big[(\frac{d}{ds})^2\zeta(s,C_j,\mathfrak{a}_\mu)\Big|_{s=0} - 2\log N(\mathfrak{a}_\mu\mathfrak{f})\frac{d}{ds}\zeta(s,C_j,\mathfrak{a}_\mu)\Big|_{s=0} + (\log N(\mathfrak{a}_\mu\mathfrak{f}))^2\zeta(0,C_j,\mathfrak{a}_\mu)\Big]. \tag{2.1}$$

By (1.17) and I, Theorem 3.1 taking account of I, Theorem 2.1, we have[1]

$$\begin{aligned}(\frac{d}{ds})^2\zeta(s,C_j,\mathfrak{a}_\mu)\Big|_{s=0} &= \sum_{z\in R(C_j,c)}\Big[n\sum_{k=1}^{n}\log\frac{\Gamma_{r(j)}^{[2]}(z^{(k)},A_j^{(k)})}{\rho_{r(j)}^{[2]}(A_j^{(k)})}\\ &+2\sum_{1\leqq i,k\leqq n, i\neq k}\Lambda_{r(j)}(A_j^{(k)},A_j^{(i)},{}^t x(z))\\ &+\frac{(-1)^{r(j)}}{n}\sum_{1\leqq i,k_1,k_2\leqq n, k_1\neq k_2, k_2\neq i, i\neq k_1}\sum_l C_{l,i,k_1,k_2}(A_j)\prod_{m=1}^{r(j)}\frac{B_{l_m}(x_m(z))}{l_m!}\\ &+\frac{2(-1)^{r(j)}}{n}\sum_{1\leqq i,k\leqq n, i\neq k}\sum_l C'_{l,i,k}(A_j)\prod_{m=1}^{r(j)}\frac{B_{l_m}(x_m(z))}{l_m!}\Big]\\ &+\frac{n(n-1)\pi^2}{6}\zeta(0,C_j,\mathfrak{a}_\mu),\end{aligned} \tag{2.2}$$

$$(\frac{d}{ds})^2\zeta^{(i)}(s,C_j,\mathfrak{a}_\mu)\Big|_{s=0} = \sum_{z\in R(C_j,c)}\log\frac{\Gamma_{r(j)}^{[2]}(z^{(i)},A_j^{(i)})}{\rho_{r(j)}^{[2]}(A_j^{(i)})}, \tag{2.3}$$

$$\begin{aligned}&(\frac{d}{ds})^2\zeta^{(i,k)}(s,C_j,\mathfrak{a}_\mu)\Big|_{s=0}\\ &= \sum_{z\in R(C_j,c)}\Big[2\,(\log\frac{\Gamma_{r(j)}^{[2]}(z^{(i)},A_j^{(i)})}{\rho_{r(j)}^{[2]}(A_j^{(i)})} + \log\frac{\Gamma_{r(j)}^{[2]}(z^{(k)},A_j^{(k)})}{\rho_{r(j)}^{[2]}(A_j^{(k)})})\\ &+2\,(\Lambda_{r(j)}(A_j^{(k)},A_j^{(i)},{}^t x(z)) + \Lambda_{r(j)}(A_j^{(i)},A_j^{(k)},{}^t x(z)))\\ &+(-1)^{r(j)}\sum_l(C'_{l,i,k}(A_j)+C'_{l,k,i}(A_j))\prod_{m=1}^{r(j)}\frac{B_{l_m}(x_m(z))}{l_m!}\Big]\\ &+\frac{\pi^2}{3}\zeta^{(i,k)}(0,C_j,\mathfrak{a}_\mu).\end{aligned} \tag{2.4}$$

[1] $\zeta^{(i)}(s,C_j,\mathfrak{a}_\mu)$ is defined by (1.13) and should not be confused with the i-th derivative.

More generally, we have

$$
\begin{aligned}
&(\frac{d}{ds})^2\zeta^{(i_1,i_2,\dots,i_k)}(s,C_j,\mathfrak{a}_\mu)\Big|_{s=0} = \sum_{z\in R(C_j,c)}\Bigg[k\sum_{p=1}^{k}\log\frac{\Gamma^{[2]}_{r(j)}(z^{(i_p)},A_j^{(i_p)})}{\rho^{[2]}_{r(j)}(A_j^{(i_p)})}\\
&+2\sum_{1\leqq p,q\leqq k,p\neq q}\Lambda_{r(j)}(A_j^{(i_p)},A_j^{(i_q)},{}^t x(z))\\
&+\frac{(-1)^{r(j)}}{k}\sum_{1\leqq p,q,r\leqq k,p\neq q,q\neq r,r\neq p}\sum_l C_{l,i_p,i_q,i_r}(A_j)\prod_{m=1}^{r(j)}\frac{B_{l_m}(x_m(z))}{l_m!}\\
&+\frac{2(-1)^{r(j)}}{k}\sum_{1\leqq p,q\leqq k,p\neq q}\sum_l C'_{l,i_p,i_q}(A_j)\prod_{m=1}^{r(j)}\frac{B_{l_m}(x_m(z))}{l_m!}\Bigg]\\
&+\frac{k(k-1)\pi^2}{6}\zeta^{(i_1,i_2,\dots,i_k)}(0,C_j,\mathfrak{a}_\mu).
\end{aligned}
\tag{2.5}
$$

We put

$$
G^{(2)}(c)=G^{(2)}(c;\{C_j\}_{j\in J},\mathfrak{a}_\mu)=n\sum_{j\in J}\frac{d^2}{ds^2}\zeta^{(1)}(s,C_j,\mathfrak{a}_\mu)\Big|_{s=0}. \tag{2.6}
$$

If a "canonical" decomposition (1.8) is possible, $X^{(2)}(c)=X^{(2)}(c;\{C_j\}_{j\in J},\mathfrak{a}_\mu)$ should be the sum of $G^{(2)}(c)$ and correction terms.

We fix a cone C_j until the end of this subsection. We put

$$
G^{(2)}(c;C_j,\mathfrak{a}_\mu)=n\frac{d^2}{ds^2}\zeta^{(1)}(s,C_j,\mathfrak{a}_\mu)\Big|_{s=0}.
$$

For $1\leqq i,k\leqq n$, we put

$$
\Lambda(i,k)=\Lambda(i,k;c,C_j,\mathfrak{a}_\mu)=\sum_{z\in R(C_j,c)}\Lambda_{r(j)}(A_j^{(i)},A_j^{(k)},{}^t x(z)), \tag{2.7}
$$

$$
\begin{aligned}
C'(i,k)&=C'(i,k;c,C_j,\mathfrak{a}_\mu)\\
&=(-1)^{r(j)}\sum_{z\in R(C_j,c)}\sum_l C'_{l,i,k}(A_j)\prod_{m=1}^{r(j)}\frac{B_{l_m}(x_m(z))}{l_m!}.
\end{aligned}
\tag{2.8}
$$

For $1\leqq i,k_1,k_2\leqq n$, we put

$$
\begin{aligned}
&C(i,k_1,k_2)=C(i,k_1,k_2;c,C_j,\mathfrak{a}_\mu)\\
&=(-1)^{r(j)}\sum_{z\in R(C_j,c)}\sum_l C_{l,i,k_1,k_2}(A_j)\prod_{m=1}^{r(j)}\frac{B_{l_m}(x_m(z))}{l_m!}.
\end{aligned}
\tag{2.9}
$$

Then (2.2) can be written as

$$
\begin{aligned}
(\frac{d}{ds})^2\zeta(s,C_j,\mathfrak{a}_\mu)\Big|_{s=0} =& n\sum_{l=1}^{n} G^{(2)}(c^{\sigma_l};C_j^{\sigma_l},\mathfrak{a}_\mu^{\sigma_l}) \\
&+2\sum_{1\leqq i,k\leqq n, i\neq k}\Lambda(i,k)+\frac{2}{n}\sum_{1\leqq i,k\leqq n,i\neq k} C'(i,k) \\
&+\frac{1}{n}\sum_{1\leqq i,k_1,k_2\leqq n, i\neq k_1,k_1\neq k_2,k_2\neq i} C(i,k_1,k_2) \\
&+\frac{n(n-1)\pi^2}{6}\zeta(0,C_j,\mathfrak{a}_\mu).
\end{aligned}
\tag{2.10}
$$

To simplify our notation further, we put

$$
\begin{aligned}
\zeta_j^{(i_1,i_2,\dots,i_k)} =& \zeta^{(i_1,i_2,\dots,i_k)}(0,C_j,\mathfrak{a}_\mu), \\
\dot\zeta_j^{(i_1,i_2,\dots,i_k)} =& \frac{d}{ds}\zeta^{(i_1,i_2,\dots,i_k)}(s,C_j,\mathfrak{a}_\mu)\Big|_{s=0}, \\
\ddot\zeta_j^{(i_1,i_2,\dots,i_k)} =& (\frac{d}{ds})^2\zeta^{(i_1,i_2,\dots,i_k)}(s,C_j,\mathfrak{a}_\mu)\Big|_{s=0}.
\end{aligned}
$$

By (2.5), we have

$$
\begin{aligned}
\ddot\zeta_j^{(i,k)}-2\ddot\zeta_j^{(i)}-2\ddot\zeta_j^{(k)} =& 2(\Lambda(i,k)+\Lambda(k,i))+(C'(i,k)+C'(k,i)) \\
&+\frac{\pi^2}{6}(\zeta_j^{(i)}+\zeta_j^{(k)}),
\end{aligned}
\tag{2.11}
$$

$$
\begin{aligned}
&\ddot\zeta_j^{(i,k,p)}-3\ddot\zeta_j^{(i)}-3\ddot\zeta_j^{(k)}-3\ddot\zeta_j^{(p)} \\
&=2(\Lambda(i,k)+\Lambda(k,i)+\Lambda(k,p)+\Lambda(p,k)+\Lambda(p,i)+\Lambda(i,p)) \\
&+\frac{2}{3}(C'(i,k)+C'(k,i)+C'(k,p)+C'(p,k)+C'(p,i)+C'(i,p)) \\
&+\frac{1}{3}(C(i,k,p)+C(i,p,k)+C(k,i,p)+C(k,p,i)+C(p,i,k)+C(p,k,i)) \\
&+\frac{\pi^2}{3}(\zeta_j^{(i)}+\zeta_j^{(k)}+\zeta_j^{(p)}).
\end{aligned}
\tag{2.12}
$$

LEMMA 2.1. *We have*

$$
\begin{aligned}
&\sum_{l=1}^{n}\sum_{k=2}^{n}\Lambda(1,k;c^{\sigma_l},C_j^{\sigma_l},\mathfrak{a}_\mu^{\sigma_l})=\sum_{l=1}^{n}\sum_{k=2}^{n}\Lambda(k,1;c^{\sigma_l},C_j^{\sigma_l},\mathfrak{a}_\mu^{\sigma_l}) \\
&=\sum_{1\leqq i,k\leqq n,i\neq k}\Lambda(i,k;c,C_j,\mathfrak{a}_\mu),
\end{aligned}
$$

$$
\sum_{1\leqq i,k\leqq n,i\neq k}\Lambda(i,k;c^{\sigma_l},C_j^{\sigma_l},\mathfrak{a}_\mu^{\sigma_l})=\sum_{1\leqq i,k\leqq n,i\neq k}\Lambda(i,k;c,C_j,\mathfrak{a}_\mu),\quad 1\leqq l\leqq n,
$$

$$\begin{aligned}&\sum_{l=1}^{n}\sum_{k=2}^{n} C'(1,k;c^{\sigma_l},C_j^{\sigma_l},\mathfrak{a}_\mu^{\sigma_l}) = \sum_{l=1}^{n}\sum_{k=2}^{n} C'(k,1;c^{\sigma_l},C_j^{\sigma_l},\mathfrak{a}_\mu^{\sigma_l})\\ &= \sum_{1\leqq i,k\leqq n, i\neq k} C'(i,k;c,C_j,\mathfrak{a}_\mu),\end{aligned}$$

$$\sum_{1\leqq i,k\leqq n,i\neq k} C'(i,k;c^{\sigma_l},C_j^{\sigma_l},\mathfrak{a}_\mu^{\sigma_l}) = \sum_{1\leqq i,k\leqq n,i\neq k} C'(i,k;c,C_j,\mathfrak{a}_\mu), \quad 1\leqq l\leqq n,$$

$$\begin{aligned}&\sum_{l=1}^{n}\sum_{2\leqq k_1,k_2\leqq n,k_1\neq k_2} C(1,k_1,k_2;c^{\sigma_l},C_j^{\sigma_l},\mathfrak{a}_\mu^{\sigma_l})\\ &= \sum_{1\leqq i,k_1,k_2\leqq n,i\neq k_1,k_1\neq k_2,k_2\neq i} C(i,k_1,k_2;c,C_j,\mathfrak{a}_\mu),\\ &\sum_{1\leqq i,k_1,k_2\leqq n,i\neq k_1,k_1\neq k_2,k_2\neq i} C(1,k_1,k_2;c^{\sigma_l},C_j^{\sigma_l},\mathfrak{a}_\mu^{\sigma_l})\\ &= \sum_{1\leqq i,k_1,k_2\leqq n,i\neq k_1,k_1\neq k_2,k_2\neq i} C(i,k_1,k_2;c,C_j,\mathfrak{a}_\mu), \quad 1\leqq l\leqq n.\end{aligned}$$

PROOF. We have $C_j^{\sigma_l} = C(v_{j1}^{\sigma_l}, v_{j2}^{\sigma_l}, \dots, v_{jr(j)}^{\sigma_l})$ and the mapping $R(C_j,c)\ni z\longrightarrow z^{\sigma_l}\in R(C_j^{\sigma_l},c^{\sigma_l})$ is a bijection such that $x_m(z^{\sigma_l}) = x_m(z)$. Let $A_{j,l}$ be the $n\times r(j)$-matrix determined by the cone $C_j^{\sigma_l}$. When we extend $\sigma_l\in J_F$ to an element $\tilde\sigma_l$ of $\mathrm{Gal}(\overline{\mathbf{Q}}/\mathbf{Q})$, we have $J_{F^{\sigma_l}} = \{\tilde\sigma_l^{-1}\sigma_i\tilde\sigma_l \mid 1\leqq i\leqq n\}$ and the (i,m)-component of $A_{j,l}$ is $v_{jm}^{\sigma_i\tilde\sigma_l}$. Thus the first row of $A_{j,l}$ does not depend on the choice of $\tilde\sigma_l$ and the other rows are obtained by a permutation on the rows of A_j which depends on the choice of $\tilde\sigma_l$. By this observation, we see that

$$\sum_{k=2}^{n}\Lambda(1,k;c^{\sigma_l},C_j^{\sigma_l},\mathfrak{a}_\mu^{\sigma_l}) = \sum_{k=1,k\neq l}^{n}\Lambda(l,k;c,C_j,\mathfrak{a}_\mu).$$

Hence we obtain the equality

$$\sum_{l=1}^{n}\sum_{k=2}^{n}\Lambda(1,k;c^{\sigma_l},C_j^{\sigma_l},\mathfrak{a}_\mu^{\sigma_l}) = \sum_{1\leqq i,k\leqq n,i\neq k}\Lambda(i,k;c,C_j,\mathfrak{a}_\mu).$$

The remaining equalities can be shown similarly. This completes the proof.

2.2. We first consider the following quantity attached to C_j:

$$\begin{aligned}X^{(2)}(c;C_j,\mathfrak{a}_\mu) =& n\ddot\zeta_j + \alpha\sum_{(i,k)\in S_1}(\ddot\zeta_j^{(i,k)} - 2\ddot\zeta_j^{(i)} - 2\ddot\zeta_j^{(k)})\\ &+\beta\sum_{(i,k,p)\in S_2}(\ddot\zeta_j^{(i,k,p)} - 3\ddot\zeta_j^{(i)} - 3\ddot\zeta_j^{(k)} - 3\ddot\zeta_j^{(p)})\end{aligned} \tag{2.13}$$

where α, $\beta\in\mathbf{Q}$ and (i,k) (resp. (i,k,p)) extends over a certain set of indices S_1 (resp. S_2).

We assume $S_1 = \{1, k \mid 2 \leqq k \leqq n\}$ or $S_1 = \{i, k \mid 1 \leqq i, k \leqq n,\ i \neq k\}$; $S_2 = \{1, k, p \mid 2 \leqq k, p \leqq n,\ p \neq k\}$ or $S_2 = \{i, k, p \mid 1 \leqq i, k, p \leqq n,\ i \neq k,\ k \neq p,\ p \neq i\}$. Put $a = \alpha|S_1|$, $b = \beta|S_2|$. By (2.10), (2.11), (2.12) and Lemma 2.1, we see that

$$\sum_{l=1}^{n} X^{(2)}(c^{\sigma_l}; C_j^{\sigma_l}, \mathfrak{a}_\mu^{\sigma_l}) = (\frac{d}{ds})^2 \zeta(s, C_j, \mathfrak{a}_\mu)\Big|_{s=0}$$

holds if

$$(4a + 12b)n = 2n(n-1), \qquad (2a + 4b)n = 2(n-1),$$
$$2bn = (n-1)(n-2), \qquad (\frac{1}{3}a + b)n = \frac{n(n-1)}{6}$$

are satisfied. These equations hold with

$$a = \frac{(n-1)(3-n)}{n}, \qquad b = \frac{(n-1)(n-2)}{2n}. \tag{2.14}$$

Now we put

$$\begin{aligned} X^{(2)}(c;\{C_j\}_{j\in J}, \mathfrak{a}_\mu) &= G^{(2)}(c) \\ &+ \sum_{j\in J}\Big[c_1 \sum_{k=2}^{n} (\ddot{\zeta}_j^{(1,k)} - 2\ddot{\zeta}_j^{(1)} - 2\ddot{\zeta}_j^{(k)}) \\ &+ c_2 \sum_{1\leqq i,k\leqq n, i\neq k} (\ddot{\zeta}_j^{(i,k)} - 2\ddot{\zeta}_j^{(i)} - 2\ddot{\zeta}_j^{(k)}) \\ &+ c_3 \sum_{2\leqq k,p\leqq n, k\neq p} (\ddot{\zeta}_j^{(1,k,p)} - 3\ddot{\zeta}_j^{(1)} - 3\ddot{\zeta}_j^{(k)} - 3\ddot{\zeta}_j^{(p)}) \\ &+ c_4 \sum_{1\leqq i,k,p\leqq n, i\neq k, k\neq p, p\neq i} (\ddot{\zeta}_j^{(i,k,p)} - 3\ddot{\zeta}_j^{(i)} - 3\ddot{\zeta}_j^{(k)} - 3\ddot{\zeta}_j^{(p)})\Big] \\ &- 2\log N(\mathfrak{a}_\mu\mathfrak{f}) \sum_{j\in J}\Big[\dot{\zeta}_j^{(1)} + \frac{2}{n}\sum_{k=2}^{n}(\dot{\zeta}_j^{(1,k)} - \dot{\zeta}_j^{(1)} - \dot{\zeta}_j^{(k)}) \\ &- \frac{1}{n^2} \sum_{1\leqq i,k\leqq n, i\neq k} (\dot{\zeta}_j^{(i,k)} - \dot{\zeta}_j^{(i)} - \dot{\zeta}_j^{(k)})\Big] \\ &+ \frac{1}{n}(\log N(\mathfrak{a}_\mu\mathfrak{f}))^2 \sum_{j\in J} \zeta_j^{(1)}, \end{aligned} \tag{2.15}$$

with c_1, c_2, c_3, $c_4 \in \mathbf{Q}$. The terms after $-2\log N(\mathfrak{a}_\mu\mathfrak{f})$ can be written as (cf. III, (3.27) $\sim$ (3.31))

$$-2\log N(\mathfrak{a}_\mu\mathfrak{f})\Big[G(c) + V(c) + \frac{1}{2}W(c)\Big].$$

We abbreviate $X^{(2)}(c; \{C_j\}_{j\in J}, \mathfrak{a}_\mu)$ to $X^{(2)}(c; \{C_j\}, \mathfrak{a}_\mu)$, or to $X^{(2)}(c)$. If $n = 1$, we have $X^{(2)}(c) = \zeta_F^{(2)}(0, c)$. We assume $n \geqq 2$ and are going to determine c_i, $1 \leqq i \leqq 4$ by the demand that the expression (2.15) should be as canonical as possible. By III, Lemma 3.8, (2.1) and (2.14), we see that

$$\zeta_F^{(2)}(0, c) = \sum_{l=1}^{n} X^{(2)}(c^{\sigma_l}; \{C_j^{\sigma_l}\}, \mathfrak{a}_\mu^{\sigma_l}) \tag{2.16}$$

holds if

$$c_1 + nc_2 = \frac{3-n}{n}, \qquad c_3 + nc_4 = \frac{1}{2n}. \tag{2.17}$$

The latter condition in (2.17) can be discarded when $n = 2$.

2.3. We are going to calculate the variation of $X^{(2)}(c; \{C_j\}, \mathfrak{a}_\mu)$ when we change $\mathfrak{a}_\mu$ to $(\alpha)\mathfrak{a}_\mu$ and $\{C_j\}_{j\in J}$ to $\{\alpha^{-1}C_j\}_{j\in J}$ with $\alpha \in F$, $\alpha \gg 0$. As in III, §3.7, we see that

$$\zeta^{(i_1,i_2,\ldots,i_k)}(s, \alpha^{-1}C_j, (\alpha)\mathfrak{a}_\mu) = (\alpha^{(i_1)}\alpha^{(i_2)}\cdots\alpha^{(i_k)})^s \zeta(s, C_j, \mathfrak{a}_\mu).$$

Hence $G^{(2)}(c) = n\sum_{j\in J}\ddot{\zeta}_j^{(1)}$ changes to

$$G^{(2)}(c) + n\sum_{j\in J}\left\{2(\log\alpha^{(1)})\dot{\zeta}_j^{(1)} + (\log\alpha^{(1)})^2\zeta_j^{(1)}\right\}.$$

We put

$$\begin{aligned} u_{i,k}^{(j)} &= \ddot{\zeta}_j^{(i,k)} - 2\ddot{\zeta}_j^{(i)} - 2\ddot{\zeta}_j^{(k)}, \\ v_{i,k,p}^{(j)} &= \ddot{\zeta}_j^{(i,k,p)} - 3\ddot{\zeta}_j^{(i)} - 3\ddot{\zeta}_j^{(k)} - 3\ddot{\zeta}_j^{(p)}. \end{aligned}$$

We see that $u_{i,k}^{(j)}$ changes to

$$\begin{aligned} &u_{i,k}^{(j)} + 2(\log(\alpha^{(i)}\alpha^{(k)}))(\dot{\zeta}_j^{(i)} + \dot{\zeta}_j^{(k)}) - 4(\log\alpha^{(i)})\dot{\zeta}_j^{(i)} - 4(\log\alpha^{(k)})\dot{\zeta}_j^{(k)} \\ &+ 2(\log(\alpha^{(i)}\alpha^{(k)}))(\dot{\zeta}_j^{(i,k)} - \dot{\zeta}_j^{(i)} - \dot{\zeta}_j^{(k)}) \\ &+ (\log(\alpha^{(i)}\alpha^{(k)}))^2\zeta_j^{(i,k)} - 2(\log\alpha^{(i)})^2\zeta_j^{(i)} - 2(\log\alpha^{(k)})^2\zeta_j^{(k)}, \end{aligned} \tag{2.18}$$

and that $v_{i,k,p}^{(j)}$ changes to

$$\begin{aligned} &v_{i,k,p}^{(j)} + 2(\log(\alpha^{(i)}\alpha^{(k)}\alpha^{(p)}))(\dot{\zeta}_j^{(i)} + \dot{\zeta}_j^{(k)} + \dot{\zeta}_j^{(p)}) \\ &- 6(\log\alpha^{(i)})\dot{\zeta}_j^{(i)} - 6(\log\alpha^{(k)})\dot{\zeta}_j^{(k)} - 6(\log\alpha^{(p)})\dot{\zeta}_j^{(p)} \\ &+ 2(\log(\alpha^{(i)}\alpha^{(k)}\alpha^{(p)}))(\dot{\zeta}_j^{(i,k,p)} - \dot{\zeta}_j^{(i)} - \dot{\zeta}_j^{(k)} - \dot{\zeta}_j^{(p)}) \\ &+ (\log(\alpha^{(i)}\alpha^{(k)}\alpha^{(p)}))^2\zeta_j^{(i,k,p)} \\ &- 3(\log\alpha^{(i)})^2\zeta_j^{(i)} - 3(\log\alpha^{(k)})^2\zeta_j^{(k)} - 3(\log\alpha^{(p)})^2\zeta_j^{(p)}. \end{aligned} \tag{2.19}$$

By I, Theorem 2.2, we see that the terms $\dot{\zeta}_j^{(i,k)} - \dot{\zeta}_j^{(i)} - \dot{\zeta}_j^{(k)}$ and $\dot{\zeta}_j^{(i,k,p)} - \dot{\zeta}_j^{(i)} - \dot{\zeta}_j^{(k)} - \dot{\zeta}_j^{(p)}$ can be expressed elementarily without using the multiple gamma function. We put

$$\delta_{i,k}^{(j)} = 2(\log(\alpha^{(i)}\alpha^{(k)}))(\dot{\zeta}_j^{(i)} + \dot{\zeta}_j^{(k)}) - 4(\log\alpha^{(i)})\dot{\zeta}_j^{(i)} - 4(\log\alpha^{(k)})\dot{\zeta}_j^{(k)} \tag{2.20}$$

$$\begin{aligned} \delta_{i,k,p}^{(j)} = &2(\log(\alpha^{(i)}\alpha^{(k)}\alpha^{(p)}))(\dot{\zeta}_j^{(i)} + \dot{\zeta}_j^{(k)} + \dot{\zeta}_j^{(p)}) \\ &- 6(\log\alpha^{(i)})\dot{\zeta}_j^{(i)} - 6(\log\alpha^{(k)})\dot{\zeta}_j^{(k)} - 6(\log\alpha^{(p)})\dot{\zeta}_j^{(p)}. \end{aligned} \tag{2.21}$$

Then $u_{i,k}^{(j)}$ (resp. $v_{i,k,p}^{(j)}$) changes to $u_{i,k}^{(j)}+\delta_{i,k}^{(j)}$ (resp. $v_{i,k,p}^{(j)}+\delta_{i,k,p}^{(j)}$) modulo elementary terms. Fix j and put

$$\Delta_1=\sum_{k=2}^{n}\delta_{1,k}^{(j)},\qquad \Delta_2=\sum_{1\leqq i,k\leqq n,i\neq k}\delta_{i,k}^{(j)},$$
$$\Delta_3=\sum_{2\leqq k,p\leqq n,k\neq p}\delta_{1,k,p}^{(j)},\qquad \Delta_4=\sum_{1\leqq i,k,p\leqq n,i\neq k,k\neq p,p\neq i}\delta_{i,k,p}^{(j)}.$$

Then we have

$$\begin{aligned}\Delta_1=&\,2(n-1)(\log\alpha^{(1)})\dot{\zeta}_j^{(1)}+2\sum_{k=2}^{n}(\log\alpha^{(k)})\dot{\zeta}_j^{(1)}+2\log\alpha^{(1)}\sum_{k=2}^{n}\dot{\zeta}_j^{(k)}\\&+2\sum_{k=2}^{n}\log\alpha^{(k)}\dot{\zeta}_j^{(k)}-4(n-1)(\log\alpha^{(1)})\dot{\zeta}_j^{(1)}-4\sum_{k=2}^{n}(\log\alpha^{(k)})\dot{\zeta}_j^{(k)}.\end{aligned}$$

Hence we obtain

$$\begin{aligned}\Delta_1=&-2n(\log\alpha^{(1)})\dot{\zeta}_j^{(1)}+2(\log N(\alpha))\dot{\zeta}_j^{(1)}+2\log\alpha^{(1)}\sum_{k=1}^{n}\dot{\zeta}_j^{(k)}\\&-2\sum_{k=1}^{n}(\log\alpha^{(k)})\dot{\zeta}_j^{(k)}.\end{aligned}\tag{2.22}$$

By similar direct calculations, we get

$$\Delta_2=4(\log N(\alpha))\sum_{k=1}^{n}\dot{\zeta}_j^{(k)}-4n\sum_{k=1}^{n}(\log\alpha^{(k)})\dot{\zeta}_j^{(k)},\tag{2.23}$$

$$\begin{aligned}\Delta_3=&-4n(n-3)(\log\alpha^{(1)})\dot{\zeta}_j^{(1)}+4(n-3)(\log N(\alpha))\dot{\zeta}_j^{(1)}\\&+4(n-3)\log\alpha^{(1)}\sum_{k=1}^{n}\dot{\zeta}_j^{(k)}+4(\log N(\alpha))\sum_{k=1}^{n}\dot{\zeta}_j^{(k)}\\&-8(n-\frac{3}{2})\sum_{k=1}^{n}(\log\alpha^{(k)})\dot{\zeta}_j^{(k)},\end{aligned}\tag{2.24}$$

$$\Delta_4=12(n-2)(\log N(\alpha))\sum_{k=1}^{n}\dot{\zeta}_j^{(k)}-12n(n-2)\sum_{k=1}^{n}(\log\alpha^{(k)})\dot{\zeta}_j^{(k)}.\tag{2.25}$$

Therefore we obtain

$$
\begin{aligned}
X^{(2)}(c;\{\alpha^{-1}C_j\},(\alpha)\mathfrak{a}_\mu) \equiv\ & X^{(2)}(c;\{C_j\},\mathfrak{a}_\mu) \\
&+ \sum_{j\in J}\Big[\{2n-2nc_1-4n(n-3)c_3\}(\log\alpha^{(1)})\dot{\zeta}_j^{(1)} \\
&+ \{2c_1+4(n-3)c_3-2\}(\log N(\alpha))\dot{\zeta}_j^{(1)} \\
&+ \{2c_1+4(n-3)c_3\}(\log\alpha^{(1)})\sum_{k=1}^{n}\dot{\zeta}_j^{(k)} \\
&+ \{4c_2+4c_3+12(n-2)c_4\}(\log N(\alpha))\sum_{k=1}^{n}\dot{\zeta}_j^{(k)} \\
&+ \{-2c_1-4nc_2-8(n-\frac{3}{2})c_3-12n(n-2)c_4\}\sum_{k=1}^{n}(\log\alpha^{(k)})\dot{\zeta}_j^{(k)}\Big].
\end{aligned}
\tag{2.26}
$$

Here $\equiv$ means that both sides are equal when we discard elementary terms explained above. Notice that the term $-2\log N(\mathfrak{a}_\mu\mathfrak{f})\sum_{j\in J}\dot{\zeta}_j^{(1)}$ in (2.15) contributes $-2\log N(\alpha)\sum_{j\in J}\dot{\zeta}_j^{(1)}$ to the right hand side of (2.26) since $\mathfrak{a}_\mu$ is replaced by $(\alpha)\mathfrak{a}_\mu$.

It is impossible to make

$$X^{(2)}(c;\{\alpha^{-1}C_j\},(\alpha)\mathfrak{a}_\mu) \equiv X^{(2)}(c;\{C_j\},\mathfrak{a}_\mu)$$

choosing c_i, $1 \leqq i \leqq 4$ suitably. It seems best to impose the condition

$$c_1 + 2(n-3)c_3 = 1. \tag{2.27}$$

Then, by (2.17) and (2.27), (2.26) simplifies to

$$
\begin{aligned}
X^{(2)}(c;\{\alpha^{-1}C_j\},(\alpha)\mathfrak{a}_\mu) \equiv\ & X^{(2)}(c;\{C_j\},\mathfrak{a}_\mu) \\
&+ 2(\log\alpha^{(1)} - \frac{1}{n}\log N(\alpha))\sum_{j\in J}\sum_{k=1}^{n}\dot{\zeta}_j^{(k)}.
\end{aligned}
\tag{2.28}
$$

If $n=2$, we have

$$2\sum_{k=2}^{n}(\ddot{\zeta}_j^{(1,k)} - 2\ddot{\zeta}_j^{(1)} - 2\ddot{\zeta}_j^{(k)}) = \sum_{1\leqq i,k\leqq n, i\neq k}(\ddot{\zeta}_j^{(i,k)} - 2\ddot{\zeta}_j^{(i)} - 2\ddot{\zeta}_j^{(k)}).$$

Hence we may take $c_2 = 0$. We can take

$$c_1 = \frac{1}{2}, \qquad c_2 = 0 \qquad (\text{when} \quad n=2)$$

without losing any generality. If $n=3$, we have

$$
\begin{aligned}
&3\sum_{2\leqq k_1,k_2\leqq n, k_1\neq k_2}(\ddot{\zeta}_j^{(1,k_1,k_2)} - 3\ddot{\zeta}_j^{(1)} - 3\ddot{\zeta}_j^{(k_1)} - 3\ddot{\zeta}_j^{(k_2)}) \\
&= \sum_{1\leqq i,k_1,k_2\leqq n, i\neq k_1, k_1\neq k_2, k_2\neq i}(\ddot{\zeta}_j^{(i,k_1,k_2)} - 3\ddot{\zeta}_j^{(i)} - 3\ddot{\zeta}_j^{(k_1)} - 3\ddot{\zeta}_j^{(k_2)}).
\end{aligned}
$$

Hence we may take $c_4 = 0$. We can take

$$c_1 = 1, \qquad c_2 = -\frac{1}{3}, \qquad c_3 = \frac{1}{6}, \qquad c_4 = 0 \qquad (\text{when} \quad n=3)$$

without losing any generality. If $n > 3$, we cannot determine c_i, $1 \leqq i \leqq 4$ by (2.17) and (2.27).

2.4. Concerning the elementary terms discarded above, we can prove that their contributions vanish under mild conditions. We take $\nu \in \mathfrak{O}_F$ so that $\nu \equiv 1 \mod \mathfrak{f}$, $-\nu \gg 0$. Let $s_0 \in C_\mathfrak{f}$ be the class of (ν). In the notation of III, §5.1, we have $s_0 = s_1 s_2 \cdots s_n$. The narrow ideal class of s_0 is trivial. We can choose the same $\mathfrak{a}_\mu$ for c and cs_0, $c \in C_\mathfrak{f}$.

LEMMA 2.2. *Let $C = C(v_1, v_2, \ldots, v_r)$ be an r-dimensional open simplicial cone with $0 \ll v_i \in \mathfrak{O}_F$, $1 \leqq i \leqq r$. Let $c \in C_\mathfrak{f}$ and define $R(C,c)$ by* IV, (6.4). *For $z = \sum_{i=1}^r x_i v_i \in R(C,c)$, $x_i \in \mathbf{Q}$, $0 < x_i \leqq 1$, we put*

$$x_i^* = \begin{cases} 1 - x_i & \text{if } \ 0 < x_i < 1, \\ 1 & \text{if } \ x_i = 1, \end{cases}$$

$z^* = \sum_{i=1}^r x_i^* v_i$. Then $z^* \in R(C, cs_0)$ and the mapping $z \longrightarrow z^*$ gives a bijection between $R(C,c)$ and $R(C, cs_0)$.

PROOF. We have $z^* = \sum_{i=1}^r v_i - z + \sum_{i, x_i = 1} v_i$. Hence we see that $z^* \in (\mathfrak{a}_\mu \mathfrak{f})^{-1}$. To prove $z^* \in R(C, cs_0)$, it suffices to show

$$(z^*)\mathfrak{a}_\mu \mathfrak{f} \equiv cs_0 \qquad \text{in} \quad C_\mathfrak{f}.$$

Since $(z)\mathfrak{a}_\mu \mathfrak{f} \equiv c$ in $C_\mathfrak{f}$ and z, z^*, $-\nu \gg 0$, it suffices to show that

$$\frac{z^*}{z} \equiv -1 \mod^\times \mathfrak{f}. \tag{2.29}$$

Since $(z)\mathfrak{a}_\mu \mathfrak{f}$ is relatively prime to $\mathfrak{f}$, we have $(z) = \mathfrak{a}_\mu^{-1} \mathfrak{f}^{-1} \mathfrak{b}$ with a fractional ideal $\mathfrak{b}$ which is relatively prime to $\mathfrak{f}$. We have

$$\frac{z^*}{z} = -1 + \frac{1}{z}\Big(\sum_{i=1}^r v_i + \sum_{i, x_i = 1} v_i\Big), \qquad \Big(\frac{1}{z}\Big) = \mathfrak{a}_\mu \mathfrak{f} \mathfrak{b}^{-1}.$$

Hence (2.29) holds. The inverse mapping from $R(C, cs_0)$ to $R(C,c)$ can be defined similarly. Therefore $R(C,c) \ni z \longrightarrow z^* \in R(C, cs_0)$ is a bijection. This completes the proof.

Let $C_j = C(v_1, v_2, \ldots, v_r)$, $r = r(j)$ be a cone as in Lemma 2.2. To stress the dependence on $c \in C_\mathfrak{f}$, we write $\zeta_j^{(i)}$, $\dot{\zeta}_j^{(i)}$, $\ldots$ as $\zeta_j^{(i)}(c)$, $\dot{\zeta}_j^{(i)}(c)$, $\ldots$. By I, Theorem 2.1 and II, Lemma 3.2, we have

$$\zeta_j^{(i)}(c) = (-1)^r \sum_l \sum_{z \in R(C_j, c)} \prod_{m=1}^r \frac{B_{l_m}(x_m(z))}{l_m!} (v_m^{(i)})^{l_m - 1}. \tag{2.30}$$

By I, Theorem 2.2 and II, Lemma 3.2, we have

$$\begin{aligned} &\dot{\zeta}_j^{(i,k)}(c) - \dot{\zeta}_j^{(i)}(c) - \dot{\zeta}_j^{(k)}(c) \\ &= \frac{(-1)^r}{2} \sum_l (C_{l,i,k}(A_j) + C_{l,k,i}(A_j)) \sum_{z \in R(C_j, c)} \prod_{m=1}^r \frac{B_{l_m}(x_m(z))}{l_m!}. \end{aligned} \tag{2.31}$$

In (2.30) and (2.31), $l = (l_1, l_2, \ldots, l_r)$ extends over all r-tuples of non-negative integers such that $l_1 + l_2 + \cdots + l_r = r$. Suppose that $\omega \in \widehat{G}_\mathfrak{f}^{(2)}$, $\omega \neq 1$. Then ω is unramified at one of the infinite places $\infty_1, \infty_2, \ldots, \infty_n$ and is ramified at all the other infinite places. Therefore we have

$$\omega(s_0) = (-1)^{n-1}. \tag{2.32}$$

By Lemma 2.2, (2.30) and (2.31), we have

$$\begin{aligned} \omega(c)\zeta_j^{(i)}(c) + \omega(cs_0)\zeta_j^{(i)}(cs_0) &= \omega(c)(-1)^r \sum_l \prod_{m=1}^r (v_m^{(i)})^{l_m-1} \\ \times \sum_{z\in R(C_j,c)} &\left\{ \prod_{m=1}^r \frac{B_{l_m}(x_m(z))}{l_m!} - (-1)^n \prod_{m=1}^r \frac{B_{l_m}(x_m(z^*))}{l_m!} \right\}, \end{aligned} \tag{2.33}$$

$$\begin{aligned} &\omega(c)(\dot{\zeta}_j^{(i,k)}(c) - \dot{\zeta}_j^{(i)}(c) - \dot{\zeta}_j^{(k)}(c)) \\ &+\omega(cs_0)(\dot{\zeta}_j^{(i,k)}(cs_0) - \dot{\zeta}_j^{(i)}(cs_0) - \dot{\zeta}_j^{(k)}(cs_0)) \\ &=\omega(c)\frac{(-1)^r}{2}\sum_l (C_{l,i,k}(A_j) + C_{l,k,i}(A_j)) \\ &\times \sum_{z\in R(C_j,c)} \left\{ \prod_{m=1}^r \frac{B_{l_m}(x_m(z))}{l_m!} - (-1)^n \prod_{m=1}^r \frac{B_{l_m}(x_m(z^*))}{l_m!} \right\}. \end{aligned} \tag{2.34}$$

LEMMA 2.3. *Assume $r = n$. Let S and T denote the left-hand sides of* (2.33) *and* (2.34) *respectively. For $1 \leqq p_1 < p_2 < \ldots < p_s \leqq n$, put*

$$R(C_j, c; \{p_1, p_2, \ldots, p_s\}) = \{z \mid z \in R(C_j, c),\ x_{p_i}(z) = 1 \quad \text{for} \quad 1 \leqq i \leqq s\},$$

$$\begin{aligned} S_{p_1,p_2,\ldots,p_s} &= (-1)^n\omega(c) \sum_{l,l_{p_i}=1 \text{ for } 1\leqq i\leqq s} \prod_{m=1}^n (v_m^{(i)})^{l_m-1} \\ &\times \sum_{z\in R(C_j,c;\{p_1,p_2,\ldots,p_s\})} \left\{ \prod_{m=1}^r \frac{B_{l_m}(x_m(z))}{l_m!} - (-1)^n \prod_{m=1}^r \frac{B_{l_m}(x_m(z^*))}{l_m!} \right\}, \end{aligned}$$

$$\begin{aligned} T_{p_1,p_2,\ldots,p_s} &= \frac{(-1)^n}{2}\omega(c) \sum_{l,l_{p_i}=1 \text{ for } 1\leqq i\leqq s} (C_{l,i,k}(A_j) + C_{l,k,i}(A_j)) \\ &\times \sum_{z\in R(C_j,c;\{p_1,p_2,\ldots,p_s\})} \left\{ \prod_{m=1}^r \frac{B_{l_m}(x_m(z))}{l_m!} - (-1)^n \prod_{m=1}^r \frac{B_{l_m}(x_m(z^*))}{l_m!} \right\}. \end{aligned}$$

Then we have

$$\begin{aligned} S = \sum_{p=1}^n S_p - \sum_{1\leqq p_1<p_2\leqq n} S_{p_1,p_2} + \cdots + (-1)^{s+1} \sum_{1\leqq p_1<p_2<\ldots<p_s\leqq n} S_{p_1,p_2,\ldots,p_s} \\ + \cdots + (-1)^n \sum_{1\leqq p_1<p_2<\ldots<p_{n-1}\leqq n} S_{p_1,p_2,\ldots,p_{n-1}} + (-1)^{n+1} S_{1,2,\ldots,n}. \end{aligned} \tag{2.35}$$

$$(2.36)\quad \begin{aligned} T = \sum_{p=1}^{n} T_p - \sum_{1\leqq p_1<p_2\leqq n} T_{p_1,p_2} + \cdots + (-1)^{s+1} \sum_{1\leqq p_1<p_2<\ldots<p_s\leqq n} T_{p_1,p_2,\ldots,p_s} \\ + \cdots + (-1)^n \sum_{1\leqq p_1<p_2<\ldots<p_{n-1}\leqq n} T_{p_1,p_2,\ldots,p_{n-1}}. \end{aligned}$$

For $1 \leqq p_1 < p_2 < \ldots < p_s \leqq n$, define an $(n-s)$-dimensional open simplicial cone $C_{j,\{p_1,p_2,\ldots,p_s\}}$ by

$$C_{j,\{p_1,p_2,\ldots,p_s\}} = C(v_1, \ldots, \check{v}_{p_1}, \ldots, \check{v}_{p_s}, \ldots, v_n),$$

the cone obtained from C_j by deleting $v_{p_1}, \ldots, v_{p_s}$ from the basis. Put $\check{C} = C_{j,\{p_1,p_2,\ldots,p_s\}}$. Then we have

$$(2.37)\quad S_{p_1,p_2,\ldots,p_s} = (-1)^s(\frac{1}{2})^s(\omega(c)\zeta^{(i)}_{\check{C}}(c) + \omega(cs_0)\zeta^{(i)}_{\check{C}}(cs_0)),$$

$$(2.38)\quad \begin{aligned} T_{p_1,p_2,\ldots,p_s} = (-1)^s(\frac{1}{2})^s \times \Big[\omega(c)(\dot{\zeta}^{(i,k)}_{\check{C}}(c) - \dot{\zeta}^{(i)}_{\check{C}}(c) - \dot{\zeta}^{(k)}_{\check{C}}(c)) \\ + \omega(cs_0)(\dot{\zeta}^{(i,k)}_{\check{C}}(cs_0) - \dot{\zeta}^{(i)}_{\check{C}}(cs_0) - \dot{\zeta}^{(k)}_{\check{C}}(cs_0))\Big]. \end{aligned}$$

Here $\zeta^{(i)}_{\check{C}}$ and $\zeta^{(i,k)}_{\check{C}}$ denote the zeta functions attached to the cone $\check{C}$ by (1.13) and the dots denote their derivatives at $s = 0$.

PROOF. Let $z \in R(C_j, c)$. Suppose that $0 < x_m(z) < 1$ for every m, $1 \leqq m \leqq n$. Since $B_k(1-x) = (-1)^k B_k(x)$, we have

$$(2.39)\quad \prod_{m=1}^{n} B_{l_m}(x_m(z^*)) = (-1)^n \prod_{m=1}^{n} B_{l_m}(x_m(z)).$$

Therefore such a z gives no contribution to (2.33) and (2.34). Suppose that $x_m(z) = 1$ for some m, $1 \leqq m \leqq n$. Then $x_m(z^*) = 1$. If l_m is odd and $l_m \geqq 3$, then we have $B_{l_m}(1) = -B_{l_m}(0) = 0$. Therefore the contribution from such a z vanishes. If l_m is even for every m such that $x_m(z) = 1$, then (2.39) holds and the contribution from z vanishes. Therefore we have

(2.40)

$$\begin{aligned} S = (-1)^n\omega(c) \sum_{l} \prod_{m=1}^{r} (v^{(i)}_m)^{l_m-1} \\ \times \sum_{z\in R(C_j,c),\exists m, x_m(z)=1, l_m=1} \left\{ \prod_{m=1}^{r} \frac{B_{l_m}(x_m(z))}{l_m!} - (-1)^n \prod_{m=1}^{r} \frac{B_{l_m}(x_m(z^*))}{l_m!} \right\}, \end{aligned}$$

(2.41)

$$\begin{aligned} T = \frac{(-1)^n}{2}\omega(c) \sum_{l} (C_{l,i,k}(A_j) + C_{l,k,i}(A_j)) \\ \times \sum_{z\in R(C_j,c),\exists m, x_m(z)=1, l_m=1} \left\{ \prod_{m=1}^{r} \frac{B_{l_m}(x_m(z))}{l_m!} - (-1)^n \prod_{m=1}^{r} \frac{B_{l_m}(x_m(z^*))}{l_m!} \right\}. \end{aligned}$$

By (2.40) and (2.41), we obtain (2.35) and (2.36). Here note that $C_{l,i,k}(A_j) = C_{l,k,i}(A_j) = 0$ if $l = (1, 1, \ldots, 1)$.

Next let us prove (2.37) and (2.38). For

$$z = \sum_{i=1}^{n} x_i v_i \in R(C_j, c; \{p_1, p_2, \dots, p_s\}),$$

put $z' = \sum_{i, i \neq p_j \text{ for } 1 \leqq j \leqq s} x_i v_i$. Then we can verify that $z \longrightarrow z'$ is a bijection between $R(C_j, c; \{p_1, p_2, \dots, p_s\})$ and $R(\check{C}, c)$. Put $l' = (l'_1, l'_2, \dots, l'_{n-s}) = (l_1, \dots, \check{l}_{p_1}, \dots, \check{l}_{p_s}, \dots, l_n)$, $(w_1, w_2, \dots, w_{n-s}) = (v_1, \dots, \check{v}_{p_1}, \dots, \check{v}_{p_s}, \dots, v_n)$. Then we have

$$\begin{aligned} S_{p_1, p_2, \dots, p_s} &= (-1)^n \omega(c) \sum_{l'} \prod_{m=1}^{n-s} (w_m^{(i)})^{l'_m - 1} \\ &\times \sum_{z \in R(\check{C}, c)} (\frac{1}{2})^s \left\{ \prod_{m=1}^{n-s} \frac{B_{l'_m}(x_m(z))}{l'_m!} - (-1)^n \prod_{m=1}^{n-s} \frac{B_{l'_m}(x_m(z))}{l'_m!} \right\}, \end{aligned}$$

since $B_1(1) = 1/2$. Here $l' = (l'_1, l'_2, \dots, l'_{n-s})$ extends over all $(n-s)$-tuples of non-negative integers such that $l'_1 + l'_2 + \dots + l'_{n-s} = n - s$. In view of (2.33), we obtain (2.37). Let A' denote the $n \times (n-s)$-matrix whose (i, l)-component is $w_l^{(i)}$. By the definition I, (2.13), we obtain $C_{l,i,k}(A_j) = C_{l',i,k}(A')$ if $l_{p_1} = \dots = l_{p_s} = 1$. Therefore we obtain (2.38) in view of (2.34). This completes the proof.

LEMMA 2.4. *Let $C = C(v_1, v_2, \dots, v_r)$ be an r-dimensional open simplicial cone with basis in $\mathfrak{O}_F$. We assume that $r < n$, $v_i \in \mathfrak{O}_F$ for $1 \leqq i \leqq r$ and that $\mathfrak{f}$ is divisible by a prime ideal $\mathfrak{p}$ of degree 1 which satisfies the following conditions* (i) $\sim$ (iii). *Let $p\mathbf{Z} = \mathfrak{p} \cap \mathbf{Z}$ and put $p\mathfrak{O}_F = \mathfrak{p}\mathfrak{q}$ with an ideal $\mathfrak{q}$.*

(i) *p is unramified in $F/\mathbf{Q}$.*

(ii) *Let a be the least integer such that $\mathfrak{a}_\mu \mathfrak{f} \otimes_{\mathbf{Z}} \mathbf{Z}_p \supset p^a \mathfrak{O}_F \otimes_Z \mathbf{Z}_p$. Then the fractional ideal $(p^a)(\mathfrak{a}_\mu \mathfrak{f})^{-1}$ is not divisible by $\mathfrak{p}$ but divisible by $\mathfrak{q}$.*

(iii) *Let $\tilde{F}$ be the normal closure of F over $\mathbf{Q}$. There exist $\tau_1, \tau_2, \dots, \tau_r \in J_F$, $\tau_i \neq \mathrm{id}$ for $1 \leqq i \leqq r$ such that*

$$p \nmid N_{\tilde{F}/\mathbf{Q}}(\det \begin{pmatrix} v_1^{\tau_1} & \dots & v_r^{\tau_1} \\ \vdots & & \vdots \\ v_1^{\tau_r} & \dots & v_r^{\tau_r} \end{pmatrix}).$$

Then $R(C, c) = \emptyset$.

PROOF. Let $\{\omega_1, \omega_2, \dots, \omega_n\}$ be an integral basis of $\mathfrak{O}_F$ over $\mathbf{Z}$. Put

$$v_i = \sum_{j=1}^{n} a_{ij} \omega_j, \quad 1 \leqq i \leqq r, \quad a_{ij} \in \mathbf{Z}, \qquad A = (a_{ij}) \in M(r, n, \mathbf{Z}).$$

By (iii), we see that $\operatorname{rank}(A \bmod p) = r$. Hence we can find $B = (b_{ij}) \in M(n - r, r, \mathbf{Z})$ such that $\operatorname{rank}(\begin{pmatrix} A \\ B \end{pmatrix} \bmod p) = n$. Put $v_{r+i} = \sum_{j=1}^{n} b_{ij} \omega_j$, $1 \leqq i \leqq n - r$. Then we have

$$\mathfrak{O}_F \otimes_{\mathbf{Z}} \mathbf{Z}_p = \mathbf{Z}_p v_1 \oplus \mathbf{Z}_p v_2 \oplus \dots \oplus \mathbf{Z}_p v_n. \tag{2.42}$$

Now assume $R(C,c) \neq \emptyset$ and take any $z = \sum_{j=1}^{r} x_j v_j \in R(C,c)$. We have $x_j \in \mathbf{Q}$, $0 < x_j \leqq 1$ for $1 \leqq j \leqq r$. Since $z \in (\mathfrak{a}_\mu \mathfrak{f})^{-1}$, we have $z \in (\mathfrak{a}_\mu \mathfrak{f})^{-1} \otimes_{\mathbf{Z}} \mathbf{Z}_p \subset p^{-a}\mathfrak{O}_F \otimes_{\mathbf{Z}} \mathbf{Z}_p$. By (2.42), we see that z can be written as

$$z = \frac{1}{p^a} \sum_{j=1}^{r} y_j v_j, \qquad y_j \in \mathbf{Q} \cap \mathbf{Z}_p, \quad 1 \leqq j \leqq r.$$

Since $(z)\mathfrak{a}_\mu \mathfrak{f} \in C_{\mathfrak{f}}$, $(z)\mathfrak{a}_\mu \mathfrak{f} = (p^a)^{-1}\mathfrak{a}_\mu \mathfrak{f}(\sum_{j=1}^{r} y_j v_j)$ is an integral ideal which is relatively prime to $\mathfrak{f}$. Therefore, by (ii), we obtain

$$\sum_{j=1}^{r} y_j v_j \notin \mathfrak{p}(\mathfrak{O}_F)_{\mathfrak{p}}, \tag{2.43}$$

$$\sum_{j=1}^{r} y_j v_j \equiv 0 \mod \mathfrak{q}. \tag{2.44}$$

Let $p\mathfrak{O}_F = \mathfrak{p}_1 \mathfrak{p}_2 \cdots \mathfrak{p}_t$ be the prime ideal decomposition. We have $t \geqq 2$ by (i). We choose the numbering so that $\mathfrak{p} = \mathfrak{p}_1$. Let $\mathfrak{P}_1$ be a prime ideal of $\tilde{F}$ lying over $\mathfrak{p}_1$. By (2.44), we have

$$\sum_{j=1}^{r} y_j v_j \equiv 0 \mod \mathfrak{p}_i, \qquad 2 \leqq i \leqq t.$$

Hence we get

$$\sum_{j=1}^{r} y_j v_j \equiv 0 \mod \mathfrak{P}$$

for any prime ideal $\mathfrak{P}$ of $\tilde{F}$ which lies over p and does not divide $\mathfrak{p}_1$. Put $G = \mathrm{Gal}(\tilde{F}/\mathbf{Q})$, $H = \mathrm{Gal}(\tilde{F}/F)$. Since $\mathfrak{p}$ is of degree 1, we see that $\mathfrak{P}_1^\sigma$, $\sigma \in G$ does not divide $\mathfrak{p}_1$ if and only if $\sigma \notin H$. Therefore we have

$$\sum_{j=1}^{r} y_j v_j \equiv 0 \mod \mathfrak{P}_1^\sigma \qquad \text{if} \quad \sigma \in G \setminus H.$$

This implies

$$\sum_{j=1}^{r} y_j v_j^\sigma \equiv 0 \mod \mathfrak{P}_1 \qquad \text{if} \quad \sigma \in G \setminus H. \tag{2.45}$$

Choosing σ to be extensions of τ_i to G, we see that $y_1 \equiv y_2 \equiv \ldots \equiv y_r \equiv 0 \mod \mathfrak{P}_1$ by (iii). This contradicts (2.43) and we complete the proof.

Remark. In (ii), if $(p^a)(\mathfrak{a}_\mu \mathfrak{f})^{-1}$ is divisible by $\mathfrak{q}$, it is not divisible by $\mathfrak{p}$. The condition (ii) is satisfied if $\mathfrak{a}_\mu$ is relatively prime to (p) and $\mathfrak{f}$ is not divisible by $\mathfrak{q}$.

PROPOSITION 2.5. *Let $C = C(v_1, v_2, \dots, v_r)$, $v_i \in \mathfrak{O}_F$, $1 \leqq i \leqq r$ be an r-dimensional open simplicial cone. Let $c \in C_{\mathfrak{f}}$ and choose $\mathfrak{a}_\mu$ so that $\mathfrak{a}_\mu \mathfrak{f}$ and c belong to the same narrow ideal class. Define $\zeta^{(i_1,i_2,\dots,i_k)}(s, C, \mathfrak{a}_\mu)$ by* (1.13) *and put*

$$\zeta^{(i_1,i_2,\dots,i_k)}(c) = \zeta^{(i_1,i_2,\dots,i_k)}(0, C, \mathfrak{a}_\mu),$$
$$\dot{\zeta}^{(i_1,i_2,\dots,i_k)}(c) = \frac{d}{ds}\zeta^{(i_1,i_2,\dots,i_k)}(s, C, \mathfrak{a}_\mu)\bigg|_{s=0}.$$

Let $\omega \in \widehat{G}_{\mathfrak{f}}^{(2)}$, $\omega \neq 1$. We assume that $\mathfrak{f}$ is divisible by a prime ideal $\mathfrak{p}$ of F of degree 1 which satisfies (i) *and* (ii) *of Lemma 2.4. If $r < n$, we assume the condition* (iii). *If $r = n$, we assume that for every $1 \leqq s < n$ and for every $1 \leqq i_1 < i_2 < \dots < i_s \leqq n$, there exist $\tau_1, \tau_2, \dots, \tau_s \in J_F$ such that $\tau_i \neq \mathrm{id}$ for $1 \leqq i \leqq s$ and that*

$$p \nmid N_{\tilde{F}/\mathbf{Q}}(\det \begin{pmatrix} v_{i_1}^{\tau_1} & \dots & v_{i_s}^{\tau_1} \\ \vdots & & \vdots \\ v_{i_1}^{\tau_s} & \dots & v_{i_s}^{\tau_s} \end{pmatrix}),$$

where $\tilde{F}$ denotes the normal closure of F over $\mathbf{Q}$. Then we have

$$\sum_{c \in C_{\mathfrak{f}}} \omega(c)\zeta^{(i)}(c) = 0 \qquad \textit{for} \quad 1 \leqq i \leqq n, \tag{2.46}$$

$$\sum_{c \in C_{\mathfrak{f}}} \omega(c)(\dot{\zeta}^{(i,k)}(c) - \dot{\zeta}^{(i)}(c) - \dot{\zeta}^{(k)}(c)) = 0 \qquad \textit{for} \quad 1 \leqq i < k \leqq n, \tag{2.47}$$

$$\sum_{c \in C_{\mathfrak{f}}} \omega(c)(\dot{\zeta}^{(i,k,p)}(c) - \dot{\zeta}^{(i)}(c) - \dot{\zeta}^{(k)}(c) - \dot{\zeta}^{(p)}(c)) = 0 \qquad \textit{for} \quad 1 \leqq i < k < p \leqq n, \tag{2.48}$$

$$\sum_{c \in C_{\mathfrak{f}}} \omega(c)(\dot{\zeta}^{(1,2,\dots,n)}(c) - \sum_{i=1}^{n} \dot{\zeta}^{(i)}(c)) = 0. \tag{2.49}$$

PROOF. We will prove (2.46) and (2.47). The proofs of (2.48) and (2.49) are similar to that of (2.47). By Lemma 2.4, the assertions hold when $r < n$. Assume $r = n$. We use Lemma 2.3. For $1 \leqq p_1 < p_2 < \dots < p_s \leqq n$, $s \leqq n-1$, put

$$\check{C} = C(v_1, \dots, \check{v}_{p_1}, \dots, \check{v}_{p_s}, \dots v_n).$$

By Lemma 2.4, we have $R(\check{C}, c) = \emptyset$. Hence we get $S_{p_1,\dots,p_s} = 0$, $T_{p_1,\dots,p_s} = 0$. Now (2.47) follows from (2.36). We see that $R(C, c; \{1, 2, \dots, n\}) = \emptyset$ since $(v_1 + v_2 + \dots + v_n)\mathfrak{a}_\mu \mathfrak{f}$ cannot be prime to $\mathfrak{f}$. Hence $S_{1,2,\dots,n} = 0$ and (2.46) follows from (2.35). This completes the proof.

2.5. We define $g_K^{(2)}(\mathrm{id},\tau;\{C_j\}_{j\in J},\{\mathfrak{a}_\mu\})$ by (1.9), i.e., we put

$$\begin{aligned}&g_K^{(2)}(\mathrm{id},\tau;\{C_j\}_{j\in J},\{\mathfrak{a}_\mu\})\\ &=\exp(\frac{1}{|G|}\sum_{\mathfrak{f}|\tilde{\mathfrak{f}}}\sum_{1\neq\omega\in\hat{G}_{\mathfrak{f}}^{(2)}}\frac{\omega(\tau)}{L'(0,\omega)}\sum_{c\in C_{\mathfrak{f}}}\omega(c)X^{(2)}(c;\{C_j\}_{j\in J},\mathfrak{a}_\mu).\end{aligned}\tag{2.50}$$

We abbreviate $g_K^{(2)}(\mathrm{id},\tau;\{C_j\}_{j\in J},\{\mathfrak{a}_\mu\})$ to $g_K^{(2)}(\mathrm{id},\tau;\{C_j\},\{\mathfrak{a}_\mu\})$, or to $g_K^{(2)}(\mathrm{id},\tau)$.

PROPOSITION 2.6. *We assume that the conditions of Proposition* 2.5 *hold for every C_j and for every $\mathfrak{f}$ such that $\mathfrak{f}$ divides $\tilde{\mathfrak{f}}$ and that $\widehat{G}_{\mathfrak{f}}^{(2)}$ contains a non-trivial character. Then we have*

$$g_K^{(2)}(\mathrm{id},\tau;\{\alpha^{-1}C_j\},\{(\alpha)\mathfrak{a}_\mu\})=(\frac{\alpha^{(1)}}{\sqrt[n]{N(\alpha)}})^a g_K^{(2)}(\mathrm{id},\tau;\{C_j\},\{\mathfrak{a}_\mu\}),$$

$$a=\frac{2}{|G|}\sum_{\mathfrak{f}|\tilde{\mathfrak{f}}}\sum_{1\neq\omega\in\hat{G}_{\mathfrak{f}}^{(2)}}\omega(\tau).$$

PROOF. Put $X^{(2)}(c)=X^{(2)}(c;\{C_j\}_{j\in J},\mathfrak{a}_\mu)$. By (2.28) and Proposition 2.5, we get

$$\begin{aligned}&g_K^{(2)}(\mathrm{id},\tau;\{\alpha^{-1}C_j\},\{(\alpha)\mathfrak{a}_\mu\})\\ &=\exp(\frac{1}{|G|}\sum_{\mathfrak{f}|\tilde{\mathfrak{f}}}\sum_{1\neq\omega\in\hat{G}_{\mathfrak{f}}^{(2)}}\frac{\omega(\tau)}{L'(0,\omega)}\sum_{c\in C_{\mathfrak{f}}}\omega(c)(X^{(2)}(c)\\ &+2(\log\alpha^{(1)}-\frac{1}{n}\log N(\alpha)))\sum_{j\in J}\dot{\zeta}_j^{(1,2,\dots,n)}(c)))\\ &=g_K^{(2)}(\mathrm{id},\tau;\{C_j\},\{\mathfrak{a}_\mu\})\\ &\times\exp(\frac{1}{|G|}\sum_{\mathfrak{f}|\tilde{\mathfrak{f}}}\sum_{1\neq\omega\in\hat{G}_{\mathfrak{f}}^{(2)}}\frac{\omega(\tau)}{L'(0,\omega)}\sum_{c\in C_{\mathfrak{f}}}\omega(c)(X^{(2)}(c)\\ &+2(\log\alpha^{(1)}-\frac{1}{n}\log N(\alpha))\zeta_F'(0,c)))\\ &=g_K^{(2)}(\mathrm{id},\tau;\{C_j\},\{\mathfrak{a}_\mu\})(\frac{\alpha^{(1)}}{\sqrt[n]{N(\alpha)}})^a.\end{aligned}$$

This completes the proof.

Next we consider the variations of $X^{(2)}(c;\{C_j\},\mathfrak{a}_\mu)$ and $g_K^{(2)}(\mathrm{id},\tau;\{C_j\},\{\mathfrak{a}_\mu\})$ when one cone C_j is replaced by ϵC_j with $\epsilon\in E_F^+$. As in III, §3.6, we see that

$$\zeta^{(i_1,i_2,\dots,i_k)}(s,\epsilon C_j,\mathfrak{a}_\mu)=(\epsilon^{(i_1)}\epsilon^{(i_2)}\cdots\epsilon^{(i_k)})^{-s}\zeta^{(i_1,i_2,\dots,i_k)}(s,C_j,\mathfrak{a}_\mu).$$

Hence (2.18) and (2.19) hold with $-\log\epsilon^{(i)}$ in place of $\log\alpha^{(i)}$. Therefore, assuming (2.17) and (2.27), we obtain

$$X^{(2)}(c;\{\{\epsilon C_j\}\cup\{C_k\}_{k\neq j}\},\mathfrak{a}_\mu)\equiv X^{(2)}(c;\{C_j\},\mathfrak{a}_\mu)-2\log\epsilon^{(1)}\sum_{k=1}^{n}\dot{\zeta}_j^{(k)}.\tag{2.51}$$

By (2.51) and Proposition 2.5, we obtain

PROPOSITION 2.7. *We assume that the conditions of Proposition* 2.5 *hold for every* C_j *and for every* $\mathfrak{f}$ *such that* $\mathfrak{f}$ *divides* $\tilde{\mathfrak{f}}$ *and that* $\widehat{G}_{\mathfrak{f}}^{(2)}$ *contains a non-trivial character. Then we have*

$$g_K^{(2)}(\mathrm{id},\tau;\{\{\epsilon C_j\}\cup\{C_k\}_{k\neq j}\},\{\mathfrak{a}_\mu\}) = (\epsilon^{(1)})^b g_K^{(2)}(\mathrm{id},\tau;\{C_j\},\{\mathfrak{a}_\mu\}),$$

$$b = \frac{-2}{|G|}\sum_{\mathfrak{f}|\tilde{\mathfrak{f}}}\ \sum_{1\neq\omega\in\hat{G}_{\mathfrak{f}}^{(2)}} \frac{\omega(\tau)}{L'(0,\omega)}\sum_{c\in C_{\mathfrak{f}}}\omega(c)\dot{\zeta}_j^{(1,2,\dots,n)}(c).$$

2.6. In the previous subsections, we examined the variations of $g_K^{(2)}(\mathrm{id},\tau)$ under the operations:

(I) $\mathfrak{a}_\mu \to (\alpha)\mathfrak{a}_\mu$, $\{C_j\}_{j\in J}\to\{\alpha^{-1}C_j\}_{j\in J}$ with $\alpha\in F$, $\alpha\gg 0$.

(II) A member C_j of $\{C_j\}_{j\in J}$ is replaced by ϵC_j with $\epsilon\in E_F^+$.

Propositions 2.6 and 2.7 show that $g_K^{(2)}(\mathrm{id},\tau)$ transforms under these operations very similarly to $g_K(\mathrm{id},\tau)$. In particular, Proposition 2.6 shows that $g_K^{(2)}(\mathrm{id},\tau)$ taken mod $\overline{\mathbf{Q}}^\times$ does not change under the operation (I) if we assume mild conditions on $\{C_j\}$ and $\tilde{\mathfrak{f}}$.[2] However, in Proposition 2.7, $(\epsilon^{(1)})^b$ seems to be transcendental. This is a serious drawback and it seems extremely difficult to pursue further analogy to $g_K(\mathrm{id},\tau)$.

To overcome this difficulty, we introduce a hierarchy for (transcendental) numbers. The first class is the field $\mathfrak{K}_0$ generated over $\overline{\mathbf{Q}}$ by the logarithms of the absolute values of all algebraic numbers. The second class is the field generated over $\mathfrak{K}_0$ by all the division values of the logarithmic multiple gamma function $\log\dfrac{\Gamma_r(x,\omega)}{\rho_r(\omega)}$ with $\omega\in(\overline{\mathbf{Q}}\cap\mathbf{R}_+)^r$; here, by x is a division value, we mean that a nonzero integral multiple of x belongs to the lattice $\mathbf{Z}\omega_1+\cdots+\mathbf{Z}\omega_r$ and $x>0$. Then $g_K^{(2)}(\mathrm{id},\tau)$ is well defined modulo the exponential of an element of $\mathfrak{K}_1$. The geometric meaning of this symbol is not clear. However we can expect that this symbol will satisfy functorial properties similar to those enjoyed by p_K. In fact, by the same considerations as in III, §3.8, we see that $g_K^{(2)}$ satisfies such properties in certain cases. It would be very interesting to know whether this is true for cases not covered by the considerations as in III, §3.8.

[2]Proposition 2.6 concerns the *simultaneous* replacement $\mathfrak{a}_\mu \longrightarrow (\alpha)\mathfrak{a}_\mu$, $1\leqq\mu\leqq h_0$. If only one $\mathfrak{a}_\mu$ is replaced by $(\alpha)\mathfrak{a}_\mu$, we must take account of the operation (II) also, as in III, §3.7.

APPENDIX III

TRANSCENDENTAL PROPERTY OF CM-PERIODS

Though we did not deal with the transcendental property of periods in this book, the reader may naturally be curious about the transcendal nature of the CM-period $p_K(\sigma, \tau)$. In this appendix, we will make a supplement on this point taking accout of conjectures of Shimura and Deligne (actually they stated them as problems).

Let K be a CM-field of degree $2n$. In [Sh8], p. 319, Shimura conjectured (or questioned) that

$$\text{(S)} \qquad \begin{gathered} p_K(\sigma_i, \mathrm{id}),\ 1 \leqq i \leqq n \text{ are algebraically independent over } \mathbf{Q} \\ \text{if } \sum_{i=1}^{n} \sigma_i \text{ is a CM-type of } K. \end{gathered}$$

If Φ is a CM-type of K and L is a CM-field containing K, then $\mathrm{Inf}_{L/K}(\Phi)$ is a CM-type of L. By Shimura's theorem (III, Theorem 1.2), we see that if (S) holds for L, then it holds for K.

In [Del2], Proposition 1.6 and Corollary 6.4, Deligne showed the following. Let A be an abelian variety of dimension n defined over a subfield k of $\mathbf{C}$. Let $H^1_B(A)$ and $H^1_{\mathrm{DR}}(A)$ denote the first Betti cohomology group and the first (algebraic) de Rham cohomology group of A respectively. Let $\{a_i\}$ (resp. $\{\alpha_i\}$) be a basis of $H^1_B(A)$ over $\mathbf{Q}$ (resp. $H^1_{\mathrm{DR}}(A)$ over k). We have the canonical isomorphism

$$H^1_{\mathrm{DR}}(A) \otimes_k \mathbf{C} \cong H^1_B(A) \otimes_{\mathbf{Q}} \mathbf{C}.$$

Write $\alpha_i = \sum_j p_{ji} a_j$ using this isomorphism. Let G be the Mumford-Tate group of A; G is the subgroup of $GL(H^1_B(A)) \times \mathbf{G}_m$ fixing all Hodge classes of $H^*_B(A)$. Deligne showed that the transcendence degree of $k(p_{ij})$ over k is not greater than $\dim G$ and conjectured[1] (or quetioned) that

$$\text{(D)} \qquad \mathrm{tr.\ deg}_k k(p_{ij} \mid 1 \leqq i, j \leqq 2n) = \dim G.$$

To clarify the relation between (S) and (D), let us begin by recalling a few basic notions on a CM-type. Let K be a CM-field and Φ be a CM-type of K. Let M be a CM-field which is normal over $\mathbf{Q}$ and contains K. Put

$$G = \mathrm{Gal}(M/\mathbf{Q}), \qquad H = \mathrm{Gal}(M/K).$$

Let S be the full inverse image of Φ by the canonical map $G \longrightarrow H\backslash G$. We call Φ *primitive* if

$$\text{(1)} \qquad H = \{g \in G \mid gS = S\}.$$

[1]For this conjecture to make sense, we should assume that $k = \overline{\mathbf{Q}}$.

This condition is equivalent to that an abelian variety of type (K, Φ) is simple (cf. [Sh12], p. 60-61). Put

$$S' = \{\sigma^{-1} \mid \sigma \in S\}, \qquad H' = \{g \mid g \in G,\ gS' = S'\}.$$

Let K' be the subfield of M corresponding to H' and let Φ' be the subset of $J_{K'}$ consisting of all isomorphisms obtained from the elements of S'. Then Φ' is a primitive CM-type of the CM-field K'; (K', Φ') is called *the reflex* of (K, Φ) ([Sh12], p. 62).

Let $r(\Phi)$ be the rank of the subgroup of I_K generated by Φg, $g \in \mathrm{Gal}(M/\mathbf{Q})$. We call $r(\Phi)$ *the rank* of the CM-type Φ (cf. [Sh12], p. 199). It can be shown that ([Sh12], 32.9, 32.10)

$$(2) \qquad r(\Phi) = r(\Phi'), \qquad r(\Phi) \leqq [K : \mathbf{Q}]/2 + 1.$$

A CM-type Φ is called *non-degenerate* if $r(\Phi) = [K : \mathbf{Q}]/2 + 1$. These notions are defined independently of the choice of M. If Φ is not primitive, then we have $r(\Phi') < [K : \mathbf{Q}]/2 + 1$. In view of (2), we see that if Φ is non-degenerate, then Φ is primitive. If K is the composite of two imaginary quadratic fields, then K does not have any primitive CM-types. For detailed discussions of these notions on a CM-type with many examples, we refer the reader to Dodson [Dod1], [Dod2].

Now let A be an abelian variety of type (K, Φ) defined over $\overline{\mathbf{Q}}$. Then we have

$$(3) \qquad \overline{\mathbf{Q}}(p_{ij} \mid 1 \leqq i, j \leqq 2n) = \overline{\mathbf{Q}}(\pi,\ p_K(\sigma, \Phi) \mid \sigma \in J_K).$$

For $x \in K'$, put

$$(\det \Phi')(x) = \prod_{\sigma \in \Phi'} x^{\sigma}.$$

Then $(\det \Phi')(x) \in K$. The mapping $K'^{\times} \ni x \longrightarrow (\det \Phi')(x) \in K^{\times}$ extends to the homomorphism of algebraic groups:

$$\det \Phi' : R_{K'/\mathbf{Q}}(\mathbf{G}_m) \longrightarrow R_{K/\mathbf{Q}}(\mathbf{G}_m).$$

Here $\mathbf{G}_m = GL(1)$ and R denotes the restriction of scalars functor. The Mumford-Tate group of A is given by

$$(4) \qquad G = \mathrm{Image}(\det \Phi', N_{K/F} \circ \det \Phi').$$

Here F denotes the maximal real subfield of K. By (4), we see without difficulty that

$$(5) \qquad \dim G = r(\Phi').$$

Hence conjecture D implies that

$$(6) \qquad \text{tr. deg.}\, \overline{\mathbf{Q}}(\pi,\ p_K(\sigma, \Phi) \mid \sigma \in J_K) = r(\Phi).$$

We see that (S) and (D) have the air of compatibility. In fact we can show:

THEOREM 1. *Let K be a CM-field normal over $\mathbf{Q}$. Let Φ be a CM-type of K. Then conjecture* (D) *implies that*

$$\text{tr.deg.}\, \overline{\mathbf{Q}}(\pi,\ p_K(\mathrm{id}, \sigma) \mid \sigma \in \Phi) = [K : \mathbf{Q}]/2 + 1.$$

In view of (6) and III, Theorem 1.2, Theorem 1 follows from

PROPOSITION 2. *Let K be a CM-field. Then there exists a CM-field L normal over $\mathbf{Q}$ which contains K and which has a non-degenerate CM-type.*

For the proof, let us prepare a lemma. Let K be a CM-field which is normal over $\mathbf{Q}$. Put $2n = [K : \mathbf{Q}]$, $G = \mathrm{Gal}(K/\mathbf{Q})$. Let $\widehat{G}$, $\widehat{G}_-$ and $\widehat{G}_+$ be the same as in III, §3.1. Let $\Phi = \sum_{i=1}^n \sigma_i$ be a CM-type. We regard Φ as an element of the group ring $\mathbf{C}[G]$. For $\pi \in \widehat{G}$, we put $\pi(\Phi) = \sum_{i=1}^n \pi(\sigma_i)$.

LEMMA 3. *For Φ to be non-degenerate, it is necessary and sufficient that* $\det \pi(\Phi) \neq 0$ *for every* $\pi \in \widehat{G}_-$.

PROOF. Let ψ be the homomorphism as abelian groups defined by $\mathbf{Z}[G] \ni \xi \longrightarrow \Phi\xi \in \mathbf{Z}[G]$. Then we have $r(\Phi) = 2n - \mathrm{rank}(\ker \psi)$. We use the same letter ψ for its extension to the linear mapping of $\mathbf{C}[G]$ to $\mathbf{C}[G]$. Then we have $r(\Phi) = 2n - \dim_{\mathbf{C}} \ker \psi$. We see that Φ is non-degenerate if and only if $\dim_{\mathbf{C}} \ker \psi = n - 1$.

Put $\widehat{G} = \{\pi_1, \pi_2, \ldots, \pi_h\}$, $n_i = \dim \pi_i$, where h is the number of conjugacy classes of G. We have the isomorphism $\mathbf{C}[G] \cong \oplus_{i=1}^h M(n_i, \mathbf{C})$ by $\sum_{j=1}^{2n} c_j g_j \longrightarrow (\sum_{j=1}^{2n} c_j \pi_i(g_j))_{1 \leqq i \leqq h}$, $c_j \in \mathbf{C}$, $G = \{g_1, \ldots, g_{2n}\}$. Hence we see that

$$\dim_{\mathbf{C}} \ker \psi = \sum_{i=1}^{h} n_i(n_i - \mathrm{rank}\, \pi_i(\Phi)).$$

Suppose that $\pi \in \widehat{G}_+$. If π is not trivial, we obviously have $\pi(\Phi) = 0$. Since G_+ can be identified with the set of the equivalence classes of all irreducible representations of $\mathrm{Gal}(F/\mathbf{Q})$, F being the maximal real subfield of K, we have

$$\sum_{\pi \in \hat{G}_+} \dim \pi(\dim \pi - \mathrm{rank}\, \pi(\Phi)) = \sum_{\pi \in \hat{G}_+} (\dim \pi)^2 - 1 = n - 1.$$

Therefore Φ is non-degenerate if and only if

$$\sum_{\pi \in \hat{G}_-} \dim \pi(\dim \pi - \mathrm{rank}\, \pi(\Phi)) = 0.$$

This completes the proof.

PROOF OF PROPOSITION 2. We may assume that K is normal over $\mathbf{Q}$. Let F_0 be a totally real field normal over $\mathbf{Q}$ such that

$$F_0 \cap K = \mathbf{Q}. \tag{7}$$

We put $L = F_0 \vee K$. We are going to show that L has a non-degenerate CM-type by choosing F_0 suitably. We have

$$\mathrm{Gal}(L/\mathbf{Q}) \cong \mathrm{Gal}(F_0/\mathbf{Q}) \times \mathrm{Gal}(K/\mathbf{Q})$$

canonically. If $\tau \in \mathrm{Gal}(F_0/\mathbf{Q})$, the element $(\tau, 1)$ corresponds to the automorphism of L which induces τ on F_0 and trivial on K; we identify τ with $(\tau, 1)$. Similarly we identify $\sigma \in \mathrm{Gal}(K/\mathbf{Q})$ with $(1, \sigma)$. Put $m = [F_0 : \mathbf{Q}]$, $\mathrm{Gal}(F_0/\mathbf{Q}) = \{\tau_1, \tau_2, \ldots, \tau_m\}$. For $1 \leqq j \leqq m$, we take a CM-type Φ_j of K. Then $\Phi = \sum_{j=1}^m \tau_j \Phi_j$ gives a CM-type of L.

Put $G = \mathrm{Gal}(K/\mathbf{Q})$, $G_0 = \mathrm{Gal}(F_0/\mathbf{Q})$. Let π be an irreducible odd representation of $\mathrm{Gal}(L/\mathbf{Q})$. Then π can be written as $\pi_1 \otimes \pi_2$ with $\pi_1 \in \widehat{G}_0$, $\pi_2 \in \widehat{G}_-$. By Lemma 3, Φ is non-degenerate if and only if

$$(8) \qquad \det(\sum_{j=1}^{m} \pi_1(\tau_j) \otimes \pi_2(\Phi_j)) \neq 0, \qquad \text{for every} \quad \pi_1 \in \widehat{G}_0,\ \pi_2 \in \widehat{G}_-.$$

Put $2n = [K : \mathbf{Q}]$. Let l be an odd prime which does not divide n. We take a prime q so that

$$q \equiv 1 \mod l, \qquad (q, D_K) = 1.$$

Let F_0 be the unique subfield of degree l of the cyclotomic field $\mathbf{Q}(e^{2\pi i/q})$. Then F_0 is totally real and $\mathrm{Gal}(F_0/\mathbf{Q}) \cong \mathbf{Z}/l\mathbf{Z}$; since no prime can ramify in $F_0 \cap K$, (7) is satisfied. Let τ be a generator of $\mathrm{Gal}(F_0/\mathbf{Q})$ and put $\tau_j = \tau^j$, $1 \leqq j \leqq l$. Put $\zeta = e^{2\pi i/l}$. Since any character of $\mathrm{Gal}(F_0/\mathbf{Q})$ can be determined by assigning the value at τ as ζ^a with $0 \leqq a \leqq l-1$, (8) can be writte as

$$(9) \qquad \det(\sum_{j=1}^{l} \zeta^{aj}\pi(\Phi_j)) \neq 0 \qquad \text{for every} \quad 0 \leqq a \leqq l-1,\ \pi \in \widehat{G}_-.$$

Let $\{\sigma_1, \sigma_2, \dots, \sigma_n\}$ be a CM-type of K. For $1 \leqq j \leqq l$, we write Φ_j in the form

$$\Phi_j = \{\sigma_1\rho^{p_j(1)}, \sigma_2\rho^{p_j(2)}, \dots, \sigma_n\rho^{p_j(n)}\}$$

where $p_j(i) = 0$ or 1. Let $q_j(i)$ be 1 (resp. -1) if $p_j(i) = 0$ (resp. $p_j(i) = 1$). Then we have

$$\pi(\Phi_j) = \sum_{i=1}^{n} q_j(i)\pi(\sigma_i), \qquad \pi \in \widehat{G}_-.$$

Hence (9) can be written as

$$(10) \qquad \det(\sum_{i=1}^{n}(\sum_{j=1}^{l} \zeta^{aj} q_j(i))\pi(\sigma_i)) \neq 0 \qquad \text{for every} \quad 0 \leqq a \leqq l-1,\ \pi \in \widehat{G}_-.$$

By a theorem of Brauer, $\pi \in \widehat{G}$ is realized over the field T generated by a primitive $2n$-th root of unity over $\mathbf{Q}$. Since l does not divide $2n$, $\mathbf{Q}(\zeta)$ and T are linearly disjoint over $\mathbf{Q}$. By taking conjugates, we see that if (10) holds for $a = 0$, 1 and every $\pi \in \widehat{G}_-$, then it holds in general.

For $a = 0$, 1, put

$$A_i(a) = \sum_{j=1}^{l} \zeta^{aj} q_j(i).$$

The idea of the proof is to make $\sum_{i=1}^{n} A_i(a)\sigma_i$ sufficiently close to a scalar multiple of σ_1, for $a = 0$ and 1, by choosing l and $q_j(i)$ suitably. For $2 \leqq i \leqq n$, we put

$$q_j(i) = \begin{cases} (\frac{j}{l}), & 1 \leqq j \leqq l-1, \\ 1, & j = l. \end{cases}$$

Let $W = \sum_{j=1}^{l-1} \zeta^j (\frac{j}{l})$ be the Gauss sum. Then we see immediately that

$$(11) \qquad A_i(0) = 1, \qquad 2 \leqq i \leqq n,$$

$$(12) \qquad A_i(1) = W + 1, \qquad 2 \leqq i \leqq n.$$

For $q_j(1)$, we first choose $q_j(1) = 1$ for $1 \leqq j \leqq (l-1)/2$ or $j = l$ and $q_j(1) = -1$ for $(l+1)/2 \leqq j \leqq l-1$. Then we have $A_1(0) = 1$,

$$A_1(1) = 2(1 + \zeta + \zeta^2 + \cdots + \zeta^{(l-1)/2}) = 2 \cdot \frac{1 - \zeta^{(l+1)/2}}{1 - \zeta}.$$

We see that

$$\frac{1}{1-\zeta} \sim -\frac{l}{2\pi i}, \qquad 1 - \zeta^{(l+1)/2} = 2 + O(\frac{1}{l}), \qquad l \to \infty.$$

Hence we have

$$A_1(1) \sim \frac{2i}{\pi} l, \qquad l \to \infty.$$

Take any small $\epsilon > 0$. We modify the definition of $q_j(1)$ so that $q_j(1) = 1$ for $1 \leqq j \leqq (l-1)/2 + \epsilon l$ or $j = l$, and $q_j(1) = -1$ otherwise. Then we have

$$A_1(0) \sim 2\epsilon l, \qquad |A_1(1)| > (\frac{2}{\pi} - 3\epsilon)l, \qquad l \to \infty.$$

Since $|W| = \sqrt{l}$, we see that $A_i(0)/A_1(0)$ and $A_i(1)/A_1(1)$ can be made arbitrarily small for $2 \leqq i \leqq n$ by taking sufficiently large l. Then (10) holds for $a = 0$, 1 and every $\pi \in \widehat{G}_-$. This completes the proof.

If K is an imaginary quadratic field, (S) is a concequence of a theorem of Siegel-Schnider. For an elliptic curve defined over $\overline{\mathbf{Q}}$, Chudnowsky showed the following ([Chu], p. 302). With the usual notation in the theory of elliptic functions, it holds that

$$(13) \qquad \text{tr. deg.}\, \mathbf{Q}(\omega_1, \omega_1, \eta_1, \eta_2) \geqq 2.$$

Hence if an elliptic curve has complex multiplication by K, we have

$$\text{tr. deg.}\, \mathbf{Q}(\pi,\ p_K(\text{id}, \text{id})) = 2$$

in view of Legendre's relation. Not much is known about (S) and (D) for the higher dimensinal case. The reader is referred to [Chu], p. 9 and Waldschmidt [Wal], p. 86. It is known that, in the notation of (S), at least one of $\pi p_K(\sigma_i, \text{id})$, $1 \leqq i \leqq n$ is transcendental.

If a CM-field K, $[K : \mathbf{Q}] \geqq 4$ is a composite of imaginary quadratic fields, then K does not have any non-degenerate CM-types; this can be seen very easily using Lemma 3. Hence there exists a CM-field of arbitrarily high degree which does not have any non-degenerate CM-types. On the other hand, we can show that a CM-field K has a primitive CM-type whenever $[K : \mathbf{Q}] \geqq 14$, which is given as an exercise in the next page.

Exercises

Let K be a CM-field of degree $2n$. Show that K has a primitive CM-type if $n \geqq 7$ by the following steps.

Let F be the maximal real subfield of K and let L be the normal closure of K over $\mathbf{Q}$. Put

$$G = \mathrm{Gal}(L/\mathbf{Q}), \qquad H = \mathrm{Gal}(L/K), \qquad H_0 = \mathrm{Gal}(L/F).$$

Let

$$G = \sqcup_{i=1}^{n} H_0 \sigma_i = \sqcup_{i=1}^{n} (H\sigma_i \sqcup H\sigma_i \rho)$$

be coset decompositions with $\sigma_1 = 1$. For $1 \leqq i, j \leqq n$, put

$$\sigma_j \sigma_i = h_{ij} \sigma_{\underline{i}(j)} \rho^{l_i(j)}, \qquad h_{ij} \in H$$

where $1 \leqq \underline{i}(j) \leqq n$, $l_i(j) = 0$ or 1. The mapping $j \longrightarrow \underline{i}(j)$ is a permutation on n letters. Take a mapping

$$p : \{1, 2, \ldots, n\} \longrightarrow \{0, 1\}$$

such that $p(1) = 0$ and put

$$\Phi_p = \{\sigma_i \rho^{p(i)} \mid 1 \leqq i \leqq n\}, \qquad S_p = \sqcup_{i=1}^{n} H\sigma_i \rho^{p(i)}, \qquad H_p = \{g \in G \mid gS_p = S_p\}.$$

Then Φ_p is a CM-type of K. We assume $n \geqq 2$.

1. Show that if $\sigma_j \rho^{\alpha}$, $\alpha = 0, 1$ belongs to H_p, then $\alpha = p(j)$ and
$$p(\underline{i}(j)) \equiv p(i) + p(j) + l_i(j) \mod 2$$
holds for $2 \leqq i \leqq n$.
2. Put $\mathbf{F}_2 = \mathbf{Z}/2\mathbf{Z}$, $x_j = p(j) \mod 2 \in \mathbf{F}_2$, $\bar{l}_i(j) = l_i(j) \mod 2 \in \mathbf{F}_2$. Regard $(x_2, \ldots, x_n) \in \mathbf{F}_2^{n-1}$ as a variable point, while $x_1 = 0$. For $2 \leqq j \leqq n$, let V_j be the linear subset of $\mathbf{F}_2^{n-1}$ defined by
$$x_i + x_j + x_{\underline{i}(j)} + \bar{l}_i(j) = 0, \qquad 2 \leqq i \leqq n.$$
Show that Φ_p is primitive for some choice of p if $\mathbf{F}_2^{n-1} \supsetneqq \cup_{j=2}^{n} V_j$.
3. Let k be a field and $A \in M(m, k)$. Suppose that all the diagonal components of A are 1 and that every row of A contains at most one nonzero off diagonal component. Show that $\mathrm{rank}(A) \geqq [\frac{m+1}{2}]$.
4. Show that $|V_j| \leqq 2^{n-1-[(n-1)/2]}$ if $n \geqq 3$. Show that K has a primitive CM-type if $n \geqq 7$.

REFERENCES

[An] G. W. Anderson, Logarithmic derivatives of Dirichlet L-functions and the periods of abelian varieties, Comp. Math. 45(1982), 315–332.

[Ar] E. Artin, Zur Theorie der L-Reihen mit allgemeinen Gruppencharakteren, Hamb. Abh. 8(1930), 292–306 (=Collected Papers, No. 7).

[As] T. Asai, On a certain function analogous to $\log|\eta(z)|$, Nagoya Math. J. 40 (1970), 193–211.

[Bak] A. Baker, Linear forms in the logarithms of algebraic numbers II, Mathematika 14(1967), 102–107.

[Bar1] E. W. Barnes, The theory of the double gamma function, Philos. Trans. Roy. Soc. (A), 196 (1901), 265–388.

[Bar2] E. W. Barnes, On the theory of the multiple gamma function, Trans. Cambridge Philos. Soc. 19 (1904), 374–425.

[Be] W. E. H. Berwick, Algebraic number-fields with two independent units, Proc. London Math. Soc. Ser. 2, 34(1932), 360–378.

[Bi] K. K. Billevič, On units of algebraic fields of third and fourth degree, Math. Sb. N. S., 40, (1956), 123–136 (in Russian).

[Bl] D. Blasius, On the critical values of Hecke L-series, Ann. of Math. 124(1986), 23–63.

[Bo1] N. Bourbaki, Topologie Générale, Hermann, 1971.

[Bo2] N. Bourbaki, Algèbre Commutative, Chapitre 5, 6, 7, Masson, 1985.

[Che] C. Chevalley, Deux théorèmes d'arithmétique, J. Math. Soc. Japan 3(1951), 36–44.

[Chu] G. V. Chudnovsky, Contributions to the theory of transcendental numbers, Mathematical Surveys and Monographs, Vol. 19, 1984, Amer. Math. Soc.

[Co] P. Colmez, Périods des variétés abéliennes à multiplication complexe, Ann. of Math. 138(1993), 625–683.

[Da1] H. Darmon, Stark-Heegner points over real quadratic fields, Contemp. Math. 210 (1998), 41–69.

[Da2] H. Darmon, Integration on $\mathcal{H}_p \times \mathcal{H}$ and arithmetic applications, Ann. of Math. 154(2001), 589–639.

[Ded] R. Dedekind, Gesammelte mathematische Werke, Druck und Verlag von Friedr. Vieweg & Sohn Akt. -Ges., Braunschweig, 1930–32.

[Del1] P. Deligne, Valeurs de fonctions L et périodes d'intégrales, Proc. Symposia Pure Math. 33(1979), part 2, 313–346.

[Del2] P. Deligne, Hodge cycles on abelian varieties (Notes by J. S. Milne), 9–100, Lecture notes in mathematics 900, Springer-Verlag, 1982.

[Dod1] B. Dodson, The structure of Galois groups of CM-fields, Trans. of Amer. Math. Soc., 283(1984), 1–32.

[Dod2] B. Dodson, Solvable and nonsolvable CM-fields, Amer. J. Math., 108(1986), 75–93.

[F] G. Fujisaki, Introduction to algebraic number theory (in two volumes), Shôkabô, 1975 (in Japanese).

[G] B. Gross, On the periods of abelian integrals and a formula of Chowla and Selberg (with an appendix by D. E. Rohrlich), Inv. Math. 45(1978), 193–211.

[GR] I. S. Gradshteyn and I. M. Ryzhik, Table of integrals, series and products, Academic Press, 1980.

[He] E. Hecke, Mathematische Werke, Vandenhoeck & Ruprecht, Göttingen, 1959.

[Hi1] H. Hida, Elementary theory of L-functions and Eisenstein series, London Math. Soc. Student Texts 26, Cambridge University Press, 1993.

[Hi2] H. Hida, Zariski density of CM points and Hecke L-values, preprint 2002.

[HW] G. H. Hardy and E. M. Wright, An introduction to the theory of numbers, Oxford University Press, fifth edition, 1979.

[J] H. Jacquet, Automorphic forms on $GL(2)$, part II, Lecture notes in mathematics 278, Springer-Verlag, 1972.

[JL] H. Jacquet and R. P. Langlands, Automorphic forms on $GL(2)$, Lecture notes in mathematics 114, Springer-Verlag, 1970.

[JKS] U. Jannsen, S. Kleiman and J-P. Serre (eds.), Motives, Proc. Symposia Pure Math. 55(1994), Part 1 and 2.

[Kli] H. Klingen, Über die Werte der Dedekindschen Zetafunktionen, Math. Ann. 145(1962), 265–272.

[Klo] H. D. Kloostermann, Theorie der Eisensteinschen Reihen von mehreren Veränderlichen, Abh. Math. Sem. Hamb. 6(1928), 163–188.

[Kr] L. Kronecker, Werke, Chelsea, New York, 1968.

[Ku] E. E. Kummer, Collected papers, Springer-Verlag, 1975.

[La] S. Lang, Complex multiplication, Grundlehren der mathematischen Wissenshaften, 225, Springer-Verlag, 1983.

[Ll] R. P. Langlands, On the functional equation satisfied by Eisenstein series, Lecture notes in mathematics 544, Springer-Verlag, 1976.

[PZ] M. Pohst and H. Zassenhaus, Algorithmic algebraic number theory, Encyclopedia of Mathematics and its applications, Cambridge University Press, 1989.

[Ri] B. Riemann, Ueber die Anzahl der Primzahlen unter einer gegebenen Grösse, Werke, No. VII.

[Ru] K. Rubin, A Stark conjecture "over **Z**" for abelian L-functions with multiple zeros, Ann. Institut Fourier 46(1996), 33–62.

[SC] A. Selberg and S. Chowla, On Epstein's zeta-function, Jour. für reine und angew. Math. 227(1967), 86–110 (=Collected Papers of A. Selberg, No. 34).

[Sh1] G. Shimura, Introduction to the arithmetic theory of automorphic functions, Publ. Math. Soc. Japan 11, Iwanami Shoten and Princeton University Press, 1971.

[Sh2] G. Shimura, On the holomorphy of certain Dirichlet series, Proc. London Math. Soc. 3rd ser. 31(1975), 79–98 (=Collected Papers II, [75a]).

[Sh3] G. Shimura, On some arithmetic properties of modular forms of one and several variables, Ann. of Math. 102(1975), 491–515 (=Collected Papers II, [75c]).

[Sh4] G. Shimura, On abelian varieties with complex multiplication, Proc. London Math. Soc. 3rd ser. 34(1977), 65–86 (=Collected Papers II, [77a]).

[Sh5] G. Shimura, On certain reciprocity-laws for theta functions and modular forms, Acta Math. 141(1978), 35–71 (=Collected Papers III, [78a]).

[Sh6] G. Shimura, The special values of the zeta functions associated with Hilbert modular forms, Duke Math. J. 45(1978), 637–679 (=Collected Papers III, [78c]).

[Sh7] G. Shimura, Automorphic forms and periods of abelian varieties, J. Math. Soc. Japan 31(1979), 561–592 (=Collected Papers III, [79a]).

[Sh8] G. Shimura, The arithmetic of certain zeta functions and automorphic forms on orthogonal groups, Ann. of Math. 111(1980), 313–375 (=Collected Papers III, [80]).

[Sh9] G. Shimura, Confluent hypergeometric functions on tube domains, Math. Ann. 260(1982), 269–302 (=Collected Papers III, [82c]).

[Sh10] G. Shimura, On Eisenstein series, Duke Math. J. 50(1983), 413–476 (=Collected Papers III, [83b]).

[Sh11] G. Shimura, Nearly holomorphic functions on hermitian symmetric spaces, Math. Ann. 278(1987), 1–28 (=Collected Papers III, [87a]).

[Sh12] G. Shimura, Abelian varieties with complex multiplication and modular functions, Princeton Mathematical Series 46, Princeton University Press, 1998.

[Sh13] G. Shimura, Arithmeticity in the theory of automorphic forms, Mathematical Surveys and Monographs, Vol. 82, 2000, Amer. Math. Soc.

[Sh14] G. Shimura, Generalized Bessel functions on symmetric spaces, Jour. für reine und angew. Math. 509(1999), 35–66 (=Collected Papers IV, [99c]).

[Sh15] G. Shimura, The relative regulator of an algebraic extension, Collected Papers, IV, 720–736 (=[01c]).

[Shi1] T. Shintani, On evaluation of zeta functions of totally real algebraic number fields at non-positive integers, J. Fac. Sci. Univ. Tokyo 23(1976), 393–417.

[Shi2] T. Shintani, On a Kronecker limit formula for real quadratic fields, J. Fac. Sci. Univ. Tokyo 24(1977), 167-199.

[Shi3] T. Shintani, On values at $s = 1$ of certain L functions of totally real algebraic number fields, in Algebraic Number Theory, Proc. International Symp., Kyoto, 1976, 201–212, Kinokuniya, 1977.

[Shi4] T. Shintani, On certain ray class invariants of real quadratic fields, J. Math. Soc. Japan 30(1978), 139–167.

[Shi5] T. Shintani, On special values of L-functions of algebraic number fields, Sugaku 29 (1977), 204–216 (in Japanese).

[Si1] C. L. Siegel, Discontinuous groups, Ann. of Math. 44(1943), 674–689 (=Gesammelte Abhandlungen III, No. 43).

[Si2] C. L. Siegel, Lectures on advanced analytic number theory, Tata Institute, 1961.

[Si3] C. L. Siegel, Berechnung von Zetafunktionen an ganzzahligen Stellen, Göttingen Nachrichten, 1969, Nr. 10, 87–102 (=Gesammelte Abhandlungen IV, No. 89).

[Si4] C. L. Siegel, Über die Fourierschen Koefficienten von Modulformen, Göttingen Nachrichten, 1970, Nr. 3, 15–56 (=Gesammelte Abhandlungen IV, No. 90).

[Sp] E. H. Spanier, Algebraic topology, McGraw-Hill, 1966.

[St1] H. M. Stark, Class number problems in quadratic fields, Actes, Congès intern. math., 1970. Tome 1, 511–518.

[St2] H. M. Stark, L-functions at $s = 1$, I, II, III, IV, Advances in Math. 7(1971), 301–343; 17(1975), 60–92; 22(1976), 64–84; 35(1980), 197–235.

[St3] H. M. Stark, Class fields for real quadratic fields and L-series at $s = 1$, Algebraic number fields, Proc. Sympos. Durham, 1975, 355–375, Academic Press, London, 1977.

[T] J. Tate, Les conjectures de Stark sur les fonctions L d'Artin en $s = 0$ (Notes by D. Bernardi et N. Schappacher), Progress in Math. 47, Birkhäuser, 1984.

[TV] E. Thomas and A. T. Vasquez, On the resolution of cusp singularities and the Shintani decomposition in totally real cubic number fields, Math. Ann. 247(1980), 1–20.

[Wal] M. Waldschmidt, Nombres transcendents et group algébriques, Astérisque, 69–70, 1979.

[Wat] G. N. Watson, A treatese on the theory of Bessel functions, Cambridge University Press, second edition, 1944.

[We1] A. Weil, Basic number theory, Die Grundlehren der mathematischen Wissenschaften, 144, Springer-Verlag, 1967.

[We2] A. Weil, Elliptic functions according to Eisenstein and Kronecker, Ergebnisse der Mathematik und ihrer Grenzgebiete 88, Springer-Verlag, 1976.

[We3] A. Weil, Sur les périodes des intégrales abéliennes, Comm. on Pure and Appl. Math. XXIX (1976), 813–819 (=Collected Papers III, [1976b]).

[WW] E. T. Whittaker and G. N. Watson, A course of Modern Analysis, fourth edition, Cambridge University Press, 1927.

[Y1] H. Yoshida, Hecke characters and models of abelian varieties with complex multiplication, J. Fac. Sci. Univ. Tokyo 28(1981), 633–649.

[Y2] H. Yoshida, On the zeta functions of Shimura varieties and periods of Hilbert modular forms, Duke Math. J. 75(1994), 121–191.

[Y3] H. Yoshida, On absolute CM–periods, Proc. Symposia Pure Math. 66, Part 1, 1999, 221–278.

[Y4] H. Yoshida, On absolute CM-periods II, Amer. J. Math. 120(1998), 1199–1236.

Index

$\mathbf{a}$, 171
$\mathbf{a}_K$, 43
$\mathbf{a}_r$, 215
$\mathbf{a}_c$, 215
$B_m(x)$, 16
$B(p,q)$, 63
$\mathcal{B}(\chi_1,\chi_2)$, 222
$C_{\mathfrak{f}}$, 52
$C_l(A)$, $C_{l,j,k}(A)$, 22
$\mathbf{e}_{\mathbf{a}}(z)$, 172
$E(g,f)$, 218
$E_k(z,s;\lambda)$, 171
$E_\chi(z,s;\mathfrak{a})$, 175
$\eta(z)$, 178
$\widehat{G}$, $\widehat{G}_+$, $\widehat{G}_-$, 74
$G(c)$, 78, 85
$g_K(\mathrm{id},\tau)$, $g_K(\mathrm{id},\tau;\{C_j\}_{j\in J},\mathfrak{a}_\mu)$, 79
$g_{K/F}(\mathrm{id},\tau;\{C_j\}_{j\in J},\mathfrak{a}_\mu)$, 89
$g_K^{(p)}(\mathrm{id},\tau)$, 251
γ_k, 169
$\gamma_k(c)$, 170
$\Gamma_r^{[i]}(x,\omega)$, 16
$\Gamma_r(x,\omega)$ $(=\Gamma_r^{[1]}(x,\omega))$, 16
$\mathbf{h}$, 171
$h_\chi(z;\mathfrak{a})$, 176
$\Lambda_r(\mathfrak{a},\mathfrak{b},\mathfrak{X})$, 27
$p_K(\xi,\eta)$, 62
$R(C_j,c)$, 53
$R(\chi,f)$, 44
$\rho_r^{[i]}(x,\omega)$, 15
$\rho_r(x,\omega)$ $(=\rho_r^{[1]}(x,\omega))$, 16
$\mathcal{S}(E_v)$, $\mathcal{S}(E_A)$, 234
T_w, 222, 242
$V(c)$, 79, 85
$W(c)$, 79, 85
$X(c)$, $X(c;\{C_j\}_{j\in J},\mathfrak{a}_\mu)$, 79
$X^{(2)}(c;\{C_j\}_{j\in J},\mathfrak{a}_\mu)$, 257
$\xi(g,h;\alpha,\beta)$, 172
$\zeta_F(s,c)$, 52
$\zeta_r(s,\omega,x)$, 13
$\zeta(s,A,x)$, 17

analytic class number formula, 43
absolute period symbol, 79
Artin's unit theorem, 46
asymptotic expansion of the r-ple gamma function, 35

Bernoulli polynomial, 16
beta function, 63
big cell, 147
big cone, 147

Chowla-Selberg formula, 63, 74
  real case, 126
cone decomposition, 52
confluent hypergeometric function, 172
Conjecture A=Conjecture III, 3.2, 75
Conjecture B=Conjecture III, 3.9, 81
Conjecture C=Conjecture III, 3.10, 83

Dedekind η-function, 178

exact Chowla-Selberg formula, 184
  generalization of, 183
Eisenstein series, 171, 218

face, 136
Fermat curve, 65
fine sequence, 149

generalized regulator, 44

Hurwitz's zeta function, 17

K-Bessel function, 173, 227
Kronecker's first limit formula, 178

lattice, 179
Liouville's theorem, 38
limit formula of Kronecker's type, 177

non-degenerate CM-type, 270

odd representation, 74
open simplex, 135
open simplicial cone, 52
order of an algebraic number field, 189

Poisson summation formula, 236
precise form of Stark's conjecture, 50
primitive CM-type, 269
proper ideal, 189

rank of a CM-type, 270
reciprocity map, 83
refinement, 84
reflex, 270
r-ple Riemann zeta function, 13
r-ple gamma function, 16

Schwartz-Bruhat function, 171, 234
self-dual measure, 236
Shimura's period symbol, 62
Shimura's reciprocity law, 205
Shintani's formula
  for $\zeta(0,A,x)$, 20
  for $\zeta'(0,A,x)$, 23

for $\zeta_F(0, c)$, 54
for $\zeta'_F(0, c)$, 54
special decomposition of I^n, 139
Stark's general conjecture, 44

Theorem of Thomas-Vasquez, 147
Theorem of Shimura
on CM-periods, 62
on critical values of Hecke's L-function, 65

zeta integral, 235

Titles in This Series

106 **Hiroyuki Yoshida,** Absolute CM-periods, 2003

105 **Charalambos D. Aliprantis and Owen Burkinshaw,** Locally solid Riesz spaces with applications to economics, second edition, 2003

104 **Graham Everest, Alf van der Poorten, Igor Shparlinski, and Thomas Ward,** Recurrence sequences, 2003

103 **Octav Cornea, Gregory Lupton, John Oprea, and Daniel Tanré,** Lusternik-Schnirelmann category, 2003

102 **Linda Rass and John Radcliffe,** Spatial deterministic epidemics, 2003

101 **Eli Glasner,** Ergodic theory via joinings, 2003

99 **Philip S. Hirschhorn,** Model categories and their localizations, 2003

98 **Victor Guillemin, Viktor Ginzburg, and Yael Karshon,** Moment maps, cobordisms, and Hamiltonian group actions, 2002

97 **V. A. Vassiliev,** Applied Picard-Lefschetz theory, 2002

96 **Martin Markl, Steve Shnider, and Jim Stasheff,** Operads in algebra, topology and physics, 2002

95 **Seiichi Kamada,** Braid and knot theory in dimension four, 2002

94 **Mara D. Neusel and Larry Smith,** Invariant theory of finite groups, 2002

93 **Nikolai K. Nikolski,** Operators, functions, and systems: An easy reading. Volume 2: Model operators and systems, 2002

92 **Nikolai K. Nikolski,** Operators, functions, and systems: An easy reading. Volume 1: Hardy, Hankel, and Toeplitz, 2002

91 **Richard Montgomery,** A tour of subriemannian geometries, their geodesics and applications, 2002

90 **Christian Gérard and Izabella Łaba,** Multiparticle quantum scattering in constant magnetic fields, 2002

89 **Michel Ledoux,** The concentration of measure phenomenon, 2001

88 **Edward Frenkel and David Ben-Zvi,** Vertex algebras and algebraic curves, 2001

87 **Bruno Poizat,** Stable groups, 2001

86 **Stanley N. Burris,** Number theoretic density and logical limit laws, 2001

85 **V. A. Kozlov, V. G. Maz′ya, and J. Rossmann,** Spectral problems associated with corner singularities of solutions to elliptic equations, 2001

84 **László Fuchs and Luigi Salce,** Modules over non-Noetherian domains, 2001

83 **Sigurdur Helgason,** Groups and geometric analysis: Integral geometry, invariant differential operators, and spherical functions, 2000

82 **Goro Shimura,** Arithmeticity in the theory of automorphic forms, 2000

81 **Michael E. Taylor,** Tools for PDE: Pseudodifferential operators, paradifferential operators, and layer potentials, 2000

80 **Lindsay N. Childs,** Taming wild extensions: Hopf algebras and local Galois module theory, 2000

79 **Joseph A. Cima and William T. Ross,** The backward shift on the Hardy space, 2000

78 **Boris A. Kupershmidt,** KP or mKP: Noncommutative mathematics of Lagrangian, Hamiltonian, and integrable systems, 2000

77 **Fumio Hiai and Dénes Petz,** The semicircle law, free random variables and entropy, 2000

76 **Frederick P. Gardiner and Nikola Lakic,** Quasiconformal Teichmüller theory, 2000

75 **Greg Hjorth,** Classification and orbit equivalence relations, 2000

74 **Daniel W. Stroock,** An introduction to the analysis of paths on a Riemannian manifold, 2000

73 **John Locker,** Spectral theory of non-self-adjoint two-point differential operators, 2000

72 **Gerald Teschl,** Jacobi operators and completely integrable nonlinear lattices, 1999

71 **Lajos Pukánszky,** Characters of connected Lie groups, 1999

70 **Carmen Chicone and Yuri Latushkin,** Evolution semigroups in dynamical systems and differential equations, 1999

69 **C. T. C. Wall (A. A. Ranicki, Editor),** Surgery on compact manifolds, second edition, 1999

68 **David A. Cox and Sheldon Katz,** Mirror symmetry and algebraic geometry, 1999

67 **A. Borel and N. Wallach,** Continuous cohomology, discrete subgroups, and representations of reductive groups, second edition, 2000

66 **Yu. Ilyashenko and Weigu Li,** Nonlocal bifurcations, 1999

65 **Carl Faith,** Rings and things and a fine array of twentieth century associative algebra, 1999

64 **Rene A. Carmona and Boris Rozovskii, Editors,** Stochastic partial differential equations: Six perspectives, 1999

63 **Mark Hovey,** Model categories, 1999

62 **Vladimir I. Bogachev,** Gaussian measures, 1998

61 **W. Norrie Everitt and Lawrence Markus,** Boundary value problems and symplectic algebra for ordinary differential and quasi-differential operators, 1999

60 **Iain Raeburn and Dana P. Williams,** Morita equivalence and continuous-trace C^*-algebras, 1998

59 **Paul Howard and Jean E. Rubin,** Consequences of the axiom of choice, 1998

58 **Pavel I. Etingof, Igor B. Frenkel, and Alexander A. Kirillov, Jr.,** Lectures on representation theory and Knizhnik-Zamolodchikov equations, 1998

57 **Marc Levine,** Mixed motives, 1998

56 **Leonid I. Korogodski and Yan S. Soibelman,** Algebras of functions on quantum groups: Part I, 1998

55 **J. Scott Carter and Masahico Saito,** Knotted surfaces and their diagrams, 1998

54 **Casper Goffman, Togo Nishiura, and Daniel Waterman,** Homeomorphisms in analysis, 1997

53 **Andreas Kriegl and Peter W. Michor,** The convenient setting of global analysis, 1997

52 **V. A. Kozlov, V. G. Maz′ya, and J. Rossmann,** Elliptic boundary value problems in domains with point singularities, 1997

51 **Jan Malý and William P. Ziemer,** Fine regularity of solutions of elliptic partial differential equations, 1997

50 **Jon Aaronson,** An introduction to infinite ergodic theory, 1997

49 **R. E. Showalter,** Monotone operators in Banach space and nonlinear partial differential equations, 1997

48 **Paul-Jean Cahen and Jean-Luc Chabert,** Integer-valued polynomials, 1997

47 **A. D. Elmendorf, I. Kriz, M. A. Mandell, and J. P. May (with an appendix by M. Cole),** Rings, modules, and algebras in stable homotopy theory, 1997

46 **Stephen Lipscomb,** Symmetric inverse semigroups, 1996

45 **George M. Bergman and Adam O. Hausknecht,** Cogroups and co-rings in categories of associative rings, 1996

44 **J. Amorós, M. Burger, K. Corlette, D. Kotschick, and D. Toledo,** Fundamental groups of compact Kähler manifolds, 1996

For a complete list of titles in this series, visit the AMS Bookstore at **www.ams.org/bookstore/**.